LIMIT STATE DESIGN OF STEEL STRUCTURES as per IS:800/2007

S. Kanthimathinathan
Tiruchy Engineering Consultancy House
Srirangam, Tiruchy
India

TECHSAR™
NEW DELHI

Published and Printed by

Techsar Pvt. Ltd.
4435-36/7, Ansari Road, Daryaganj
New Delhi – 110 012 (India)
E-mail: tech_info@techsarworld.com
Website: www.techsarworld.com

ISBN 978-93-82332-78-7

Reprint 2025

This book is dedicated to
My father Mr. B. Sreenivasan, (late)
Superintendent (Retired). Sri. Renganathaswamy
Devasthanan, Srirangam.
My mother Mrs. S. Mythili, (late)

My thanks goes to my
"Daughter Mrs. K. Vishnu Priya Prem Kumar, M.Sc"
"Son Mr. K. Srinivasa Rengan, B.Tech"
and
Wife Mrs. K. Ragini,
Without whose cooperation I could not have completed this task.

I was inspired by
Mr. K. Veeraragavachary, Retd., AGM/BHEL/Tiruchy.
Mr. J.J.R. Muthayya, Retd., DGM/BHEL/Tiruchy.
Mr. G. Srinivasan, Retd., DGM/BHEL/Tiruchy.
Dr. V. Gopalakrishnan, Ex. Executive Director BHEL/Tiruchy.
Dr. M. Shanmugam, R.E.C./Tiruchy. (USA)
Dr. C. Natarajan, N.I.T./Tiruchy.
I thank all of them for inspiring me.

Foreword

This book is unique in the sense that it has been written, not by an academic, but by a practising engineer. Since there are very few books in the market, for solving structural engineering problems based on Indian standards, Mr. S. Kanthimathinathan's work will certainly be welcomed with open hands by the students and the practising engineers, who strongly feel the need for such a book.

With infrastructure getting the kind of attention it does, it is imperative that the teaching and design practices in India get updated to keep up with international trends. It is generally felt that they have remained obsolescent so far and among the many reasons is the lack of appropriate textbooks. This volume is a welcome addition in that respect.

The author has worked as a design engineer for several years in both public and private organizations dealing with design of supporting structures for steam generators. In that process, he has become quite familiar with the current European, British, American and Australian design codes and he has adopted them well for Indian standards. The coverage is comprehensive, including the latest advancements in several related areas, which are normally not found in the usual textbooks.

The book exhaustively covers both bolted and welded constructions and a large number of worked examples have been added to enhance the value of the book. Special emphasis is laid on the various options for efficient connections and their detailing.

A good discussion has been done on the assessment of loading and its importance in preventing premature failure and ensuring the safety of structures.

With India's growing population, there is a dire need for building another 100 million homes in the shortest possible time, using the most modern techniques. Similarly, India needs to build urgently more infrastructure facilities, bridges, port and harbour facilities, industrial and commercial structures and mass rapid transit systems.

The structural engineers of India thus face a massive challenge at the moment. It is imperative that a person like S. Kanthimathinathan, with his vast experience in the industry, comes up with appropriate books to help the Indian engineers in this hour of need.

I wish to express my admiration for the untiring efforts put in by the author in bringing out this book amidst his regular, tiresome work and I hope this volume becomes a forerunner for many more such books to come in the future.

Dr. V. Gopalakrishnan
Principal, Saranathan College of Engineering, Tiruchy
& Former Executive Director, Bharat Heavy Electricals Limited, Tiruchy

Preface

This book is based on Limit State method as per IS:800/2007.

Any structure from concept to utilization should pass through the following stages:

(1) Planning: Based on functional requirements, available land and fund.

(2) Structural Configuration: Forming the frames of the building based on functional requirements and architecture.

(3) Structural Analysis: Application of external loads and its combinations and finding the effects of the same on structural components (Bending moment, shear, etc.).

(4) Structure Design: Sizing the members and connections to withstand the forces and moments.

(5) Detailed Engineering: This is the activity converting the design into fabrication drawings. This is the communication between the designer and the manufacturer. In the drawings, the size of members, orientation of members, connection details with bill of materials will be provided for fabrication.

(6) Fabrication: This is a process by which the finished structural members will be produced as per the fabrication drawing. The fabrication or manufacturing process includes cutting the raw material economically, drilling, edge preparation, welding and non-destructive testing. This also includes surface preparation [Mechanical cleaning (or) Shot blastering] and surface protection (painting).

The revised Indian Standards IS:800/2007 includes Limit State method. This book deals with structural design based on IS:800/2007, IS:875, IS:1893, SP6(1), IS:816 and IS:1367. The readers are advised to go through these standards for better and clear understanding.

The thrust area of this book is to give more examples which will be useful for the students as well as practising design engineers. The relevant clause numbers of the standards are given then and there so as to enable the reader to go through the clause of the standards.

There are 13 chapters in this book. Each chapter gives the rationale behind the design of particular component, worked examples with figures and tables.

Chapter 1 gives the list of symbols. The symbols used elsewhere in the book but not covered in the list of symbols is described then and there.

Chapter 2 gives introduction to structural steel design. This chapter gives various sections and shapes used as tension members, compression members and flexural members, types of connections, definitions of loads. This chapter also includes the difference between limit state design and working stress design. Examples of calculating centre of gravity for unsymmetrical sections, the formula for finding equivalent uniformly distributed load to obtain bending moment for triangular and trapezoidal loading is given. The effect of a concentrated load on the corner columns of a particular floor is given for quickly calculating the load on columns.

The method of joints and method of sections are explained in detail with figures.

Chapter 3 is for tension members. Design procedure for design of tension member with suitable worked examples including design of lug angle are given.

Chapter 4 deals with the design of compression members. Step-by-step design procedure to find section strength and overall strength of the compression members is given. A number of worked examples on axially loaded column (though not in existence) uniaxial bending and biaxial bending of columns including design of batten and lacing is covered in this chapter.

Chapter 5 covers design of flexural members. This chapter includes two examples on design of beams of platform subjected to imposed load and platform beams carrying a vessel. These examples give the design of secondary and main beams for a complete platform. This chapter includes design of crane gantry for single crane and more than one crane. Design of plate girder is also discussed in depth. Design of purlin with and without sag rod as an example of biaxial bending of a flexural member is given in this chapter.

Chapter 6 is for design of column bases. This covers slab base, gusseted base, the design of anchor bolts, for columns with and without moment. This also includes design of stiffeners for column bases.

Chapter 7 deals with welded connections. Types of welding along with weld symbols are given in this chapter. Fillet welded connections and butt welded connections are covered with worked examples and figures. Weld design for eccentric connections and splice connections are also covered with worked examples.

Chapter 8 deals with bolted connections. Simple connections and moment connections are discussed with figures and worked examples. This chapter starts with how to designate a bolt and different classes of bolt. Ready to use tables are given with gross area, net area, shear capacity and bearing capacity for various diameters and classes. Worked examples on bolted splices are also given.

Chapter 9 is devoted to the calculation of wind load on a structure as per IS:875-Part-3. This will enable the reader to completely understand the provision in IS:875-Part-3 and also to use the Indian standard for calculating wind load on any other type of structure.

Chapter 10 is for design of pitched roof truss. An example on analyzing a pitched roof using graphical statics is dealt with elaborately. The forces in each member are tabulated and grouping is done. Design of the tension and compression members of the truss including tie member between the trusses is provided. This chapter gives various configurations of trusses for various spans and empirical relation for calculating self weight.

Chapter 11 is devoted for calculating the force due to earthquake at various elevations of a structure. Worked example explains how to calculate the force due to earthquake. The worked examples are used to explain how the elevation of the structure affect the horizontal load due to seismic force.

Chapter 12 gives a draft design basis report which will be useful for consultants.

Chapter 13 gives templates for column and base plate with anchor bolts for use in design offices between design engineer and drafting section.

I am thankful to Dr. V. Gopalakrishnan for writing the foreword for this book.

I thank all my friends, students and family members who have encouraged me in writing this book.

S. Kanthimathinathan

Contents

Chapter 1

List of Symbols

Symbols used are based on IS: 800/2007 and shall have the following meanings with respect to the structure or member or condition, unless otherwise defined elsewhere:

A – Area of cross section

A_c – Area at root of threads

A_e – Effective cross-sectional area

A_{ef} – Reduced effective flange area

A_f – Total flange area

A_g – Gross cross-sectional area

A_{gf} – Gross cross-sectional area of flange

A_{go} – Gross cross-sectional area of outstanding (not connected) leg of a member

A_n – Net area of the total cross section

A_{nb} – Net tensile cross-sectional area of bolt

A_{nc} – Net cross-sectional area of the connected leg of a member

A_{nf} – Net cross-sectional area of each flange

A_{no} – Net cross-sectional area of outstanding (not connected) leg of a member

A_{pb} – Nominal bearing area of bolt on any plate

A_q – Cross-sectional area of a bearing (load carrying) stiffener in contact with the flange

A_s – Tensile stress area

A_{sb} – Gross cross-sectional area of a bolt at the shank

A_{tg} – Gross sectional area in tension from the centre of the hole to the toe of the angle section/Channel section, etc. (block shear failure, see 6.4 of IS: 800/2007) perpendicular to the line of force

A_{tn} – Net sectional area in tension from the centre of the hole to the toe of the angle perpendicular to the line of force (block shear failure, see 6.4 of IS: 800/2007)

A_V – Shear area

A_{Vg} – Gross cross-sectional area in shear along the line of transmitted force (block shear failure, see 6.4 of IS: 800/2007)

A_{Vn} – Net cross-sectional area in shear along the line of transmitted force (block shear failure, see 6.4 of IS: 800/2007)

a, b – Larger and smaller projection of the slab base beyond the rectangle circumscribing the column, respectively (see 7.4 of IS: 800/2007)

a_o – Peak acceleration

a_1 – Unsupported length of individual elements being laced between lacing points

B – Length of side of cap or base plate of a column

b – Outstand/width of the element

b_1 – Stiff bearing length, Stiffener bearing length

b_e – Effective width of flange between pair of blots

b_f – Width of the flange

b_i – Width of flange as an internal element

b_o – Width of flange outstand

b_p – Panel zone width between column flange at beam–column junction

b_s – Shear lag distance

b_t – Width of tension field

b_w – Width of outstanding leg

C – Centre to centre longitudinal distance of battens

C_m – Coefficient of thermal expansion

C_{my}, C_{mz} – Moment amplification factor about respective axes

c – Spacing of transverse stiffener

c_b – Moment amplification factor for braced member

c_m – Moment reduction factor for lateral torsional buckling strength calculation

c_s – Moment amplification factor for sway frame

D – Overall depth/diameter of the cross section

d – Depth of web, nominal diameter

d_2 – Twice the clear distance from the compression flange angles, plates or tongue plates to the neutral axis

d_h – Diameter of a bolt/rivet hole

d_o – Nominal diameter of the pipe column or the dimensions of the column in the depth direction of the base plate

d_p – Panel zone depth in the beam–column junction

E – Modulus of elasticity for steel

$E(T)$ – Modulus of elasticity for steel at T°C

$E(20)$ – Modulus of elasticity for steel at 20°C

E_p – Modulus of elasticity of the panel material

F_{cdw} – Buckling strength of un-stiffened beam web under concentrated load

F_d – Factored design load
F_n – Normal force
F_o – Minimum proof pretension in high strength friction grip bolts
F_{psd} – Bearing capacity of load carrying stiffener
F_q – Stiffener force
F_{qd} – Stiffener buckling resistance
F_{test} – Test load
$F_{\text{test, a}}$ – Load for acceptance test
$F_{\text{test.min}}$ – Minimum test load from the test to failure
$F_{\text{test.R}}$ – Test load resistance
$F_{\text{test, s}}$ – Strength test load
F_w – Design capacity of the web in bearing
F_x – External load, force or reaction
F_{xd} – Buckling resistance of load carrying web stiffener
f – Actual normal stress range for the detail category
f_1 – Frequency for a simple supported one way system
f_2 – Frequency of floor supported on steel girder perpendicular to the joist
f_a – Calculated stress due to axial force at service load
f_{abc} – Permissible bending stress in compression at service load
f_{ac} – Permissible compressive stress at service load
f_{abt} – Permissible bending stress in tension at service load
f_{apb} – Permissible bending stress of the bolt at service load
f_{asb} – Permissible bending stress of the bolt in shear at service load
f_{at} – Permissible tensile stress at service load
f_{atb} – Permissible tensile stress of the bolt at service load
f_{aw} – Permissible stress of the weld at service load
f_b – Actual bending stress at service load
f_{bc} – Actual bending stress in compression at service load
f_{bd} – Design bending compressive stress corresponding to lateral buckling
f_{br} – Actual bearing stress due to bending at service load
f_{bt} – Actual bending stress in tension at service load
f_{bs} – Permissible bending stress in column base at service load
f_c – Actual axial compressive at service load
f_{cc} – Elastic buckling stress of a column, Euler buckling stress
f_{cd} – Design compressive stress
$f_{cr,\,b}$ – Extreme fibre compressive stress corresponding elastic lateral buckling moment

f_e – Equivalent stress at service load
f_f – Fatigue stress range corresponding to 5×10^6 cycles of loading
f_{feq} – Equivalent constant amplitude stress
f_{fd} – Design normal fatigue strength
$f_{f\max}$ – Highest normal stress range
f_{fn} – Normal fatigue stress range
f_{nw} – Normal stress in weld at service load
f_o – Proof stress
f_p – Actual bearing stress at service load
f_{pb} – Actual bearing stress in bending at service load
f_{psd} – Bearing strength of the stiffeners
f_r – Frequency
f_{sb} – Actual shear stress in bolt at service load
f_t – Actual tensile stress at service load
f_{tb} – Actual tensile stress of the bolt at service load
f_u – Characteristic ultimate tensile stress
f_{ub} – Characteristic ultimate tensile stress of the bolt
f_{um} – Average ultimate stress of the material as obtained from test
f_{up} – Characteristic ultimate tensile stress of the connected plate
f_v – Applied shear stress in the panel designed utilizing tension field action
f_w – Actual stress of weld at service load
f_{wd} – Design stress of weld at service load
f_{wn} – Nominal strength of fillet weld
f_x – Maximum longitudinal stress under combined axial force and bending
f_y – Characteristic yield stress
f_y – (T) Yield stress of steel at T°C
f_y – (20) Yield stress of steel at 20°C
f_{yb} – Characteristic yield stress of bolt
f_{yf} – Characteristic yield stress of flange
f_{ym} – Average yield stress as obtained from test
f_{yp} – Characteristic yield stress of connected plate
f_{yq} – Characteristic yield stress of stiffener material
f_{yw} – Characteristic yield stress of web material
G – Modulus of rigidity for steel
g – Gauge length between centre of the holes perpendicular to the load direction, acceleration due to gravity

h – Depth of the section
h_b – Total height from the base to the floor level concerned
h_c – Height of the column
h_e – Effective thickness
h_f – Centre to centre distance of flanges
h_i – Thickness of fire protection material
h_L – Height of the lip
h_s – Storey height
h_y – Distance between shear centre of the two flange of a cross section
I – Moment of inertia of the member about an axis perpendicular to the plane of the frame
I_{fc} – Moment of inertia of the compression flange of the beam about the axis parallel to the web
I_{ft} – Moment of inertia of the tension flange of the beam about minor axis
I_q – Moment of inertia of a pair of stiffener about the centre of the web, or a single stiffener about the face of the web
I_s – Second moment of inertia
I_{so} – Second moment of inertia of the stiffener about the face of the element perpendicular to the web
I_T – Transformed moment of inertia of the one way system (in terms of equivalent steel, assuming the concrete flange of width equal to the spacing of the beam to be effective)
I_t – St. Venant's torsion constant
I_w – Warping constant
I_y – Moment of inertia about the minor axis of the cross section
I_z – Moment of inertia about the major axis of the cross section
K_b – Effective stiffness of the beam and column
K_h – Reduction factor to account for the high strength friction grip connection bolts in oversized and slotted holes
KL – Effective length of the member
KL/r – Appropriate effective slenderness ratio of the section
KL/r_y – Effective slenderness ratio of the section about the minor axis of the section
KL/r_z – Effective slenderness ratio of the section about the major axis of the section
$\left(\frac{KL}{r}\right)_o$ – Actual maximum effective ratio of the laced column
$\left(\frac{KL}{r}\right)_e$ – Effective slenderness ratio of the laced column accounting for shear deformation

K_v – Shear buckling coefficient
K_w – Warping restraint factor
k – Regression coefficient
K_{sm} – Exposed surface area to mass ratio
L – Actual length unsupported length, length centre to centre distance of the intersecting members, cantilever length
L_c – Length of end connection in bolted and welded members, taken as the distance between outermost fasteners in the end connection, or the length of the end weld, measured along the length of the member
L_{LT} – Effective length for lateral torsional buckling
L_m – Maximum distance from the restraint to the compression flange at the plastic hinge to an adjacent restraint (limiting distance)
L_o – Length between points of zero moment (inflecting) in the span
l – Centre to centre length of the supporting member
l_e – Distance between prying force and bolt centre line
l_g – Grip length of bolts in a connection
l_j – Length of the joint
l_s – Length between points of lateral support to the compression flange in a beam
l_v – Distance from bolts centre line to the toe of fillet weld or to half the root radius for a rolled section
l_w – Length of weld
M – Bending moment
M_a – Applied bending moment
M_{cr} – Elastic critical moment corresponding to lateral torsional buckling of the beam
M_d – Design flexural strength
M_{dv} – Moment capacity of the section under high shear
M_{dy} – Design bending strength about the minor axis of the cross section
M_{dz} – Design bending strength about the major axis of the cross section
M_{eff} – Reduced effective moment
M_{fr} – Reduced plastic moment capacity of the flange plate
M_{fd} – Design plastic resistance of the flange alone
M_{nd} – Design bending strength under combined axial force and uniaxial moment
M_{ndy}, M_{ndz} –Design bending strength under combined axial force and the respective uniaxial moment acting alone
M_p – Plastic moment capacity of the section
M_{pb} – Moment in the beam at the intersection of the beam and column centre lines
M_{pc} – Moment in the column above and below the beam surfaces

M_{pd} – Plastic design strength
M_{pdf} – Plastic design strength of flanges only
M_q – Applied moment on the stiffener
M_s – Moment at service (working) load
M_{tf} – Moment resistance of tension flange
M_y – Factored applied moment about the minor axis of the cross section
M_{yq} – Moment capacity of the stiffener based on its elastic modulus
M_z – Factored applied moment about the major axis of the cross section
N – Number of parallel planes of battens
N_d – Design strength in tension or in compression
N_f – Axial force in the flange
N_{sc} – Number of stress cycles
n – Number of bolts in the bolt group/critical section
n_e – Number of effective interfaces offering frictional resistance to slip
n_n – Number of shear planes with the threads intercepting the shear plane in the bolted connection
n_s – Number of shear planes without threads intercepting the shear plane in the bolted connection
P – Factored applied axial force
P_{cc} – Elastic buckling load
P_d – Design axial compressive strength
P_{dy}, P_{dz} – Design compression strength as governed by flexural buckling about the respective axis
P_e – Elastic Euler buckling load
P_{min} – Minimum required strength for each flange splice
P_r – Required compressive strength
P_s – Actual compressive at service load
P_y – Yield strength of the cross section under axial compression
p – Pitch length between centres of holes parallel to the direction of the load
p_s – Staggered pitch length along the direction of the load between lines of the bolt holes (see Figure 5 of IS: 800/2007)
Q – Prying force
Q_a – Accidental load (Action)
Q_c – Characteristic loads (Action)
Q_d – Design load (Action)
Q_p – Permanent loads (Action)
Q_v – Variable loads (Action)

q – Shear stress at service

R – Ratio of the mean compressive stress in the web (equal to stress at mid depth) to yield stress of the web; reaction of the beam at support

R_d – Design strength of the member at room temperature

R_i – Net shear in bolt group at bolt "i"

R_r – Response reduction factor

R_{tf} – Flange shear resistance

R_u – Ultimate strength of the member at room temperature

r – Appropriate radius of gyration

r_1 – Minimum radius of gyration of the individual element being laced together

r_f – Ratio of the design action on the member under fire to the design capacity

rvv – Radius of gyration about the minor axis (v-v) of angle section

r_y – Radius of gyration about the minor axis

r_z – Radius of gyration about the major axis

S – Minimum transverse distance between the centroid of the bolt group or weld group

S_c – Constant stress range

S_d – Design strength

S_o – Original cross section area of the test specimen

S_p – Spring stiffness

S_u – Ultimate strength

s_c – Anchorage length of tension field along the compression flange

s_t – Anchorage length of tension field along the tension flange

s_a – Actual stiffener spacing

T – Temperature in degree Celsius; Factored tension

T_b – Applied tension in bolt

T_{cf} – Thickness of compression flange

T_d – Design strength under axial tension

T_{dg} – Yielding strength of gross section under axial tension

T_{dn} – Rupture strength of net section under axial tension

T_{db} – Design strength of bolt under axial tension, Block shear strength at end connection

T_e – Externally applied tension

T_f – Factored tension force of the friction type bolt

T_i – Limiting temperature of the steel

T_{nb} – Nominal strength of bolt under axial tension

T_{nd} – Design tension capacity

T_{ndf} – Design tension capacity of friction type bolt
T_{nf} – Nominal tensile strength of friction type bolt
T_s – Actual tension under service load
t – Thickness of element/angle, time in minutes
t_f – Thickness of flange
t_p – Thickness of plate
t_{pk} – Thickness of packing
t_q – Thickness of stiffener
t_s – Thickness of base slab
t_t – Effective throat thickness of welds
t_w – Thickness of web
V – Factored applied shear force
V_b – Shear in batten plate
V_{bf} – Factored frictional shear force in friction type connection
V_{cr} – Critical shear strength corresponding to web buckling
V_d – Design shear strength
V_{db} – Block shear strength
V_{nb} – Nominal shear strength of bolt
V_{nbf} – Bearing capacity of bolt for friction type connection
V_p – Plastic shear resistance under pure shear
V_n – Nominal shear strength
V_{npb} – Nominal bearing strength of bolt
V_{nsb} – Nominal shear capacity of bolt
V_{nsf} – Nominal shear capacity of bolt as governed by slip in friction type connection
V_s – Transverse shear at service load
V_{sb} – Factored shear force in the bolt
V_{sd} – Design shear capacity
V_{sdf} – Design shear strength in friction type bolt
V_{sf} – Factored design shear force of friction type bolts
V_t – Applied transverse shear
V_{tf} – Shear resistance in tension field
W – Total load
w – Uniform pressure from below on the slab base due to axial compression under the factored load
w_{tf} – Width of tension field
x_t – Torsional index

Z_e – Elastic section modulus

Z_{ec} – Elastic section modulus of the member with respect to extreme compression fibre

Z_{et} – Elastic section modulus of the member with respect to extreme tension fibre

Z_p – Plastic section modulus

Z_v – Contribution to the plastic section modulus of the total shear area of the cross section

y_g – Distance between point of application of the load and shear centre of the cross section

y_s – Co-ordinate of the shear centre in respect to centroid

α – Imperfection factor for buckling strength in columns and beams

α_t – Coefficient of the thermal expansion

β_M – Ratio of smaller to the larger bending moment at the ends of a beam column

β_{My}, β_{MZ} – Equivalent uniform moment factor for flexural buckling for *y–y* and *z–z* axes respectively

β_{MLT} – Equivalent uniform moment factor for lateral torsional buckling

χ – Strength reduction factor to account for buckling under compression

χ_m – Strength reduction factor, χ, at f_{ym}

χ_{LT} – Strength reduction factor to account for later torsional buckling of beams

δ – Storey deflection

δ_L – Horizontal deflection of the bottom of storey due to combined gravity and notional load

δ_p – Load amplification factor

δ_U – Horizontal deflection of the top of storey due to combined gravity and notional load

Φ – Inclination of the tension field stress in web

γ – Unit weight of steel

γ_f – Partial safety factor for load

γ_m – Partial safety factor for material

γ_{mo} – Partial safety factor against yield stress and buckling

γ_{m1} – Partial safety factor against ultimate stress

γ_{mb} – Partial safety factor for bolts connection with bearing type bolts

γ_{mf} – Partial safety factor for bolts connection with high strength friction grip bolts

γ_{fft} – Partial safety factor for fatigue load

γ_{mft} – Partial safety factor for fatigue strength

γ_{mv} – Partial safety factor against shear failure

γ_{mw} – Partial safety factor for strength of weld

ε – Yield stress ratio, $\sqrt{250/f_y}$
λ – Non-dimensional slenderness ratio = $\sqrt{f_y\,(KL/r)^2/\pi^2 E} = \sqrt{f_y/f_{cc}} = \sqrt{P_y/P_{cc}}$
λ_{cr} – Elastic buckling load factor
λ_e – Equivalent slenderness ratio
λ_{LT} – Non-dimensional slenderness ratio in lateral bending
λ_{scr} – Elastic buckling load factor of each storey
μ – Poisson's ratio
μ_c – Correction factor
μ_f – Coefficient of friction (slip factor)
μ_r – Capacity reduction factor
θ – Ratio of the rotation at the hinge point to the relative elastic rotation of the far end of the beam segment containing plastic hinge
ρ – Unit mass of steel
τ – Actual shear stress range for the detail category
τ_b – Buckling shear stress
τ_{ab} – Permissible shear stress at the service load
$\tau_{cr,\,e}$ – Elastic critical shear stress
τ_f – Fatigue shear stress range
τ_{fmax} – Highest shear stress range
τ_{fd} – Design shear fatigue strength
τ_{fn} – Fatigue shear stress range at Nsc cycle for the detail category
τ_v – Actual shear stress service load
ψ – Ratio of the moments at the ends of the laterally unsupported length of a beam
Γ – Frame buckling load factor

Chapter 2

Introduction

Structures are required to support loads, and to transfer these loads to the foundations of the structures. Structural engineer's responsibility is to frame a structural system which will transfer the loads acting on it safely, meeting the requirements of serviceability. The structural engineer should consult the owner, the architect, the construction team, mechanical and process engineers before finalising the arrangement of structural components.

The arrangement of structure for a building is based on many factors. Some such factors are listed below.

(1) Functional requirement including lighting and ventilation.

(2) Process requirement

(3) Environmental factors

(4) Elevations of platforms based on equipment disposition.

(5) Maintenance requirement

(6) Accessibility

(7) Soil condition

(8) Utility services like gas ducts, pipes, etc. The structural component should not interfere with these service lines.

(9) Head room requirement for erecting and maintaining equipment.

(10) Crane capacity and span.

Apart from other requirements specific to the process, keeping all the above in mind, the structural engineer should bring out a suitable arrangement. Such arrangement should meet the functional and process requirements. It should also result in economy, safety and durability.

Advantages of Using Steel Structure

(1) It is possible to have a smaller section since steel is capable of carrying higher load compared to other materials. Steel shapes and plates of different grades with different yield/ultimate stresses are available.

(2) It is homogeneous and has long life if protected from corrosion.

(3) Pre-rolled sections are available for use. If the designer requires, he can arrange for built-up section of his choice by adding members (or) plates to rolled shapes.

(4) It is possible to strengthen (reinforce) by adding members (or) plates even after fabrication and erection.

(5) Steel structures can be easily dismantled and relocated. While relocating, the designer's responsibility is to check its adequacy for the environmental forces (wind and earthquake forces) at the proposed site.

Expectation from a Steel Structural Designer

(1) Steel structure should have adequate load carrying capacity.

(2) It should be stable.

(3) Minimum weight. This is possible only when the load coming over the structural member is estimated correctly and the orientation of the structural element is correct.

(4) Labour required for fabrication and erection is minimum. Here a conflicting situation arises. For welded structures, the labour at the shop and at erection site is more. For bolted structure, the work at the shop is more but the labour at erection site is less. It is better to have all the shop joints welded and site joints bolted.

(5) Erection is to be done at shortest time.

(6) Members can be identical. Alternatively, we can say variety reduction. This reduces the variety of sections, thickness of plates required for a project. Due to this reduction, buying becomes easier. At times it may not be possible to buy small quantity of quality material since suppliers will insist to buy minimum order quantity. But there is another dimension to this. Unless variety reduction is done judiciously by the designer, there is a possibility of sacrificing minimum weight.

Identical members help interchangeability at site.

Analysis

Preliminary selection of size of member should be done with utmost care. Here the experience of the designer on similar structures will help.

Checking the capacity (strength) of the selected section (member), stability and rigidity comes under analysis. The designer may have to perform many iterations before arriving at the most suitable and economical member size.

Before analysis, the designer should decide the following:

(1) Type of connection–welded or bolted.

(2) End condition–fixed or simply supported

(3) Support condition at base–fixed or hinged. For this, the designer should decide the type of foundation.

(4) Method of analysis–working stress method or limit state method or plastic analysis.

The designer will have to use a number of formulae with many factors and coefficients while designing a structure. He has to select more appropriate factor and coefficient to get the correct/best/optimal design.

If a structural member no longer meets the service requirements, that is, its limiting stage, this means the member cannot withstand the load coming over (or) the member may undergo local damage (or) excessive deformation. The first limiting state is determined by its strength (or) load carrying capacity. The second is development of excessive deformation.

Design force in a member due to the worst combination should be less than the strength of the member (or) load carrying capacity of the member. This state is the first limiting state. The design load used in the limit state method is not the actual load (working or service load). The design load is arrived at by multiplying the actual (working or service) load by a factor called partial safety factor for loads while designing a structure using the limit state method. The actual load multiplied by partial safety factor for loads is called the factored load. The partial safety factors for loads used for arriving at design load from actual or working or service load is more than unity and the same is given in Table 4 of IS:800:2007.

The design strength of the member shall be more than (or) equal to the design force (factored load) to meet the strength requirement. The design strength is calculated based on geometry, length, section modulus, radius of gyration, yield stress and ultimate stress. While considering the yield stress (or) ultimate stress, a factor called partial safety factor for materials is used. This is presented in Table 5 of IS:800:2007.

The second limiting state is determined by the deflection of the member which depends on service load, span of the member, Young's modulus and moment of inertia. To meet this limiting state, the member should have sufficient stiffness. Many times the designer may not have the options to reduce the load, span or change the material with higher Young's modulus. But the designer can increase the moment of inertia by increasing the width/thickness/depth of the member. The best option is to increase the depth since the moment of inertia depends on depth to the power of three. The deflection limit for members used as various structural elemets are given in Table 6 of IS:800:2007.

In the allowable (or) working stress method of design, the actual load acting on the structure is considered. The actual load is not multiplied by a factor as in the case of limit state method. The stress induced in the structural members by the external actual load should not exceed the allowable stress as in the codes and standards. The allowable stress is a fraction of yield stress of the material. This fraction is to account for variation in loads, variation in the geometrical property and mechanical property of the material. This fraction varies with respect to type of load and mode of failure.

For example, let us consider a tension member. The member is subjected to a tensile force of T. The permissible (or) allowable stress in the steel member under direct tension = $0.6\, f_y$

$$\text{Stress} = \frac{\text{Load}}{\text{Area}} = \frac{T}{\text{Area}} = 0.6\, f_y$$

$$\therefore \qquad \text{Area} = \frac{T}{0.6\, f_y} = 1.67 \times \left(\frac{T}{f_y}\right)$$

This 1.67 is the factor of safety.

In the limit state method, the same tensile force will be multiplied by the partial safety factor Therefore, factored tensile load or forssce = 1.5 T

$$\text{Area} = \frac{\text{Load}}{\text{Stress}} = \frac{1.5T}{f_y/(\gamma_{mo})} = 1.5 \times 1.1\left(\frac{T}{f_y}\right) \text{ where } \gamma_{mo} = 1.1$$

$$= 1.65\left(\frac{T}{f_y}\right)$$

In the limit state method, the load is given a separate factor called partial safety factor for loads as presented in Table 4 of IS:800:2007 and the stress is given a separate factor called partial safety factor for materials as presented in Table 5 of IS:800:2007.

Finally, it can be noticed that the factor is 1.67 in the case of allowable (or) working stress design and 1.65 in the case of limit state design.

But this is not same for compression member. Let us consider a compression member for which the slenderness ratio (= KL/r) = 100 and f_y = 250 MPa. The permissible compressive stress as in working stress method (IS:800/1984 Table 5.1) = 80 MPa.

Let us refer to IS:800/2007 Table 9(a), 9(b), 9(c) and 9(d) for limit state method. For the same value of the slenderness ratio (= KL/r) = 100 and f_y = 250 MPa, we get

As per Table 9(a), f_{cd} = 132 MPa

As per Table 9(b), f_{cd} = 118 MPa

As per Table 9(c), f_{cd} = 107 MPa

As per Table 9(d), f_{cd} = 92.6 MPa

In Working Stress Method

Load = Area × 80 MPa

∴ Area = (Load/80) mm^2

In Limit State

(a) Buckling class a

$$1.5 \times \text{Load} = \text{Area} \times 132$$

$$\therefore \quad \text{Area} = \left(\frac{1.5 \times \text{load}}{132}\right) = \left(\frac{\text{load}}{88}\right) \text{mm}^2$$

(b) Buckling class b

$$1.5 \times \text{Load} = \text{Area} \times 118$$

$$\therefore \quad \text{Area} = \left(\frac{1.5 \times \text{load}}{118}\right) = \left(\frac{\text{load}}{78.7}\right) \text{mm}^2$$

(c) Buckling class c

$$1.5 \times \text{Load} = \text{Area} \times 107$$

$$\therefore \quad \text{Area} = \left(\frac{1.5 \times \text{load}}{107}\right) = \left(\frac{\text{load}}{71.3}\right) \text{mm}^2$$

(d) Buckling class d

$$1.5 \times \text{Load} = \text{Area} \times 92.6$$

$$\therefore \quad \text{Area} = \left(\frac{1.5 \times \text{load}}{92.6}\right) = \left(\frac{\text{load}}{61.7}\right) \text{mm}^2$$

Now it can be noticed that the area required (or) load carrying capacity of the member in compression is different for different buckling classes under the limit state method of design. The required area calculated using the working stress method does not take into account the buckling class. The buckling class for various cross sections are given in Table 10 of IS:800:2007. This depends on the height to width ratio and thickness of the flange. From the above working, we can understand that a member with buckling class will be economical under the limit state method when compared to the working stress method. For other buckling classes, it is different.

In the design of members using allowable (or) working stress method, the structural member is considered in working state. In the case of the limit state method, the member is considered in limit state. Hence, it is necessary to find the limiting states based on failure mode of the structural member. This has resulted in various safety factors as against a single factor of safety in the case of working stress method. In working the stress method, the factor of safety varies based on type of load. The partial safety factor for load, partial safety factor for material, buckling class, effective length for simply supported beams taking into account torsional restraint and working restraint imperfection factor and other factors are more scientifically chosen for the design of structural members using limit state method.

Many engineers have started asking whether the limit state method will bring economy compared to working stress method. To put it in simple terms "will the weight of the structural members will come down as compared to working stress method"?

Limit state method takes care of all the factors scientifically as explained above and the same will be safe. Achieving economy rests with the design engineer who applies the optimum load and optimises the design.

Structural Steel

Eight grade designations are available as per IS: 2062: 2006. There are four sub-qualities for grade designations E250 to E450 and two sub-qualities are available for grade designations E 550 to E650. The grade designations are based on yield stress. For grade designation E250, the yield stress is 250 N/mm^2.

The sub-qualities *A*, *BR*, *BO* and *C* are generally used for welding process.

Sub-quality *A* stands for "Impact test not required, semi-killed/killed. Sub-quality *BR* stands for "impact test optional, at room temperature (25 ± 2°C) if required, killed.

Sub-quality *BO* stands for "Impact test mandatory at 0°C, killed".

Sub-quality *C* stands for "Impact test mandatory at – 20°C, killed".

Normally the Indian rolled shapes *I*, channel and angle are of grade designation E250 sub-quality *A*.

Poor hardening, high plasticity and weldability are the properties of low carbon steel. The carbon content should be less than or equal to 0.23% for low carbon steel.

2.1 STRUCTURAL COMPONENTS OF AN INDUSTRIAL SHED

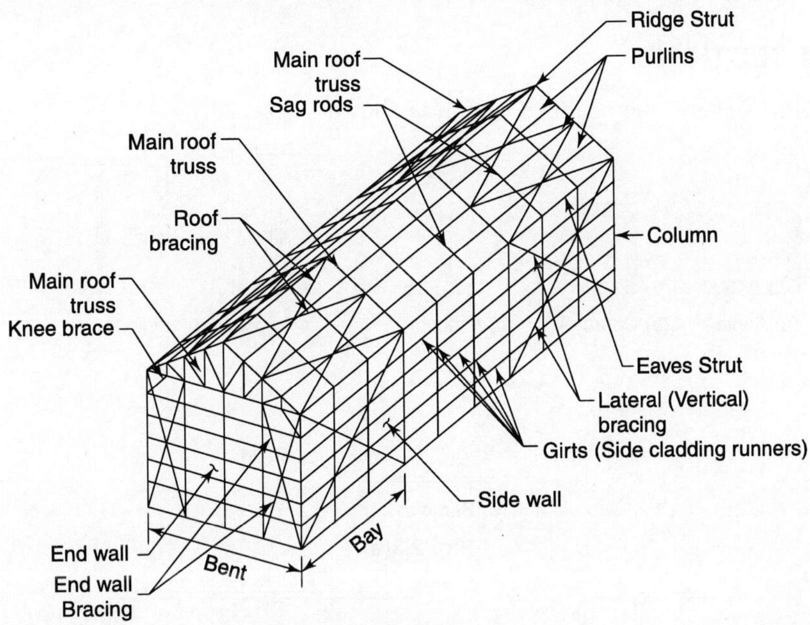

Fig. 2.1

2.2 ROLLED SECTIONS USED FOR STRUCTURES

(i) Angle section
(ii) Channel section
(iii) *I*-section
(iv) Z-section (thin gauge)
(v) Solid circular section (rods)
(vi) Hollow circular section (tubes and pipes)
(vii) Plate section
(viii) Solid square section
(ix) Hollow square and rectangular section

Some Compound and Built-up Sections

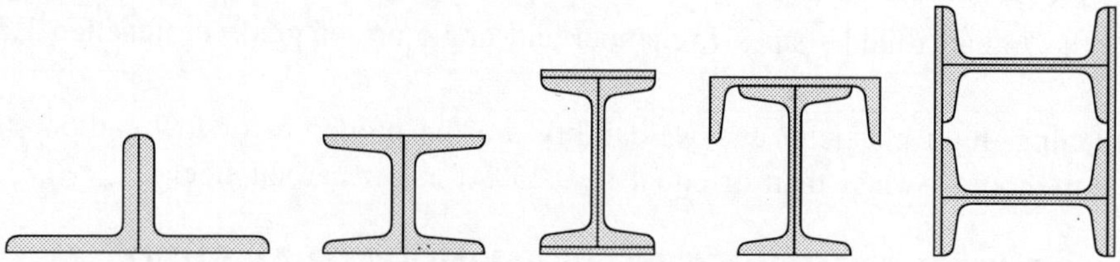

Fig. 2.2 Compound and built-up sections

2.3 BEAM SECTIONS

Sections used for flexural members (beams) are shown below.

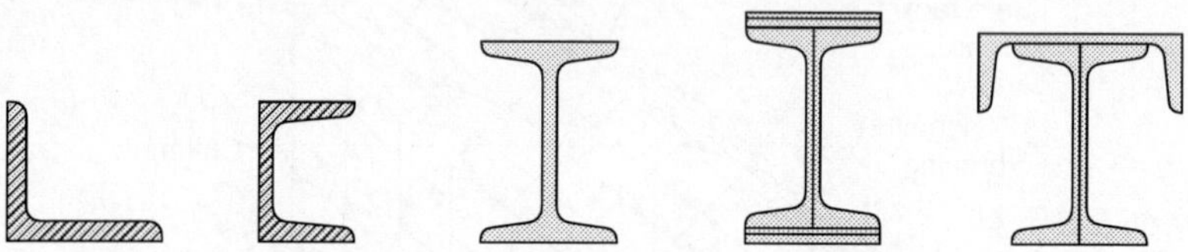

(a) Angle (b) Channel (c) I-section (d) Built-up sections

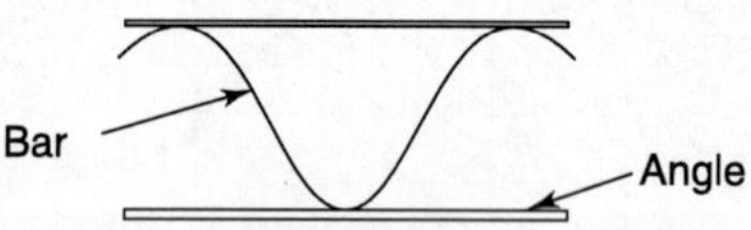

(e) Bar joist

Fig. 2.3(a)

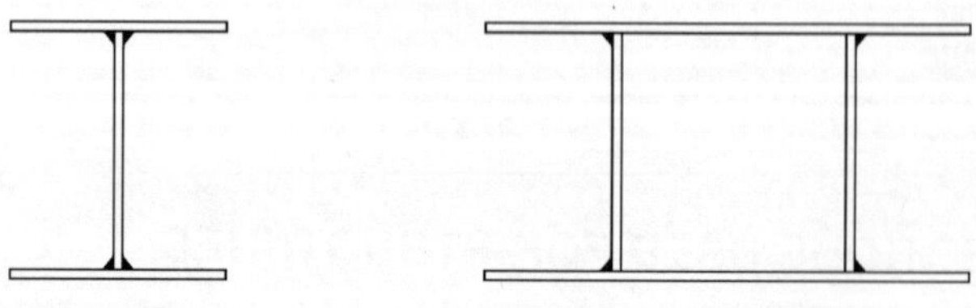

(f) Plate girder (g) Double web box girder

Fig. 2.3(b)

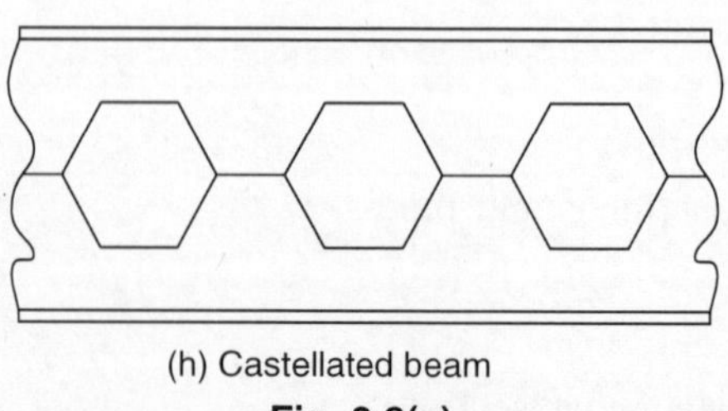

(h) Castellated beam

Fig. 2.3(c)

2.4 TYPICAL TENSION MEMBERS

Sections used for tension members are shown below.

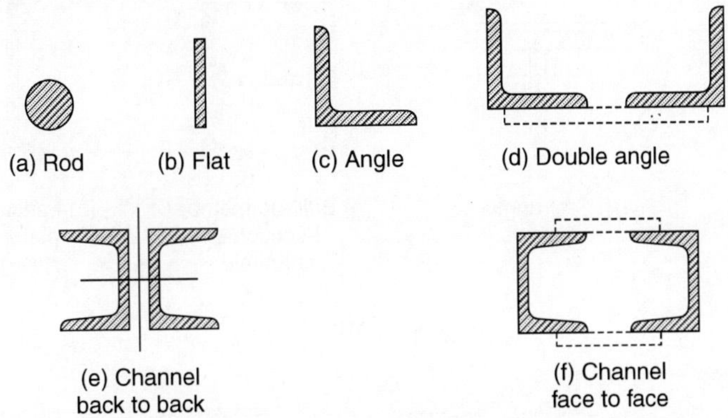

(a) Rod (b) Flat (c) Angle (d) Double angle (e) Channel back to back (f) Channel face to face

Fig. 2.4(a)

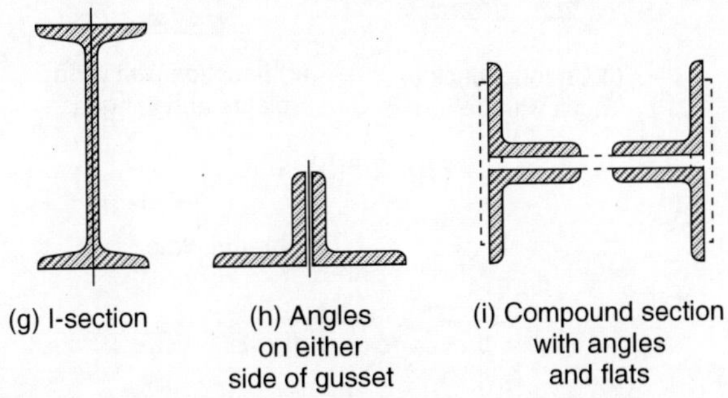

(g) I-section (h) Angles on either side of gusset (i) Compound section with angles and flats

Fig. 2.4(b)

2.5 TYPICAL COMPRESSION MEMBERS

Sections used for compression members, columns and beam-columns are shown below:

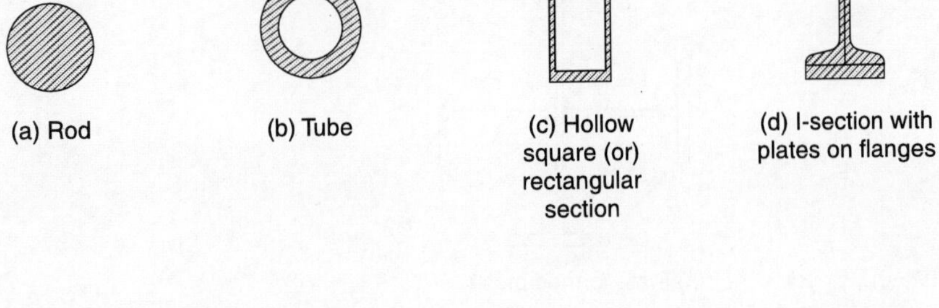

(a) Rod (b) Tube (c) Hollow square (or) rectangular section (d) I-section with plates on flanges

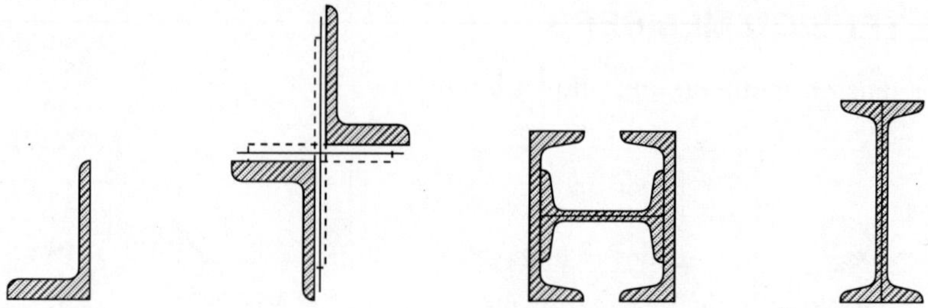

(e) Angle

(f) Star angle

(g) Built-up member i-section & channels

(h) I-section with plates on flanges

Fig. 2.5(a)

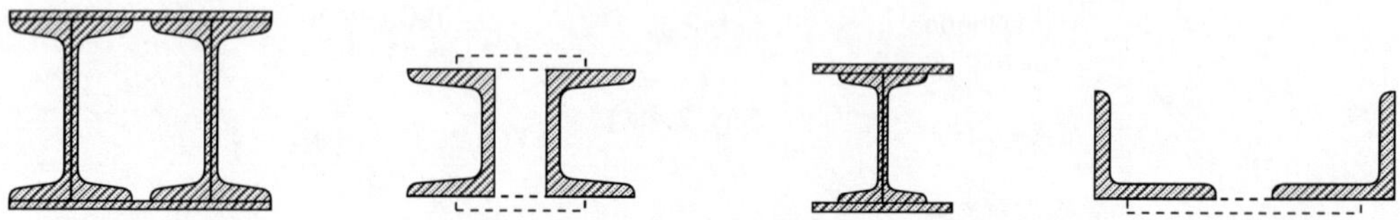

(i) Double I-section with cover plate

(j) Channel back to back with batten

(k) I-section with wide plates and flanges

(l) Double angle with flat lacing

Fig. 2.5(b)

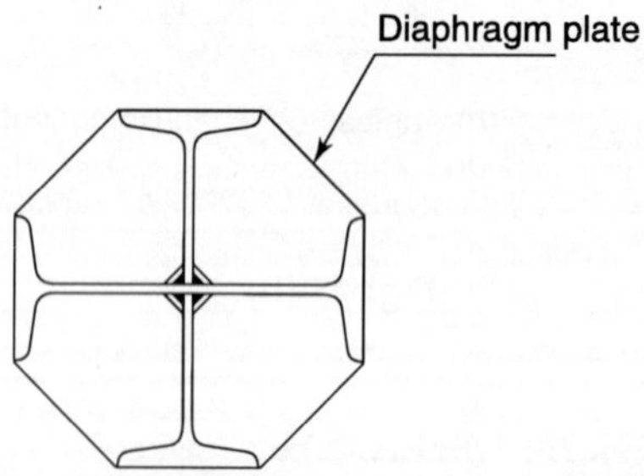

(m) Plus-i-using rolled i-section

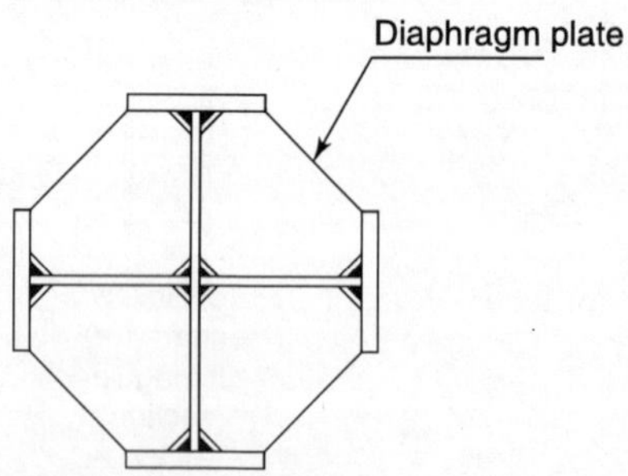

(n) Plate formed plus-i

Fig. 2.5(c)

2.6 DESIGNATION FOR PHYSICAL PROPERTY OF SECTION

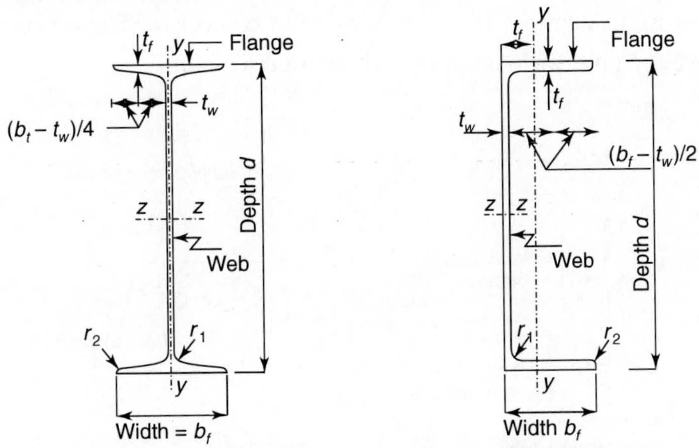

Rolled steel beam section Rolled steel channel section

Fig. 2.6(a)

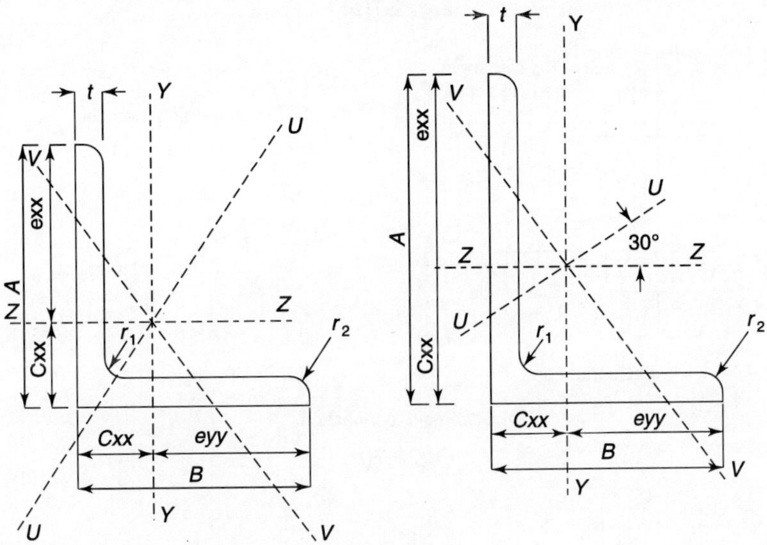

Fig. 2.6(b)

2.7 TYPES OF CONNECTION AND BEHAVIOUR

(a) Shear connection (or) simple connection: This will transfer only shear. Cleat angles or fin plate connected to web of beam is an example (Fig. 2.7(a)).

(b) Semi-rigid connection: This will transfer shear and partial moment (Fig. 2.7(b)).

(c) Rigid connection: This will transfer shear and moment. End plate connection is an example (Fig. 2.7(c)).

Figures showing typical arrangement of these connections using bolts are provided below and in chapter relevant to connection design. These connections can also be made by welding in place of bolts.

Rotation of the members is also shown against relevant figures to give an idea. It can be noticed that the supporting member is not rotating with respect to simple connection.

Figures 2.8 *a* to 2.8 *j* give various types of bolted connections.

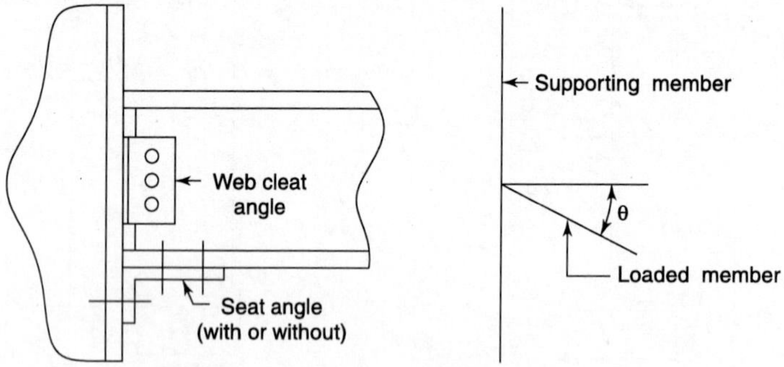

(i) Connection (ii) Simple joint

(a) Shear connection

Fig. 2.7(a)

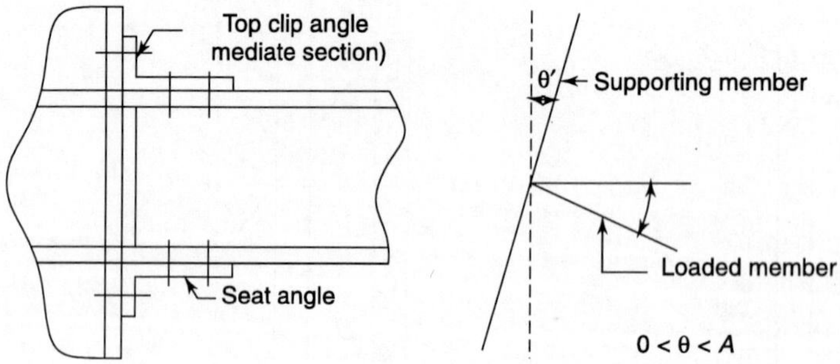

(i) Connection (ii) Semi-rigid joint

(b) Semi-rigid connection

Fig. 2.7(b)

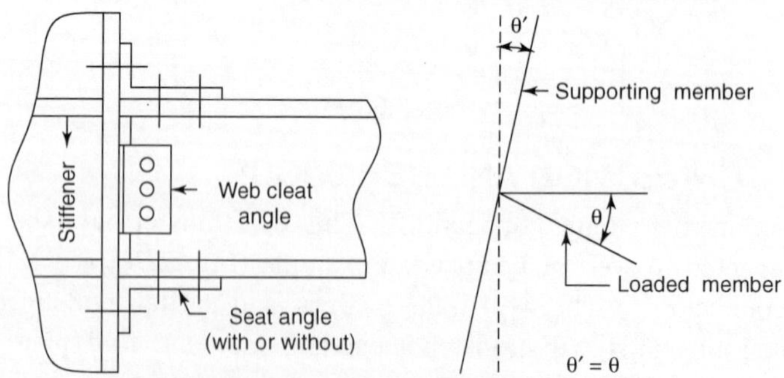

(i) Connection (ii) Rigid joint

(c) Rigid connection

Fig. 2.7(c)

2.8 TYPES OF BOLTED CONNECTIONS

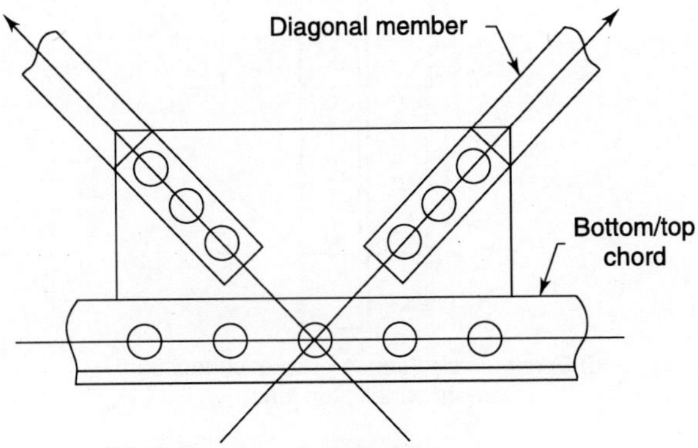

(a) Connection at a truss joint

Fig. 2.8(a)

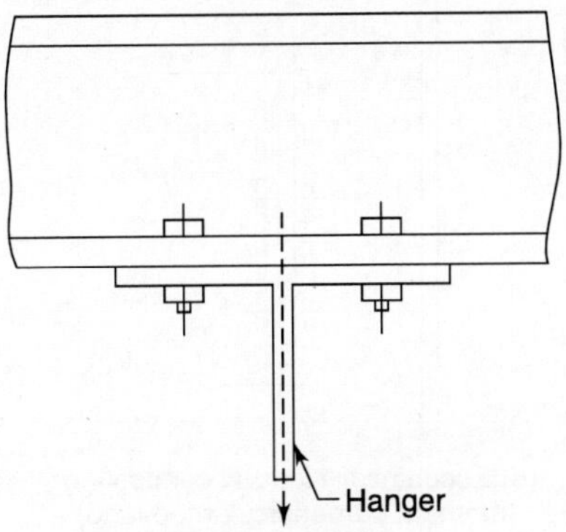

(b) Hanger connection

Fig. 2.8(b)

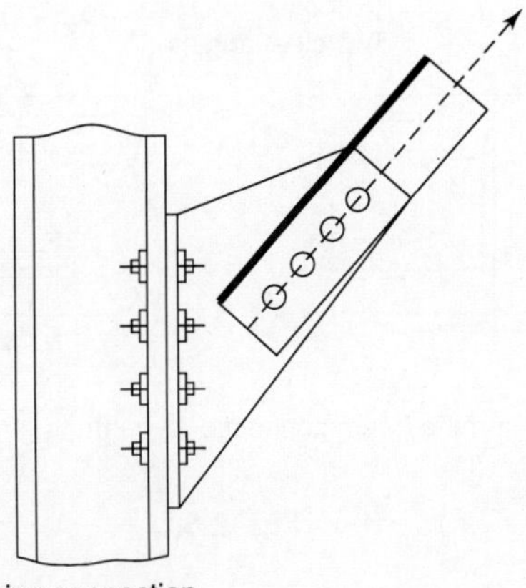

(c) Bracing connection

Fig. 2.8(c)

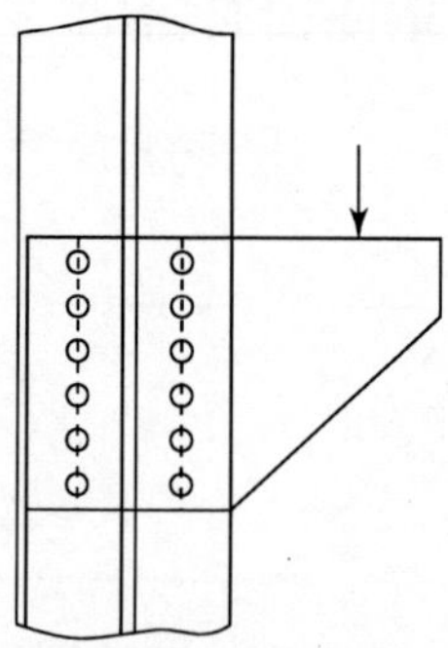

(d) Eccentrically loaded bracket connection
(moment parallel top lane)

Fig. 2.8(d)

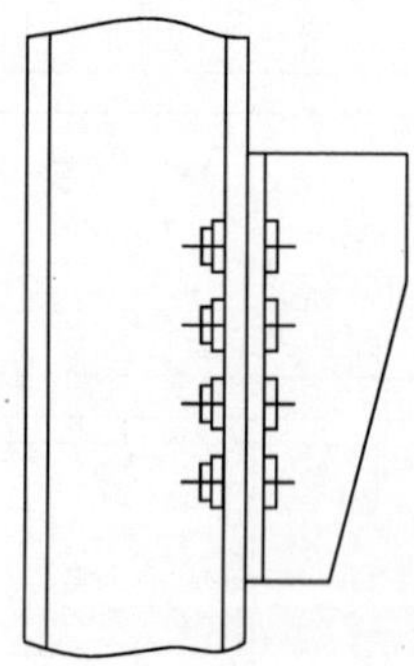

(e) Eccentrically loaded connection
(moment perpendicular top lane)

Fig. 2.8(e)

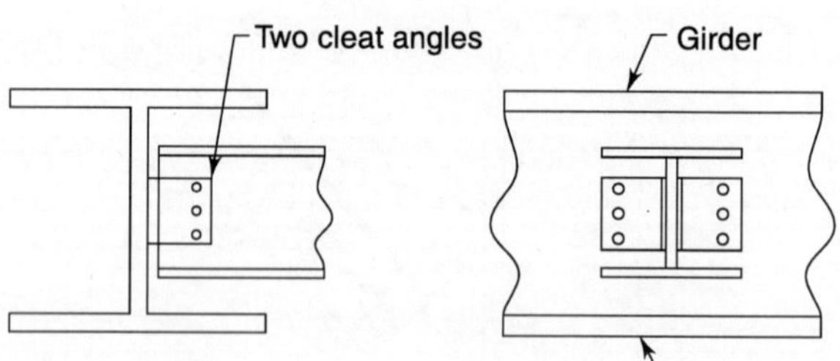

(f) Simple beam connections–beam to girder

Fig. 2.8(f)

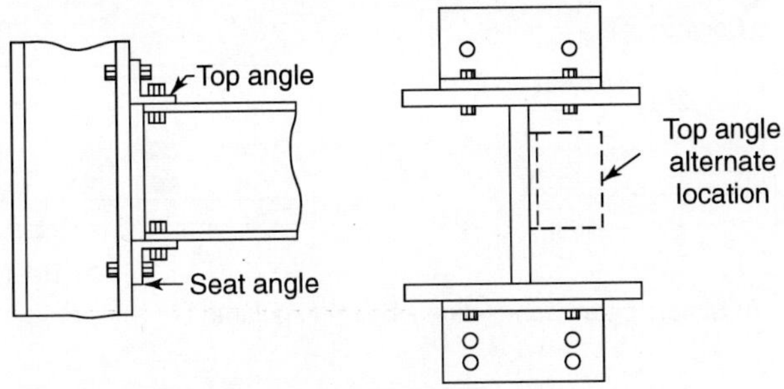

(g) Unstiffened seated beam connection

Fig. 2.8(g)

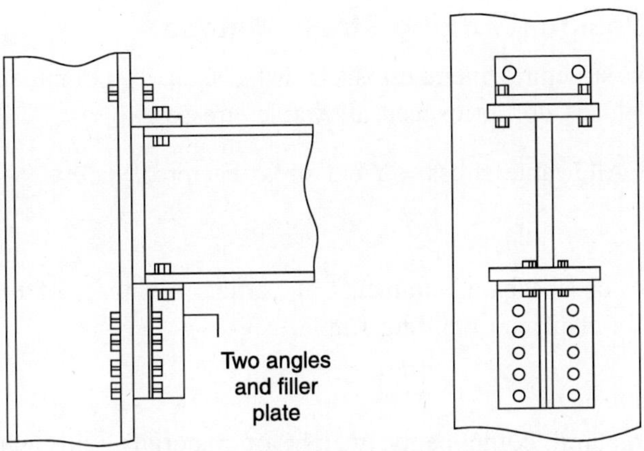

(h) Stiffened seated beam connection

Fig. 2.8(h)

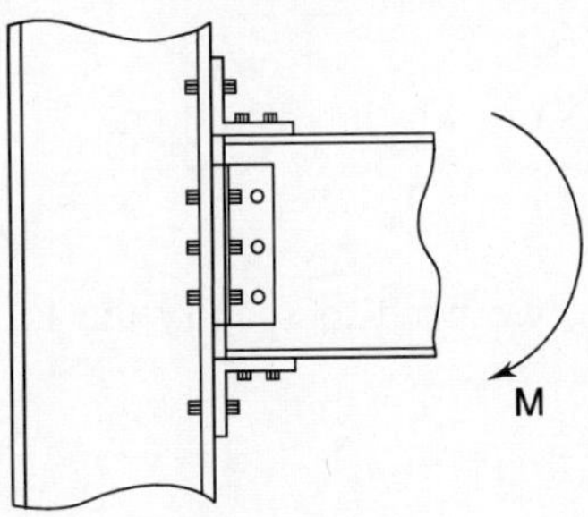

(i) Moment connection

Fig. 2.8(i)

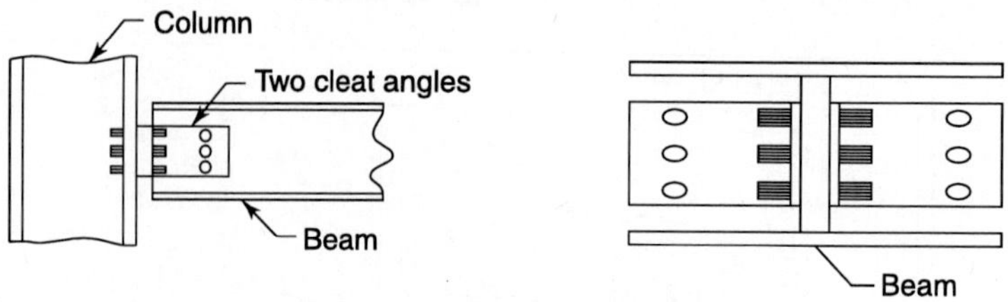

(j) Simple beam connections–beam to column

Fig. 2.8(j)

2.9 DEFINITIONS OF CERTAIN TERMS

Allowable Stress Design (Working Stress Method)

A method of sizing the structural members such that elastically computed stress produced in the members by actual loads do not exceed allowable stresses.

Allowable stress = Yield stress/Factor of safety

P-Delta Effect

The second order effect on shear and moments of vertical members of the frame induced by axial loads on a laterally displaced building frame.

Forces

Self weight of the structural components, weight of materials with which the building is constructed (example: walls, wall cladding), weight of the occupants, weight of stored materials, wind pressure are called forces (or) loads on the structure. The force due to earthquake is also to be considered.

Unit of Measurement of Force

Forces are measured in newtons (N) or kN in SI system.

External Forces

To fully describe a load (or) force, we need to specify the following characteristics of load:

The magnitude

The direction

The location

Resultants and Components

A single force replacing two or more forces which has the same effect, i.e., which is equivalent to the sum of the effects of other forces can be calculated. This single force is called the resultant (Fig. 2.9).

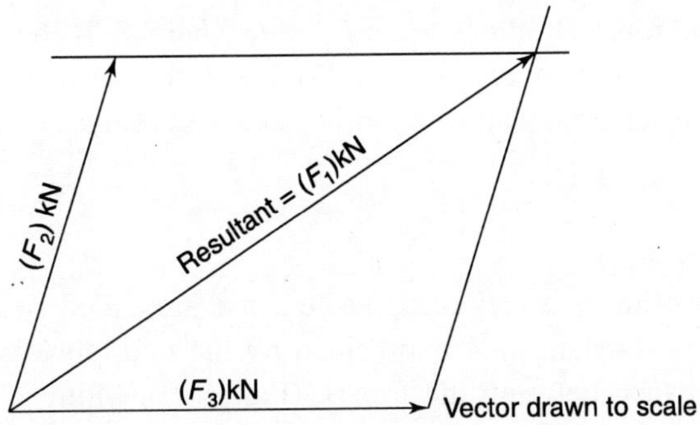

Fig. 2.9

To Replace a Single Force F_1 by Two Forces F_2 and F_3

The forces F_2 and F_3 are called the components of the single force F_1. Force F_1 can be replaced by two mutually perpendicular components F_2 and F_3 acting in directions '*a*' and '*b*' at point *A* as below.

$$F_2 = F_1 \cos \theta \text{ and } F_3 = F_1 \cos \alpha$$

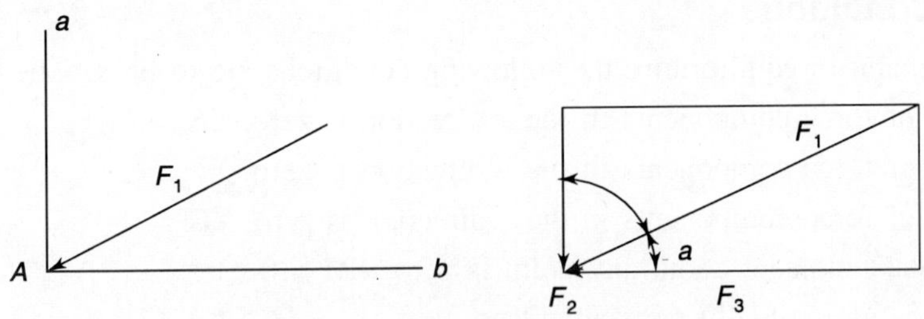

Fig. 2.10

Moments

The moment of a force about a point is obtained by multiplying the force and the perpendicular distance from the point to the line of force (Fig. 2.11). The perpendicular distance is called the 'lever arm'. Moments have units of Newton meters, Nm or kNm, in SI units. Let us adopt a convention that clockwise moments are positive, and anticlockwise are negative.

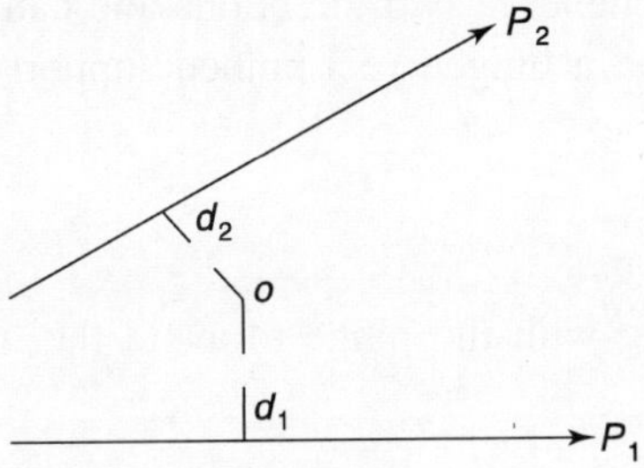

Fig. 2.11

The moment due to the force P_1 about 'o' is $P_1 \times d_1$ where d_1 is the lever arm. The rotation is anticlockwise and hence the moment is –ve. The moment due to the force P_2 about 'o' is $P_2 \times d_2$, where d_2 is the lever arm. The rotation is clockwise. Hence, it is +ve. Net moment at $o = P_2d_2 - P_1d_1$.

Couple of Forces

Two equal, parallel, non-collinear forces of opposite sense are called couple of forces. Moment at any point in its plane is constant, and is obtained by the multiplication of one of the forces and the perpendicular distance between the forces, (Fig. 2.12). Moment = $Q \times a$.

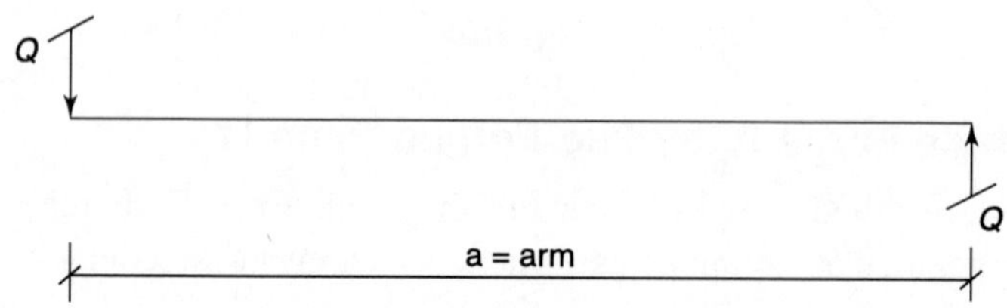

Fig. 2.12

Equilibrium Conditions

For a body to remain in equilibrium, the following conditions are to be satisfied:

(a) The sum of force components in the x direction is zero, $\Sigma F_x = 0$

(b) The sum of force components in the y direction is zero, $\Sigma F_y = 0$

(c) The sum of force components in the z direction is zero, $\Sigma F_z = 0$

(d) The resultant moment about any point is zero, $\Sigma M = 0$

The x, y, z axes are mutually perpendicular.

Reactions at Supports

For the structural component to remain in equilibrium, the supports should be capable of carrying the forces. These forces at the support are called reactions, and they must be calculated before the structure is analysed.

In single-plane structures, we are concerned with three types of supports:

(*a*) A support which permits rotation of a structural component about its axis, capable of providing reaction components in two directions and can prevent motion in two perpendicular directions is called a hinged (or) pinned support (Fig. 2.13(a)).

P_r is the resultant.

P_1 is the shear.

P_2 is the vertical reaction.

(*b*) A roller is provided along with the hinge to avoid the force P_1 (Fig. 2.13(b)).

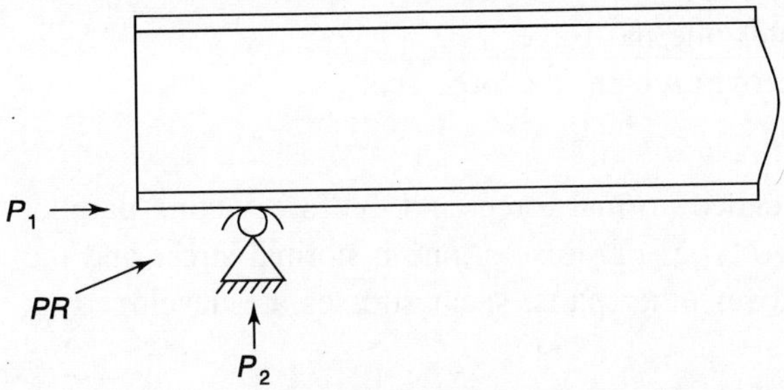

Fig. 2.13(a)

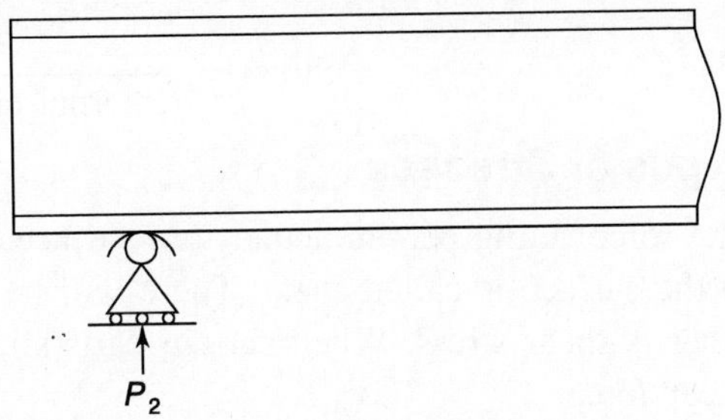

Fig. 2.13(b)

(*c*) A support which prevents rotation, prevents movements in two perpendicular directions and gives three reaction components is called a fixed support (Figure 2.13(c). M = moment.

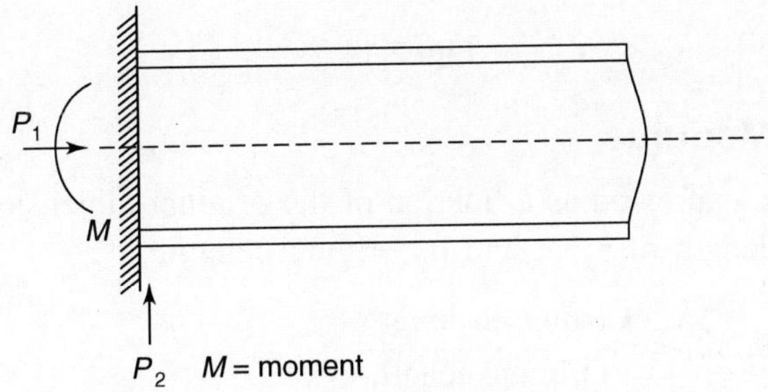

Fig. 2.13(c)

In short, a roller gives a single reaction component. *A* pin gives two components. A fixed support gives three components (two forces and a moment).

The resultant of load is a force equal in magnitude to the total load, acting at the centre of gravity of the loads.

Intensity of Pressure

Load is distributed over an area. When a load is distributed over the external surface of a structure, the load per unit area is referred to as pressure. Pressure is defined as force per unit area:

$$\text{Pressure } (P) = \frac{\text{Total load}}{\text{Surface area over which the load acts}}$$

Stress

Stress is defined as internal force per unit area. Pressure refers to external forces and surface areas, the term stress is used for internal forces and internal areas.

$$\text{Stress} = \frac{\text{Total internal force}}{\text{Internal area over which the force acts}}$$

Types of Stresses

The force acting perpendicular to a surface is called normal force, and the force acting parallel to the surface is called shear. There will be two types of stresses, one is normal stress and the other is shear stress. When certain parts slide over other parts, shear stresses are developed by shear force.

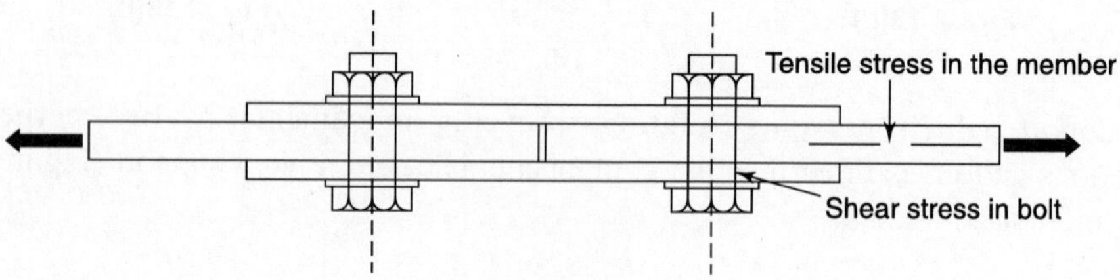

Fig. 2.14

Strain and Young's Modulus

The change in dimension, expressed as a fraction of the original dimension, is called the strain. Example, the change in length of a bar and its original length.

$$\text{Strain} = \frac{\text{Change in length}}{\text{Original length}}$$

The ratio between stress and strain is called Young's modulus of elasticity (E).

$$\text{Stress} = E \times \text{Strain}$$

Young's modulus is a measure of the stiffness of a material. Materials having a large value of E can be used to form stiff structures.

Permissible Stress (or) Allowable Stress

Safety is the first and foremost requirement of a structure. Every structure is designed to have a reserve strength above that required to take up the externally applied loads.

$$\text{Permissible Stress} = \frac{\text{Yield stress}}{\text{Factor of safety}}$$

For example, the yield stress of a particular type of steel may be 250 MPa. With a safety factor of safety of 1.67, the permissible stress in tension will be (250/1.67) = 150 MPa.

Ties and Struts

A structural member carrying an axial tensile force is called a tie. A structural member carrying an axial compressive force is called a strut or a column.

Portal Frame

A rigid frame in a single plane fixed at base is called a portal frame or rigid frame.

2.10 LOADS ON STRUCTURES

The loads acting on steel structures are dead load, imposed load including both gradually applied and dynamic load, wind load, seismic load, earth or groundwater load and those due to temperature changes and foundation settlement.

Dead Load

Dead loads are those loads which will be applied permanently during the life of the buildings. They are the gravity forces. They are from the mass of the structure, finishes, cladding and all other permanent parts of the building. They can be accurately estimated of all loads, since they are calculated from the known densities of the materials (IS:875:Part-1) used in construction.

Imposed Load

Imposed loads are the loads applied to the floor of a building as a result of the activities carried on within the building based on functional requirement. Consequently, they result from the type of occupancy of a particular part of a building, and are usually expressed in kN per square metre of floor area. (IS: 875: Part-2). Imposed load on a roof are those produced as follows:

(a) During maintenance and (b) During the life of the structure by movable objects.

Wind Loads

Air in motion is called wind. When the flow of wind is obstructed by an object; the wind exerts a pressure on the object. This pressure depends on velocity of wind and shape and orientation of the object. The design wind velocity depends on many factors. They are basic wind speed in the region and the type of terrain and topography. From design wind speed (or) velocity; design wind pressure can be calculated (IS: 875-Part-3). External pressure coefficients and internal pressure coefficients depend on the height, width, length, shape and opening of the object (structure). By utilising the design wind pressure, pressure coefficients; and area resisting, wind force on the structure can be calculated.

Earth or Groundwater Load

The loads due to earth or groundwater act as pressure normal to the contact surface of the structure. These loads are considered static.

Seismic Loads

Ground in motion creates force on the structure and this is called seismic force (or) load on the structure. Earthquake load is horizontal, it can act in any direction at a time. One direction only is to be considered at a time while analysing. The following factors are considered in calculating the seismic force on a structure.

(i) Location of the structure
(ii) Type of soil
(iii) Materials of construction
(iv) Size of the structure
(v) Method of construction of the structure

India is divided into four seismic zones considering the past earthquake and the seismic coefficients for each zone is given in IS: 1893-Part-1.

2.11 LIMIT STATE DESIGN VS WORKING STRESS DESIGN

Limit State Design

The strength limit state design criteria are based on the possible methods of failure of the structure concerned with yielding, buckling, brittle fracture (rupture) and fatigue. In other words, any limiting condition beyond which the structure ceases to meet its intended function is called limit state. Every load has its own probability of occurrence with different degrees of variation. Due to this, the partial safety factors for loads will also differ for different loads (Table 4 of IS: 800).

In the limit state designs, the design load effect shall not exceed the design strength of the structure. The following are the design load effects caused by external application of load and its appropriate combination.

(a) Bending moment
(b) Torsion
(c) Axial force
(d) Shear force

The actual loads will be multiplied by the appropriate partial safety factors for loads, (Table 4 of IS: 800). The design strength is calculated from the strength divided by partial safety factor for materials (Table 5 of IS:800) which takes care of variation in strength.
Design load effect ≤ Design strength.

Working Stress Design

This method of design ensures that the calculated stresses from the most worst load combination shall not exceed the permissible stresses.

$$\text{Working Stress} \leq \text{Permissible Stress} = \frac{\text{Yield Stress}}{\text{Factor of Safety (F.S.)}}$$

In load combinations which includes wind or seismic loads, the permissible stresses in steel structures may be increased by 33%. For anchor bolts and foundation, this increase shall be limited to 25%. Such an increase in allowable stresses should not be considered if the wind or seismic load is the major load in the load combination. This is to ensure that the structure should be capable of carrying load due to combinations excluding wind or seismic loads without increase in stresses.

Errors and Uncertainties

Deliberate errors

Assumptions made (i) to simplify the analysis, (ii) in the assessment of loading and (iii) in the structural behaviour

Accidental errors

Lack of precision in calculating the load, behaviour of the structure and in the method of analysis.

Uncertainties

- Material property (Example: yield strength)
- Variation in dimensions (Physical property) of the sections (Tolerances are specified)
- Manufacturing (Tolerances are specified)
- Erection (Tolerances are specified)
- Density of material
- Imposed loads
- Wind and seismic loads (By probability theory)

Material

The value of Young's modulus of elasticity, $E = 2 \times 10^5$ N/mm^2. (Clause 2.2.4.1 of IS:800).

$$\text{Strain at yield} = f_Y/E$$

The yield stress also varies with the heat treatment used and with the amount of working which occurs during the rolling process. For design purpose, minimum yield stress is identified.

$$\text{Shear yield stress} = f_y/\sqrt{3}$$

Analysis of Structures

The term analysis of structure is used to denote the analytical process by which the information of the response of the structure for various loads and load combinations can be obtained. The basis for this process is the knowledge of the behaviour of the material, and this is used to analyse the behaviour of the individual components and joints of the structure. Response of the structure depends on material specification, shape, physical property, span, end/support condition and location, direction and magnitude of the load.

2.12 CENTRE OF GRAVITY (C.G.) AND MOMENT OF INERTIA

The Centre of Gravity

The centre of gravity of a body is the point through which the resultant attraction of the earth acts for all orientations of the body.

Second Moment of Area

The second moment of area (or the moment of inertia) is a property of area of great importance in mechanics.

2.13 EXAMPLES ON CENTRE OF GRAVITY CALCULATION

Example 2.1 *Find the centre of gravity of the cross section as shown in Fig. 2.15.*

AREA ID	a	y	$a \times y$	h	h^2	$a \times h^2$	$bd^3/12$
1	5	0.25	1.25	1.37	1.87	9.35	0.104
2	3.25	3.75	12.18	2.13	4.53	14.72	11.44
Total	$\Sigma a = 8.25$	–	$\Sigma ay = 13.43$	–	–	$\Sigma ah^2 = 24.07$	$\Sigma I_g = 11.544$

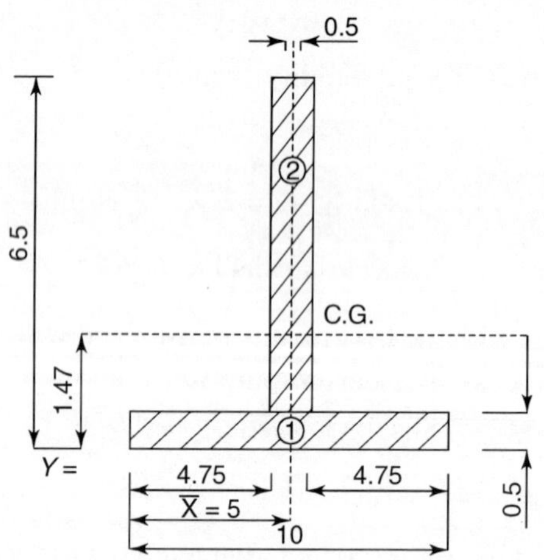

Fig. 2.15

a = area, y = distance of C.G. from base.

C = distance of C.G. of combined area from base.

$C = \Sigma ay/\Sigma a = 13.43/8.25 = 1.63$ cm

Moment of inertia about $xx = I = ah^2 + I_g = 24.07 + 11.544 = 35.614$ cm^4

Area (1) I_g

$$b_1d_1^3/12 = 10 \times 0.5^3/12 = 0.104 \text{ cm}^4$$

Area (2) I_g

$$b_2d_2^3/12 = 0.5 \times 6^3/12 = 11.44 \text{ cm}^4$$

Example 2.2 *Find the centre of gravity of the cross section as shown in Fig. 2.16.*

Area ID	a	y	$a \times y$	h	h^2	$a \times h^2$	$bd^3/12$
1	7.5	0.25	1.875	2	4	30	0.15625
2	6	3.5	21	1.25	1.56	9.36	18
3	6	3.5	21	1.25	1.56	9.36	18
Total	$\Sigma a = 19.5$	–	$\Sigma ay = 43.875$	–	–	$\Sigma ah^2 = 48.72$	$\Sigma I_g = 36.16$

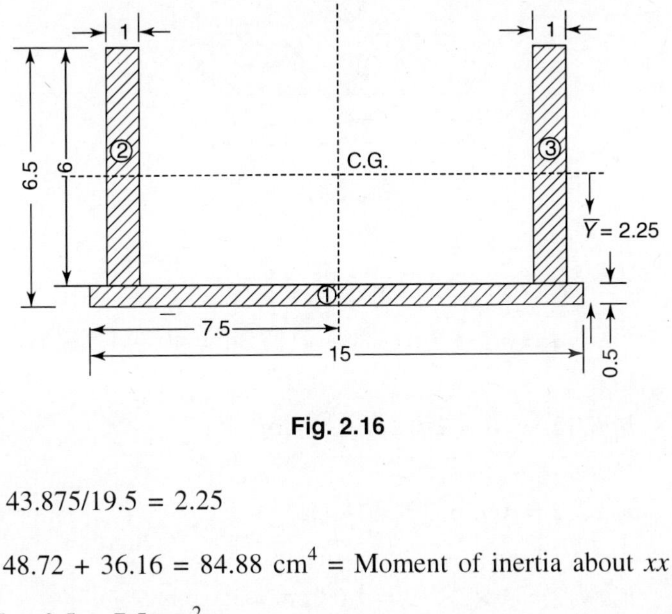

Fig. 2.16

$C = \Sigma ay/\Sigma a = 43.875/19.5 = 2.25$

$I = ah^2 + I_g = 48.72 + 36.16 = 84.88 \text{ cm}^4$ = Moment of inertia about xx axis

Area $= a_1 = 15 \times 0.5 = 7.5 \text{ cm}^2$

$$a_2 = a_3 = 6 \times 1 = 6 \text{ cm}^2$$

$$I_g = b_1d_1^3/12 = 15 \times 0.5^3/12 = 0.15625 \text{ cm}^4$$

$$I_{g2} = I_{g3} = 2 \times b_2d_2^3/12 = 1.0 \times 6^3/12 = 18 \text{ cm}^4$$

Example 2.3 *Find the centre of gravity of the cross section as shown in Fig. 2.17.*

Area ID	a	y	a × y	h	h^2	a × h^2	bd^3 / 12
1	20	1	20	2.59	6.7	134	6.67
2	6	5	30	1.41	1.99	11.94	18
3	8	9	72	5.41	29.27	234.16	2.67
Total	Σa = 34		Σay = 122			Σah^2 = 380.1	ΣI_g = 27.34

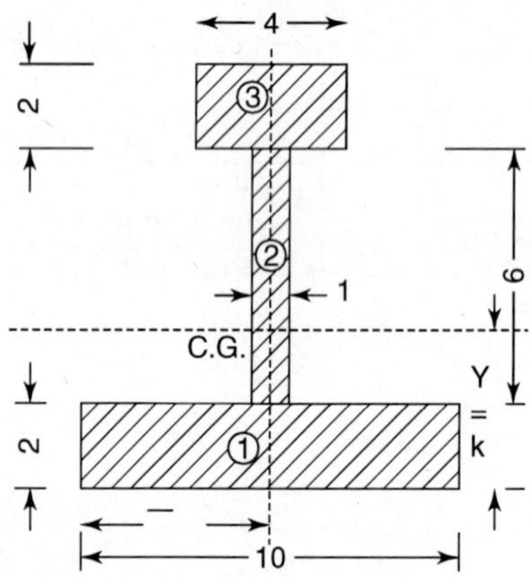

Fig. 2.17

$$C = \Sigma ay/\Sigma a = 122/34 = 3.59$$

$$I = ah^2 + l_g = 380.1 + 27.34 = 407.44 \text{ cm}^4$$

ea (1) I_g

$$b_1 d_1^3/12 = 10 \times 2^3/12 = 6.67 \text{ cm}^4$$

Area (2) I_g

$$b_2\, d_2^3/12 = 1 \times 6^3/12 = 18 \text{ cm}^4$$

Area (3) I_g

$$b_3 d_3^3/12 = 4 \times 2^3/12 = 2.67 \text{ cm}^4$$

Example 2.4 *Find the centre of gravity of the cross section as shown in Fig. 2.18.*

Area ID	a	y	a × y	h	h^2	a × h^2	bd^3/12
1	3.2	0.1	0.32	3.1	9.61	30.75	0.01
2	0.76	4	3.04	0.5	0.25	0.19	3.65
3	0.76	4	3.04	0.5	0.25	0.19	3.65
4	2.4	7.9	18.96	4.4	19.36	46.46	0.0
Total	Σa = 7.12	–	Σay = 25.36	–	–	Σah^2 = 77.59	ΣI_g = 7.4

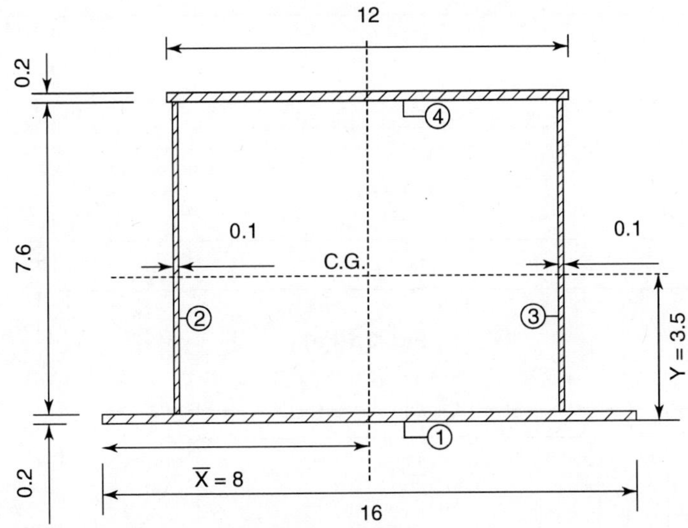

Fig. 2.18

$$C = \Sigma ay / \Sigma a = 25.36/7.12 = 3.56$$

$$I = ah^2 + I_g = 77.59 + 7.4 = 84.99 \text{ cm}^4$$

$$a_1 = 16 \times 0.2 = 3.2 \text{ cm}^2 \; a_2 = 0.1 \times 7.6 = 0.76 \text{ cm}^2$$

$$a_3 = 0.1 \times 7.6 = 0.76 \text{ cm}^2 \; a_4 = 12 \times 0.2 = 2.4 \text{ cm}^2$$

Area (1) I_g; $b_1d_1^3/12 = 16 \times 0.2^3/12 = 0.01$ cm^4

Area (2) I_g; $b_2d_2^3/12 = 0.1 \times 7.6^3/12 = 3.65$ cm^4

Area (3) I_g; $b_2d_2^3/12 = 0.1 \times 7.6^3/12 = 3.65$ cm^4

Area (4) I_g; $b_3d_3^3/12 = 12 \times 0.2^3/12 = 0.00$ cm^4

2.14 EQUIVALENT UNIFORMLY DISTRIBUTED LOAD FOR TRIANGULAR LOAD AND TRAPEZOIDAL LOADS

Equivalent uniformly distributed load for

(1) Short span = $wS/3$ w = load in kN/sqm

S and L are in metre

$m = S/L$

(2) Long span = $(wS/3) \times [(3 - m^2)/2]$

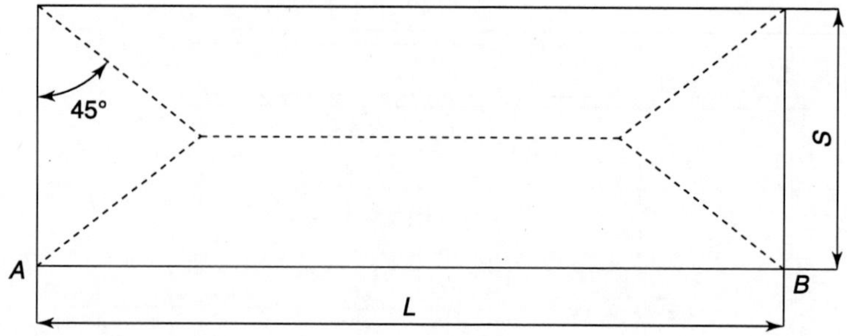

Fig. 2.19(a)

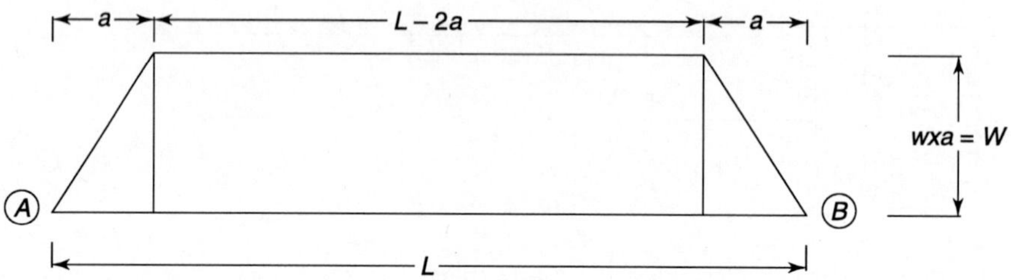

Fig. 2.19(b)

(1) For simply supported beam

Maximum *BM* at mid span = $M = (w/24) \times (3L^2 - 4a^2)$

(2) For fixed beam

$$M_A = \{-wL^2/12(L - a)\}\ [1 - 2(a/L)^2 + (a/L)^3] = M_B$$

$$BM \text{ at mid span} = M_A - M$$

Equivalent UDL Comparison

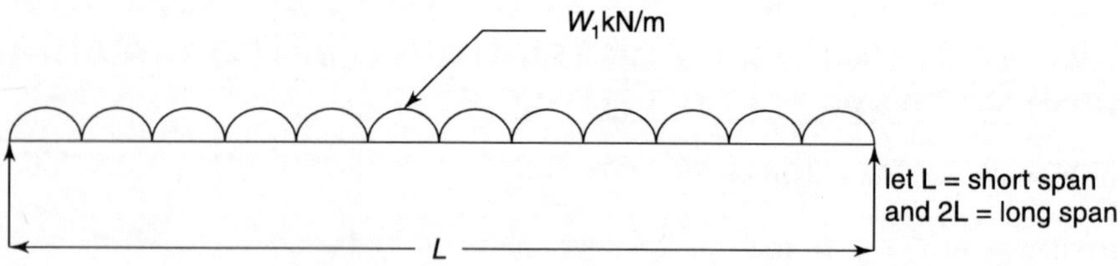

Fig. 2.20

- **For Uniformly Distributed Load**

$$\delta_1 = 5w_1L^4/384EI \ \&\ M_l = w_1L^2/8$$

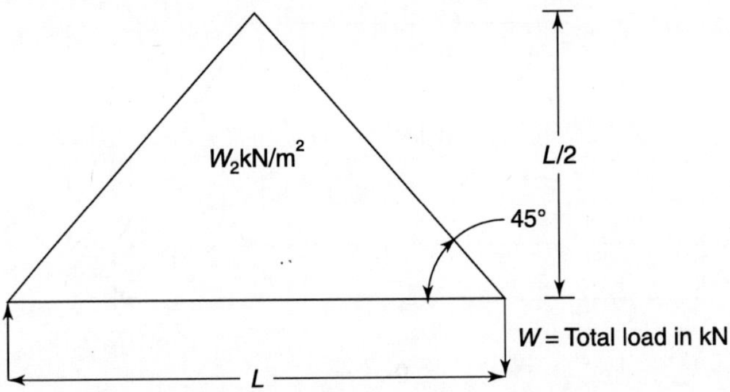

Fig. 2.21

- **For Triangular Load**

$$\delta_2 = WL^3/60EI$$
$$W = 1/2 \times L \times L/2 \times = (w_2L^2/4)$$
$$\delta_2 = (w_2L^2/4) \times (L^3/60EI) = w_2L^5/240EI$$
$$M_2 = WL/6 = (w_2L^2/4) \times L/6 = w_2L^3/24$$

Equating Deflection

Deflection due to uniformly distributed load = $5w_1L^4/384EI$

Deflection due to triangular load = $w_2L^5/240EI$

Deflection due to triangular load = $w_2L^5/240EI$

Equating the two $5w_1L^4/384EI = w_2L^5/240EI$

$$w_1 = (w_2 \times L/240) \times (384/5) = w_2L \times 384/240 \times 5 = \frac{w_2L}{3.125}$$

Equating BM

Bending moment due to uniformly distributed load = $w_1L^2/8$

Bending moment due to triangular load = $w_2L/24$

Equating these two, $w_1L^2/8 = w_2L^3/24$

$$w_1 = w_2L/3$$

Shear Force

Total load on short span = $2 \times (w_2L/3) \times L = 2 \times (w_2L^2/3)$ (1)

$$\text{Say } L_1 = 2L_2$$
$$m = S/L = L/2L = 0.5$$

Equivalent uniformly distributed load on long span

$$(w_2L/3) \times ((3 - m^2)/2) = (w_2L/3) \times ((3 - 0.5^2/2) = (w_2L)/(2.75/6) = (w_2L) \times (11/24)$$

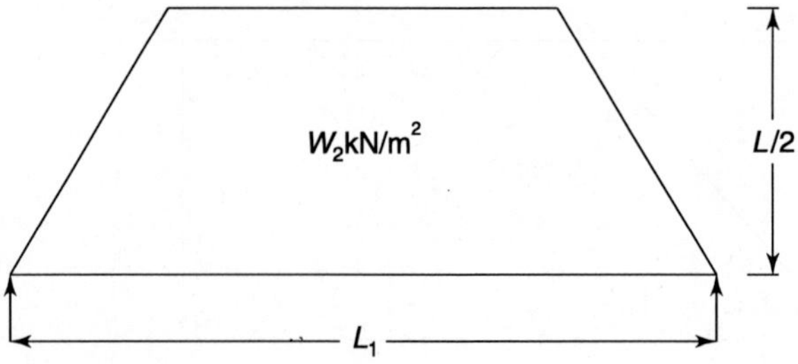

Fig. 2.22

Total load on long span

$$2L \times \{(11/24) \times w_2L = 11/12 \times w_2L^2 \qquad (2)$$

Grand total of (1) & (2)

$$= (2/3)w_2L^2 + (11/12)w_2L^2 = w_2L^2(2/3 + 11/12)$$

$$= w_2L^2(24 + 33)/36) = 57/36\ w_2L^2 = 1.6w_2L^2$$

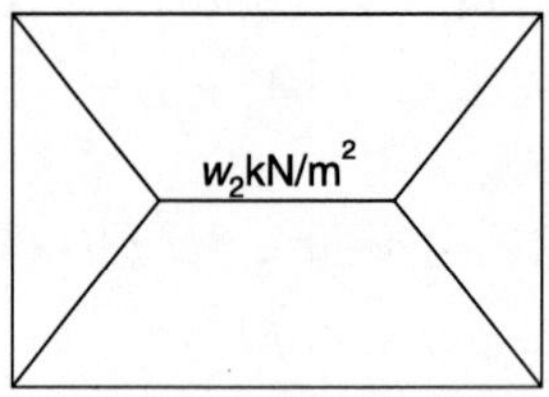

Fig. 2.23

$$L \times 2L \times w_2 = 2w_2L^2$$

Difference of $0.4w_2L^2$

This difference is because the equivalent uniformly distributed load is based on the maximum moment due to the triangle/trapezoidal load equated to the maximum moment due to the uniformly distributed load. From the above example, it can be seen that the equivalent uniformly distributed load calculated equating bending moment, deflection and shear force are not equal.

2.15 FRAMES

Refer to Clause 4.1.2 of IS: 800.

Building structure consists of a number of frames. Mostly these frames are mutually perpendicular. A number of longitudinal frames and cross (transverse) frames form a complete structure. These frames consist of columns and beams with or without bracing members. When a frame is with bracing, it is called a braced structure and when it is not braced, it is called a rigid structure.

In some structures, the frames may not be in orthogonal directions. The frames may meet at less or more than 90°.

(i) **Braced Structure** *A* simple braced frame is shown in Figs. 2.24 and 2.28. The structure system (frame) consists of columns and beams to transfer loads to the foundation. These frames are provided with diagonal members which are called bracing or diagonal bracing. The joints will be having a thin flexible plate called gusset plate to create pin joints at ends. These bracing members are expected to take only axial forces. These bracing members give stability to the frames against the action of horizontal or lateral loads. Bracing members are provided in frames to reduce or prevent sway of the frame. A frame with bracing members is very efficient in taking lateral load due to wind/seismic/ other loads. These bracing members are subjected to tension or compression depending on the direction of horizontal loads/forces. But these members may interfere with pipes, ducts and other openings which has to be considered at the initial planning stage.

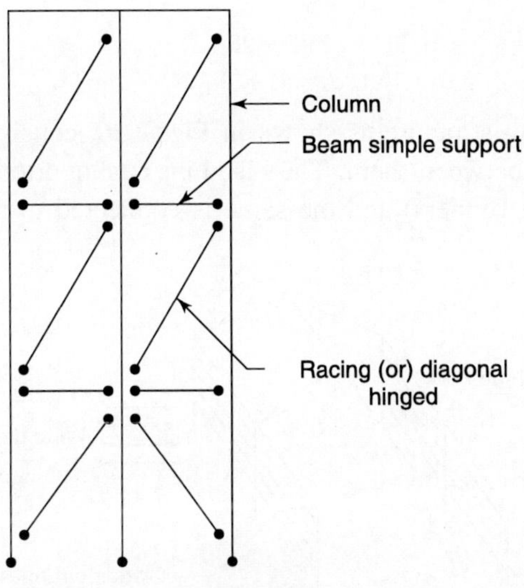

Fig. 2.24

(ii) **Rigid Frames** If the frames are not provided with bracing members and the beams are pin jointed; columns will be heavy to take up horizontal loads and to prevent or reduce the sway. In such cases, the joints between beam and columns will be made rigid or fixed. Such frame is called a rigid frame. Bending moments will be more because of fixity and to resist horizontal loads. This system is not advisable for tall structures and structures subjected to high lateral (or) horizontal loads. Also refer to Figs. 2.25 and 2.28.

(iii) **Shear** If the structure consists of walls, these walls will resist lateral loads. When the applied lateral load lies in the same plane as that of the wall, the wall is subjected to shear and is called a shear wall. This system is the most economic of all.

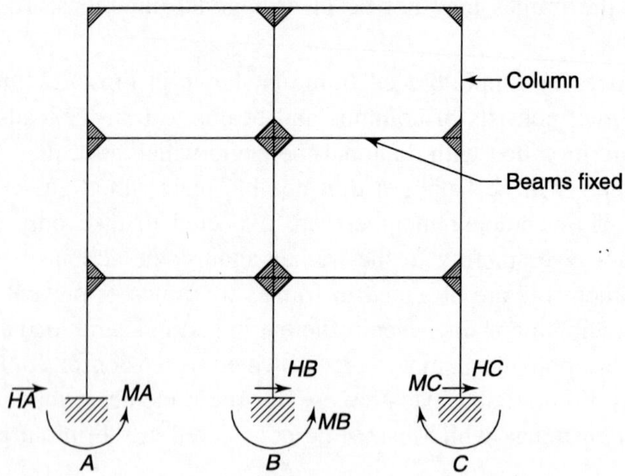

Fig. 2.25

Consider a simple building structure as shown in Fig. 2.26, consisting of three rigid frames and the section of cladding between them. The cladding is supported by the horizontal beams (called cladding runners/side runners) and the same is connected to the columns.

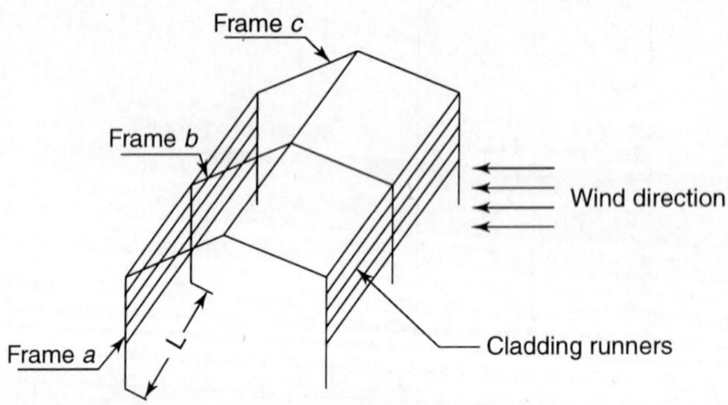

Fig. 2.26

The cladding material must be strong enough to span between cladding runners or side runners which in turn acts as a beam spanning between vertical members (columns). This should be capable of taking wind force acting horizontally but perpendicular to the surface of cladding. The force on frame '*b*' will be twice that of end frames '*a*' or '*c*'. however, for uniformity, the end frames are also kept same as that of middle frame. But if the ends of the building are also cladded; then, the corner columns will be subjected to horizontal forces due to wind on mutually perpendicular directions.

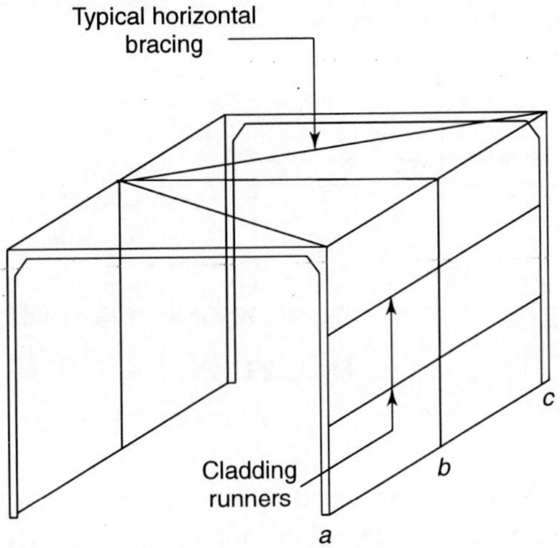

Fig. 2.27

We can form a horizontal truss in the plane of the roof as shown in Fig. 2.27 connecting top of columns to share the horizontal load.

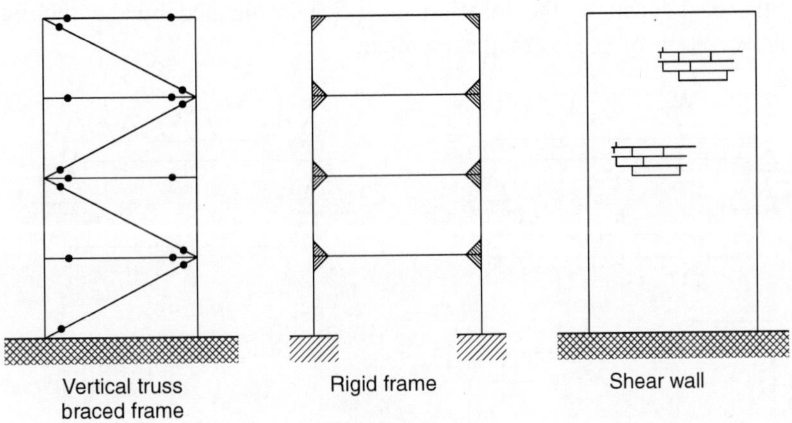

Fig. 2.28

2.16 SHEAR FORCE, BENDING MOMENT AND POINT OF CONTRAFLEXURE

Shear Force

The shear force at any cross section *AA* in a beam loaded as in Fig. 2.29 is the algebraic sum of all forces acting perpendicular to the axis of the beam either to left (or) right side of section *AA*.

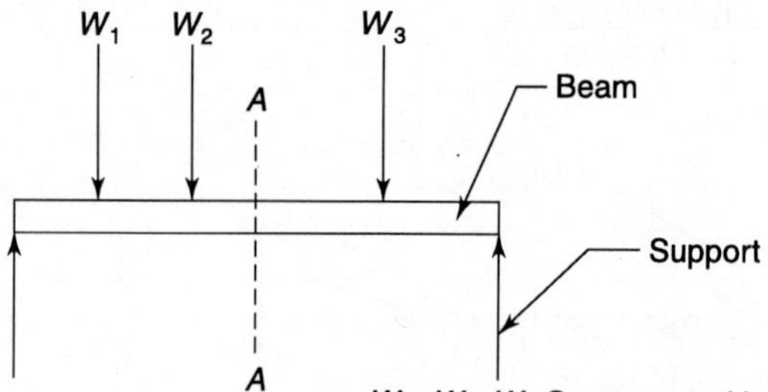

W_1, W_2, W_3-Concentrated loads

Fig. 2.29

Bending Moment

The moment of a force about a point is obtained by multiplying the force and the perpendicular distance from the point to the force. The bending moment at cross section *AA* in a beam is the algebraic sum of the moments about *AA* of all forces to one side of *AA*.

Point of Contraflexure

The point at which bending moment is zero is called the point of contraflexure. This is the best location to have splices to join two pieces for getting a beam to get the required length. Though this is the best location; the location of splice is dictated by the required length of beam and available length of pieces of the member.

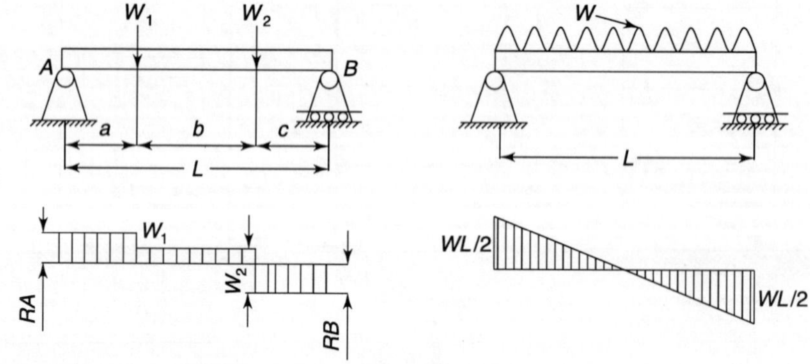

$$RA = \frac{W_1(b+c)}{L} + \frac{W_2 x c}{L}$$

$$RB = \frac{W_1(a) + W_2(a+b)}{L}$$

Fig. 2.30 Shear force diagrams

Theorem of Parallel Axes

The second moment or moment of inertia of an area *A* about an axis *XX* away from its centroid equals the second moment or moment of inertia about a parallel axis passing through its centre of gravity plus area times the square of the distance between the axes (Fig. 2.31).

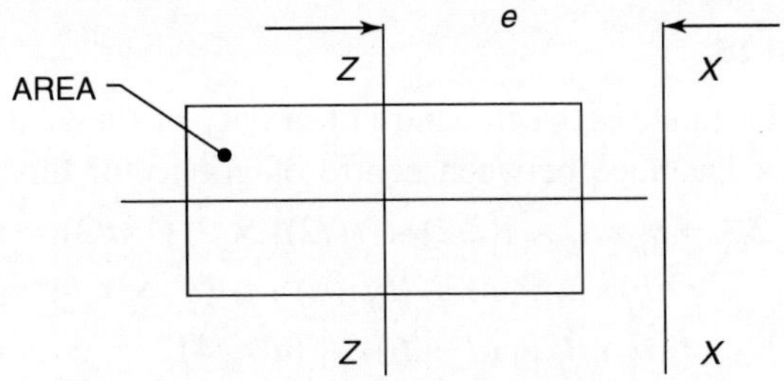

Fig. 2.31

$$I_{xx} = I_{zz} + Ae^2$$

Elastic Section Modulus (Z_e)

The elastic section modulus is a property of area that is important in flexural structural members. The section modulus of a shape about an axis *zz* is the second moment of area about that axis I_{zz}; divided by the distance from the axis to the furthermost part of the shape (*y*).

$$\text{Section modulus } Z_{ez} = I_{zz}/y$$

where *y* is the distance from the axis *zz* to the top or the bottom of the shape.

The unit of Z_z is mm^3.

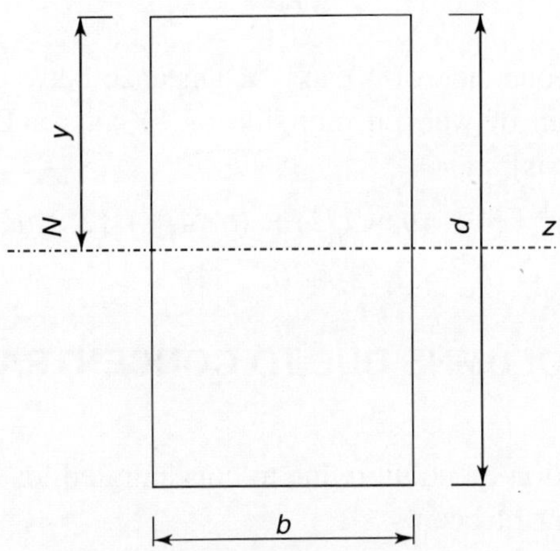

Fig. 2.32

For a rectangle *b* wide and *d* deep,

$$I_{xx} = bd^3/12 \text{ and } y = d/2$$

and $$Z_{ez} = bd^2/6$$

i.e., $$Z_{ez} \text{ (rect)} = bd^2/6$$

Plastic Section Modulus

Z_{pz} = 2 × [Area of flange × Distance between centre of gravity of flange and *ZZ* axis] + 2 × [Area of web above the *ZZ* axis × Distance between centre of gravity of this area and *ZZ* axis].

$$Z_{pz} = b_f \times t_f \times [(d/2) + (t_f/2)] \times 2 + (d/2) \times t_w \times (d/4) \times 2$$
$$= (b_f \times t_f \times d) + (b_f \times t_f^2) + (d^2 \times t_w/4)$$
$$= (b_f \times t_f \times (d + tf)) + (d^2 t_w/4)$$

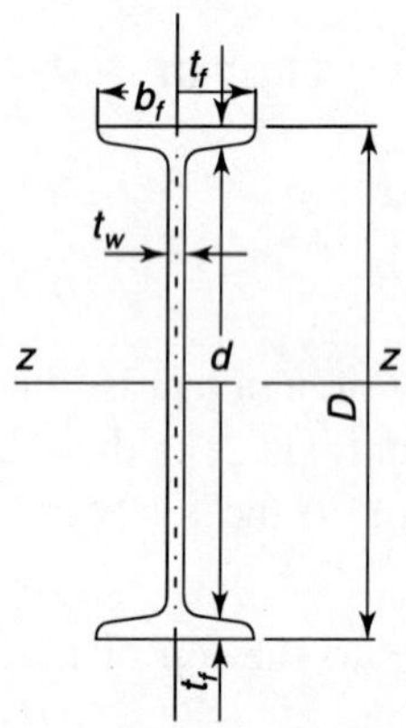

Fig 2.33

Z_{py} = 4 × [Area of flange on one side of *YY* axis × Distance between centre of gravity of this area and *YY* axis] + 2 × [Area of web on one side of *YY* axis × Distance between centre of gravity of this area and *YY* axis].

$$Zpy = [4 \times (b_f \times t_f/2) \times (b_f/4)] + [2 \times (d \times t_w/2) \times (t_w/4)]$$
$$= (b_f^{\ 2} \times t_f/2) + (dt_w^{\ 2}/4)$$

2.17 REACTION ON COLUMNS DUE TO CONCENTRATED LOAD "W" ON FLOOR

This will help to find the reaction on column due to concentrated load anywhere on the platform irrespective of the arrangement of beams.

2.18 BUILT-UP MEMBERS

Many options are available to a design engineer to choose built-up members. Different built-up sections are shown in Figs. 2.2, 2.3, 2.4 and 2.5.

The longitudinal spacing of fasteners connecting components of built-up compression members, for example, battens or lacings must be such that the effective slenderness ratio of the individual shape does not exceed 70% of the slenderness ratio of the whole member about the axis parallel to the batten or lacings or 50 whichever is less. Minimum two such intermediate

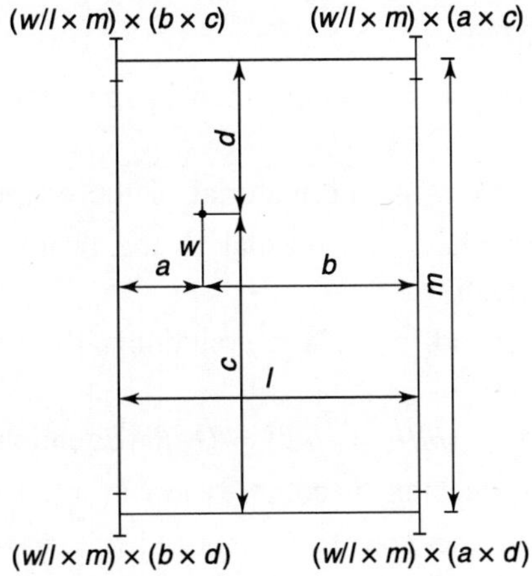

Fig. 2.34 *Plan of platform*

connectors must be used along the length of the built-up member. Also refer to Clause 7.8 of IS: 800.

2.19 STRESSES AND STRENGTH

Axial Tensile Strength (Direct Tension)

Section 6 of IS: 800

The factored design tension T in the members shall satisfy the following requirement:

$$T < T_d$$

where T_d = The design strength of a member under axial tension,

T_d is the lowest of the design strength due to yielding of gross section (T_{dg}), rupture strength of critical section (T_{dn}), and block shear (T_{db}) given in Clauses 6.2, 6.3 and 6.4, of IS: 800 respectively.

Axial Compressive Strength (Direct Compression)

The members under axial compression (P) will fail by flexural buckling before yielding. The buckling strength of these members depends on effective length and least radius of gyration and is also influenced by residual stresses, initial deformation, and accidental eccentricities of load. To take care of all these factors, the strength of members subjected to axial compression is defined by buckling class a, b, c, or d as given in Table 7 of IS: 800.

The design compressive strength P_d, of a member is given by: $P_d = A_e f_{cd}$

where A_e = effective sectional area and f_{cd} = design compressive stress.

Axial load P shall be less than P_d.

Bending Stresses & Stength

A French engineer by name Navier, based his theory upon three principal assumptions as follows:

(1) Strain is proportional to stress for the material of the beam.

(2) Cross sections which are plane and normal to the beam axis before bending remain plane and normal after bending.

(3) The beam is assumed to bend in a plane containing the vertical centroid axis and the beam axis.

From the theory of simple bending $(M/I) = (f/y/) = (E/R)$ Equation 1.0

where f = bending stress (tension or compression in a particular fibre), N/mm^2

Y = distance of extreme fibre from centroid of cross-sectional area, mm

M = applied B.M. at the cross section being considered, Nmm

I = second moment of area of the cross section, mm^4

E = Young's modulus of the material, N/mm^2

R = radius of curvature of the beam.

By rearrangement of Equation 1.0, we have

$f = (M \times y)/I = M/z$ Equation 2.0

i.e., the stress in any fibre is proportional to the bending moment and inversely proportional to the second moment of area of the cross section of the beam. The above three assumptions are important for understanding the behaviour of beams.

For beams of constant cross section, the most highly stressed points are the extreme fibres at the location of maximum bending moment, because at these points both M and Y are maximum.

$$f_{max} = M_{max}/Z_{min}$$

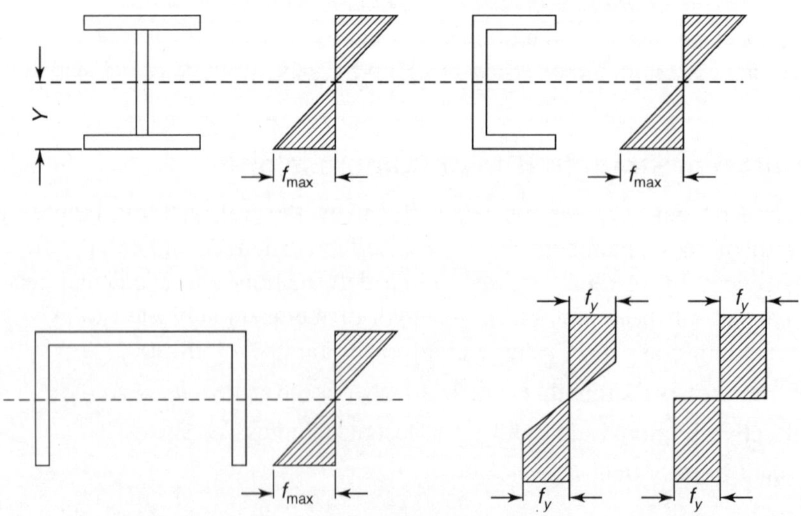

Fig. 2.35

The extreme fibre will reach yield stress. $f_{max} = f_y$. Then on further increase in load, the interior fibres will reach f_y as shown in Fig. 2.35 (b). Any further increase in load will form plastic hinge when all the fibres reach f_y as shown in 2.35 (c).

Bending strength = $M_d = \beta_p Z_p f_y / \gamma_{mo} \leq .2 Z_e f_y / \gamma_{mo}$ (Clause 8.2.1.2 of IS: 800) for simply supported beams and when $(d/t_w) < 67\ \varepsilon$ and $V \leq 0.6\ V_d$. Other cases are dealt with in this book at appropriate worked examples.

Shear Stress

Shear stress for a rectangular beam is: $f_s = (V \times A \times Y)/(I \times b)$ in N/mm^2

where

f_s is the longitudinal shear stress acting on a particular plane in N/mm^2

V is the external shear force action in *N*. *A* is the area of that part of the beam cross section between the particular plane and the extreme fibre in mm^2.

Y is the distance from the neutral surface to the centroid A in mm

I is the second moment of area of the beam cross section in mm^4

b is the width of the beam at the particular plane in mm

Torsion

Maximum shear: $V = M_T \times r/J$

M_T = torsional moment, Nmm

r = radius of gyration, mm

J = polar moment of inertia mm^4 = sum of I about any two perpendicular axes through the shear centre.

Polar moment of inertia $\rho^2 d_A$

ρ = radius from shear centre to any point in section

d_A = difference in area at a point

For hollow circular section

$J = (D^4 - d^4)/32$

For rectangular box

$r = T/kb^2d$ (K as in Table 2.1)

b = short side of rectangle

d = long side of rectangle

Table 2.1

d/b	1.0	1.5	2.0	2.5	3.0	4.0	5.0	10.0	∞
K	0.208	0.231	0.246	0.258	0.267	0.282	0.291	0.312	0.330

For channel and I beam $T = [(1/3)b^2d)]$

$J = (1/3)\ \Sigma b^3 d$

$r = Tb'/J$

where b' is the thickness of web or flange of the member. Maximum shear will occur at centre of one of the long sides of the rectangular part that has the greater thickness.

2.20 PROPERTIES OF STRUCTURAL STEEL

(a) $E = 2 \times 10^5$ MPa (N/mm^2)

(b) G = Shear modulus (modulus of rigidity) = 0.769×10^5 MPa

(c) Poisson's ratio = 0.3

(d) Coefficient of thermal expansion 0.000012 mm/mm/°C

(e) From 50° to + 150°F, density of carbon steel 7850 kg/cum

2.21 GRAPHICAL STATICS

Vectors

A force is represented by a vector in graphical statics (Fig. 2.36). A vector is shown graphically as an arrow, for which:

(1) the length is proportional to the magnitude of the force to a scale; and

(2) the direction is same as the direction of the force with respect to reference.

Graphical solutions require two diagrams:

Free body diagram or space diagram showing all forces acting on a body in their correct location.

A vector diagram or force diagram shows the vectors by which the forces are represented. These two diagrams must be drawn accurately and to scale.

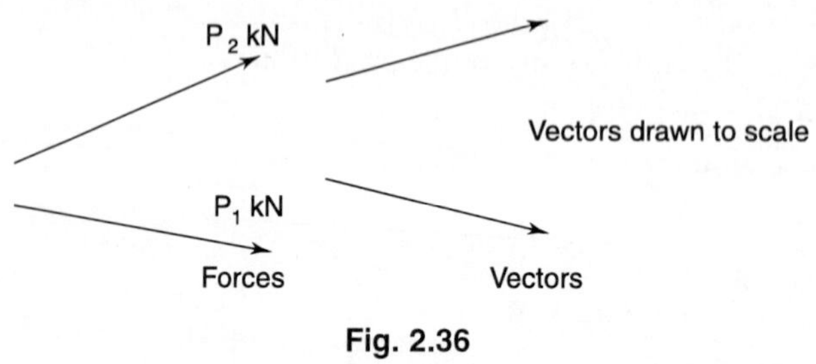

Fig. 2.36

Bow's Notation

The space between the forces is named using alphabets (capital letters) in a free body diagram. This naming is done by reading the spaces in clockwise direction. The forces are named using the names of spaces on either side of the force.

The vectors are named by lower case alphabet letters, with the earlier letter of the alphabet being attached to the tail of the vector, and the latter to the head of the vector in the vector diagram.

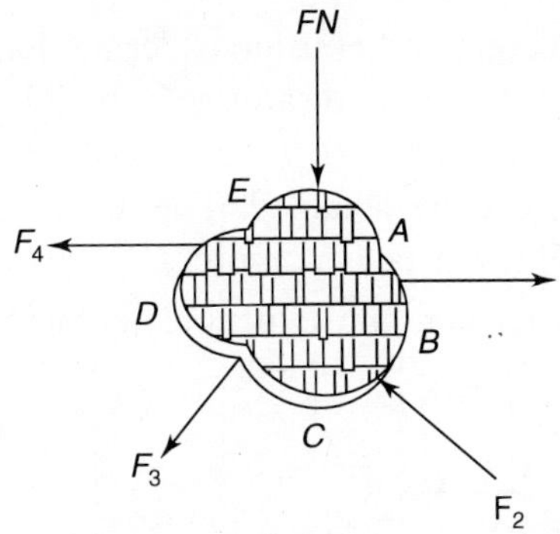

Fig. 2.37

2.22 PIN-JOINTED STRUCTURES

Trusses and braced structures are designed as pin joined frames. All the joints between members are assumed to have frictionless pins, which cannot transfer moment. The following two assumptions are made.

(i) All loads are assumed to be applied at the joints (nodes).

(ii) Each member will be subjected to axial forces only (pure tension or pure compression).

The Maxwell Diagram and Method of Joints

The Maxwell diagram is a graphical method for analysing determinate plane frames. It has a force diagram which contains within the one figure the polygons for every joint in the framework.

First, it is essential to determine the external reaction forces for constructing Maxwell diagram. This is most easily done by manual calculation.

Bow's notation, using letters of the alphabet in clockwise order around the frame is used to identify external forces. Closed spaces formed by the members of the frame are given names using by numerals 1, 2, 3 (Fig. 2.38).

Commence the force diagram by representing the external forces by vectors *ab*, *bc*, *cd* is prepared to scale. The forces on the pin at joint (*P*) (clockwise around the joint) are *A*1, 1*E* and *EA*. Vectors can be drawn parallel to these forces to form a triangle for forces. Draw lines through point '*a*' parallel to *A*1 and through '*e*' parallel to 1*E*. The intersection point of these lines is point 1, the third point on the triangle of forces for this joint (*P*). Let us consider the joint *Q* adjacent to spaces *A* and *B*. The forces acting at this joint are *AB*, *B*2, 2-1 and 1-A. From to the force diagram, locate the existing lines l*a*, *ab* and determine the location of point 2 such that *b*2 is parallel to force *B*2 and *ab* is parallel to force 2-1. Proceed in the same way with joints *T*, *U*, *R*, *V*, *S* and *W*, and plot the points 3, 4, 5 and 6.

The lines drawn represent the magnitudes to scale and directions of the forces at the joints. The force in member 2-1 will be the length of 2-1 in vector diagram to scale. Thus, we can find the magnitude of force in all members. Let us consider the joint (*T*). The members are *E*-1, 1-2, 2-3 and 3-*E* from vector diagram *e*-1 is towards left. At joint (*T*) mark an arrow towards left for member *E*-1. From vector diagram, 1-2 is upward. At *T*, mark the arrow for member 1-2. Thus, the directions of forces can be obtained. The arrows away from the joint denotes tension. The arrows towards the joint indicates compression.

In the above diagram *T* denotes tension and *C* denotes compression.

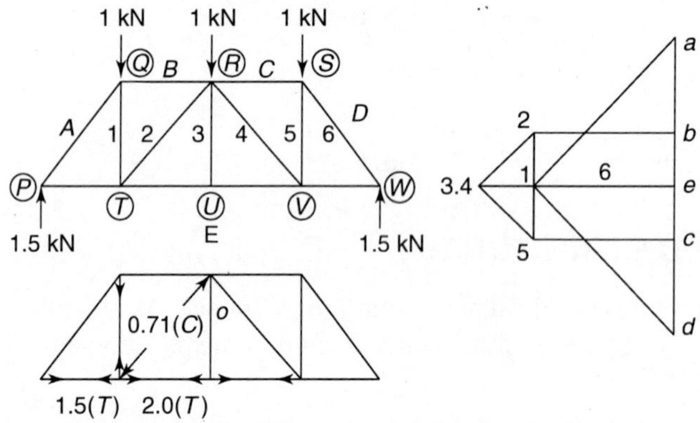

Fig. 2.38 Vector diagram and force diagram

In this method of joints, the following points are important:

(a) Reactions must be obtained before drawing the Maxwell diagram.

(b) Forces at a joint are always read in clockwise order.

(c) The order in which the joints are to be analysed is dictated by unknown forces, only two unknown forces per joint can be analysed.

(d) All forces are applied to the joint and not to the member.

(e) If two points in the Maxwell diagram coincide (e.g., 3 and 4 in Fig. 2.38), the force in that member is zero.

(f) The results of the analysis should be shown on the diagram of the truss. *A* member appearing thus ←→ is in compression. Similarly ——→←—— indicates tensile force in the member.

Maxwell's diagram is used to analyse the pin-jointed frame and is called method of joints. Two equations of equilibrium $\Sigma H = 0$ and $\Sigma V = 0$ are established for forces in two mutually perpendicular directions. Those equations are solved to get the forces in the members. Thus, only two unknown forces can be solved at each joint.

Method of Section

Equilibrium of the section to the left of *XX* or to the right of *XX* is considered for finding the forces in members. To determine the force in the member *GH* in Fig. 2.39; let us cut the truss through section *XX*. The members *BC*, *BH* and *GH* are cut by section *XX*. To keep the portion

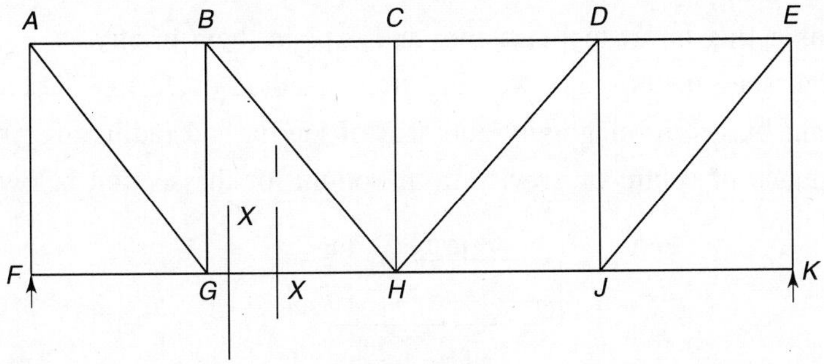

Fig. 2.39

of the truss on the left of section *XX* in equilibrium; let us assume that forces F_{BC}, F_{BH} and F_{GH} are applied.

The method of sections is explained below.

Cut three members for which forces are to be determined and apply forces that are applied at the ends of cut members.

Take the moment about a point of intersection so that the force in the other member can be found out by using the equilibrium equations.

Method of joints gives speedy solution for parallel chord trusses. The method of sections is more useful for pitched roof and north light roofs (saw-tooth roofs).

EXERCISE

1. Draw a neat sketch of an industrial shed with steel members and show the main components.
2. List the rolled shapes used as structural members.
3. Draw the sketches showing channels face to face and back to back and show the two mutually perpendicular axes.
4. List the types of connections used to connect a beam to column.
5. Draw a sketch showing moment connection between a beam and column flange.
6. What is the unit of measurement of force in SI unit?
7. What are all the requirements to fully specify a force?
8. What is meant by resultant of forces?
9. Explain with a sketch, the law of parallelogram of forces.
10. Explain couple of forces with a sketch.
11. What are all the conditions to be satisfied to keep a body in equilibrium?
12. Explain the difference between pressure and stress.
13. Define strain, permissible stress and yield stress.

14. List the loads acting on a steel structure and explain them briefly.
15. Compare limit state method and working stress method of design of steel structures.
16. What is meant by centre of gravity, moment of inertia and radius of gyration?
17. Find the distance of centre of gravity from bottom for the section below:

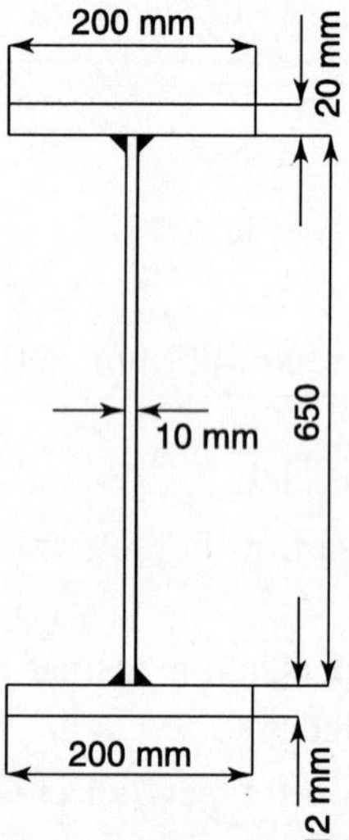

18. Why vertical bracing is used in a structural frame?
19. Define point of contraflxure.
20. Write down two commonly used properties of structural steel.
21. Explain the following three methods to analyse pin-jointed frames: (a) Graphical, (b) Method of joints, (c) Method of section.
22. Find the elastic section modulus and plastic section modulus for the sections below.

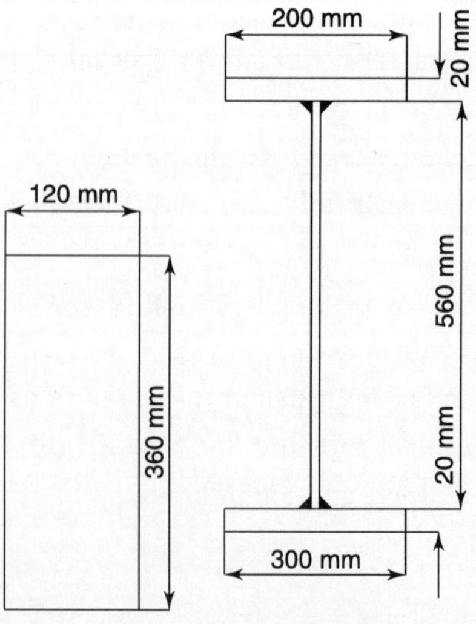

Chapter 3

Design of Tension Members

General

We will discuss the members subjected to axial tension in this chapter. Examples for tension members are tie member, diagonal member of roof, and bracing members of braced structure. The diagonal member of trusses, and bracing members are subjected to reversal of stresses. They are designed for compressive force and checked for tensile force. The hangers for platforms and hanger rods to suspend some components are tension members.

The tension members can be round, square rods, angles, channels and box channels. A single angle section will have eccentricity between the end connections and the centre of gravity of the angle. Double angle (or) channel sections will have less eccentricity.

Tension members are designed based on net sectional area. If the connection is welded; the centre of gravity of weld and the centre of gravity of the member should coincide as far as practicable. If the connection is bolted; the centre line of bolt should fall in line as indicated in SP6 (1). Section 6 of IS: 800 gives the method of designing the tension members. The net area after deducting bolt hole shall be considered. If the connection is by welding, no deduction in cross-sectional area is made.

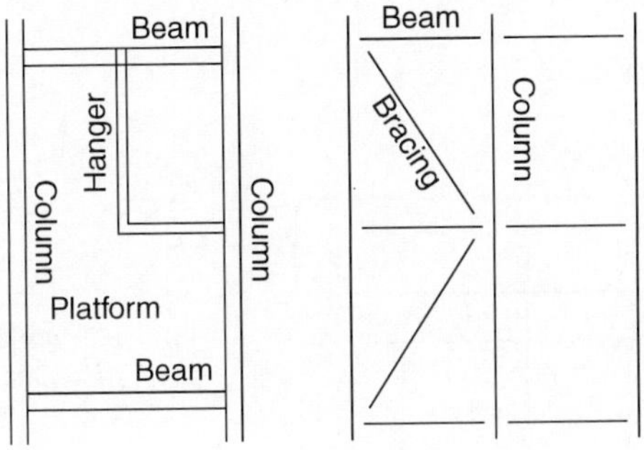

Fig. 3.1

3.1 DESIGN PROCEDURE

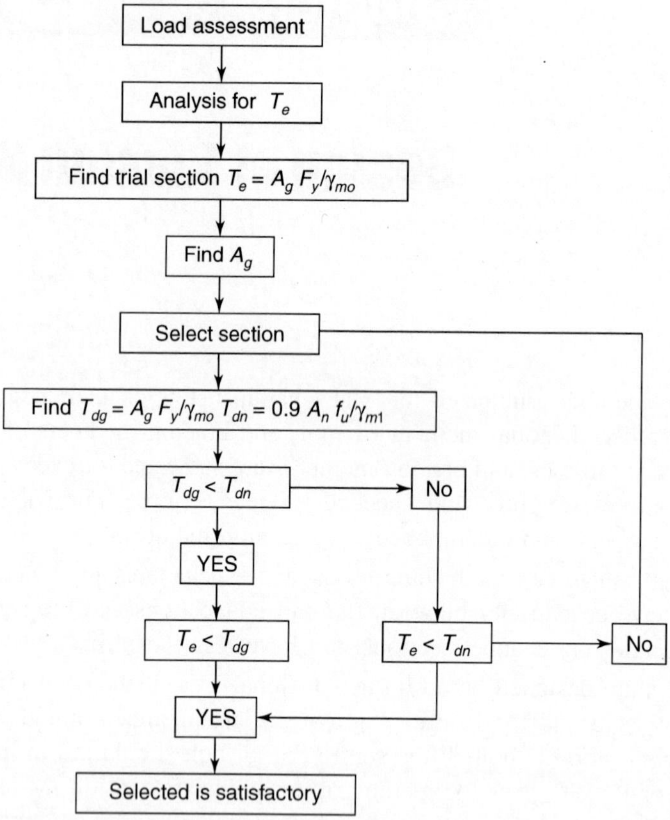

Note: Select a section means choose a section which has gross cross-sectional area greater than calculated A_g.

Members without Holes

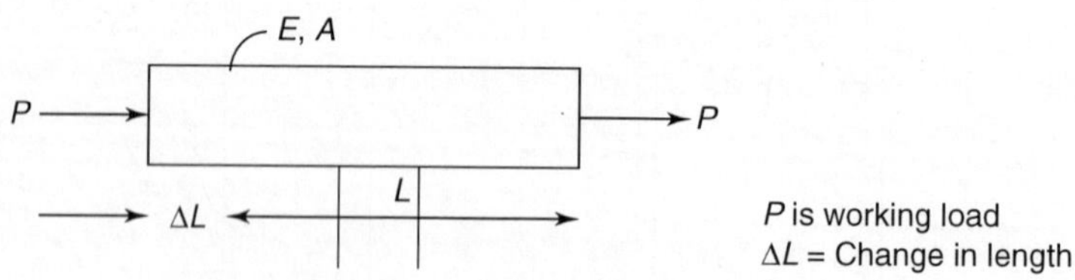

$$\text{Average strain} = \frac{\Delta L}{L}$$

$$\text{Strain} = \text{Stress}/E = P/AE = \Delta L/L$$

$$\text{Average stress} = f_a = \frac{P}{A}$$

$$\Delta L = \frac{PL}{AE}$$

$$\text{Load at yield} = \frac{Af_y}{\gamma_{mo}} = T_{dg}$$

Beyond ultimate load ($f_u \times A$); the neck is formed and the load P reduces until the member fails.

Tensile Strength

Tensile strength based on the following are to be calculated.

(1) As governed by yielding of gross section (Clause 6.2 of IS: 800)

$$T_{dg} = A_g f_y / \gamma_{mo}$$

(2) As governed by rupture of net section (Clause 6.3 of IS: 800)

(a) For plates

$$T_{dn} = 0.9\ A_n f_u / \gamma_{m1}$$

The presence of small local holes in a tension member (such as small bolt holes used for connecting the member) causes early yielding around the holes. When the holes are large, the member may fail before the yielding of gross section. In such case, the local rupture load will govern.

$$T_{dn} = 0.9\ A_n f_u / \gamma_{m1} \text{ where } A_n = A_g - \Sigma\, d_h\, t$$

(b) Plates with staggered holes under tension

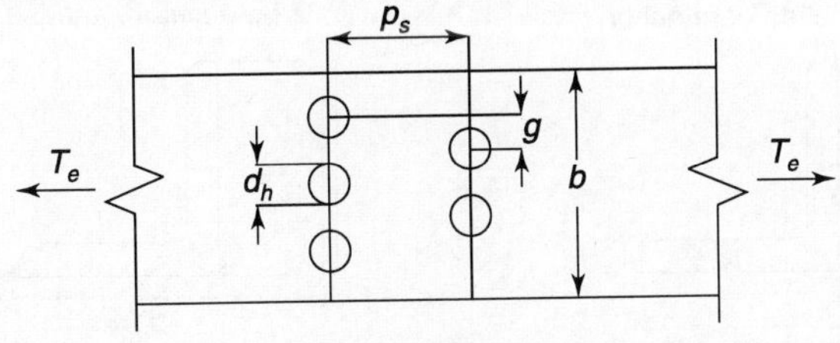

Fig. 3.2

$$A_n = \left(b - nd_h + \Sigma\left(\frac{p_{s,i}^2}{4g_i}\right)\right) \times t \qquad \text{(Clause 6.3.1 of IS: 800)}$$

T_{dn} = rupture design strength in tension of respective plate/angle calculated in accordance with Clause 6.3 of IS: 800/2007

$$T_{dn} = 0.9\ A_n f_u / \gamma_{m1}$$

A_n = net effective area of the member given by $A_n = \left(b - nd_h + \Sigma\left(\frac{Psi^2}{4gi}\right)\right) \times t$

b = width of plate

t = thickness of plate

d_h = diameter of the bolt hole (Table 19 of IS: 800)

p_s = staggered pitch length between the bolt holes as shown in Fig. 3.2

n = number of bolt holes in the critical section

i = subscript for summation of all the inclined legs

g = gauge length between the bolt holes, as shown in Fig. 3.2.

(c) For threaded rods

$$T_{dn} = 0.9\, A_n f_u / \gamma_{m1} \text{ where } A_n = \text{root area of thread.}$$

Angle under Tension

Single angle (Clause 6.3.3 of IS: 800)

The rupture strength of an angle connected through one leg is affected by shear lag. The design strength, T_{dn} as governed by rupture at net section is given by

$$T_{dn} = (0.9\, A_{nc}\, f_u/\gamma_{m1}) + (\beta\, A_{go}\, f_y/\gamma_{mo})$$

$$\beta = 1.4 - 0.076\, (w/t)(f_y/f_u)\, (b_s/L_c) \leq (f_u\, \gamma_{mo}/f_y\, \gamma_{m1}) \geq 0.7$$

w = outstand leg width

b_s = shear lag width as shown in Fig. 3.3

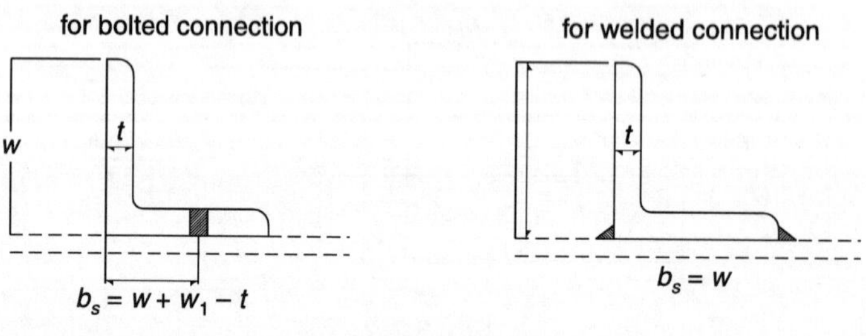

Fig. 3.3

L_c = length of the end connection, i.e., distance between the outermost bolts in the end joint measured along the load direction or length of the weld along the load direction. This is known only after designing the connection which in turn depends on the size of the angle. Hence, choose the size of the angle using the following approximate relation. Number of bolts can be decided based on load and the capacity as in Table 8.8.

Approximate $\qquad T_{dn} = \alpha A_n f_u / \Delta\gamma_{m1}$

α = 0.6 for one or two bolts 0.7 for three bolts and 0.8 for four or more bolts along the length in the end connection or equivalent weld length.

γ_{m1} = partial safety factor against ultimate stress.

γ_{mo} = partial safety factor against yield stress and buckling.

As Governed by Block Shear

Bolted connection (Clause 6.4.1 of IS:800)

$$T_{db} = ((A_{vg}\, f_y)/(\gamma_{mo}\sqrt{3})) + (0.9\, A_{tn}\, f_u/\gamma_{m1})$$

A_{vg}, A_{vn} = minimum gross and net area in shear along bolt line parallel to external force, (1-2 and 3-4 as shown in Fig. 3.4(a), 1-2 as shown in Fig. 3.4(b).

A_{tg}, A_{tn} = minimum gross and net area in tension from the bolt line perpendicular to the line of force, respectively (2-3 as shown in Fig. 3.4(b).

Welded connection (Clause 6.4.2 of IS:800)

The block shear strength (T_{db}) shall be checked for welded end connections by taking an appropriate section in the member around the end weld, which can shear off as a block. In most of the cases, block shear strength in welded connections is higher than the strength of welded connection itself.

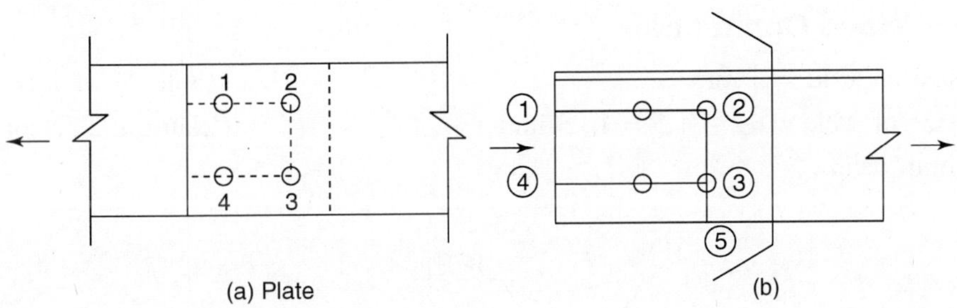

(a) Plate (b)

Fig. 3.4

Refer Fig. 3.4(b). Two different types of failures can occur. One along 1-2-3-5. For this to happen bolt nearer to 4 will have to fail. Other possible failure is along 1-2-3-4. This is similar to Fig. 3.4(a).

3.2 DESIGN EXAMPLES

Plate/Flat under Tension

Example 3.1 *Design a plate/flat tension member to take a tensile force of 100 kN (working load due to dead load and imposed load) using IS:2062 material for which f_y = 250 MPa with welded connection.*

(1) As governed by yielding of cross section (Clause 6.2 of IS: 800)

Factored load = 1.5 × 100 = 150 kN (Table 4 of IS: 800)

Approximate cross-sectional area required = $P/(f_y/\gamma_{mo})$ = (150 × 1000)/(250/1.1) = 660 mm^2

Let t = 12 mm

$$\therefore \quad b = \text{width} = \frac{660}{12} = 55 \text{ mm}$$

Provide flat = 60 × 12 mm

Strength in yielding = $T_{dg} = A_g f_y/\gamma_{mo}$ = 60 × 12 × 250/1.1 = 163.64 kN (1)

(2) As governed by rupture of net section (Clause 6.3.1 of IS:800)

$$T_{dn} = 0.9 A_n f_u/\gamma_{m1}$$

Since the connection is proposed to be welded, $A_n = A_g = bt$ = 60 ×12 = 720 mm^2

f_u = 410 MPa (Table 1 of IS: 800)

γ_{m1} = 1.25 (Table 5 of IS: 800/2007)

T_{dn} = (0.9 × 720 × 410/1.25) = 212544 N = 212.544 kN (2)

(3) Block shear strength (Clause 6.4.2 of IS:800)

Design of Welded Connection

Minimum size of weld = 5 mm (Table 21 of IS: 800/2007)

Maximum size of weld = 12 – 1.5 = 10.5 mm (Clause 6.2.5.1 of IS: 816)

Provide 10 mm weld

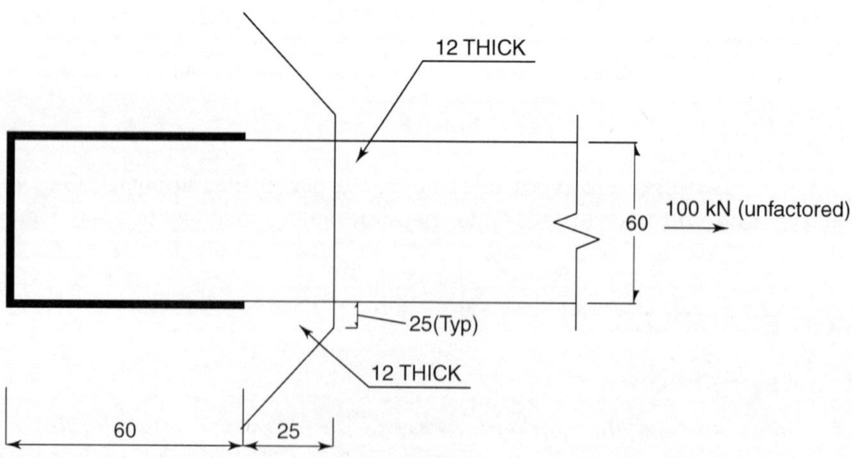

Fig. 3.5

Effective throat thickness = 0.7071 × 10 = 7.071 mm (Clause 6.2.3.1 of IS: 816)

Strength of shop weld per mm^2

$$f_{wd} = f_u/\sqrt{3}(\gamma_{mw}) = 410/\sqrt{3}(1.25) = 189.37 \text{ N/mm}^2 \text{ (Clause 10.5.7.1.1 of IS: 800).}$$

Strength of 10 mm weld per mm length = (7.071 × 189.37) N/mm

$$l_w \times 7.071 \times 189.37 = 150 \times 1000$$

$$l_w = 150 \times 1000/(7.071 \times 189.37) = 112 \text{ mm}$$

Required minimum length of weld = 112 mm.

At the end, weld is provided for full width of plate/flat.

Effective length of weld at end of plate

= Actual length of weld – twice the size of weld. (Clause 6.2.4 of IS:816)

= 60 – 2 × 10 = 40 mm.

Balance will be on sides of plate.

Required length of weld per side = (112 – 40)/2 = 36 mm (effective length).

Actual length required = 36 + 2 × 10 = 56 mm.

Provide 60 mm length of 10 mm fillet weld on sides.

Total effective length of weld = 40 + 2 (60 – 2 × 10) = 40 + 80 = 120 mm > 112 mm.

Block Shear

At the end of the plate, block shear failure will be by tension and on sides by shear.

$$A_{vg} = A_{vn} = \text{Area under shear} = (180 - 60) \times 12 = 1440 \text{ mm}^2$$

$$A_{tg} = A_{tn} = \text{Area under tension} = 60 \times 12 = 720 \text{ mm}^2$$

$$T_{db} = (A_{vg} f_y/\sqrt{3}\ \gamma_{mo}) + (0.9\ A_{tn} f_u/\gamma_{m1})$$

$$= [(1440 \times 250)/(1.1 \times \sqrt{3})] + [(0.9 \times 720 \times 410)/1.25] = 401.5 \text{ kN} \quad (3)$$

$$T_{db} = (0.9\ A_{vn} f_u/\sqrt{3}\ \gamma_{m1}) + (A_{vg} f_y/\gamma_{mo})$$

$$= [(0.9 \times 1440 \times 410)/(1.25 \times \sqrt{3})] + [(720 \times 250)/1.1] = 409.1 \text{ kN} \quad (4)$$

Strength = 163.64 kN (least of the four)

This is greater than the factored load 150 kN.

Hence, adequate.

Example 3.2 *Design a single angle tension member for an axial tension of 100 kN using IS: 2062 material having yield stress 250 MPa. The ends are welded. Also check the angle if the connection is bolted using M20. Class 4.6 bolts. The tension is due to dead and imposed loads only.*

Solution: Factored load = 1.5 × 100 = 150 kN (Table 4 of IS: 800)

Approximate area of cross section required = $\dfrac{150 \times 1000}{(250/1.1)}$ = 660 mm^2

Try 75 × 75 × 6; A_g = 866 mm^2

Design strength due to yielding

$$T_{dg} = A_g f_y/\gamma_{mo} \text{ (Clause 6.2 of IS: 800)}$$

$$= 866 \times 250/(1.1 \times 1000) = 196.82 \text{ kN} \qquad (1)$$

Design strength due to rupture

$$T_{dn} = 0.9\, A_{nc}\, f_u/\gamma_{m1} + \beta\, A_{go}\, f_y/\gamma_{mo} \text{ (Clause 6.3.3 of IS: 800)}$$

The connection is welded.

$$A_{nc} = \frac{866}{2} = 433 \text{ mm}^2 = A_{go};\ f_y = 410 \text{ Mpa};\ \gamma_{m1} = 1.25;\ \gamma_{mo} = 1.1$$

$$\beta = 1.4 - 0.076\left(\frac{w}{t}\right)\left(\frac{f_y}{f_u}\right)\left(\frac{bs}{Lc}\right) \le (f_u \times \gamma_{mo})/(f_y \times \gamma_{m1}) \ge 0.7$$

(Refer Fig. 3.7)

$$= 1.4 - 0.076\left(\frac{75}{6}\right)\left(\frac{250}{410}\right)\left(\frac{75}{150}\right) = 1.1 \text{ (Clause 6.3.3 of IS: 800)}$$

$$(f_u \times \gamma_{mo})/(f_y \times \gamma_{m1}) = (410 \times 1.1)/(250 \times 1.25) = 1.4432$$

Keeping length of weld as 150 mm on each side. L_c = 150 mm.

$$T_{dn} = (0.9 \times 433 \times 410/1.25) + \left(\frac{1.11 \times 433 \times 250}{1.1}\right)$$

$$= 127822 + 109270 = 237092\text{N} = 237 \text{ kN} \qquad (2)$$

Welded Connection

Length of weld

Thickness of angle 6 mm

Thickness of gusset plate 8 mm

Minimum size of weld = 3/4 × 6 = 4.5 mm (Clause 10.5.8.2 of IS: 800)

Provide 4 mm weld.

Strength/mm = 0.7071 × 4 × 189.37 = 535.6 N (Refer Table 7.2)

Required length of weld = 150 ×1000/535.6 = 280 mm.

End of angle (75 mm) is welded. Effective length = 75 – 2 × 4 = 67 mm.

Required effective length on each side = $\dfrac{280 - 67}{2}$ = 106.5 mm

Provide 150 mm on sides

Total effective length of weld provided = 2 × (150 – 2 × 4) + (75 – 2 × 4) = 351 mm > 280 mm.

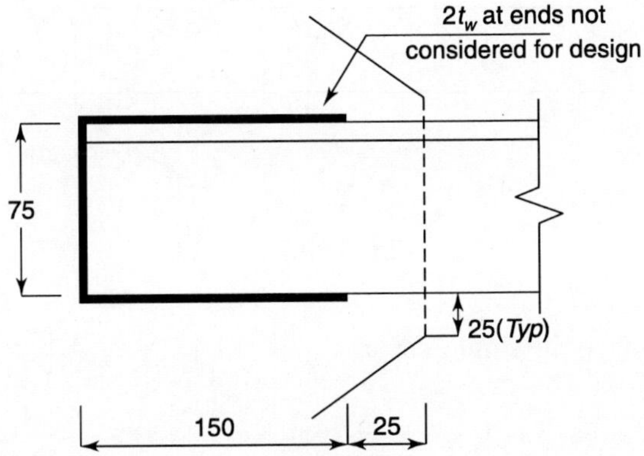

Fig. 3.6

Single Angle Bolted Connection

M20 bolts of class 4.6 are to be used.

Hole diameter = 20 + 2 = 22 mm (Table 19 of IS: 800)

Try ISA75 × 75 × 6 (already designed with welded connection)

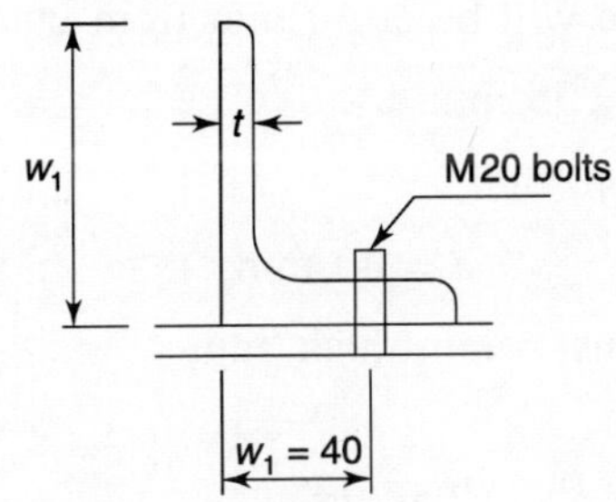

Fig. 3.7

To Find L_c

Shear Strength of bolt

Nominal shear capacity of the bolt = $V_{dsb} = (f_u/\sqrt{3})\ (n_n \times A_{nb} + n_s A_{sb})/\gamma_{mb}$

(Clause 10.3.3 of IS: 800)

f_u = 400 MPa

n_n = 0, considering threads are not in shear plane.

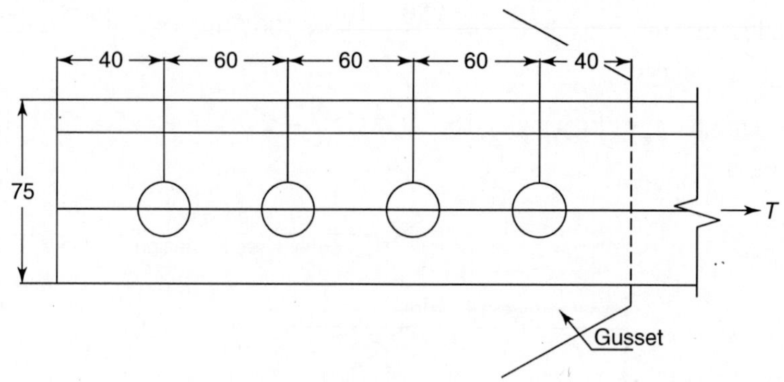

Fig. 3.8(a)

N_s = 1, since the bolt is in single shear.

A_{sb} = 314 mm^2

$V_{dsb} = (f_u/\sqrt{3})\ (n_n \times A_{nb} + n_s A_{sb})/\gamma_{mb} = (400/\sqrt{3}) \times (0 \times 314.29 + 1 \times 314.28)/1.25$

$= 230.94 \times 314.29/1.25 = 58012$ N = 58.06 kN (Refer Table 8.8b).

Bearing strength of bolt (Clause 10.3.4 of IS: 800)

Minimum pitch = 2.5 × Diameter of bolt = 2.5 × 20 = 50 mm.

Provide pitch = P = 60 mm.

For ISA75 × 75 × 6, bolt centre will be at 40 mm from outer face of angle. (SP: 6(1))

Edge distance will be 75 – 40 = 35 mm.

Let us provide end distance as 40 mm.

e = 40 mm.

Strength of bolt in bearing against 6 mm thick angle

$$= V_{dpb} = \left(\frac{2.5 k_b d t f_u}{\gamma_{mb}}\right)$$

k_b is the least of (Clause 10.2.4.2 and Table 19 of IS: 800)

(1) $e/3do = 40/3 \times 22 = 0.61$

(2) $p/3 \times do = (60/3 \times 22) - 0.25 = 0.66$

(3) $f_{ub}/f_u = 400/410 = 0.975$

(4) 1.0

Strength of bolt in bearing = V_{dpb} = 2.5 × 0.61 × 20 × 6 × 410/(1.25 × 1000) = 60 kN.

Bolt value = least of strength in shear and strength in bearing = 58.06 kN.

Number of bolts = $n = \dfrac{150}{58.06} \approx 2.5835$ bolts. Provide four bolts.

For four bolts there will be three pitches. Hence, L_c = 3 × 60 = 180 mm

Design tensile strength of angle in rupture

$$T_{dn} = 0.9\, A_{nc}\, f_u/\gamma_{m1} + \beta\, A_{go}\, f_y/\gamma_{mo}$$

$$A_{nc} = (866/2) - 22 \times 6 = 301^2 \text{ mm}$$

$$A_{go} = 433^2 \text{ mm}$$

$$\beta = 1.4 - 0.076 \left(\frac{w}{t}\right)\left(\frac{f_y}{f_u}\right)\left(\frac{bs}{L_c}\right)$$

$bs = w + w_1 - t = 75 + 40 - 6 = 109$ mm (Refer Fig. 3.3)

β = 1.4 – 0.076 (75/6) × (250/410) × (109/180 = 1.05 (Clause 6.3.3 of IS:800:2007)

$$T_{dn} = (0.9 \times 301 \times 410/1.25) + \frac{1.05 \times 433 \times 250}{1.1}$$

$$= 192185 \text{ N} = 192.2 \text{ kN} \qquad (3)$$

Block Shear Strength

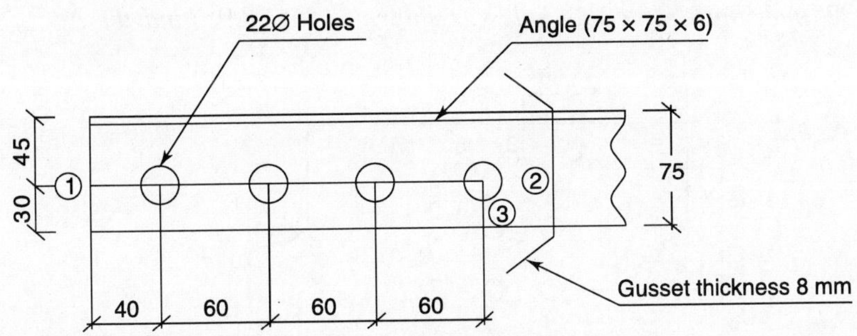

Fig. 3.8(b)

Block shear along section 1-2-3 is to be checked.

$$T_{db} = (A_{vg}\, f_y/\sqrt{3}\, \gamma_{mo}) + (0.9\, A_{tn}\, f_u/\gamma_{m1})$$

(or)

$$T_{db} = (0.9 A_{vn}\, f_u/\sqrt{3}\, \gamma_{m1}) + (A_{tg}\, f_y/\gamma_{mo})$$

$$A_{vn} = (40 + 3 \times 60 - 3.5 \times 22) \times 6 = 858 \text{ mm}^2$$

$$A_{vg} = (40 + 3 \times 60) \times 6 = 1320 \text{ mm}^2$$

$$A_{tg} = 35 \times 6 = 210 \text{ mm}^2$$

$$A_{tn} = (35 - (22/2)) \times 6 = 144 \text{ mm}^2$$

$$T_{db} = [(1320 \times 250/\sqrt{3} \times 1.1) + (0.9 \times 144 \times 410/1.25)]/1000 = 173.2 + 42.51$$

$$= 215.7 \text{ kN} \quad (4)$$

$$T_{db} = [(0.9 \times 858 \times 410/\sqrt{3} \times 1.25) + (210 \times 250/1.1)] = 146.23 + 47.72 \text{ kN} \quad (5)$$

The least of values (1) to (5) is the strength of angle which is = 192.2 kN

This is greater than the factored load 150 kN.

Hence, ISA 75 × 75 × 6 is adequate.

Example 3.3 *Both the flanges of an ISMB 600 used as hangers have holes to accommodate M25 bolts. Find the net area, f_y = 250 MPa.*

Solution: Only two holes can be made per flange in a cross section of ISMB (IS: 808 or steel tables).

Thickness of flange = 20.8 mm

Bolt diameter is 25 mm

Clearance = 3 mm (Table 19 of IS: 800)

$$d_h = 25 + 3 = 28 \text{ mm}$$

$$A_n = A_g - 4 \times 28 \times 20.8 = 15621 - 2329.6 = 13291.4 \text{ mm}^2$$

Example 3.4 *25 mm thick plate has holes as shown in Fig. 3.9. Find the A_n.*

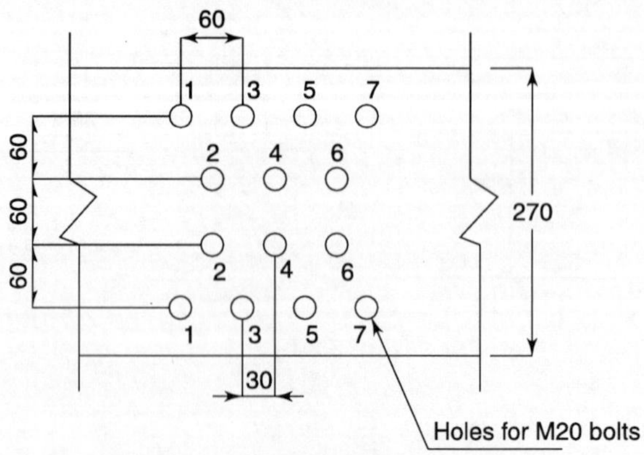

Fig. 3.9

Solution: Bolts are M20

$$d_h = 20 + 2 = 22 \text{ mm (Table 19 IS: 800: 2007)}$$

$$A_g = 270 \times 25 = 6750 \text{ mm}^2$$

Net area (A_n) (Clause 6.3.1 of IS: 800 and Fig. 3.9)

Section 1-1 = Section 2-2 = Section 3-3 = Section 4-4 = Section 5-5 = Section 6-6 = Section 7-7 = (270 – 2 × 22) × 25 = 5650 mm^2

Section 1-2-2-1 = Section 3-4-4-3 = Section 5-6-6-5 = (270 – 2 × 22 + 2 × 30^2/3 × 60) × 25 = (182 + 10) × 25 = 4800 mm^2

$$A_n = 4800 \text{ mm}^2$$

Example 3.5 *Find the design strength of the tension member in the above example if Fe490 steel is used.*

Solution: Thickness is 25 mm > 20 mm but < 40 mm.

Hence, f_y = 330 MPa; f_u = 490 MPa (Table 1 Serial number Viii of IS: 800: 2007)

Design strength due to yielding = $T_{dg} = A_g f_y / \gamma_{mo}$ = 6750 × 330/1.1 × 1000 = 2025 kN

Design strength due to rupture = $T_{dn} = 0.9 A_n f_u / \gamma_{m1}$ = 0.9 × 4800 × 490/1.25 × 1000 = 1693.44 kN.

Lesser of the two is the design strength (tension capacity) = 1693.44 kN.

Example 3.6 *An ISMC 100 section is used as tension member with M16 bolts two rows on web. Find the net area in tension.*

Solution: For M16 bolts, the diameter of bolt holes = 18 mm (Table 19 of IS: 800)

$$A_g = 1170 \text{ mm}^2$$

Thickness of web = 4.7 mm

$$A_n = 1170 - 2 \times 18 \times 4.7 = 1000.8 \text{ mm}^2$$

Net area in tension = 1000.8 mm^2

Example 3.7 *Find the tension capacity of ISMB 600 with M20 bolts on both the flanges.*

$$f_y = 250 \text{ N/mm}^2 \; f_u = 410 \text{ N/mm}^2$$

Solution: **Design strength due to yielding of gross section** (Clause 6.2 of IS: 800)

$$T_{dg} = A_g f_y / \gamma_{mo} = (15621 \times 250)/(1.1 \times 1000) = 3550 \text{ kN}$$

(Clause 6.2 & Table 5.0 of IS: 800)

Design strength due to rupture (Clause 6.3.1 of IS: 800)

Diameter of hole for M20 bolt = 22 mm

Thickness of flange = 20.8 mm

Only two rows of bolt on each flange is possible.

For two flanges there will be four holes in a cross section.

$$A_n = 15621 - 4 \times 22 \times 20.8 = 13790.6 \text{ mm}^2$$

$$T_{dn} = 0.9\, A_n f_u/\gamma_{m1} = \frac{0.9 \times 13790.6 \times 410}{1.25 \times 1000} = 4071 \text{ kN}$$

Tension capacity is the least of T_{dg} and T_{dn}

Hence, tension capacity = 3550 kN

Note: In this problem, we have calculated the strength of the section with bolt holes. But to obtain the tension capacity of the section with its connection, we have to design the connection and calculate the block shear strength also. The least will be the strength.

Example 3.8 *ISA 150 × 75 × 10 is used as a tension member. Larger leg of the angle is connected to gusset plate by using nine numbers of M20 bolts of Class 4.6 at 60 mm pitch in single row. Find the tension capacity of the angle f_y = 250 MPa.*

Solution: **Design strength due to yielding of gross section** (Clause 6.2 of IS: 800)

$$A_g = 2156 \text{ mm}^2$$

Design strength due to yielding = $T_{dg} = A_g f_y/\gamma_{mo}$ = (2156 × 250)/(1.1 × 1000) = 490 kN

Design strength due to rupture (Clause 6.3.3 of IS:800)

$$T_{dn} = 0.9 \times A_{nc}\, f_u/\gamma_{m1} + \beta\, A_{go}\, f_y/\gamma_{mo}$$

where

$$\beta = 1.4 - 0.076 \left(\frac{w}{t}\right)\left(\frac{f_y}{f_u}\right)\left(\frac{b_s}{L_c}\right) \le (f_u\, \gamma_{mo}/f_y/\gamma_{m1}) \ge 0.7$$

w = outstand leg width = 75 mm

t = thickness of the leg = 10 mm

$b_s = w + w_1 - t = 75 + 90 - 10 = 155$ mm

w_1 = 90 mm [SP:6(1)]

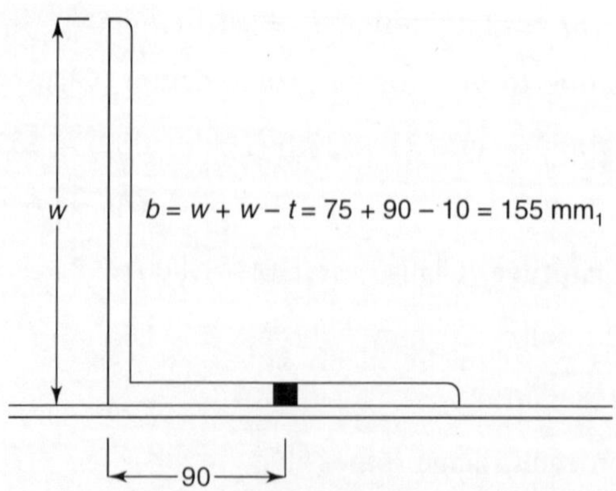

Fig. 3.10

Considering 490 kN as the capacity, number of $M20$ class 4.6 bolts required = 490/58 = 8.44 (Refer to Table 8.8(b))

Provide 9 numbers on single row

Minimum pitch = 2.5 × 20 = 50 mm. Provide 60 mm pitch.

L_c = 8 × 60 = 480 mm

w/t = 75/10 = 7.5

f_y/f_u = 250/410 = 0.61

b_s/L_c 155/480 = 0.323

$\beta = 1.4 - 0.076 \times 7.5 \times 0.61 \times 0.323 = 1.29 \leq (f_u \times \gamma_{mo})/(f_y \times \gamma_{m1}) \geq 0.7$

$(f_u \times \gamma_{mo})/(f_y \times \gamma_{m1}) = 410 \times 1.1/250 \times 1.25 = 1.443 \geq 0.7$

Note:

If β is ≥ 0.7 and ≤ 1.443; keep β = the value.

If β is ≤ 0.7; keep $\beta = 0.7$

If β is ≤ 1.443; keep $\beta = 1.443$

In this case $\beta = 1.3 \leq 1.443$ and ≥ 0.7

$A_{nc} = (150 - (10/2)) \times 10 - (22 \times 10) = 1450 - 220 = 1230 \text{ mm}^2$

$A_{go} = 75 \times 10 = 750 \text{ mm}^2$

$T_{dn} = (0.9 \times 1230 \times 410/1.25) + (1.29 \times 750 \times 250/1.1)/1000 = 583 \text{ kN}$

Lesser of the two is the capacity.

Capacity = 490 kN

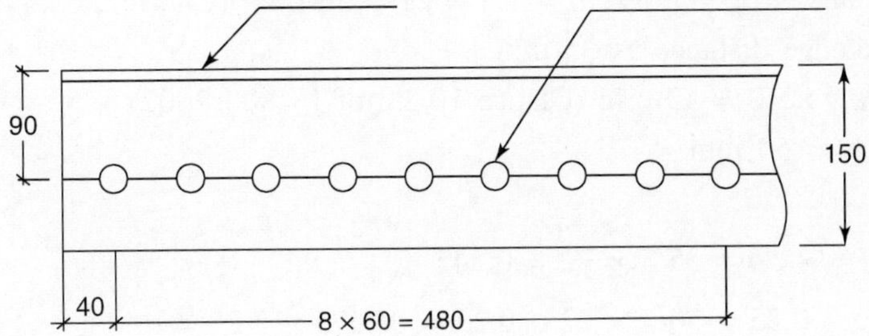

Fig. 3.11

Note: Refer Chapter 8 "Design of Bolted Connections" to check the adequacy of bolts including block shear strength.

3.3 TENSION SPLICES

Filler plate or packing plate is to be provided when two of varying thickneses are connected with cover plates. Additional bolts are to be provided if the thickness of filler plate (or packing) is more than 6 mm. For each 2 mm thickness of packing, the number of bolts required by normal calculations should be increased by 2.5%. The additional bolts should preferably be placed in an extension of the packing.

Example 3.9 *Design a tension splice to connect two plates of size 250 × 20 mm and 220 mm × 12 mm for a design load of 380 kN. (Figure 3.12). Use M20 bolts of property class 8.8.*

Solution: Thickness of packing plate = 20 – 12 = 8 mm

Let us use 10 mm thick splice plate on either side and 20 mm diameter bolts of property class 8.8 for connection.

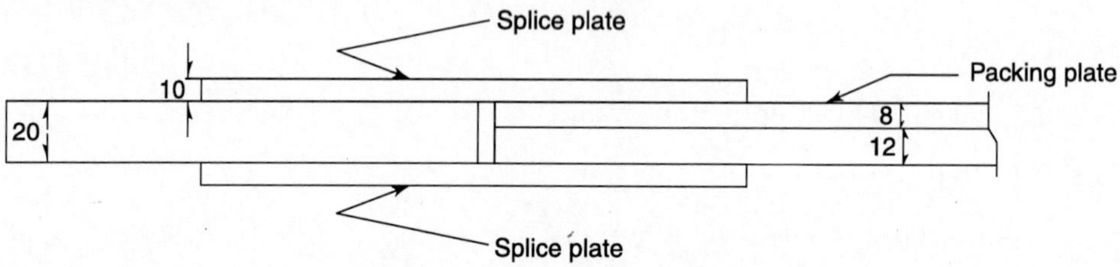

Fig. 3.12

There are two splice (cover) plates of each 10 mm thick. Total all thickness of both the plates put together is more than the thickness of main plate. Out of the main plates, least thickness = 12 mm.

Strength of bolt in bearing against 12 mm thick plate

$$= \left(\frac{2.5 k_b d t f_u}{\gamma_{mb}}\right) \qquad \text{(Clause 10.3.4 of IS: 800)}$$

Minimum edge distance = e = 1.5 × d_h = 1.5 × 22 = 33 mm. (Clause 10.2.4.2 of IS: 800)

Let us provide, end/edge distance as 35 mm.

Minimum pitch = 2.5 × 20 = 50 mm (Clause 10.2.2 of IS: 800: 2007)

Let us provide pitch as 50 mm.

k_b is the least of

$$e/3 \times do = 35/3 \times 22 = 0.545$$

$$p/3 \times \text{do} - 0.25 = 50/3 \times 22 - 0.25 = 0.7575 - 0.25 = 0.51$$

$$f_{ub}/f_u = 800/410 = 1.95$$

k_b is the least of the above three values and shall not exceed one.

Strength of bolt in bearing against 10 mm thick plate = 2.5 × 0.51 × 22 × 12 × 410/1.25 × 1000 = 110.4 kN.

Double shear strength of bolt = 232.26 kN.

Bolt value = 110.4 kN.

Factored load = 1.5 × 380 = 570 kN.

Number of bolts required on each side = 570/110.4 = 5.16 nos.

Provide 7 bolts.

Since the thickness of packing plate is more than 6 mm, additional bolts required at the rate of 2.5% for each 2 mm thickness of packing = ((10 – 6)/2) × 2.5 = 5%

Additional bolt = 7 × 5/100 = 0.35

Provide one additional bolt as shown in Fig. 3.13.

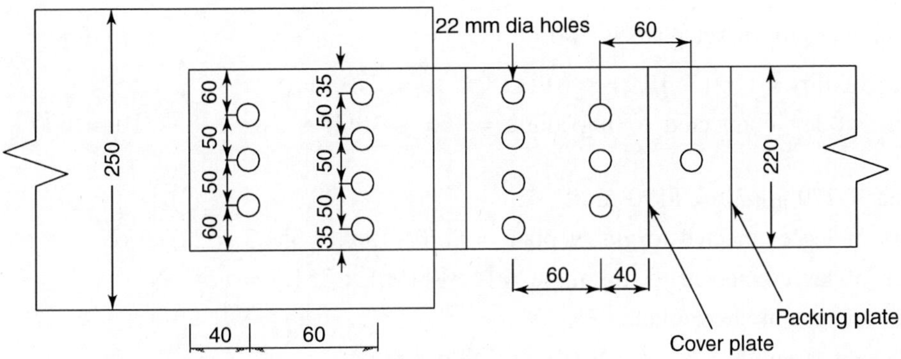

Fig. 3.13

3.4 LUG ANGLE

Lug angles are provided for tension members to reduce the length of joint and thereby the size of gusset plate.

Example 3.10 *Using a lug angle, design a suitable bolted connection using class 4.6 bolt for full tension capacity of an angle ISA 100 × 65 × 10 mm. Angle IS: 2062 material with* f_y *= 250 MPa.*

Solution: Area of cross section of angle = 1550 mm^2

Design strength due to yielding of cross section

$$T_{dg} = A_g f_y/\gamma_{mo} = 1550 \times 250/(1.1 \times 1000) = 352.3 \text{ kN. (Clause 6.2 of IS:800)}$$

Let us provide 16 mm diameter Class 4.6 bolts

Diameter of hole = 18 mm (Table 19 of IS: 800)

Minimum end distance = 1.5 × 18 = 27 mm.

Let us provide 30 mm.

Minimum pitch = 2.5 × 16 = 40 mm.

Let us provide 50 mm.

Rupture strength of single angle

$$T_{dn} = 0.9\ A_{nc}\ f_u/\gamma_{m1} + \beta\ A_{go}\ f_y/\gamma_{mo} \text{ (Clause 6.3.3 of IS: 800)}$$

One leg is connected to gusset plate and the other leg is connected to the lug angle.

A_{nc} = net area of the angle = 1550 – 2 × 18 × 10 = 1550 – 360 = 1190 mm^2

There is no outstanding leg since both are connected.

T_{dn} = (0.9 × 1190 × 410)/(1.25 × 1000) = 351.3 kN.

Tensile strength of member = 351.3 kN. (Least of the two values)

Leg connected to gusset plate is 100 mm.

Net area of this = (100 – 10/2) × 10 – 18 × 10 = 770 mm^2

Net area of leg connected to lug angle = (65 – 10/2) × 10 – 18 × 10 = 600 – 180 = 420 mm^2

Net area = 770 + 420 = 1190 mm^2

Strength of leg connected to gusset plate = (770/1190) × 351.3 = 227.3 kN

Strength of leg connected to lug angle = (420/1190) × 351.3 = 124 kN.

Threads are not in shear plane

Single shear strength = 37.16 kN (Refer Table 8.8(b)) (1)

Gusset plates are provided such that the thickness is equal to or greater than the thickness of member connected. Let us provide 10 mm thick gusset plate.

Strength of bolt in bearing against 10 mm thick angle = $V_{dpb} = \left(\dfrac{2.5 k_b d t f_u}{\gamma_{mb}}\right)$

(Clause 10.3.4 of IS: 800)

k_b is the least of (Clause 10.2.4.2 and Table 19 of IS: 800)

(1) $e/3do$ = 30/3 × 18 = 0.555 (for 65 mm leg, edge distance is 30 mm SP: 6(1))

(2) $p/3 \times do$ = (50/3 × 18) –0.25 = 0.676 (Clause 10.2.2 of IS: 800)

(3) f_{ub}/f_u = 400/410 = 0.975

(4) 1.0

f_u = ultimate tensile stress of plate.

In this case f_u for plate is 410 MPa.

Strength of bolt in bearing = 2.5 × 0.555 × 16 × 10 × 410/(1.25 × 1000) = 72.82 kN (2)

Bolt value is the least of (1) and (2)

Bolt value 37.16 kN

Number of bolts required for connecting the angle to the gusset plate = 227.3/37.16 = 6.12 Provide 7 bolts.

Design of Lug Angle

Purpose is to reduce the length of gusset plate.

Strength of leg connected to lug = 124 kN

It is a general engineering practice to design the lug angle for 20% and connection for 40% more strength.

Required strength of lug angle = 1.2 × 124 = 148.8 kN

A_{net} required = $148.8 \times 10^3/(250/1.1) = 654.72$ mm^2

Try 65 × 65 × 8 angle

Area = 976 mm^2

A_{net} = $976 - 2 \times 18 \times 8 = 688$ mm^2 > 654.72 mm^2

Strength of bolt in bearing against 8 mm thick angle

$$= V_{dpb} = \left(\frac{2.5k_b dtf_u}{\gamma_{mb}}\right) \text{ (Clause 10.3.4 of IS: 800)}$$

k_b is the least of (Clause 10.2.4.2 and Table 19 of IS: 800)

(1) $e/3 \times do = 30/3 \times 18 = 0.555$ (for 65 mm leg, edge distance is 30 mm SP:6(1))

(2) $p/3 \times do = (50/3 \times 18) - 0.25 = 0.676$ (Clause 10.2.2 of IS:800)

(3) $f_{ub}/f_u = 400/410 = 0.975$

(4) 1.0

Strength of bolt in bearing = $V_{dpb} = 2.5 \times k_b \times d \times t \times f_u/\gamma_{mb}$ (Clause 10.3.4 of IS: 800)

$= 2.5 \times 0.555 \times 16 \times 8 \times 410/(1.25 \times 1000) = 58.25$ kN.

Bolt value = 37.16 kN

Number of bolts required = 1.4 × 124/37.16 = 4.67

Provide 5 Nos. of M16-Class 4.6 bolts.

Design strength due to block shear (Clause 6.4 of IS: 800)

$$T_{db} = (A_{vg} f_y/\sqrt{3}\ \gamma_{mo}) + (0.9\ A_{tn}\ f_u/\gamma_{m1})$$

(or)

$$T_{db} = (0.9A_{vn}\ f_u/\sqrt{3}\ \gamma_{m1}) + (A_{tg}\ f_y/\gamma_{mo})$$

$$A_{vg} = 10\ (6 \times 50 + 30) = 3300 \text{ mm}^2$$

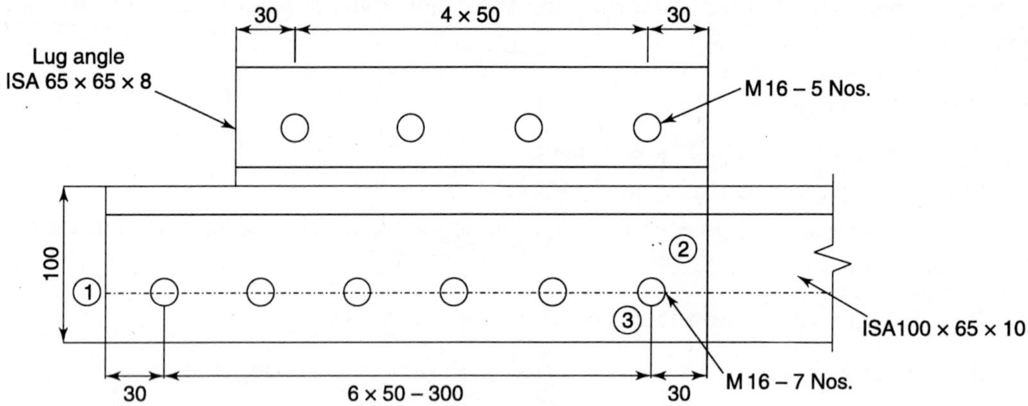

Fig. 3.14

A_{vn} = 10(6 × 50 + 30) – (6.5 × 18) = 2130 mm^2

A_{tg} = 10 × 40 = 400 mm^2

A_{tn} = 10(40 – 0.5 × 18) = 310 mm^2

T_{db} = (3300 × 250/ $\sqrt{3}$ × 1.1) + (0.9 × 310 × 410/1.25) = 524.52 kN > 351.3 kN considered

T_{db} = (0.9 × 2130 × 410/ $\sqrt{3}$ × 1.25) + (400 × 250/1.1) = 453.93 kN > 351.3 kN considered

Hence, satisfactory.

EXERCISE

1. What is meant by "tension member"?
2. Give some examples for tension member in a structure.
3. Write down the design procedure for design of tension member as per the limit state method.
4. Write down the formula used to calculate net area (A_n) of a plate with staggered holes under tension. Draw a neat sketch.
5. Draw the section along which block shear will be checked for the following connection.

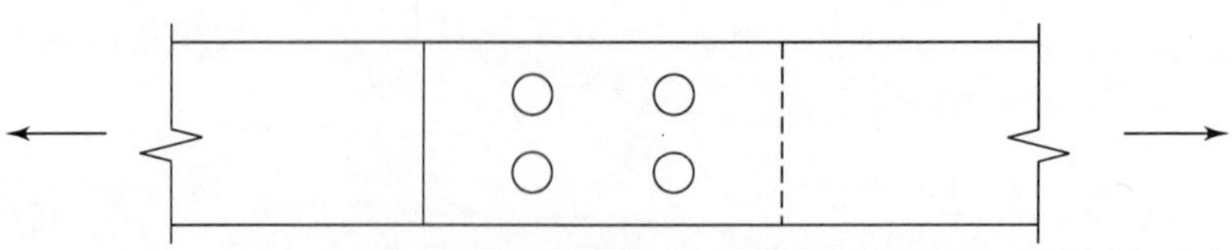

6. In what way, the design of tension member connected by welding at ends differs with the same member connected by bolting at ends?

7. Design a single angle tension member to take an axial load of 120 kN due to dead load, imposed load and wind load. Use material with f_y = 250 MPa. The end connection is by welding. Connection design and block shear check need not be made.
8. Design a tension member to carry an axial tension of 160 kN due to dead and imposed load using two angles one on each side of gusset. The end connection is made using 3 Nos. of M20 bolts with an edge distance of 40 mm and a pitch of 60 mm. Design the angle and also check for block shear. Yield stress of the material is 250 MPa.

Chapter 4

Design of Compression Members

The structural members subjected to axial compressive force are called compression members. Examples are columns, vertical and horizontal bracing members and posts in structures. Members subjected to moments in addition to axial compressive force are called beam-columns. Predominant failure mode for members in compression is due to buckling. Buckling depends on two parameters. One is the effective length of the member which depends on its end condition and the other one is the radius of gyration.

4.1 EFFECTIVE LENGTH OF ANGLE STRUTS

Table 4.1 gives the effective length and maximum allowable slenderness ratio for angle compression members depending on their end connections.

Single Angle Discontinuous

End Connections

1. Welded (or) with one bolt at each end.

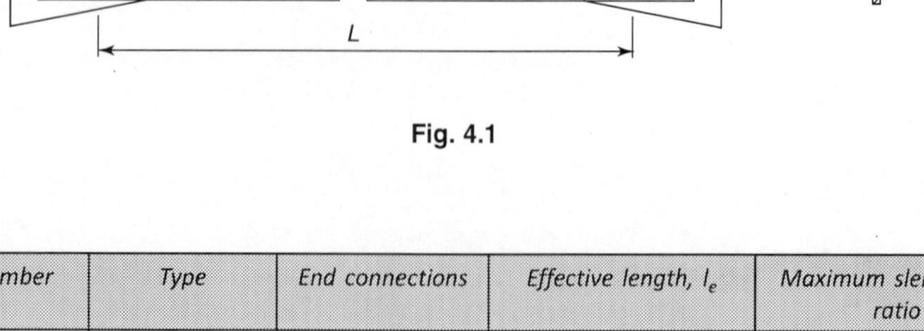

Fig. 4.1

Table 4.1

Serial number	*Type*	*End connections*	*Effective length, l_e*	*Maximum slenderness ratio*
1	Single angle discontinuous	One bolt or weld at each end	0.7 to 1.0 L. Preferably, 1.0 L. Clause 7.5.1 of IS: 800.	180

2. Welded (or) with two (or) more bolts at each end.

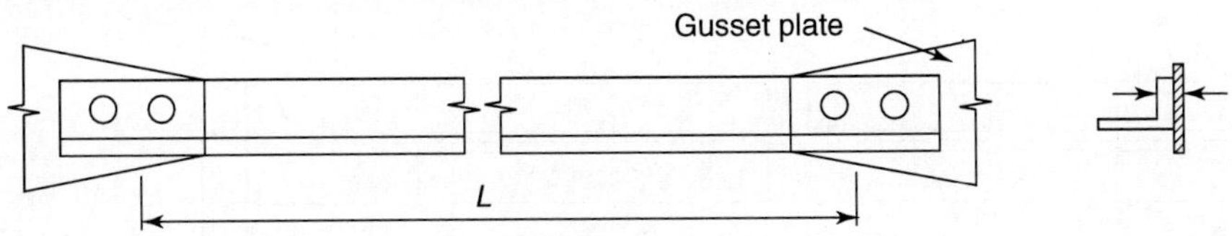

Fig. 4.2

Table 4.2

Serial number	*Type*	*End connections*	*Effective length, l_e*	*Maximum slenderness ratio*
1	Single angle discontinuous	Two or more bolts or weld at each end	Preferably l_e = 0.85 L Clause 7.5.1 of IS: 800	180

3. Two angles connected on same side of gusset plate

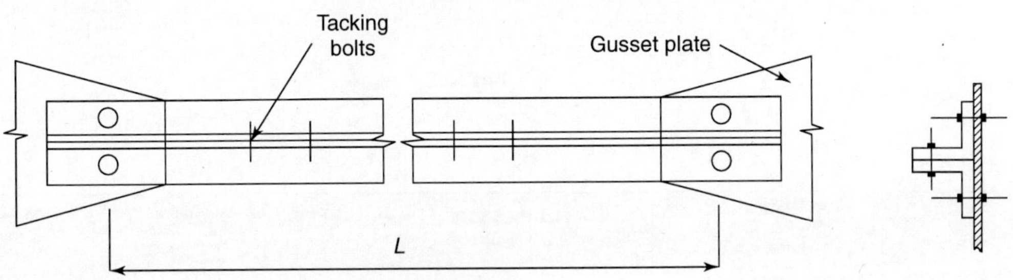

Fig. 4.3

Table 4.3

Serial number	*Type*	*End connections*	*Effective length, l_e*	*Maximum slenderness ratio*
1	Double angle discontinuous	Connected on same side of gusset plate with one bolt (or) welded	L_e = 0.7 L to 1.0 L Clause 7.5.2.2 of IS: 800	180

4. Welded (or) with two (or) more bolts at each end.

Table 4.4

Serial number	*Type*	*End connections*	*Effective length, l_e*	*Maximum slenderness ratio*
1	Double angle discontinuous	Two or more bolts or weld at each end	0.7 L to 0.85 L. Clause 7.5.2.2 of IS: 800. Preferable = 0.85 L.	180

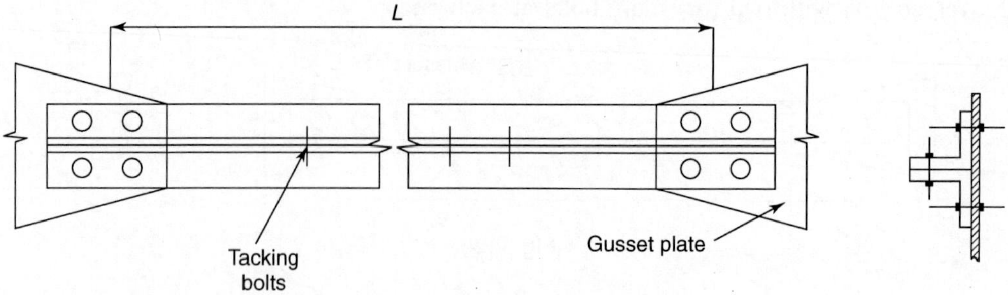

Fig. 4.4

5. Two angles connected on both sides of gusset plate by two or more bolts or welding

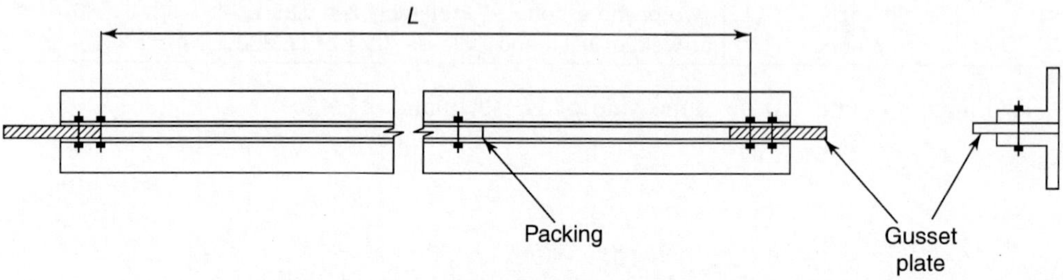

Fig. 4.5

Table 4.5

Serial number	*Type*	*End connections*	*Effective length, l_e*	*Maximum slenderness ratio*
1	Double angle discontinuous	Connected on both sides of gusset plate by one bolt or welding	0.7 L to 0.85 L. Clause 7.5.2.1 of IS: 800 for axis parallel to gusset. For the axis perpendicular to the gusset, 1.0 L.	180
		Connected on both sides of gusset plate by two or more bolts or welding.	1.0 L	
	Continuous angles of trussess and towers.	(1) In the plane of truss (2) Other plane	0.7 to 1.0 L 1.0 L	

L = Unsupported length in mm

Trusses

In the case of bolted or welded trusses and braced frames, the effective length l_e of the compression member is taken between 0.7 and 1.0 times the distance between centres of intersection. In the case of members of a truss buckling may be in the plane of truss or perpendicular to the plane of truss, for which the effective length is taken 1.0 time the distance between the points of restraint.

4.1.1 Maximum Slenderness Ratio

The slenderness ratio ($\lambda = l_e/r$) should not exceed the values given in Table 4.6 (Table 3 of IS:800)

Table 4.6

Types of member	*Maximum slenderness ratio*
For members carrying compressive load due to dead and imposed loads.	180
For members subjected to compressive load due wind/earthquake forces provided the deformation of such members does not adversely affect the stress in any part of the structure.	250
For members carrying tension but subjected to reversal of stress due to wind or earthquake forces.	350

Note:

The above are the maximum permissible slenderness ratio.

$$L_e = \text{effective length} = kL$$

$$r = \text{radius of gyration}$$

Table 4.7 Slenderness ratios for angle and channel sections

Refer sketch	*Type of section*	*End connection*	*Slenderness ratio*
(a)	Single angle	At least two bolts in line or welded	Preferable 0.85 L/$r_{min.}$ Clause 7.5.1 of IS: 800
(b)	Single angle	Single bolt	(0.7 L to 1.0 L)/$r_{min.}$ Preferable 1.0 L. Clause 7.5.1 of IS: 800.
(c)	Double angles parallel or back to back on one side of gusset	Single bolt or welded	For each main component $\lambda_c = \frac{L_c}{r_{min}} \leq 50$ r_{min} = minimum slenderness ratio of one main component. Preferable 1.0 L/$r_{min.}$ Clause 7.5.1 of IS:800.
(d)	Double angle back to back separated	At least two bolts in line connected to one leg or equivalent weld	(0.7 L to 1.0 L)/$r_{min.}$ Preferable 0.85 L. Clause 7.5.2.1 of IS: 800.
(e)	Parallel angle battened	At least two bolts in line connected to one leg or equivalent weld	(0.7 L to 1.0 L)/$r_{min.}$ Preferable 0.85 L. Clause 7.5.2.1 of IS: 800. Clause 7.5.1 of IS:800.

Contd...

Contd...

(f)	Double angle back to back in contact	At least two bolts in line connected to one leg or equivalent weld	(0.7 L to 1.0 L)/$r_{min.}$ Preferable 0.85L/$r_{min.}$ Clause 7.5.2.1 of IS: 800.
(g)	Double angle back to back separated	One bolt connected to one leg of one angle	Preferable 1.0 L/$r_{min.}$ Clause 7.5.2.2 of IS: 800.
(h)	Parallel angle battened	One bolt connected to one leg of one angle	Preferable 1.0 L/$r_{min.}$ Clause 7.5.2.2 of IS: 800.
(i)	Double angle back to back in contact	One bolt connected to one leg of one angle	Preferable 1.0 L/$r_{min.}$ Clause 7.5.2.2 of IS:800.
(j)	Double angle back to back separated	One leg of both angles connected by minimum two bolts in line or equivalent weld	Preferable 0.85 L/$r_{min.}$ Clause 7.5.2.2 of IS: 800.
(k)	Single channel	Connected to a gusset or another member by two or more rows of minimum two bolts in each row or equivalent weld	Preferable 0.85 L/$r_{min.}$
(l)	Single channel	Connected to a gusset or another member by one row of minimum two bolts	Preferable 1.0 L/$r_{min.}$

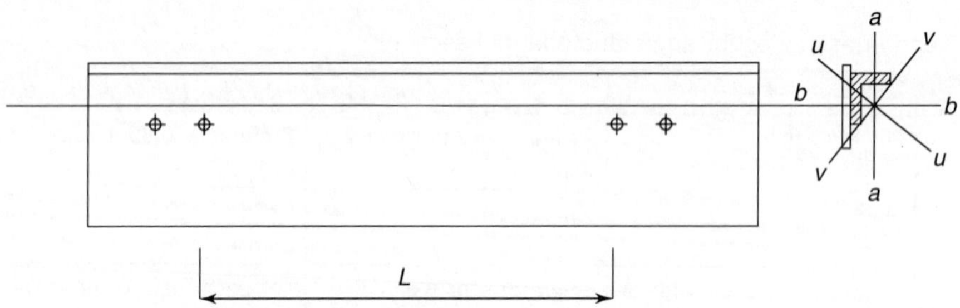

Sketch (a)

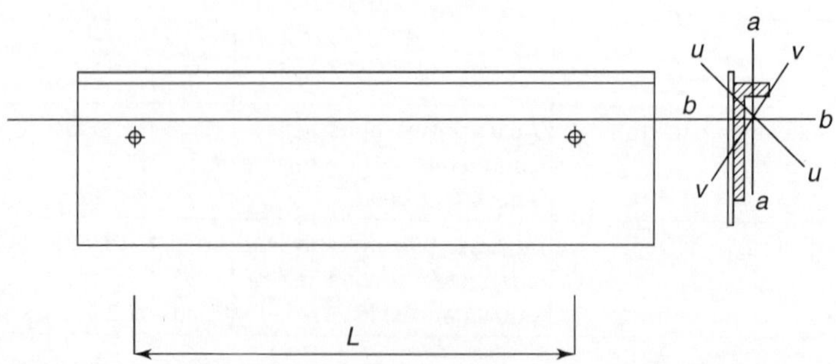

Sketch (b)

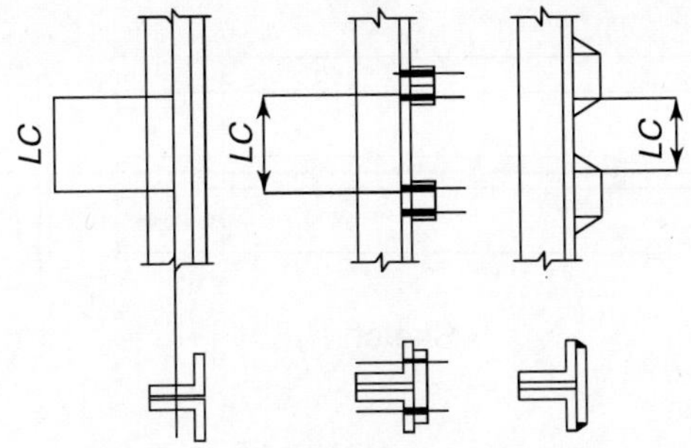

Sketch (c)

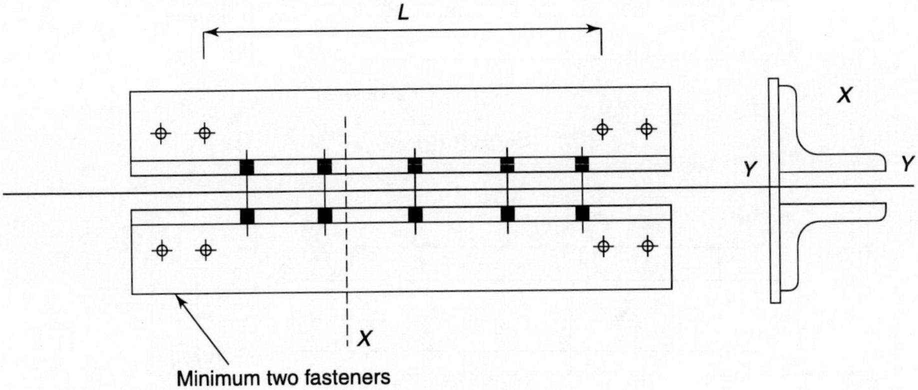

Sketch (d)

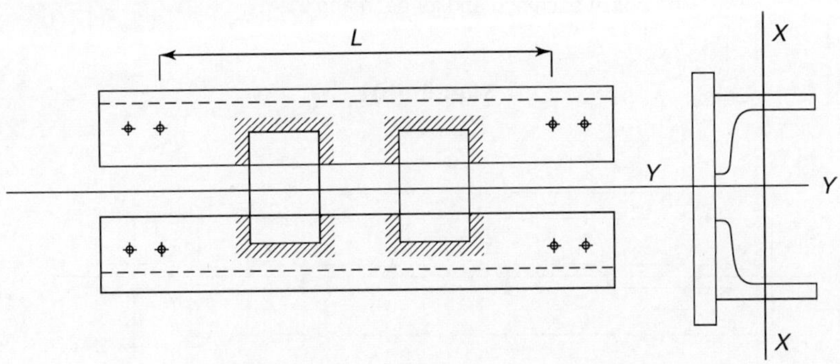

Sketch (e)

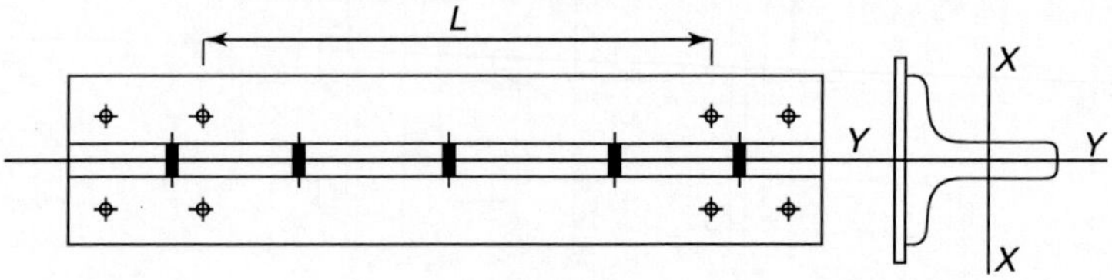

Sketch (f)

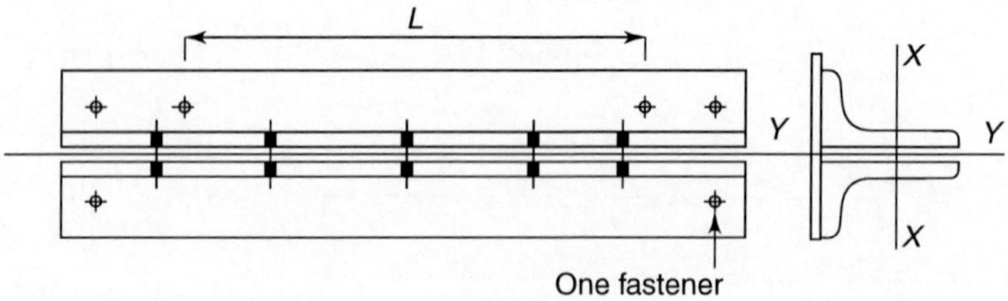

Sketch (g)

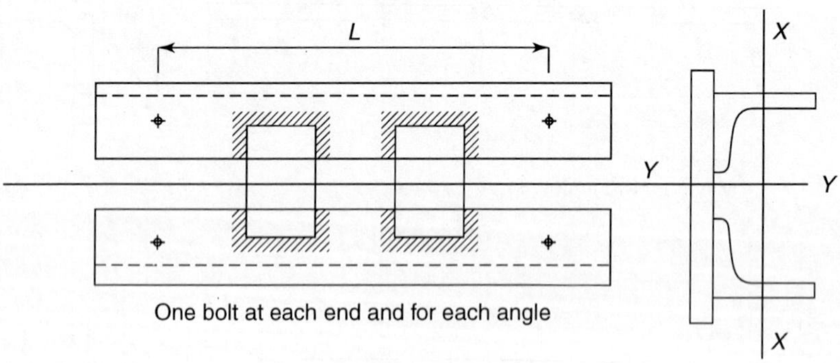

Sketch (h)

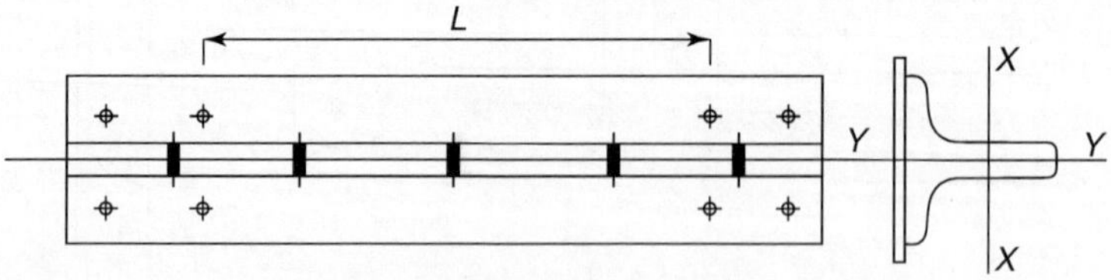

Sketch (i)

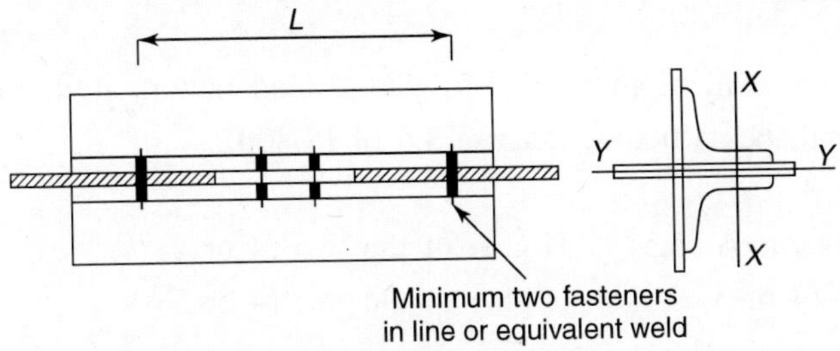

Sketch (j)

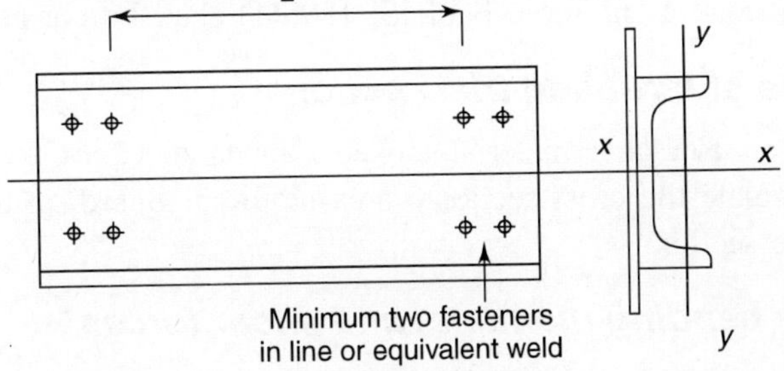

Sketch (k)

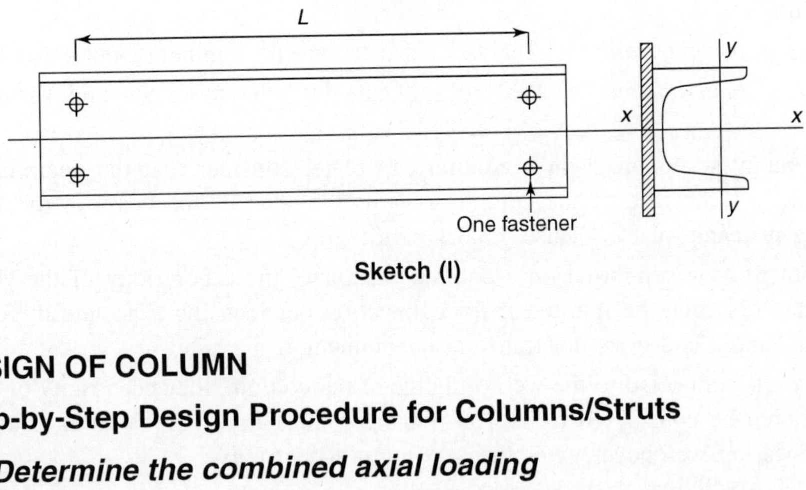

Sketch (l)

4.2 DESIGN OF COLUMN

4.2.1 Step-by-Step Design Procedure for Columns/Struts

Step 1 Determine the combined axial loading

Primary load cases

1. Self weight
2. Dead load
3. Imposed load
4. Wind load
5. Seismic load
6. Erection load (Clause 3.3 of IS:800).

Design load

The above primary loads shall be multiplied by partial load factors as in Table 4 of IS:800.

The combination shall be as below. (Clause 3.5 of IS:800).

Case (a) = 1 + 2 + 3

Case (b) = 1 + 2 + 3 + (4 or 5) Higher of the two (4 or 5)

Case (c) = 1 + 2 + (4 or 5) Higher of the two (4 or 5)

Case (d) = 1 + 2 + 6

Follow the above load combination and after the analysis, determine the factored axial compression (*P*). Reduction in imposed load as per Clause 3.2 and Appendix A of IS:875 (Part-2) and % imposed load as in Table B of IS: 1893(1) shall be considered.

Step 2 Select type of strut and trial section

Select the type based on available material (shape), appearance, architectural and frictional requirements. Then calculate the cross-sectional area required. Based on the area required and the moment, choose the section.

Step 3 Determine bending moments and shear forces

The type of construction (continuous or simple) should be decided at the beginning of the design. Calculate the bending moments in a compression member according to the construction method chosen:

Continuous (as in a rigid frame): Analyse and calculate the moments and shear forces.

Simple (as in a braced frame): Analyse and calculate the moments and shear forces as below:

(a) If the beam is supported on a column cap plate, consider that the beam-end reaction is applied on the face of the column towards the span of the beam or the edge of the bearing plate/cap plate (Clause 7.3.3.1 of IS:800).

(b) If a roof truss is supported on a column cap plate, the eccentricity of the vertical load from the truss may be ignored if the connection between the truss and the column cap plate is simple and does not transfer any moment.

(c) If a beam is connected to the web or flange of the column, the eccentricity of the vertical load should be considered to be 100 mm from the face of the column or at the centre of the bearing whichever gives the maximum eccentricity.

(d) Ignore the effect of eccentric end connections for struts, bracing members made of angles provided the slenderness ratios are limited to the values given in IS: 800.

(e) For laced struts, battened struts, star angle struts and battened parallel angle struts, allow for eccentricity of end connections. Try to avoid eccentricity by making the line passing through the C.G. of all members joining at that joint to meet at a single point called set-off point (S.O.P.) or working point (W.P.).

(f) Maximum column loads will occur when all beams are fully loaded. The moments due to eccentricity should be added to moments from fixity. This total moment will be shared between columns above and below the joint in proportion to the stiffness of the column I/L. Deflection due to self weight should be limited to 1/1000. Otherwise, design for axial load plus bending due to deformation are to be considered. (P-δ effect). Find out the following.

M_{xx} = Bending moment in ZZ direction (ZZ being major axis)

M_{yy} = Bending moment in YY direction

Step 4 Determine the local bending moments of compression members in open web girders/latticed girders/trusses

For trusses, the members are considered axially loaded members. To achieve this, the loads are applied at nodes. This helps analysis of truss by using method of joints, method of sections or graphical statics. But in reality, the purlins are not always connected at nodes. It is connected to the top chord away from the node also. This induces bending. At first, the analysis is made by assuming that the loads are at nodes and then the member is checked for bending also.

Step 5 Determine the effective length

The most important element in the design of a compression member is its effective length in both the mutually perpendicular axes. Tables 4.1 to 4.8 may be used to find the effective lengths. The effective slenderness ratio of laced columns shall be taken as 1.05 times the actual maximum slenderness ratio. (Clause 7.6.1.5 of IS:800). The effective slenderness ratio of battened columns shall be taken as 1.1 times the actual maximum slenderness ratio. (Clause 7.7.1.4 of IS: 800). For columns subjected to moment lacing is the best option.

Step 6 Calculate the required area of cross section

Assuming f_{cd} 0.5 f_y to 0.7 f_y, find approximate area required

$$\text{Area required} = \text{factored load}/f_{cd}.$$

This is very approximate because, even if the column is axially loaded, we have to consider minimum eccentricity. Hence, the column will be subjected to axial load and moment. Apart from this, the failure mode for column is by flexural buckling. The above calculation will give very less area of cross section.

Step 7 Choose a section

Choose a section for which cross section area is more than the required area as in step 6 above. More exact increase in area or the value of f_{cd} cannot be suggested since it depends on the effective length and magnitude of moment.

Step 8 Determine the slenderness ratio

The slenderness ratio is defined as the effective length with respect to an axis of buckling divided by the radius of gyration of the column section about the same axis.

Table 4.8 Effective length of compression members

Description of structure	*Effective length*	
Free to rotate at top and base effectively held in position at both ends but not restrained in ditection.	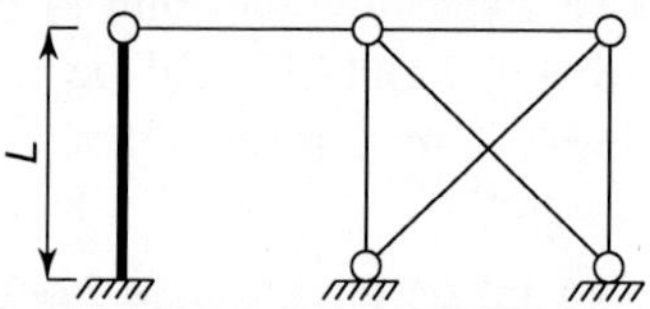	1.0 L
Free to rotate at top, restrained at base effectively held in position at both ends and in direction at one end.	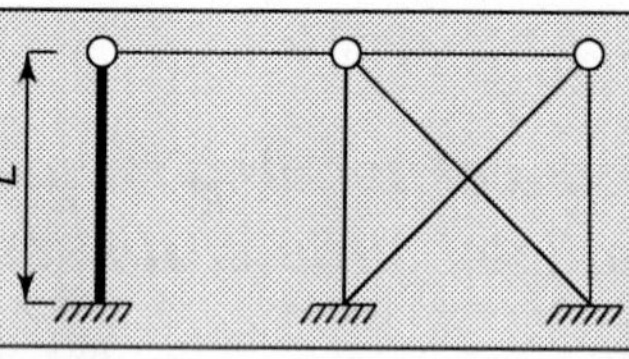	0.8 L
Unbraced, restrained at base, pinned at top. Effectively held in position and direction at one end but not restrained in position or direction at the other end.	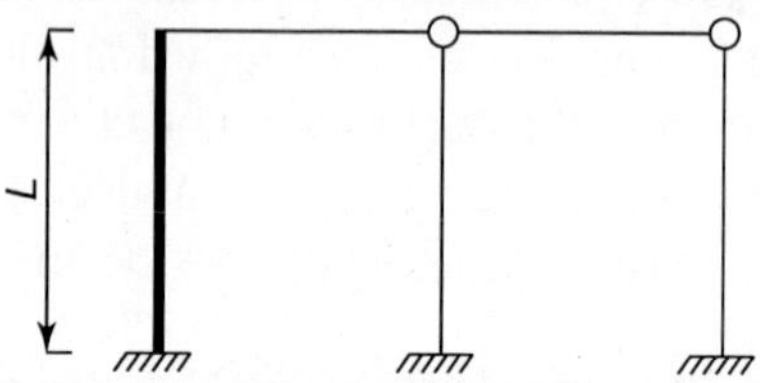	2.0 L
Braced, partially restrained by beams at both ends. Effectively held in position at both ends and partially restrained in direction at both ends.	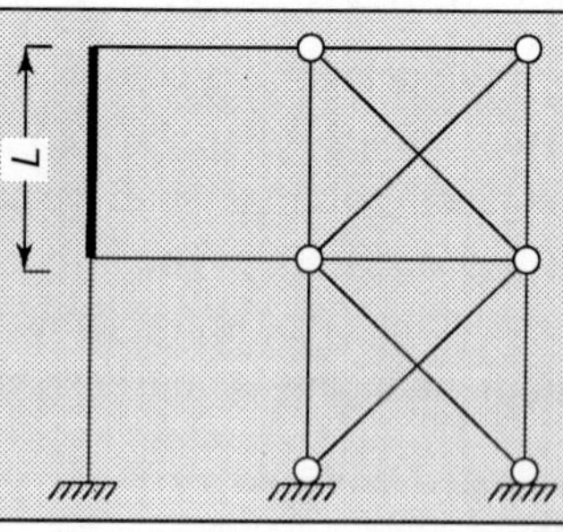	0.85 L
Braced, restrained at base and rigidly restrained by a very stiff beam at the top. Effectively held in position and direction at both ends.	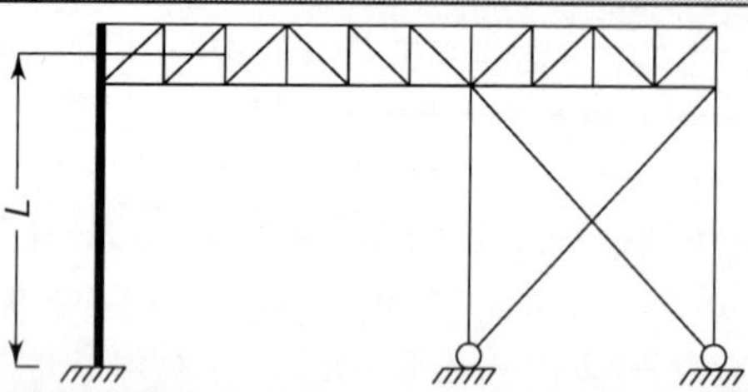	0.7 L
Unbraced, restrained base and rigidly restrained by a very stiff beam at the top. Effectively held in position and direction at one end and held in direction at the other end.	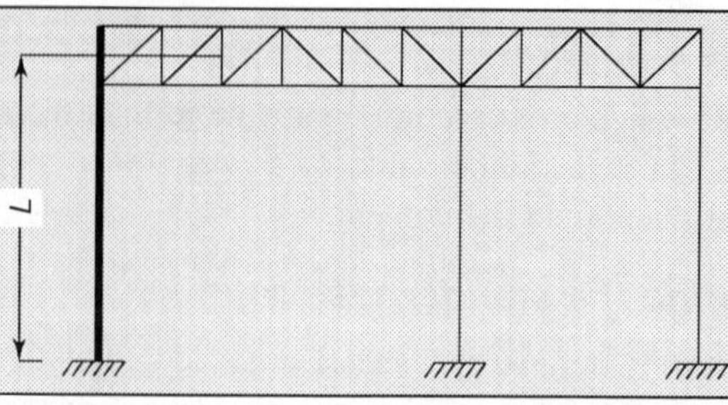	1.2 L

Contd...

Contd...

Unbraced, restrained at base and partially restrained by a beam at the top. Effectively held in position and direction at one end and partially restrained in direction at the other end.	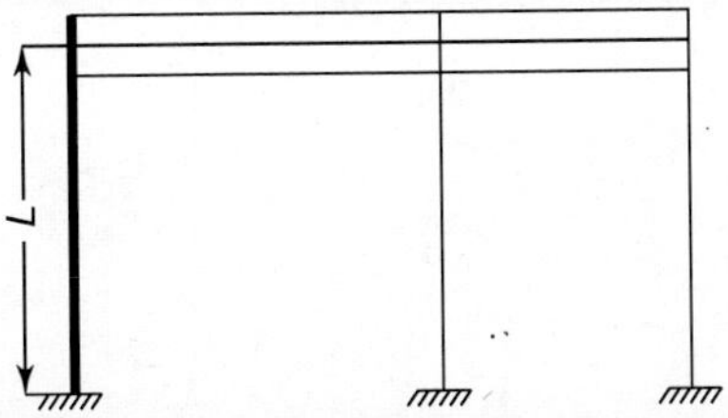	1.5 L
Columns for factory sheds effectively held in position and direction about both axes at the base, not held in position at the top along the Y-Y axis, or direction about the Z-Z axis, but effectively held in position along the Z-Z axis by bracing and also held in position at the crane girder level by braced bays. Partial frame method may be used if the column is rigidly connected to a stiff roof truss of latticed girder at the top. If at the base of the column, the directional restraint about the Y-Y axis is not available then effective length factors will be increased from 0.85 to 1.0	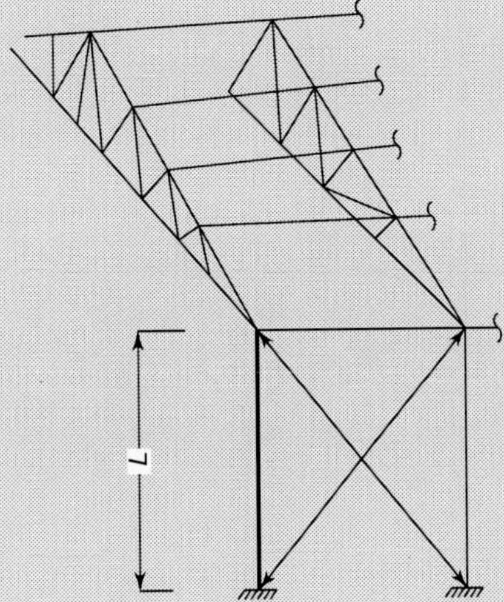	L_{ez} = 1.5 L L_{ey} = 0.85 L
The angles/channels provided as tie or side cladding runners act as restraint to Y-Y axis. To restrain both the flanges, adequate tie should be provided on the inner flange also.	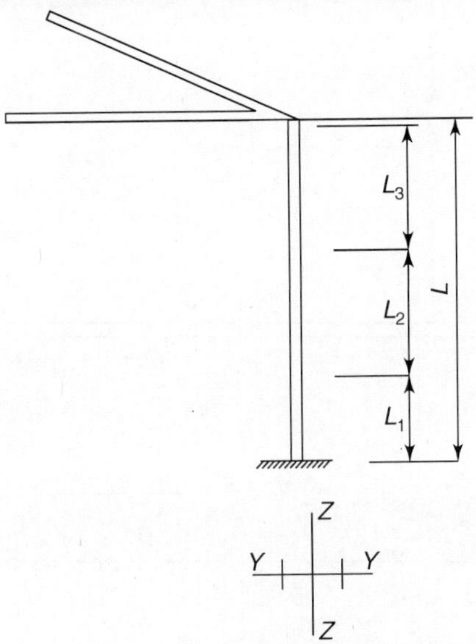	L_{ez} = 1.50 L L_{eg} = 0.85 L_2, 1.0 L_2 or 1.0 L_3

Contd...

Contd...

	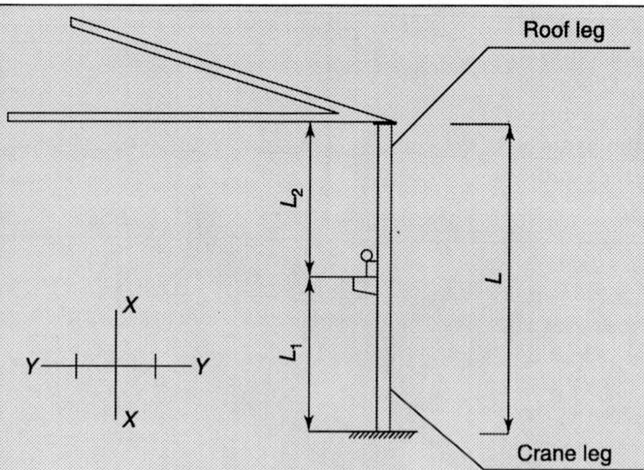		For roof leg: L_{ez} = 1.5 L_2 L_{eg} = 1.5 L_1 For crane leg L_{ez} = 1.5 L_1 L_{eg} = 0.85 L_1
	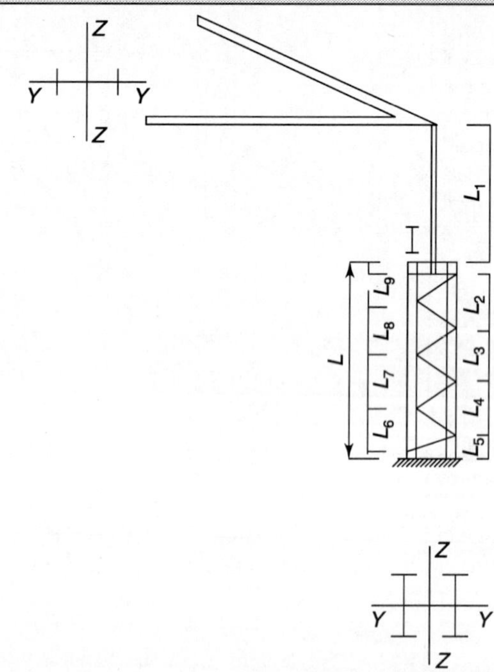		For roof leg: L_{ez} = 1.5 L_1; L_{ey} = 1.0 L_1 For crane leg L_{ez} = 1.5 L L_{ey} = 0.85 L
Braced structure Single diagonal bracing	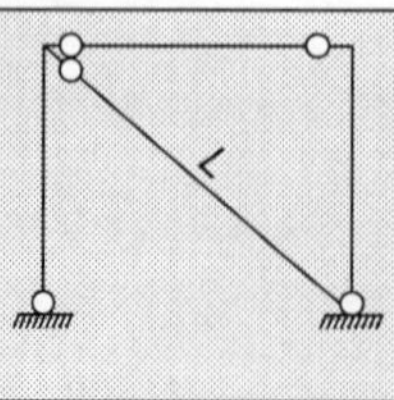	L_e = 1.0 L	Double intersection lacing, connected at intersection. 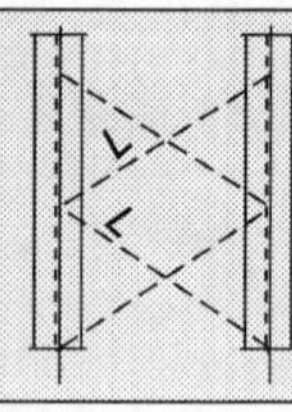 L_e = 0.7 L

Contd...

Contd...

Cross bracing, connected at intersections	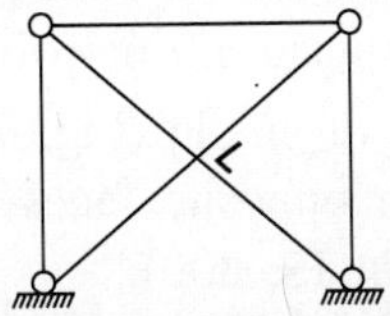 $L_e = 0.7$ L	
Single intersection lacing	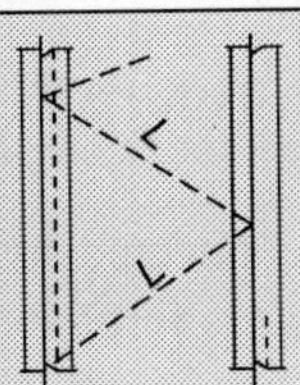 $L_e = 1.0$ L	In this case the L_e in the plane of bracing will be 0.5 L and the perpendicular plane will be 1.01. However, the frame can be considered 0.7 L in both the axes.
L = Length between liner end fastener of the lacing member (Clause 7.6.6.3 of IS:800)		

$$\lambda_z = \frac{KL_{ez}}{r_z} \text{ and } \lambda_y = \frac{KL_{ey}}{r_y}$$

where L_{ez} and L_{ey} are the unsupported lengths about the two orthogonal axes.

The values of K for single angle, double angles and single channel sections are given in Tables 4.1 to 4.8.

Step 9 Determine the compressive resistance

Calculate the f_{cd} as in Clause 7.1.2.1 of IS: 800 or obtain the value of f_{cd} from Tables 9(a) to 9(d) based on KL/r and buckling class of the section. Find $P_d = A_e \times f_{cd}$ which shall be greater than P as in step 2.

Step 10 Check the columns with uniaxial or biaxial moments for combined effect

Follow the steps in Section 9 of IS: 800 particularly Clause 9.3.1 and 9.3.2 of IS: 800 to check the columns with uniaxial or biaxial moments for combined effect.

Step 11 Design of lacing systems (see Fig. 4.25)

General rules:

(1) Columns consisting of two main components connected by a lacing system, should have a radius of gyration about the axis perpendicular to the plane of lacing more than the radius of gyration about the axis parallel to the plane of lacing if practically possible. (Clause 7.6.1.1 of IS: 800).

(2) Lacing should not vary in section or inclination throughout the length of the member.

(3) Inclination of the lacing should be between 40 and 70 to the longitudinal axis of the member (axis along the length of the member-Clause 7.6.4 of IS: 800).

(4) Lacing can be double (Fig. 10 B of IS: 800) or single (Fig. 10 C of IS: 800). Single lacing should not be opposed in direction on opposite faces of the main member to avoid torsion in the member. (Clause 7.6.1.4 of IS: 800).

(5) Tie plates (end batten) similar to battens should be provided at the ends of the lacing systems. Such end or tie plates will act as stiffeners for base plate or cap plates of the column.

(6) Slenderness ratio for individual main members connected by the lacing should not be greater than 50 or 0.7 times the higher slenderness ratio of the member as a whole. The unsupported length of individual main member is the length between the points of intersections of the lacing with the main members. Radius of gyration of the individual main member is the radius of gyration about the minor axis which should be used to compute slenderness ratio of the individual member = KL/r_y (Clause 7.6.5.1 of IS: 800).

Select a type and section of lacing (flat, angle, channel) to satisfy the general rules. Find effective length of the lacing bars as per Clause 7.6 of IS: 800. Check that the slenderness ratio of the lacing bars do not exceed 145.

Find shear force V across a plane perpendicular to the axis of the member, which is given by:

$$V = \left(\frac{2.5}{100}\right) \times F \times V_s$$

where F = maximum axial compression in the member

V_s = transverse shear in the member at any section acting with F.

The lacing bar should be checked for an axial tension or compression equal to P, which is given by

$$P = \frac{V}{\sin\varphi}\frac{1}{N}$$

where N = number of lacing bars cut by a plane perpendicular to the axis of the member.

φ = angle of inclination of the lacing bars to the axis of the member along length.

The required section for lacing bars as compression members shall be determined by assuming f_{cd}. Calculate the design strength in compression as per Clause 7.1 of IS: 800.

P_d = capacity of lacing bar in compression = $A_g \times f_{cd}$ for welded connection.

P_d = capacity of lacing bar in compression = $A_n\, f_{cd}$ for bolted connection.

A_g = gross cross-sectional area of the lacing bar

A_n = net cross-sectional area of the lacing bar

P_d shall be greater than P.

This cross section of lacing bar shall be checked as tension member as per Clause 6.1 of IS: 800. Find design strength due to yielding of gross section, due to rupture of critical section and block shear strength. These shall be more than the actual tensile force.

The tie plates should be designed as end batten, as in Step 12(8).

Step 12 Design of battens (Figs. 4.10 & 4.11)

General rules

1. Columns consisting of two components (*I* or channel) battened shall have the individual components of the same cross section and kept symmetrically about their major axis. The columns should have a radius of gyration about the axis perpendicular to the plane of the batten not less than the radius of gyration about the axis parallel to the plane of the batten (Clause 7.7.1.1 of IS: 800).
2. The joints between the individual components and the battens must be rigid. Use welded connections or connection by at least two bolts along the axis of the main member on each individual component.
3. Battens should be provided opposite to each other in each plane at the ends of the member (base and cap of columns) and at the point of lateral restraints.
4. Intermediate battens on opposite faces of main member should be uniformly spaced throughout the length of the member. (Clause 7.7.1.3 of IS: 800)
5. The maximum slenderness ratio of the individual component ($\lambda = KL/r_y$) between end welds or end fasteners of adjacent battens should not exceed 50 or 0.7 times the higher slenderness ratio of the member as a whole.
6. The thickness of the plate used as batten should not be less than 1/50 of the minimum distance between welds or fasteners on the batten connecting the individual components, i.e., the free length of batten. (Clause 7.7.2.3 of IS: 800)
7. Effective depth of the end battens (dimension measured along the length of the main member) should not be less than the distance between the centroids of the main members of the compound column. (Clause 7.7.2.3 of IS: 800)
8. Effective depth of the intermediate battens (dimension measured along the length of the main member) should not be less than 3/4th of the distance between the centroids of the individual components of the column. This shall not be less than twice the width of the member in the plane of batten. (Clause 7.7.2.3 of IS: 800)

   Effective depth of batten is the distance (measured along the length of column) between outermost bolts or welds at ends of batten.
9. Select a type and section of end batten and intermediate battens, find the spacing using the condition in 5 above and decide the thickness and effective depth as per 7 & 8 above.
10. Find transverse shear force V across a plane perpendicular to the axis of the column, which is given by: $V = 0.025\ F + V_s$ where F = maximum axial compressive force in the column and V_s = transverse shear in the column at a batten position if any (including those due to moment) acting with F. The batten should be checked as per Clause 7.7.2 of IS: 800 for shear and bending.

4.3 BEAM COLUMNS – COLUMN WITH MOMENT

The following are some patterns of loading which causes bending moment on a column.

Eccentricity of an axial force (Fig. 4.6a)

Load from brackets (Fig. 4.6b)

Fixed base condition (Fig. 4.6c)

Transverse load (due to wind or earthquake) (Fig. 4.6d)

Building frame action (Fig. 4.6e)

Portal (or) gable frame action (Fig. 4.6f)

The column and pattern are pictorially represented below.

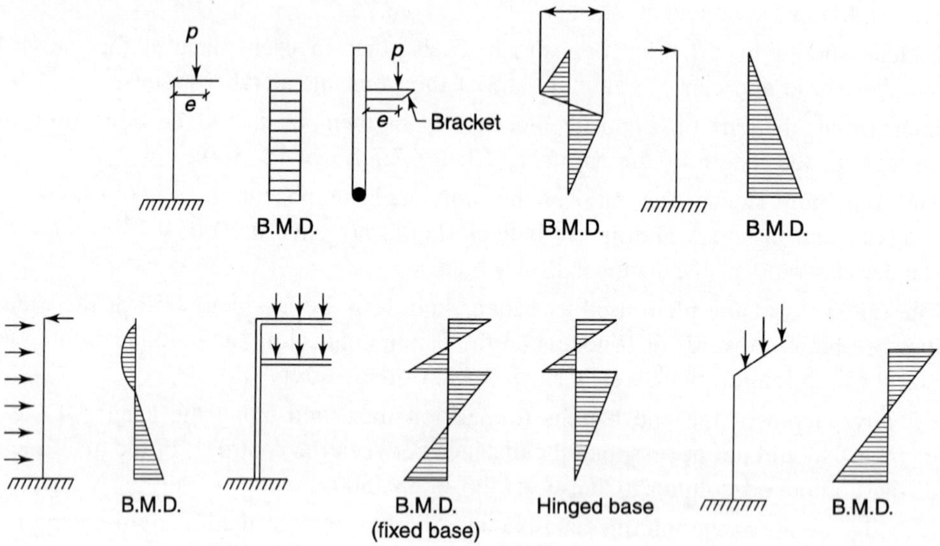

Fig. 4.6

4.4 BOUNDARY CONDITIONS AND RESTRAINTS

Euler's buckling load = critical load = $P_{cr} = \pi^2 EI/l_e^2$

Critical bucking stress $f_{cr} = P_{cr}/A$

$= \pi^2 EI/Al_e^2$

$= \pi^2 E\, r^2/l_e^2$ Since $r = (I/A)^{1/2}$ and $r^2 = I/A$

$= \pi^2 E/(le/r)^2$

$= \pi^2 E/\lambda^2$ (Clause 7.1.2 of IS: 800)

Effective length l_e depends on boundary/end/support conditions (Table 11 of IS:800)

4.5 EFFECTIVE LENGTH OF BRACED COLUMN

(Clause 7.2.2 & Annex D of IS:800)

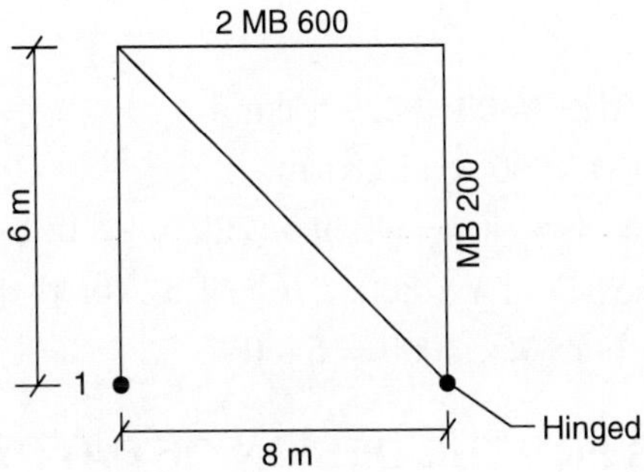

Fig. 4.7

Bracing is provided to resist/reduce sway. The frame is non-sway frame.

Let us use the graph to find the effective length.

For such braced columns, the base of the column is generally pinned.

Let us take joint 2.

Stiffness of beam = $k_b = (I/L)_b = 91813 \times 10000/8000 = 114.8 \times 10^3$

Stiffness of column = $k_c = (I/L)_c = 8603.6 \times 10000/6000 = 14.3 \times 10^3\ \text{mm}^3$

Consider joint 1. It is pinned. Hence, $\beta_1 = 1.0$. (Fig. 27 in Annex-D of IS:800).

At joint 2, the relative property of the column is β_2

$$\beta_2 = k_c/k_b + kc = 14.3/(14.3 + 114.8) = 0.11$$

From Fig. 27 in Annex-D of IS:800, $l_e/L = 0.71$ for $\beta_1 = 1.0$ & $\beta_2 = 0.11$.

$$l_e = 0.71\ L = 0.71 \times 6000 = 4260\ \text{mm}$$

The effective length of the column is $0.71 \times$ unsupported length.

4.6 EFFECTIVE LENGTH OF COLUMN IN RIGID FRAME

(Clause 7.2.2 & Annex-D of IS:800)

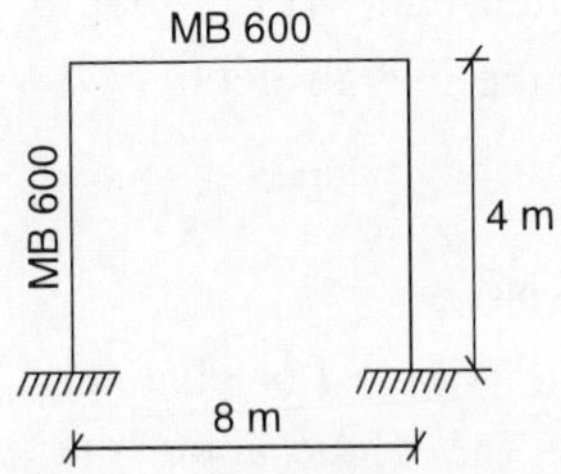

Fig. 4.8

This is a rigid frame in which the lateral forces are taken by column. It is considered sway frame.

Base of the column is fixed. Hence, $\beta_1 = 0$. (Figure 28 in Annex-D of IS:800)

Let us take joint 2.

Stiffness of column = k_c = 91813/400 = 229.5 cm^3

Stiffness of beam = k_b = 91813/800 = 114.8 cm^3

$$\beta_2 = k_c/k_c + k_b = 229.5/(229.5 + 114.8) = 0.67$$

Referring to Fig. 28 in Annex-D of IS: 800, $l_e/L = 1.37$ for $\beta_1 = 0$ and $\beta_2 = 0.67$

$$l_e = 1.37 \times 4000 = 5480$$

4.7 CODE PROVISIONS FOR THE DESIGN OF BATTEN PLATES

Plates, flats, angles, channels, *I* sections can be used as battens

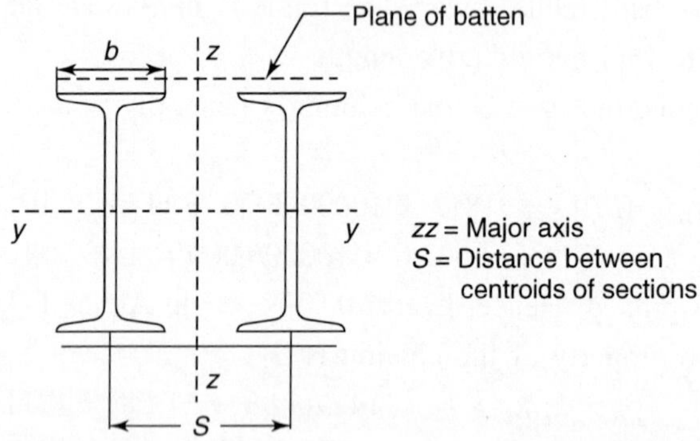

Fig. 4.9

ZZ is the major axis.

S is the distance between the centroids of the individual members.

a_i is the distance between innermost line of bolts or weld. (Fig. 4.16 of this book)

D is the overall depth of the batten plate. (Fig. 4.10 of this book)

d is the effective depth of the batten plate. (Fig. 4.10 of this book)

In the case of bolted connection, the overall depth = $d + 2 \times$ end distance.

IS Recommendations

(1) $r_z \geq r_y$ (Clause 7.7.1.1 of IS: 800)

(2) *C* should be such that λ of the (single *I* or channel) component shall not exceed 50 (or) 0.7 times λ of the column as the whole about the axis parallel to the plane of batten. (Clause 7.7.3 of IS:800)

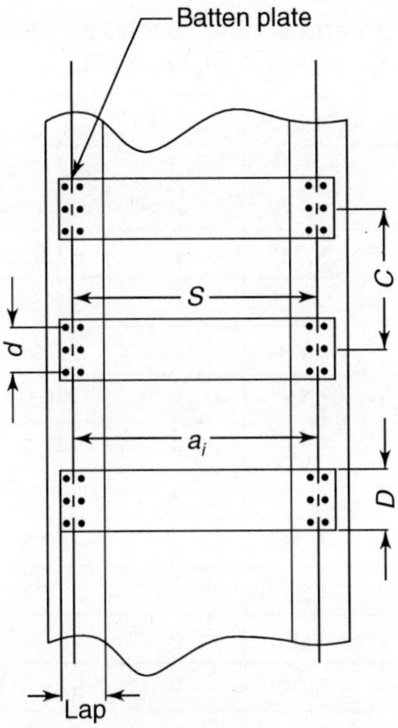

Fig. 4.10

(3) d = effective depth = Distance between end bolts/welds.

(4) $d > S$ for end battens and (3/4) of S for intermediate battens and $d > 2b$ for any battens. (Clause 7.7.2.3 of IS: 800)

(5) $t > a_i/50$ (Clause 7.7.2.3 of IS:800)

(6) lap > four times the thickness of the batten. The length of the weld in aggregate is not less than half the depth of the batten plate. The length of weld and depth of batten plate shall be measured along the longitudinal axis of the main member (Clause 7.7.4.1 of IS: 800)

(7) V = 2.5% of the total axial force as the whole compression member. (Clause 7.7.2.1 of IS: 800)

$$M = M_1 = M_2 = 1/2\ (V/N) \times C = VC/2N$$

The longitudinal shear induced due to these members is

$$V_1 = 1/S\ (M_1 + M_2) = VC/NS$$

Longitudinal shear stress = $(V_1/D \times t)$ which should be less than $(f_y/\gamma_{mo} \times \sqrt{3})$ MPa.

$$M_d = 1.2 \times Z_e \times f_y/\gamma_{mo} \qquad \text{(Clause 8.2.1.2 of IS: 800)}$$

$$Z_e = b \times D^2/6$$

$$M_d = 1.2 \times (b \times D^2/6) \times f_y/\gamma_{mo}$$

Bending stress = M/Z = $[6 \times M/(D^2 t)]$ should be less than or equal to (f_y/γ_{mo}) MPa

Table 4.9 Classification of India rolled *I* and channel sections (Clause 3.7.2 of IS: 800) class of section for Indian rolled *I* and channel sections (Clause 3.7.2 and Table 2 of IS: 800)

Designation of rolled section	*Based on outstand*				*Based on web*			
	b in mm	*t_f in mm*	*b/t_f*	*class of sec*	*d in mm*	*t_w in mm*	*d/t_w*	*class of sec*
ISMB 100	37.5	7.2	5.21	P	67.60	4.0	16.9	P
ISMB 150	40	7.6	5.26	P	116.8	4.8	24.3	P
ISMB 200	50	10.8	4.63	P	156.4	5.7	27.4	P
ISMB 250	62.5	12.5	5.00	P	199.0	6.9	28.8	P
ISMB 300	70	12.4	5.65	P	247.2	7.5	33.0	P
ISMB 350	70	14.2	4.93	P	293.6	8.1	36.3	P
ISMB 400	70	16.0	4.38	P	340.0	8.9	38.2	P
ISMB 450	75	17.4	4.31	P	385.2	9.4	41.0	P
ISMB 500	90	17.2	5.23	P	431.6	10.2	42.3	P
ISMB 600	105	20.8	5.05	P	518.4	12.0	43.2	P
ISMC 75	40	7.3	5.48	P	43.4	4.4	9.9	P
ISMC 100	50	7.5	6.67	P	67.0	4.7	14.3	P
ISMC 125	65	8.1	8.02	P	89.8	5.0	18.0	P
ISMC 150	75	9.0	8.33	P	112.0	5.4	20.7	P
ISMC 200	75	11.4	6.58	P	155.2	6.1	25.4	P
ISMC 250	80	14.1	5.67	P	197.8	7.1	27.9	P
ISMC 300	90	13.6	6.62	P	246.8	7.6	32.5	P
ISMC 350	100	13.5	7.41	P	295.0	8.1	36.4	P
ISMC 400	100	15.3	6.54	P	339.4	8.6	39.5	P

Note: (1) It can be noticed that for all rolled *I* sections $b/t_f < 9.4\varepsilon$ and $d/t_w < 84\varepsilon$. Hence, all rolled *I* sections (f_y = 250 MPa) are plastic.

(2) It can be noticed that for all rolled channel sections $b/t_f < 9.4\varepsilon$ and $d/t_w < 42\varepsilon$. Hence, all rolled channel sections (f_y = 250 MPa) are plastic.

Table 4.10 Classification of single and double angle sections (Clauses 3.7.2 & 3.7.4 and Table 2 of IS: 800)

Size of angle	*b/t = d/t*	*(b + d)/t*	*Classification of sections*	
1	*2*	*3*	*4*	*5*
ISA50 × 50 × 6	8.3	16.7	Semi-compact	Plastic
ISA60 × 60 × 6	10.0	20.0	Semi-compact	Compact
ISA60 × 60 × 8	7.5	15.0	Semi-compact	Plastic
ISA60 × 60 × 10	6	12.0	Semi-compact	Plastic
ISA65 × 65 × 6	10.8	21.6	Semi-compact	Semi-compact
ISA65 × 65 × 8	8.1	16.3	Semi-compact	Plastic

Contd...

Contd...

ISA65 × 65 × 10	6.5	13.0	Semi-compact	Plastic
ISA75 × 75 × 6	12.5	25.0	Semi-compact	Semi-compact
ISA75 × 75 × 8	9.4	18.7	Semi-compact	Plastic
ISA75 × 75 × 10	7.5	15.0	Semi-compact	Plastic
ISA80 × 80 × 6	13.3	26.7	Slender	Semi-plastic
ISA80 × 80 × 8	10.0	20.0	Semi-compact	Compact
ISA80 × 80 × 10	8.0	16.0	Semi-compact	Plastic
ISA80 × 80 × 12	6.7	13.3	Semi-compact	Plastic
ISA90 × 90 × 6	15.0	30.0	Semi-compact	Semi-compact
ISA90 × 90 × 8	11.3	22.5	Semi-compact	Semi-compact
ISA90 × 90 × 10	9.0	18.0	Semi-compact	Plastic
ISA90 × 90 × 12	7.5	15.0	Semi-compact	Plastic
ISA100 × 100 × 6	16.7	33.3	Slender	Slender
ISA100 × 100 × 8	12.5	25.0	Semi-compact	Semi-compact
ISA100 × 100 × 10	10.0	20.0	Semi-compact	Semi-compact
ISA100 × 100 × 12	8.3	16.7	Semi-compact	Plastic
ISA110 × 110 × 8	13.8	27.5	Slender	Semi-compact
ISA110 × 110 × 10	11.0	22.0	Semi-compact	Semi-compact
ISA110 × 110 × 12	9.2	18.3	Semi-compact	Plastic

Note: (1) Column 4 of the above table is applicable for single, or double angles with the components separated, axial compression.

(2) Column 5 of the above table is applicable for (a) single angle, compression due to bending (b) outstanding leg of an angle in contact back to back in a double angle member (c) outstanding leg of an angle with its back is in continuous contact with another component.

4.8 DESIGN EXAMPLES

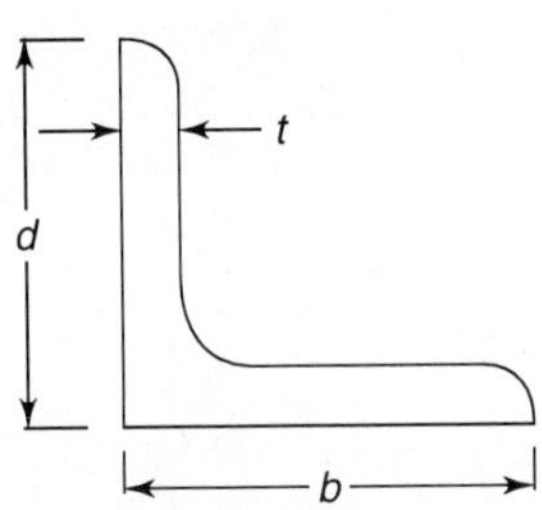

Fig. 4.11

Example 4.1 *An axial load carrying capacity of column*

Single ISMB 500 is used as a compression member and is restrained at both the ends by beams. Unsupported length is 9 m and it is laterally restrained at midpoint about minor axis. Find the load carrying capacity in compression. f_y = 250 MPa.

Solution: Design compressive strength = $P_d = A_e f_{cd}$

(Clause 7.1.2 of IS:800)

A_e = effective sectional area (Clause 7.3.2 of IS:800)

To find limiting width thickness ratio and class of section (Table 2 of IS: 800)

b = 180/2 = 90 mm; t_f = 17.2 mm.

$$\text{Yield stress ratio} = \varepsilon = \sqrt{\frac{250}{f_y}} = \sqrt{\frac{250}{250}} = 1.0$$

$$\frac{d}{t_w} = (500 - 2 \times 17.2 - 2 \times 17)/10.2 = 42.31 < 84\ \varepsilon$$

$$\frac{b}{f_f} = \frac{90}{17.2} = 5.23 < 9.4\ \varepsilon$$

Section is plastic (Table 2 of IS:800)

Gross cross-sectional area of ISMB 500 = A_g = 11074 mm^2

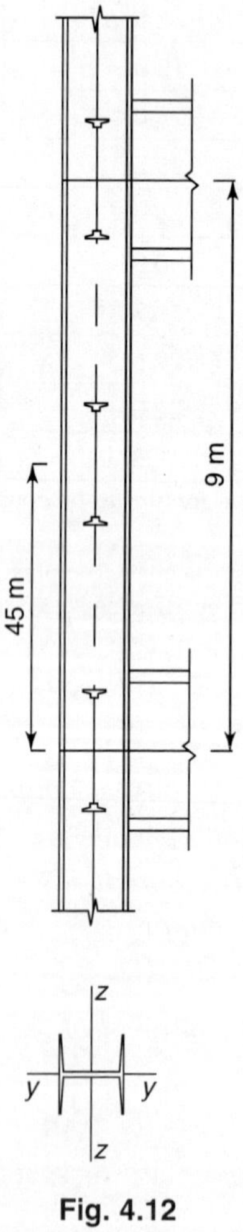

Fig. 4.12

$$f_{cd} = \frac{f_y/\gamma_{mo}}{\varphi + [\varphi^2 - \lambda^2]^{0.5}}$$

(Clause 7.1.2.1 of IS:800)

$$f_y = 250 \text{ N/mm}^2$$

Partial safety factor for material = γ_{mo} = 1.1 (Table 5 of IS:800)

$$\varphi = 0.5\ [1 + \alpha(\lambda - 0.2) + \lambda^2]$$

$$\lambda = \sqrt{\frac{f_y}{f_{cc}}}$$

$$f_{cc} = \pi^2\ E/(KL/r)^2$$

Since both the ends of column are fully restrained by beams,

$$K = 0.65. \text{ (Table 11 of IS: 800).}$$

$$KL = \ell_{ez} = 0.65 \times 9 = 5.85 \text{ m}$$

Since there is a restrained on *yy* axis at middle and the ends are fully restrained, $K = 0.65$. (Table 11 of IS: 800)

$$KL = \ell_{ey} = 0.65 \times 9/2 = 2.925 \text{ m}$$

$$r_z = 202.1 \text{ mm};\ r_y = 35.2 \text{ mm}$$

$$E = 2 \times 10^5 \text{ N/mm}^2$$

$$f_{ccz} = \pi^2\ E/(\ell_z/r_z)^2 = \pi^2 \times 2 \times 10^5/(5850/202.1)^2 = 2358 \text{ N/mm}^2$$

$$f_{ccy} = \pi^2\ E/(\ell_y/r_y)^2 = \pi^2 \times 2 \times 10^5/(2925/35.2)^2 = 286.1 \text{ N/mm}^2$$

$$\lambda_z = \sqrt{\frac{f_y}{f_{ccz}}} = \sqrt{\frac{250}{2358}} = 0.3256$$

$$\lambda_y = \sqrt{\frac{f_y}{f_{ccy}}} = \sqrt{\frac{250}{286.1}} = 0.9348$$

To find buckling class (Table 10 of IS:800)

$$\frac{h}{b_f} = \frac{500}{180} > 1.2;\ t_f < 40 \text{ mm}$$

Buckling class for major axis *a*; α_z = 0.21 (Table 7 of IS: 800)

Buckling class for minor axis *b*; α_y = 0.34 (Table 7 of IS: 800)

$$\varphi_z = 0.5\ [1 + 0.21(0.3256 - 0.2) + 0.3256^2] = 0.566$$

$$\varphi_y = 0.5\ [1 + 0.34(0.9348 - 0.2) + 0.9348^2] = 1.062$$

Stress reduction factor *x* (Clause 7.1.2.1 of IS: 800)

$$X_z = \frac{1}{\left[\varphi_z + \left(\varphi_z^2 - \lambda_z^2\right)^{0.5}\right]}$$

$$= \frac{1}{\{0.566 + (0.566^2 - 0.3256^2)^{0.5}} = 0.972$$

$$X_y = \frac{1}{\{1.062 + (1.062^2 - 0.9348^2)^{0.5}} = 0.6386$$

$$(KL/r)_z = \frac{5850}{202.1} = 28.95$$

$$(KL/r)_z = \frac{2925}{35.2} = 83.1$$

X_z from Table 8a = 0.994 – (0.025/10) × 8.95 = 0.972

The value as calculated is also 0.972.

X_y from Table 8b = 0.661 – (0.072/10) × 3.1 = 0.6386

The value as calculated is 0.6386.

Lesser of the two (value of X_z and X_y) is considered for calculating f_{cd}.

$f_{cd} = X_y f_y/\gamma_{mo} = 0.6386 \times 250/1.1 = 145.14$ MPa and $X_z f_y/\gamma_{mo} = 0.972 \times 250/1.1 = 221$ MPa.

$$P_d = 11074 \times 145.14/1000 = 1607.3 \text{ kN}.$$

Axial load carrying capacity of this column is 1607.3 kN.

Note:

(1) Value of X_z is obtained from Table 8a because the buckling class for major axis *z* is *a*. The value is interpolated.

(2) Value of X_y is obtained from Table 8b because the buckling class for major axis *y* is *b*. The value is interpolated.

(3) Even after providing restraint at mid height, the stress reduction factor for weaker axis is 0.6386. If possible more restraints can be provided.

Example 4.2 *Design of plate formed I section as column.*

Design a plate formed I section to carry a compressive load of 800 kN due to dead and imposed loads. Ends are simply supported and the available has f_y = 235 N/mm^2 for which f_u = 355 N/mm^2. The unsupported length is 9 M. At mid height, the minor axis is restrained against buckling.

Solution: Trial section

Assuming stress reduction factor $x = 0.5$

$$f_{cd} = 0.5 \times f_y/\gamma_{mo} = 0.5 \times 235/1.1 = 106.8 \text{ MPa}$$

Partial safety factor for load = 1.5 (Table 4 of IS:800)

Factored load = 1.5 × 800 = 1200 kN

$$\text{Area required} = \frac{P \times 1000}{f_{cd}} = \frac{1200 \times 1000}{106.08} = 11234 \text{ mm}^2$$

Choose overall depth 500 mm, width of flange 200 mm, thickness of flange 16 mm and that of web 10 mm.

$$\varepsilon = \sqrt{\frac{250}{235}} = 1.031$$

$$b = \frac{200 - 10}{2} = 95,\ t_f = 16,\ \frac{b}{t_f} = \frac{95}{16} = 5.94 < 8.4\ \varepsilon \text{ (Table 2 of IS: 800)}$$

$$8.4\ \varepsilon = 8.4 \times 1.031 = 8.66$$

$$d = 500 - 2 \times 16 = 468 \text{ mm};\ t_w = 10;\ d/t_w = 468/10 = 46.8 < 84\ \varepsilon$$

Section is plastic (Table 2 of IS: 800)

$$A_e = 2 \times 16 \times 200 + 468 \times 10 = 6400 + 4680 = 11080 \text{ mm}^2$$

$$r_z = I_z/A$$

$$I_z = \frac{200 \times 500^3}{12} - \frac{190 \times 468^3}{12} = 460365493 \text{ mm}^4$$

$$I_y = \frac{2 \times 16 \times 200^3}{12} + \frac{468 \times 10^3}{12} = 21372333 \text{ mm}^4$$

$$r_z = \sqrt{\frac{460365493}{11080}} = 203.84 \text{ mm}$$

$$r_y = \sqrt{\frac{21372333}{11080}} = 43.92 \text{ mm}$$

$$E = 2 \times 10^5 \text{ N/mm}^2$$

Unsupported length on *ZZ* axis = 9 m.

Both ends are simply supported. (Translation restrained and rotation free). Hence, $K = 1.0$.

Unsupported length on *YY* axis = 9/2 = 4.5 m since this is restrained at mid height. One end is simply supported and the other end is fully restrained. Hence, $K = 0.8$. (Table 11 of IS: 800).

$\ell_z = k \times L_z = 1.0 \times 9 = 9.0$ m (Since both the ends are simply supported).

$\ell_y = k \times L_y = 0.8 \times 4.5 = 3.6$ m (Since the minor axis is restrained at middle).

$$\frac{\ell_z}{r_z} = \frac{9000}{203.84} = 44.15$$

$$\frac{\ell y}{r_y} = \frac{3600}{43.9} = 81.96$$

Euler buckling stress f_{ccz} and f_{ccy} as per Clause 7.1.2.1 of IS: 800.

$$f_{ccz} = \pi^2\ E/(\ell_z/r_z)^2 = \frac{\pi^2 \times 2 \times 10^5}{44.15^2} = 1013.5\ \text{N/mm}^2$$

$$f_{ccy} = \frac{\pi^2 \times 2 \times 10^5}{81.96^2} = 294.1\ \text{N/mm}^2$$

Slenderness ratio λ_z and λ_y as per Clause 7.1.2.1 of IS: 800.

$$\lambda_z = \sqrt{\frac{f_y}{f_{ccx}}} = \sqrt{\frac{235}{1013.5}} = 0.48$$

$$\lambda_y = \sqrt{\frac{f_y}{f_{ccy}}} = \sqrt{\frac{235}{294.1}} = 0.89$$

To find buckling class of welded section (Table 10 of IS: 800) and factor a (Table 7 of IS:800).

$t_f < 40$ mm.

Buckling class for major axis *b*; $\alpha_z = 0.34$

Buckling class for minor axis *c*; $\alpha_y = 0.49$

$$\varphi_z = 0.5\ [1 + 0.34\ (0.48 - 0.2) + 0.48^2] = 0.6628$$

$$\varphi_y = 0.5\ [1 + 0.49\ (0.89 - 0.2) + 0.89^2] = 1.065$$

Stress reduction factor χ

$$\chi_z = 1/[\varphi_z + (\varphi_z^2 - \lambda_z^2)^{0.5}] = 1/[0.6628 + (0.6628^2 - 0.48^2)^{0.5}] = 0.893$$

$$f_{cdz} = \chi_z \times f_y/\gamma_{mo} = 0.893 \times 235/1.1 = 190.8\ \text{MPa}.$$

$$\chi_y = 1/[\varphi_y + (\varphi_y^2 - \lambda_y^2)^{0.5}] = 1/[1.065 + (1.065^2 - 0.48^2)^{0.5}] = 0.61$$

$$f_{cdy} = \chi_y \times f_y/\lambda_{mo} = 0.61 \times 235/1.1 = 130\ \text{MPa}.$$

From Table 9.c; for $KL/r_y = 81.96$,

$f_y = 235$ N/mm^2

KL/r	230		240
80	130		133
81, 96	127.2	128.7	130.2
90	116		119

$f_{cdy} = 128.7$ N/mm^2 as against the calculated value of 130 MPa. The variation is only 1%.

$$P_d = A_e \times f_{cdy} = 11080 \times 128.7/1000 = 1426 \text{ kN} > 1200 \text{ kN}$$

Note:

(1) In this case the capacity is higher than the factored load. Hence, we can try by reducing any one (or) two (or) all the following.

(a) The depth

(b) The thickness of web

(c) The thickness of flange

(d) The width of flange

(2) If the capacity is less than the factored load, then increase any one (or) two (or) all the above.

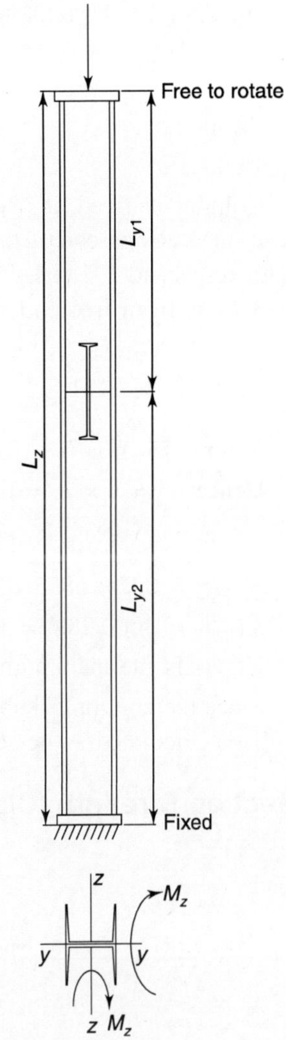

Fig. 4.13

Example 4.3 *Design of column with biaxial B.M.*

Design a column using ISMB rolled section (material specification IS: 2062; $f_y = 250$ MPa) for the following design data.

Axial force = 75 kN

Moment about major axis = M_z = 45 kNm

Moment about minor axis = M_y = 3 kNm

Column is fixed at one end and free to rotate at the other end.

Unsupported length on major axis = 6.875 m

Restraint on minor axis is at 3.75 m from free end.

Force and moments are factored.

Solution:

Assuming stress reduction factor as 0.5.

$f_{cd} = 0.5\, f_y/\gamma_{mo} = 0.5 \times 250/1.1 = 113.6$ MPa

Area required = Load/Stress = $75 \times 1000/113.6 = 660$ mm^2

This is only a rough estimate since column carries moment also.

The required area will be more.

Try ISMB 300

$$L_Z = 6875 \text{ mm. } L_Y = 3750 \text{ mm.}$$

$$A = 5626 \text{ mm}^2$$

$$Z_z = 573.6 \times 10^3 \text{ mm}^3;\ r_x = 123.7 \text{ mm}$$

$$Z_y = 64.8 \times 10^3 \text{ mm}^3;\ r_y = 28.4 \text{ mm}$$

Note: The physical property of rolled sections may slightly vary between manufacturers and tables giving the properties.

Buckling class (Table 10 of IS: 800).

$$h/b_f = 300/140 = 2.14 > 1.2 \ \&\ t_f < 40 \text{ mm}$$

ZZ axis *a*

YY axis *b*

Column is fixed at one end and free to rotate at the other end. (Translation and rotation restrained at one end at the other end translation restrained and free to rotate). Hence, $K = 0.8$ with respect to *ZZ* axis (Table 11 of IS:800). With respect to *YY* axis, the column is restrained at 3.75 m from free end.

$$L_{y1} = 3.75 \text{ m.}$$

$$L_{y2} = 3.125 \text{ M.}$$

K for 3.75 m is 0.8 (fixed-free) & K for (6.875 – 3.75 =) 3.125 m is 0.65 (fixed-fixed).

Hence, $\ell_y = k \times L_y = 0.8 \times 3.75 = 3.0$ m or $= 0.65 \times 3.125 = 2.03125$ m.

$$\ell_y = 3.0 \text{ m will govern.}$$

$KL_z/r_z = 0.8 \times 6875/123.7 = 44.5$; $f_{cd} = 209.4$ MPa (Table 9a of IS:800)

$KL_y/r_y = 3000/28.4 = 105.63$; $f_{cd} = 111$ MPa. (Table 9b of IS:800)

KL_y/r_y is the maximum for which f_{cd} is minimum. Hence, this governs the design.

Since the column is subjected to axial load and moment, section strength and overall strength will be calculated as per Clause 9.3.1 & 9.3.2 of IS: 800.

Section Strength (Class 9.3.1 of IS:800)

$$N_d = N_d = A_g \times f_y/\gamma_{mo} = 5626 \times 250/1.1 = 1278.636 \text{ kN}$$

$$b/t_f = 70/12.4 = 5.645 < 9.4\ \varepsilon$$

$$d = (300 - 2 \times 12.4 - 2 \times 14) = 247.2 \text{ mm.}$$

$$d/t_w = 247.2/7.5 = 32.96 < 84\ \varepsilon$$

Hence, section is plastic. (Table 2 of IS:800)

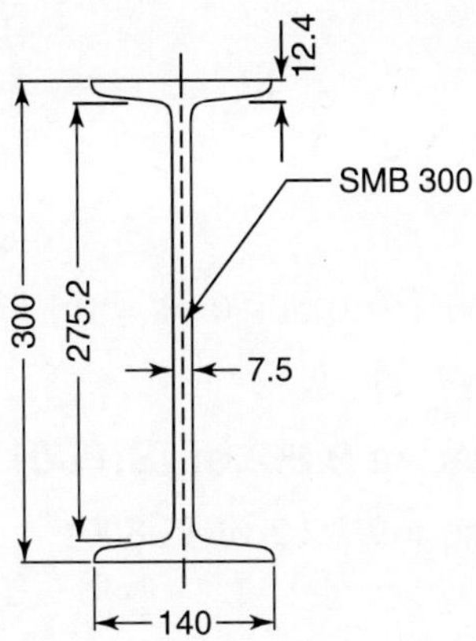

Fig. 4.14

$$n = N/N_d = 75/1278.6 = 0.06 < 0.2$$

Section is plastic and hence $\beta = 1.0$

$M_{dz} = \beta_p Z_p f_y/\gamma_{mo}$ (Clause 8.2.1.2 & Table 46 Annex-H)

$$= 1.0 \times 651.74 \times 250/1.1 \times 10^3 = 148.123 \text{ kNm}$$

$$M_{dy} = \beta_p Z_{py} f_y/\gamma_{mo}$$

$\beta_p = 1.0$ Since section is plastic.

$$Z_{py} = (4 \times 70 \times 35 \times 12.4) + \left(2 \times 275.2 \times \frac{7.5}{2} \times \frac{7.5}{4}\right) = 125390 \text{ mm}^3$$

$$M_{dy} = 1.0 \times 125390 \times 250/(1.1 \times 10^6) = 28.5 \text{ kNm}$$

$$\eta = < 0.2$$

Hence,

$$M_{ndy} = M_{dy} = 28.5 \text{ kNm}$$

For $\eta \le 0.2$, $M_{ndz} = 1.11 \times M_{dz}(1 - \eta) \le M_{dz}$

$$M_{ndz} = 1.11. \times 148.123 \times (1 - 0.06) = 154.55 \text{ kNm}$$

Should be less than *Mdz*

Hence, $M_{ndz} = 148.123$ kNm

$$\left(\frac{M_y}{M_{ndy}}\right)^{\alpha_1} + \left(\frac{M_z}{M_{ndz}}\right)^{\alpha_2} \le 1.0 \text{ (Clause 9.3.1.1 of IS: 800)}$$

$\alpha_1 = 5n \ge 1 = 5 \times 0.06 = 0.3 < 1.0$ and hence $\alpha_1 = 1.0$ (Table 17 of IS: 800)

$$\alpha_2 = 2$$

$$\left(\frac{3}{28.5}\right)^1 + \left(\frac{4.5}{148.123}\right)^2 = 0.106 < 1.0$$

$$\frac{N}{N_d} + \frac{M_y}{M_{dy}} + \frac{M_z}{M_{dz}} = 0.06 + 0.105 + 0.03 = 0.195 < 1.0$$

Overall Member Strength (Clause 9.3.2 of IS:800)

Bending and axial compression (Clause 9.3.2.2 of IS:800)

$$\frac{P}{P_{dy}} + K_y \frac{C_{my} M_y}{M_{dy}} + K_{LT} \frac{M_z}{M_{dz}} \le 1.0$$

$$P_d = A_e \times f_{cd}$$

$$P_{dy} = 111 \times 5626/1000 = 624.486 \text{ kN.}$$

$$P_{dz} = 209.4 \times 5626/1000 = 1178.1 \text{ kN.}$$

$$K_y = 1 + (\lambda_y - 0.2)\ \eta_y < 1 + 0.8\ \eta_y$$

$$\eta_y = \frac{75}{624.486} = 0.12$$

$$\eta_z = \frac{75}{1178.1} = 0.0637$$

$$\lambda_y = \sqrt{[f_y (KL/r)^2/\pi^2 E]}$$

(Clause 7.1.2.1 of IS: 800)

$$= \sqrt{[(250 \times 105.6^2)/(\pi^2 \times 2 \times 10^5)]} = 1.19$$

$$\lambda_z = \sqrt{(250 \times 44.5^2/\pi^2 \times 2 \times 10^5)} = 0.5$$

C_{my} (Table 18 of IS:800)

With respect to *ZZ* axis, column is fixed at one end and free at the other end. At free end moment will not be there. Hence, $\psi = 0$

$C_{mz} = 0.6 + 0.4 \times 0$ and this should be equal to or greater than 0.4.

With respect to *YY* axis, restraint is there at 3.75 m from free end. Already we have seen that this will govern. In this case also $\psi = 0$

Hence, $C_{my} = 0.6 = C_{mz}$

$$K_{LT} = 1 - \frac{0.1\ \lambda_{LT}\ \eta_y}{C_{mLT} - 0.25} \geq 1 - \frac{0.1\ \eta_y}{(C_{mLT} - 0.25)}$$

(Clause 9.3.2.2 of IS: 800)

$$\lambda_{LT} = \sqrt{\beta_p\ Z_p f_y / M_{cr}} \leq \sqrt{1.2 Z_e\ f_y / M_{cr}}$$ (Clause 8.2.2 of IS:800)

$$M_{cr} = \beta_p Z_p f_{crb}$$

(Clause 8.2.2.1 of IS:800)

$$f_{crb} = \frac{1.1 \times \pi^2 E}{(L_{LT}/r_y)^2} \times \left(1 + \frac{1}{20}\left(\frac{L_{LT}/r_y}{h_f/t_f}\right)^2\right)^{0.5}$$

Ends are fully restrained against torsion and warping not restrained. (Table 15 of IS: 800)

Hence, $L_{LT\,z} = 1.0 \times 6.875 = 6875$ mm (Table 15 of IS:800)

$$L_{LTy} = 3750 \text{ mm}$$

$$f_{crb} = \frac{1.1 \times \pi^2 E}{(3750/28.4)^2} \times \left(1 + (1/20)\left(\frac{3750/28.4}{287.6/12.4}\right)^2\right)^{0.5} = 201.8 \text{ MPa}$$

h_f = Distance between centre to centre of flanges = 300 – 12.4 = 287.6 mm. t_f = 12.4 mm.

From Table 14, for K_L/r_y = (3750/28.4) = 132 and h_f/t_f = (287.6/12.4) = 23.2, value of f_{crb} can be obtained by interpolation as below.

K_L/r_y	f_{crb}		
	h_f/t_f	h_f/t_f	h_f/t_f
	20	23.2	25
130	226.7		197.1
132	222.5	203.75	193.2
140	205.8		177.5

Value of f_{crb} = 203.75 and the calculated value is 201.8 MPa. We can adopt the value obtained from the table.

For ISMB 300, $Z_p = 651.71 \times 10^3$ mm^3 (Table 46 of IS: 800)

$$M_{cr} = \beta_p Z_p f_{crb} = 1.0 \times 651.71 \times 10^3 \times 203.75/10^6 = 132.8 \text{ kNm}$$

$$\lambda_{LT} = \frac{148.123}{132.8} = 1.056$$

$$K_{LT} = 1 - \frac{0.1 \times 1.056 \times 0.12}{0.6 - 0.25} = 1 - 0.0362 = 0.9638$$

This shall be greater than or equal to

$$1 - \frac{0.1\,\eta_y}{(C_{MLT} - 0.25)} = 1 - \frac{0.1 \times 0.12}{0.35} = 0.9657$$

Hence, $K_{LT} = 0.9657$

$$\frac{P}{P_{dy}} = \frac{75}{624.486} = 0.12 = \eta_y$$

To find $K_y \times \frac{C_{my} M_y}{M_{dy}}$

$$K_y = 1 + (\lambda_y - 0.2)\,\eta_y = 1 + (1.19 - 0.2) \times 0.12 = 1.1188$$

Should be less than $1 + 0.8\,\eta_y = 1 + 0.8 \times 0.12 = 1.096$

Hence, $K_y = 1.096$

$$K_y \times \frac{C_{my}\,M_y}{M_{dy}} = \frac{1.096 \times 0.6 \times 3 \times 10^6}{28.5 \times 10^6} = 0.069$$

$$K_{LT} \times \frac{M_z}{M_{dz}} = \frac{0.9657 \times 45 \times 10^6}{148.123 \times 10^6} = 0.293$$

$$\frac{P}{P_{dy}} + K_y \times \frac{C_{my}\,M_y}{M_{dy}} + K_{LT} \times \frac{M_z}{M_{dz}} = 0.12 + 0.069 + 0.293 = 0.482 << 1.0$$

The factor is less than 1.0, we can try with lower sections such as ISMB 250.

Example 4.4 *Built-up column using I sections*

Design a built-up column using I sections carrying an axial load of 2000 kN. The column having a length of 7 m is effectively held in position at both ends but restrained against rotation at one end only. Design the batten plates also. f_y = 250 MPa. The load is due to dead and imposed loads only.

Solution: Effectively held in position at both ends but restrained against rotation at one end only.

Effective length = l_e = 0.8 L = 0.8 × 7 = 5.6 m = 5600 mm (Table 11 of IS: 800)

Let the stress reduction factor be 0.5. f_{cd} = 0.5 × 250 = 125 MPa.

Factored load = 1.5 × 2000 = 3000 kN (Table 4 of IS: 800).

Area required = 3000 × 1000/125 = 24000 mm^2. Area of each I = 12000 mm^2

Try with ISMB 400.

Properties of Each ISMB 400

Area 7846 mm^2; I_z = 20458 cm^4; I_y = 622 cm^4 r_z = 16.15 cm; r_y = 2.82 cm

$$Z_z = 1023 \text{ cm}^3;\ Z_y = 88.9 \text{ cm}^3$$

The radius of gyration about the axis perpendicular to the plane of the batten not less than the radius of gyration about the axis parallel to the plane of the batten. (Clause 7.7.1.1 of IS:800).

Accordingly keeping $r_{zz} = r_{yy}$

$$\sqrt{I_{zz}/A} = \sqrt{I_{yy}/A}$$

$$\therefore \quad I_{zz} = I_{yy}$$

Let S = the distance between centroids of the two main components

$$2 \times 20458 = 2 \times \{622 + 78.46 \times (S/2)^2\}$$

$$\therefore \quad S = 31.8 \text{ cm. Provide } S = 330 \text{ mm}$$

By keeping S greater than what is required; we can meet the condition as in Clause 7.7.1.1 of IS:800

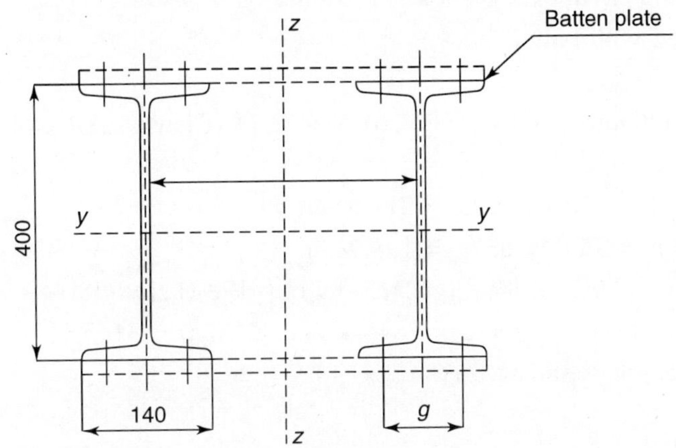

Fig. 4.15

Combined Section Properties

$$A = 2 \times 78.46 = 156.92 \text{ cm}^2$$

$$I_{yy} = 2 \times I_z = 2 \times 20458 = 40916 \text{ cm}^4$$

$$r_{yy} = 16.15 \text{ cm}$$

$$I_{zz} = 2(622 + 78.46 \times (33/2)^2) = 43965.5 \text{ cm}^4$$

$$r_{zz} = \sqrt{I_y/A} = \sqrt{(3965.5/156.92)} = 16.74 \text{ cm} > r_{yy}.$$

Hence, Clause 7.7.1.1 of IS: 800 is satisfied.

$$P = 3000 \text{ kN}$$

$$f_{ccy} = \pi^2 E/(KL_y/r_y)^2 = \pi^2 \times 2 \times 10^5/(5600/161.5)^2 = 1975510/1119 = 1643 \text{ MPa}$$

$$f_{ccz} = \pi^2 E/(KL_y/r_y)^2 = \pi^2 \times 2 \times 10^5/(5600/167.4)^2 = 1765 \text{ MPa}$$

$$\lambda_z = \sqrt{\frac{f_y}{f_{ccz}}} = \sqrt{\frac{250}{1765}} = 0.376$$

$$\lambda_y = \sqrt{\frac{f_y}{f_{ccz}}} = \sqrt{\frac{250}{1643}} = 0.39$$

$$\varphi = 0.5(1 + \alpha(\lambda - 0.2) + \lambda^2)^2 \qquad \text{(Clause 7.1.2.1 of IS: 800)}$$

To find buckling class (Table 10 of IS: 800) and imperfection factor α (Table 7 of IS: 800) This is a built-up member. Buckling class is *c*. Hence, $\alpha = 0.49$

Column is provided with batten

γ_{yy} is minimum.

KL/r_{yy} of built-up column = 1.1 × 5600/161.5 = 38.14 (Clause 7.7.1.4 of IS: 800).

From Table 9c of IS: 800,

For KL/r_{yy} 30 and f_y = 250 MPa; f_{cd} = 211 MPa.

For KL/r_{yy} 40 and f_y = 250 MPa; f_{cd} = 198 MPa.

By interpolating, for KL/r_{yy} 38.14; f_{cd} = 211 – [(211 – 198)/10] × (38.14 – 30) = 200.42 MPa.

Load carrying capacity = $A_e f_{cd}$ = 2 × 7846 × 200.42/1000 = 3145 kN > 3000 kN

Hence, section strength is satisfactory.

Design of Batten

Spacing = *C* (Clause 7.7.3 of IS:800)

C/r_{min} should not be greater than 50 (or) 0.7 times slenderness ratio of the member as a whole about its axis parallel to the batten (in this case *yy*).

0.7 times KL/r_y = 0.7 × 38.14 = 26.7

$\frac{C}{r_{min}}$ should not be greater than 26.7 which is less than 50.

$$C < 26.7 \times 28.2 = 752.94 \text{ mm}$$

Keep C = 750 mm

Thickness of Batten (Clause 7.7.2.3 of IS: 800)

The thickness of batten shall not be less than (1/50)th of the distance between the innermost connecting lines of bolts (or) welds.

Centroid distance = S = 330 mm.

Gauge distance of bolts = 80 mm.

$S - g = 330 - 80 = 250$ mm (Fig. 4.17 of this work).

$t > 250/50 = 5$ mm. Provide 8 mm.

Depth of Batten

$$S = 330 \text{ mm}$$

$$b = 140 \text{ mm}$$

For intermediate batten $d > (3/4)S = (3/4) \times 330 = 247.5$ mm and shall not be less than 2 × *. flange width = 2 × 140 = 280 mm. (Clause 7.7.2.3 of IS:800)

For end batten $d > S = 330$ mm (Clause 7.7.2.3 of IS: 800)

Let the effective depth of intermediate batten be 280 mm.

Using M20 Class 4.6 bolts, diameter of hole = 22 mm. (Table 19 of IS: 800)

Minimum edge distance for M20 bolts = 1.5 × 22 = 33 mm (Clause 10.2.4.2 of IS: 800).

Depth of intermediate batten = 280 + 2 × 33 = 346 mm. Provide 350 mm.

Depth of end batten = 330 + 2 × 33 = 396 mm. Provide 40 mm.

End battens act as stiffeners to cap/base plate of the column.

Provide 8 thick 350 mm depth battens at 750 mm C/C.

Checking the Adequacy

$$V = 0.025\ P = 0.025 \times 3000 = 75 \text{ kN (Clause 7.7.2.1 of IS: 800)}$$

$$S = 330 \text{ mm}$$

Longitudinal shear = $V_1 = VC/NS = 75 \times 0.75/2 \times 0.33 = 85.2$ kN

Design shear strength = $V_d = \dfrac{A_v f_{yw}}{\gamma_{mo}\sqrt{3}}$ (Clause 8.4 of IS: 800)

$$= \frac{350 \times 8 \times 250}{1.1 \times \sqrt{3} \times 1000} = 367 \text{ kN} > 85.2 \text{ kN}$$

$$0.6\ V_d = 0.6 \times 367 = 220.2 \text{ kN} > 85.2 \text{ kN}$$

Hence, low shear

$$M = VC/2N = \frac{75 \times 0.75}{2 \times 2} = 14.06 \text{ kNm}$$

Capacity in bending = $M_d = 1.2\ Z_e\ f_y/\gamma_{mo}$ (Clause 8.2.1.2 of IS:800)

$$= 1.2 \times \frac{8 \times 350^2}{6} \times \frac{250}{1.1 \times 10^6} = 44.54 \text{ kNm} > 14.06 \text{ kNm.}$$

Hence, satisfactory.

Combined Capacity (Clause 9.2.2 of IS: 800)

When the factored shear force is higher than 0.6 V_d, the bending strength shall be reduced as in Clause 9.2.2 of IS:800. In this example, the applied factored shear is less than 0.6 V_d (low shear).

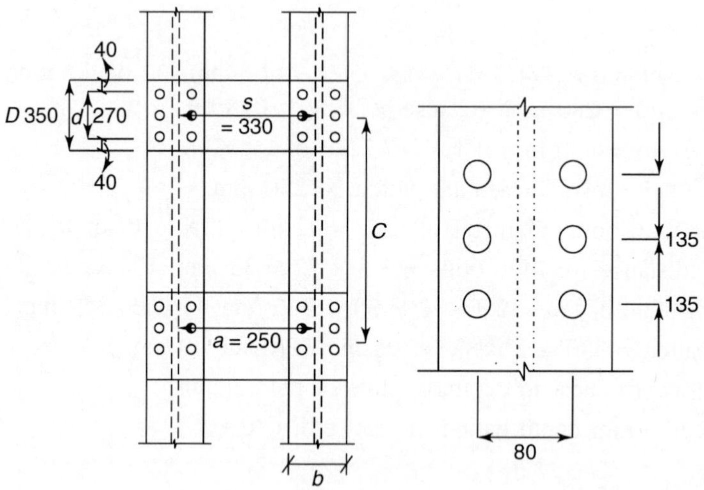

Fig. 4.16

Example 4.5 *Built-up column using channel facing each other*

Design a built-up column to carry an axial compressive load of 1500 kN. The length of column is 5 m. It is effectively held in position at both ends but restrained against rotation at one end only. f_y = 250 MPa. The load is due to dead and imposed loads only.

Solution: Factored load = 1.5 × 1500 = 2250 kN (Table 4 of IS:800)

Assuming stress reduction factor as 0.5, f_{cd} = 0.5 × 250 = 125 MPa

$$\text{Area required} = \frac{2250 \times 1000}{125} = 18000 \text{ mm}^2$$

Keeping 2 ISMC, required area of each channel section = 9000 mm^2

Try 2 ISMC 350

Let us denote the axis perpendicular to the plane of batten/lacing as *ZZ*.

$$r_{zz} > r_{yy} \qquad \text{(Clause 7.7.1.1 of IS:800)}$$

$$\sqrt{I_{zz}/A} = \sqrt{I_{yy}/A}$$

Keeping $I_{zz} = I_{yy}$; we can calculate 'S'

$$I_{yy} = 2 \times I_z = 2 \times 10008 = 20016 \text{ cm}^4$$

$$I_{zz} = 2 \times (I_y + a(S/2)^2) = 2 \times [430.6 + 53.66(S/2)^2] = 2 \times 10008$$

$$S = 26.7 \text{ cm}$$

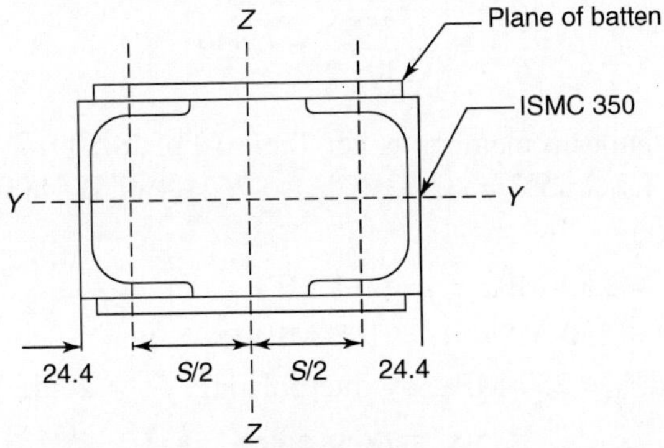

Fig. 4.17

Keeping 350 mm outer to outer

$$C_{yy} = 24.4 \text{ mm.}$$

$$S = 350 - 2 \times 24.4 = 301.2 \text{ mm}$$

Distance of centroid of each channel from ZZ axis = $S/2$ = 301.2/2 = 150.6 mm.

$$I_{zz} = 2 \times (430.6 + 53.66 \times 15.06^2) = 25201.76 \text{ cm}^4$$

$$I_{yy} = 20016 \text{ cm}^4$$

$$Z_{zz} = 25201.76/(35/2) = 1440 \text{ cm}^3$$

$$A = 2 \times a = 2 \times 53.66 = 107.32 \text{ cm}^2$$

$$r_{zz} = \sqrt{I_{zz}/A} = 15.32 \text{ cm}$$

$$r_{yy} = \sqrt{I_{yy}/A} = 13.66 \text{ cm}$$

Note that $r_{zz} > r_{yy}$ Requirement as per Clause 7.7.1.1 of IS: 800 is met.
Effective length = 0.8 × L = 0.8 × 5 = 4 m = 4000 mm (Refer Table 11 of IS:800)

$$\lambda_y = \frac{l_{ey}}{r_{yy}} = \frac{4000}{136.6} = 29.3$$

$$P = 2250 \text{ kN}$$

$$f_{ccy} = \pi^2 E/(KL_y/\lambda_y)^2 = \pi^2 E/(4000/136.6)^2 = 2303.9 \text{ MPa}$$

$$f_{ccz} = \pi^2 E/(KL_y/\lambda_y)^2 = \pi^2 \times 2 \times 10^5/(4000/153.2)^2 = 2897.9 \text{ MPa}$$

$$\lambda_z = \sqrt{\frac{f_y}{f_{ccz}}} = \sqrt{\frac{250}{2897.9}} = 0.29$$

$$\lambda_y = \sqrt{\frac{f_y}{f_{ccy}}} = \sqrt{\frac{250}{2303.9}} = 0.329$$

Buckling class = *c* (for built-up member as per Table 10 of IS:800) λ_y of the built-up column with batten = 1.1 × λ_y = 1.1 × 29.3 = 32.23 (Clause 7.7.1.4 of IS: 800)

From Table 9c of IS:800,

For Kl_y/r_{yy} = 30 and f_y = 250 MPa, f_{cd} = 211 MPa

For Kl_y/r_{yy} = 40 and f_y = 250 MPa, f_{cd} = 198 MPa

For Kl_y/r_{yy} = 32.23 and f_y = 250 MPa, by interpolating, f_{cd} = 208.1 MPa

$$\text{Load carrying capacity} = A_e\, f_{cd} = \frac{2 \times 5366 \times 208.1}{1000} = 22233.3 \text{ kN} < 2250 \text{ kN}$$

Inadequate

Note: The load carrying capacity or design strength is marginally less. Hence, the cover plates on each flange may be welded and battens can be avoided. This will be a cost-effective solution. Alternatively, we can try next section ISMC 400 as below.

To Find *S*

Keeping $I_{zz} = I_{yy}$; we can calculate '*S*'

$$2 \times 15082.8 = 2(504.8 + 62.93 \times (S/2)^2)$$

Therefore, $S = 30.44$ cm.

Providing 400 mm outer to outer;

$$S = 400 - 2 \times C_{yy} = 400 - 2 \times 24.2 = 351.6 \text{ mm}$$

Distance of centroid of each channel from *ZZ* axis = *S*/2 = 35.16/2 = 17.58 cm.

$$I_{zz} = 2 \times (504.8 + 62.93 \times 17.58^2) = 39907 \text{ cm}^4$$

$$I_{yy} = 30165.6 \text{ cm}^4$$

$$Z_{zz} = 1995.35 \text{ cm}^3$$

$$r_{zz} = 17.8 \text{ cm}$$

$$r_{yy} = 15.48 \text{ cm}$$

$$\lambda_y = \frac{4000}{154.8} = 25.84$$

$$f_{ccy} = \pi^2 E/(4000/154.8)^2 = 2958.7 \text{ MPa}$$

$$f_{ccz} = \pi^2 E/(4000/178)^2 = 3912 \text{ MPa}$$

$$\lambda_y = \sqrt{\frac{250}{3912}} = 0.253$$

$$\lambda_y = \sqrt{\frac{250}{2958.7}} = 0.29$$

It is proposed to provide battens.

$$Kl_y/r_{yy} = 1.1 \times 25.84 = 28.42 \qquad \text{(Clause 7.7.1.4 of IS:800)}$$

From Table 9c of IS:800

For $Kl_y/r_{yy} = 20$ and $f_y = 250$ MPa, $f_{cd} = 224$ MPa

For $Kl_y/r_{yy} = 30$ and $f_y = 250$ MPa, $f_{cd} = 211$ MPa

For $Kl_y/r_{yy} = 28.42$ and $f_y = 250$ MPa, by interpolating, $f_{cd} = 213$ MPa

Load carrying capacity = $A_e\, f_{cd} = 2 \times 6293 \times 213/1000 = 2680.8$ kN > 2250 kN (Clause 7.7.1.4 of IS: 800)

Hence, satisfactory.

Design of Batten

Spacing of batten C

$\frac{C}{r_{min}} > 50$ (or) 0.7 times KL/r of column (Clause 7.7.3 of IS:800)

where KL/r is the slenderness ratio of the member as a whole about its axis parallel to batten

$$KL/r = 28.42 \text{ and } 0.7 \times 28.42 = 19.9$$

$$\frac{C}{r_{min}} = \frac{C}{r_{yy}} > 19.9$$

Here r_{yy} is the radius of gyration of individual member = 28.3 mm.

$$C = 19.9 \times 28.3 = 563.17 \text{ mm}$$

Let $C = 560$ mm

Distance between connecting bolts = 400 – 2g = 400 – 2 × 60 = 280 mm

Thickness (Clause 7.7.2.3 of IS:800)

$$t >= 280/50 = 5.6 \text{ mm}$$

Provide 8 mm thick plate (Minimum is 6 mm)

$$S = 351.6 \text{ mm};\ b = 100 \text{ mm}$$

For intermediate batten $d > (3/4)\ S = (3/4) \times 351.6 = 263.7$ mm and shall not be less than $2 \times$ flange width $= 2 \times 100 = 200$ mm. (Clause 7.7.2.3 of IS:800)

For end batten $d > S = 351.6$ mm (Clause 7.7.2.3 of IS:800)

For M20 bolt, diameter of hole = 22 mm (Table 19 of IS:800). Edge distance $= 1.5 \times 22 =$ 33 mm

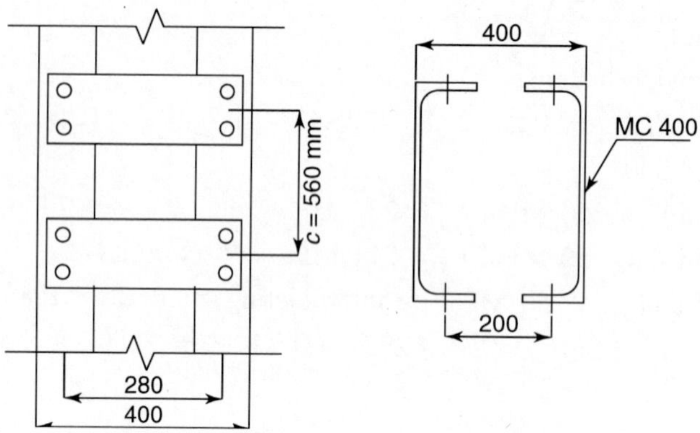

Fig. 4.18

$$D = 263.7 + 2 \times 33 = 329.7 \text{ mm}$$

Provide 330×8 mm plate

For end battens, $D = 351.6 + 2 \times 33 = 417.6$ mm. Let us provide 420×8 mm plate

Check the Adequacy

$V = 0.025\ P = 0.025 \times 2250 = 56.25$ kN (Clause 7.7.2.1 of IS:800)

Distance between inner bolts = 280 mm (Fig. 4.19)

$$V1 = VC/NS = 56.25 \times 1000 \times 560/(2 \times 280) = 56.25 \text{ kN}$$

$$M = VC/2N = 56.25 \times 1000 \times 560/(2 \times 2) = 7.875 \times 10^6 \text{ Nmm}$$

Design shear strength $= V_d = \dfrac{A_v \times f_{yw}}{\gamma_{mo}\ \sqrt{3}} = \dfrac{330 \times 8 \times 250}{1.1 \times \sqrt{3}} = 346.4$ kN (Clause 8.4 of IS: 800)

$$0.6\ V_d = 0.6 \times 346.4 = 207.84 \text{ kN} > 56.25 \text{ kN}$$

Hence, low shear.

Design strength in bending $= M_d = 1.2\ Z_e\ f_y/\gamma_{mo}$ (Clause 8.2.1.2 of IS:800)

$$M_d = 1.2\ Z_e f_{yw}/\gamma_{mo} = 1.2 \times \frac{8 \times 330^2}{6} \times \frac{250}{1.1} = 39.6 \text{ kNm} > 7.875 \text{ kNm}$$

Hence, satisfactory.

Provide 8 × 330 mm intermediate battens at 560 mm *c*/*c* and 8 × 420 mm end battens/tie plates.

Combined Capacity (Clause 9.2.2 of IS:800)

When the factored shear force is higher than 0.6 V_d, the bending strength shall be reduced as in Clause 9.2.2 of IS: 800. In this example the applied factored shear is less than 0.6 V_d (low shear).

Example 4.6 *Design of built-up column with channel back to back with lacing bars (or with* batten).

Design a built-up column using channel sections to carry an axial factored load of 2100 kN. The column has an unsupported length of 7 m and is effectively held in position at both ends but restrained against rotation at one end only. Material f_y = 250 MPa.

Solution: Assuming stress reduction factor as 0.75

Required area of cross section = 2100 × 1000/(0.75 × 250/1.1) = 12320 mm^2

Try two ISMC 400

Physical property of single ISMC 400

Area of each channel = 6293 mm^2

Width of flange = 100 mm t_f = 15.3 mm

t_w = 8.6 mm, I_y = 15032.8 × 10^4 mm^4, I_z = 504.8 × 10^4 mm^3

C_{yy} = 24.2 mm

Two Channels

Area provided = 2 × 6293 = 12586 mm^2 > 12320 mm^2

a = 6293 mm^2

To Find Spacing

Refer Clause 7.6.1.1 and 7.7.1.1 of IS: 800. The radius of gyration (r_{zz}) about the axis perpendicular to the plane of the batten (*zz* in Fig. 4.19) shall not be less than the radius of gyration (r_{yy}) about the axis parallel to the plane of the batten (*YY* in Fig. 4.20). Hence, $r_{zz} \geq r_{yy}$

$$\sqrt{\frac{I_{zz}}{A}} > \sqrt{\frac{I_{yy}}{A}}$$

$$Izz > Iyy$$

$$I_{yy} = 2 \times I_z = 2 \times 15032.8 \times 10^4 = 30065.6 \times 10^4 \text{ mm}^4$$

$$I_{yy} = (S/2)^2$$

$$I_{zz} = 2(I_z + a(S/2)^2) = 2 \times (504.8 \times 10^4 + 6293 \times (S/2)^2)$$

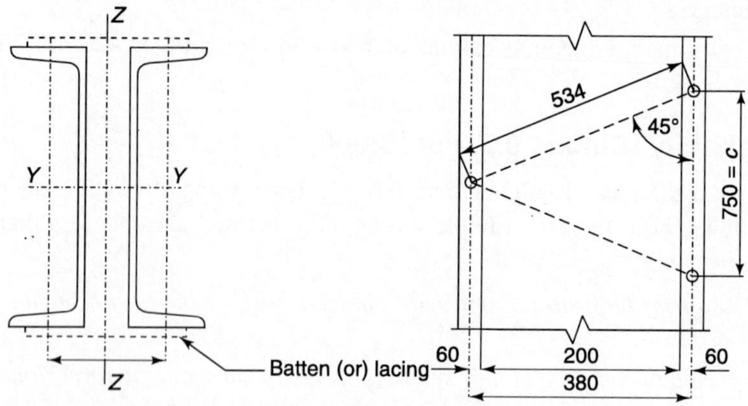

Fig. 4.19

Equating I_{zz} & I_{yy}

$$2 \times 15032.8 \times 10^4 = 2(504.8 \times 10^4 + 6293 \times (S/2)^2)$$

$$\therefore \quad S = \sqrt{[(15032.8 \times 10^4 - 504.8 \times 10^4)/6293] \times 4} = 303.88 \text{ mm}$$

Clear space $S - 2C_{yy} = 303.88 - 2 \times 24.2 = 255.48$ mm

Provide clear space = 260 mm

$$\therefore \quad X = \frac{260}{2} + 24.2 = 154.2 \text{ mm}$$

C/C of channels = 2 × 154.2 = 308.4 mm

To Find *M. I, Z* and *Y*

$$I_{yy} = 2 \times 15032.8 \times 10^4 = 30065.6 \times 10^4 \text{ mm}^4$$

$$I_{zz} = (2 \times 504.8 \times 10^4 + 6293 \times 1542^2) = 30936.1377 \times 10^4 \text{ mm}^4$$

$$r_{yy} = \sqrt{\frac{30065.6 \times 10^4}{2 \times 6293}} = \sqrt{23888.13} = 154.55 \text{ mm}$$

$$r_{zz} = \sqrt{\frac{30936 \times 10^4}{2 \times 6293}} = \sqrt{24579.69} = 156.77 \text{ mm}$$

Since the column is restrained against rotation @ one end only, $K = 0.8$ (Table 11 of IS:800:2007)

$$L_z = L_y = 0.8 \times 7 = 5.6 \text{ m}, \; P = 2100 \text{ kN}$$

$$f_{ccy} = \pi^2 E/(KL/r_y)^2 = \pi^2 \times 2 \times 10^5/(5600/154.55)^2 = 1975510/1313 = 1505 \text{ MPa}$$

$$f_{ccz} = \pi^2 E/(KL/r_z)^2 = \pi^2 \times 2 \times 10^5/(5600/156.77)^2 = 1975510/1276 = 1548 \text{ MPa}$$

$$\lambda_z = \sqrt{\frac{f_z}{f_{ccz}}} = \sqrt{\frac{250}{1548}} = 0.4$$

$$\lambda_y = \sqrt{\frac{f_y}{f_{ccy}}} = \sqrt{\frac{250}{1505}} = 0.4076$$

Buckling class of cross section for built-up member, buckling about any axis is *c* (Table 10 of IS:800).

Kl_y/r_{yy} of the built-up column with lacing = 1.05 × l/r_{yy} = 1.05 × 5600/154.55 = 38 (Clause 7.6.1.5 of IS: 800)

For Kl_y/r_{yy} = 38 and f_y = 250 MPa; f_{cd} = 201 MPa (Table 9c of IS: 800)

Load carrying capacity = $A_e f_{cd}$ = 12586 × 201/1000 = 2530 kN > 2100 kN

Hence, satisfactory.

Design of Lacing System (Single Lacing – Fig. 10 A of IS: 800)

Let us adopt single lacing system at 45 deg.

Gauge length = 60 mm (See Fig. 4.20)

Spacing of lacing bars = 2 (260 + 60 + 60) cot 45 deg = 2 × 380 × cot 45° = 760 mm

C/r_y should be < 0.7 × maximum kL/r of column (or) 50 whichever is less (Clause 7.6.5.1 of IS: 800).

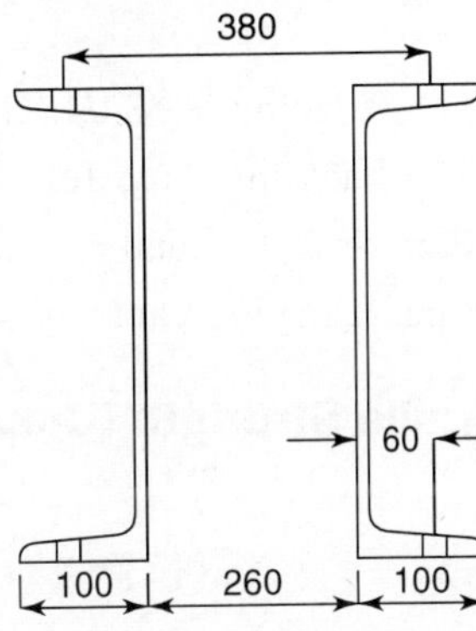

Fig. 4.20

$$0.7 \times 38 = 26.6 \text{ (or) } 50$$

$$C/r_y = 26.6$$

$$r_{min} = r_y = 28.3 \text{ mm for ISMC 400.}$$

$$C = 26.6 \times 28.3 = 753 \text{ mm}$$

Hence, provide spacing of 750 mm

$$750/2 = 375 \text{ mm}$$

$\tan\theta = \frac{375}{380}$; θ = 44.6°. This shall be between 40° and 70° (Clause 7.6.4 of IS:800)

Length of lacing bar = $\sqrt{(375^2 + 380^2)}$ = 534 mm

Maximum shear = 2.5% of total load = $\frac{2.5}{100} \times 2100$ = 52.5 kN

Compressive force in each lacing bar = $\frac{V}{N}\operatorname{cosec}\theta = \frac{52.5}{2} \times \operatorname{cosec} 44.6°$ = 36.9 kN

Using M20 class 4.6 bolt; one on each end, diameter of hole = 22 mm (Table 19 of IS: 800) and edge distance shall not be less than 1.5 × 22 = 33 for planed edges (Clause 10.2.4.2 of IS:800).

Hence, minimum width of lacing = 3 × 22 = 66 mm.

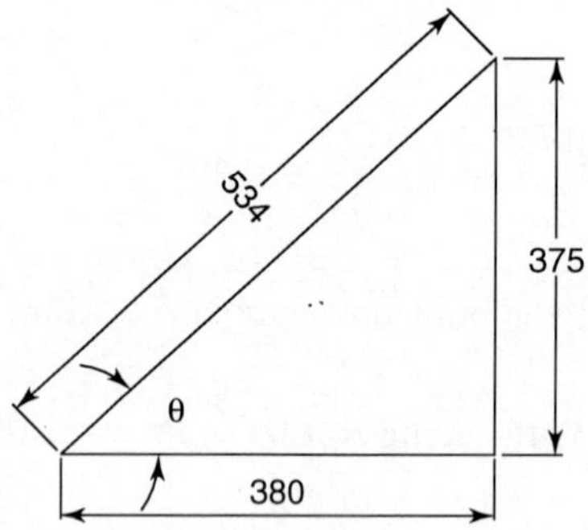

Fig. 4.21

Minimum thickness = $\frac{1}{40}$ × 534 = 13.3 mm (Clause 7.6.3 of IS: 800)

Provide 16 mm thick 70 mm wide flat as lacing.

$$r_{\min} = \frac{t}{\sqrt{12}} = \frac{16}{\sqrt{12}} = 4.62 \text{ mm}$$

L/r of lacing = 534/4.62 = 116 < 145 (Clause 7.6.6.3 of IS:800)

The buckling class for solid cross sections is c (Table 10 of IS:800).

For L/r = 116 and f_y = 250 MPa from Table 9c; f_{cd} = 88.06 MPa

Load carrying capacity = 16 × 70 × 88.06/1000 = 98.62 kN > 36.9 kN

Tensile Strength (Clause 6.1 of IS: 800)

$$A_n = 16 \times 70 - 1 \times 22 \times 16 = 768 \text{ mm}^2$$

$$A_g = 16 \times 70 = 1120 \text{ mm}^2$$

$$T_{dn} = 0.9\, A_n\, f_u/\gamma_{m1} = 0.9 \times 768 \times 410/(1.25 \times 1000) = 226.71 \text{ kN}$$

$$T_{dg} = 1120 \times 250/1.1 \times 1000 = 254.54 \text{ kN}$$

Strength = 226.71 kN > 36.9 kN

Hence, satisfactory.

Connection

Since two lacing bars are connected by a single bolt, force in this bolt = 2 × 36.9 kN = 73.8 kN.

Using M20 class 4.6 bolts. Strength of bolt in double shear keeping the lacing bar one over the other and connecting to the flange of channel and the threads are not in shear plane.

$$V_{dsb} = (f_u/\sqrt{3})\ (\eta_n\ A_{\eta bt} + \eta_s A_{sb})/\gamma_{mb} \qquad \text{(Clause 10.3.3 of IS:800)}$$

$$= (400/\sqrt{3})\ (0 + 2 \times \pi \times 20^2/4)/(1.25 \times 1000) = 116 \text{ kN}$$

Strength in bearing = $\left(\dfrac{2.5\,k_b dtf_u}{\gamma_{mb}}\right)$ (Clause 10.3.4 of IS:800)

Edge distance = e = 70/2 = 35 mm.

$$k_b = e = (35/3 \times 22) = 0.53$$

For class 4.6 bolt, f_{ub} = 400 MPa.

$$k_b = 400/410 = 0.9756$$

$$k_b = 1.0$$

Value of k_b is the least of the above = 0.53

Thickness of flange of ISMC 400 = 15.3 mm.

Strength in bearing = 2.5 × 0.53 × 20 × 15.3 × 410/(1.25 × 1000) = 133 kN

Flange thickness is less than the thickness of lacing bar and hence the thickness of flange is considered.

Bolt value = 116 kN > 73.8 kN.

Hence, one M20 class 4.6 bolt is satisfactory.

Tie Plates at Ends

Required effective depth = centroid distance = clear gap + 2 C_{yy} = 260 + 2 × 24.2 = 308.4 mm (Clause 7.7.2.3 of IS:800)

Minimum edge distance 1.5 or 1.7 times the diameter of bolt hole (Clause 10.2.4.2 of IS:800)

Let the edge distance be 2 × diameter of bolt hole = 2 × 22 = 44 mm

Required overall depth of tie plate = 308.4 + 2 × 44 = 396.4 mm

Width = 260 + 2 × (60 + 44) = 464 mm more than outer to outer dimension of built-up column which is 460 mm. Hence, keep 460 mm wide tie plate.

Thickness = (260 + 2 × 60)/50 = 7.6 mm

Let the thickness be 8 mm

Provide tie plate of 400 depth, 460 width and 8 thick with four nos. of M20 Class 4.6 bolts.

Design of Batten for the Same Column

Spacing of batten

$\dfrac{C}{r_{\min}}$ should not be greater than 50 (or) 0.7 times KL/r of column (Clause 7.7.3 of IS:800)

where KL/r is the higher slenderness ratio of the member as a whole.

$$KL/r = 1.1 \times 5600/154.55 = 39.86 \text{ (Clause 7.7.1.4 of IS: 800)}$$

$$0.7 \times KL/r = 0.7 \times 39.86 = 27.9$$

$$\frac{C}{r_{min}} > 27.9$$

$$C = 27.9 \times r_{min} = 27.9 \times 28.3 = 790 \text{ mm}$$

r_{min} is the minimum radius of gyration of single member.

Let the spacing of batten = 770 mm C/C g from steel table = 60 mm, $a_i = 260 + (2 \times 60)$ = 380 mm

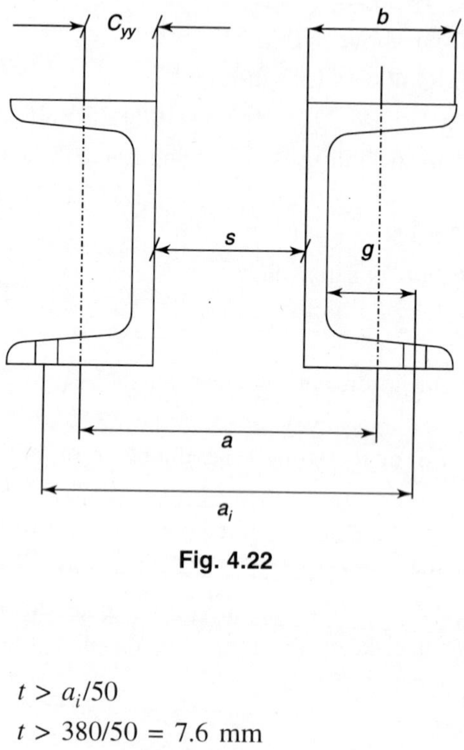

Fig. 4.22

Thickness of Batten

$$t > a_i/50 \quad \text{(Clause 7.7.2.3 of IS:800)}$$

$$t > 380/50 = 7.6 \text{ mm}$$

Provide 8 mm.

Depth of Batten (Clause 7.7.2.3 of IS:800)

$a = 260 + 2 \times 24.4 = 308.4$ mm; $b = 100$ mm

For intermediate batten $d > (3/4)a = (3/4) \times 308.4 = 231$ mm and shall not be less than 2 × flange width = 2 × 100 = 200 mm. (Clause 7.7.2.3 of IS:800)

For M20 bolt, diameter of hole = 22 mm (Table 19 of IS:800)

Minimum edge distance = 1.5 × 22 = 33 mm

Hence, overall depth of intermediate batten – 231 + 2 × 33 = 297 mm.

For end batten $d > a$ = 308.4 mm (Clause 7.7.2.3 of IS:800)

Let the effective depth of end batten be = 308.4 mm

$$D = 308.4 + 2 \times 33 = 374.4 \text{ mm.}$$

Therefore, $D = 375$ mm

Providing end batten and intermediate battens of uniform depth,

Provide 8 mm thick 375 deep battens at 770 mm *c/c*.

Check for Adequacy

$$V = 2.5\% \text{ of axial load} = \frac{2.5}{100} \times 2100 = 52.5 \text{ kN}$$

$$S = a_i = 380 \text{ mm}$$

Longitudinal shear = $V_1 = VC/NS$ = 52.5 × 770/(2 × 380) = 53.2 kN

$M = VC/2N$ = 52.5 × 1000 × 770/(2 × 2) = 10.1 × 10^6 Nmm = 10.1 kNm

Design Shear Strength of Batten Plate (Clause 8.4 of IS:800)

$$V_d = \frac{A_v f_{yw}}{\gamma_{mo}\sqrt{3}}$$

Let us provide four bolts of M20 class 4.6 on each flange.

A_v = Shear area = Net area of C.S. = 8 × 0.375 – 4 × 22 = 2912 mm

$$V_d = \frac{2912 \times 250}{1.1 \times \sqrt{3}} = 382.1 \text{ kN} > (52.5/2) = 26.25 \text{ kN}$$

Bending Strength of Batten Plate (Clause 8.2.1.2 of IS: 800)

$$\frac{d}{t} = \frac{375}{8} = 46.875 < 67\ \epsilon$$

Shear buckling need not be checked. (Clause 8.2.1.1 of IS:800)

$$0.6\ V_d = 0.6 \times 382.1 = 229.25 \text{ kN} > 26.25 \text{ kN}$$

$$V < 0.6\ V_d$$

Low shear

$$M_d = 1.2\ Z_e f_y/\gamma_{mo} = 1.2 \times \frac{8 \times 375^2}{6} \times (250/1.1) = 51.13 \text{ kNm} > 10.1 \text{ kNm.}$$

Combined Capacity (Clause 9.2.2 of IS: 800)

When the factored shear force is higher than 0.6 V_d, the bending strength shall be reduced as in Clause 9.2.2 of IS: 800. In this example the applied factored shear is less than 0.6 V_d (low shear).

Design of Connection

Try M20 class 4.6 bolts.

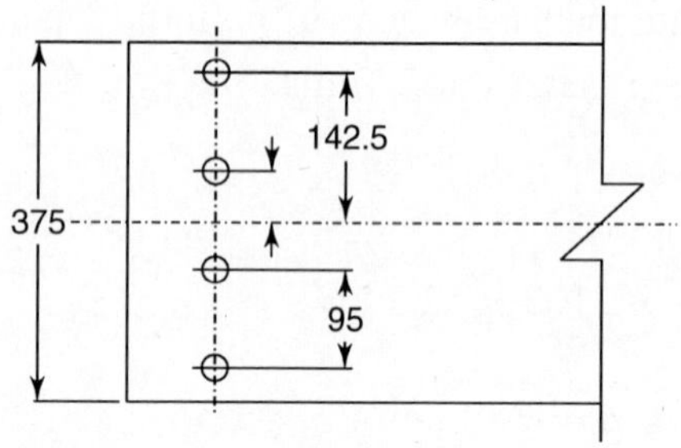

Fig. 4.23

Strength of M20 bolt in single shear = 58.06 kN ------(A) (Thread not in shear plane)

Strength in bearing (Clause 10.3.4 of IS: 800)

$$k_b = \frac{e}{3_{do}};\ \frac{p}{3_{do}} - 0.25;\ \frac{f_{ub}}{f_u};\ 1.0$$

$$\frac{e}{3_{do}} = \frac{40}{3 \times 22} = 0.61 \quad (1)$$

$$\frac{p}{3_{do}} - 0.25 = \frac{95}{3 \times 22} - 0.25 = 1.19 \quad (2)$$

$$\frac{f_{ub}}{f_u} = \frac{400}{410} = 0.9756 \quad (3)$$

$$1.0 \quad (4)$$

k_b is the least of the above four values

$\therefore \quad k_b = 0.5$

Design bearing strength = $V_{dpb} = \left(\frac{2.5\,k_b d t f_u}{\gamma_{mb}}\right)$ (Clause 10.3.4 of IS: 800) = 2.5 × 0.61 × 20 × 8 × 410/1000 = 100 kN ------*B*

Bolt value is the least of *A* and *B* = 58.06 kN

Providing 4 bolts in each row

Actual shear/bolt = 26.25/4 = 6.56 kN

Force due to moment at extreme bolt = $\dfrac{Mr}{\Sigma r^2}$

$$r = 142.5 \text{ mm}$$

$$\Sigma r^2 = 2(142.5^2 + 47.5^2) = 45125 \text{ mm}^2$$

$$\text{Force} = \frac{10.1 \times 10^6 \times 142.5}{45125 \times 10^3} = 31.9 \text{ kN}$$

$$R = \sqrt{6.56^2 + 31.9^2} = 32.56 \text{ kN} < 58.06 \text{ kN}$$

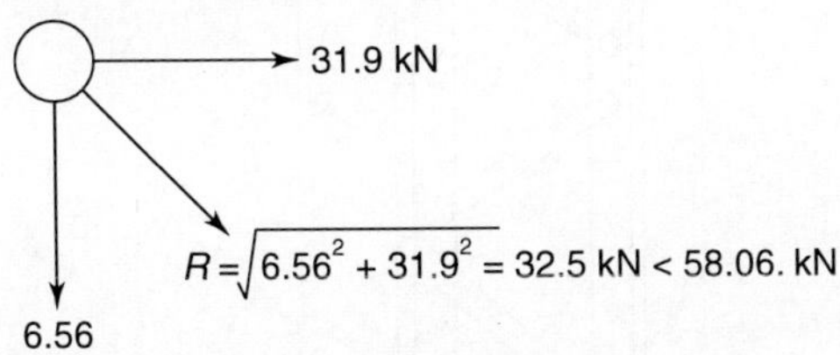

Fig. 4.24

Hence, provide 4 bolts of M20 class 4.6 on each side which is satisfactory.

Example 4.7 *Design a built-up column using channels and providing lacing. The column is subjected to an axial load of 1125 kN and moment of 112.5 kNm about the axis perpendicular to the plane of lacing. The column is free to rotate at both ends. Unsupported length is 6500 mm.*

Moment and the force is factored. Use IS:2062 material with yield stess = 250 MPa.

Solution: Assuming stress reduction factor as 0.6;

$$f_{cd} = 0.6 \times 250 = 150 \text{ MPa}$$

Area required = (1125 × 1000/150) = 7500 mm^2

It is proposed to have two channels facing each other.

Required area per channel = 3750 mm^2

Try two ISMC 350

Outer to outer of web of channel is kept at 350 mm

Section Properties of Single Channel

$$\text{Area} = 5366 \text{ mm}^2 = a$$

$$I_x = 10008 \times 10^4 \text{ mm}^4$$

$$I_y = 430.6 \times 10^4 \text{ mm}^4$$

$$r_z = 136.6 \text{ mm}; \ r_y = 28.3 \text{ mm}$$

$$C_{yy} = 24.4 \text{ mm}$$

Sectional Properties of Combined Section

$$A = 2 \times a = 10732 \text{ mm}^2$$

$$I_{yy} = 2I_x = 20016 \times 10^4 \text{ mm}$$

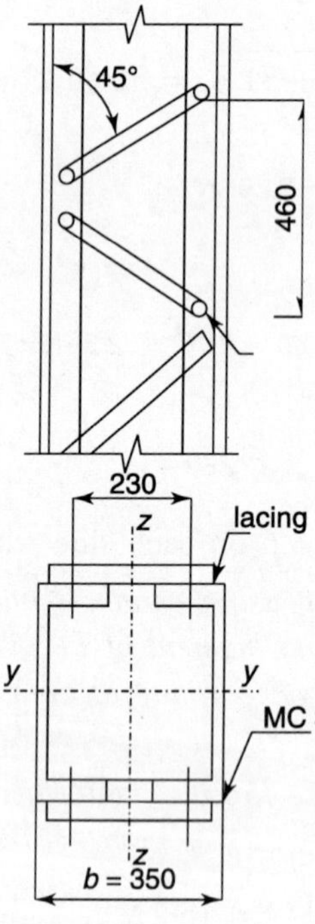

Fig. 4.25

$$I_{zz} = 2[I_y + a(b/2 - C_{yy})^2] = 2[430.6 \times 10^4 + 5366\ (175 - 24.4)^2] = 25202 \times 10^4 \text{ mm}^4$$

$$r_{yy} = \sqrt{(I_{yy}/A)} = 136.6 \text{ mm}$$

$$r_{zz} = \sqrt{(I_{zz}/A)} = 153.2 \text{ mm}$$

Since the column is hinged at both ends, $K = 1.0$ (Table 11 of IS:800)

The individual members of the whole column are proposed to be connected with lacing, effective slenderness ratio = 1.05 times KL/r. (Clause 7.6.1.5 of IS:800)

$$KL = 1 \times 6.5 = 6.5 \text{ m}$$

$f_{ccy} = \pi^2 E/(1.05 \times KL/r_{yy})^2 = \pi^2 \times 2 \times 10^5/(6825/136.6)^2 = 791.4$ MPa = 872.5 MPa

$f_{ccz} = \pi^2 E/(1.05 \times KL/r_{zz})^2 = \pi^2 \times 2 \times 10^5/(6825/153.2)^2 = 995.4$ MPa

$$\lambda_z = \sqrt{\frac{f_y}{f_{ccz}}} = \sqrt{\frac{250}{995.4}} = 0.501$$

$$\lambda_y = \sqrt{\frac{f_y}{f_{ccy}}} = \sqrt{\frac{250}{791.4}} = 0.562$$

KL/r_{yy} of built-up column with lacing

$= 1.05 \times 6500/136.6 = 50$ (Clause 7.6.1.5 of IS: 800)

For $KL/r = 50$ and $f_y = 250$ MPa; $f_{cd} = 183$ MPa (Table 9c of IS:800)

Buckling class is *c* for built of column (Table 10 of IS: 800).

Load carrying capacity = $A_e f_{cd}$. = 10732 × 183/1000 = 1964 kN > 1125 kN

Hence, satisfactory.

Design of Lacing

Lacing bar shall be inclined at an angle 40° to 70° to the longitudinal axis of the member (Clause 7.6.4 of IS: 800).

Single lacing at 45° to the vertical axis of the column is proposed.

Distance of centre of bolt hole to outer face of web = g = 60 mm.

Spacing of lacing bars = (350 – 2 × 60) × 2 = 460 mm

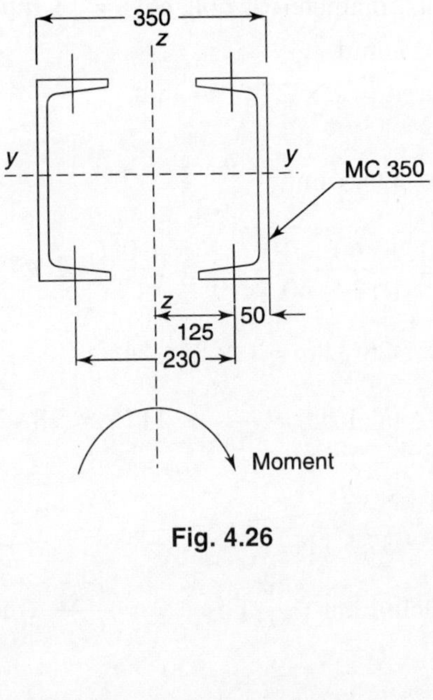

Fig. 4.26

r_{min} of ISMC 350 = 28.3 mm

λ of individual ISMC 350 between lacing intersections = $\dfrac{460}{28.3}$ = 16.25 < 50 and 0.7 × λ of built-up column = 0.7 × 50 = 35. (Clause 7.6.5.1 of IS: 800). Hence, satisfactory.

$$\text{Minimum thickness of lacing bar} = \frac{\text{Length of lacing between bolts}}{40}$$

(Clause 7.6.5.1 of IS: 800)

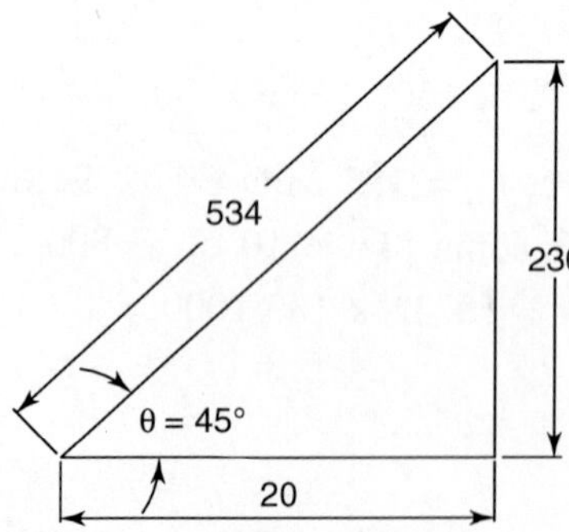

Fig. 4.27

Length of lacing between bolts

= 230 × $\sqrt{2}$ = 325.3 mm. Hence, minimum thickness is $\dfrac{325.3}{40}$ = 8.13 mm

Proposing M16 class 4.6 bolts, diameter of bolt hole = 18 mm (Table 19 of IS: 800)

Edge distance = 1.5 × 18 = 27 mm.

Minimum width of bar required = 2 × 27 = 54 mm.

Try 60 × 10 lacing bar.

$$L = 325.3 \text{ mm}$$

$$r_{min} = \sqrt{\frac{60 \times 10^3}{12 \times 60 \times 10}} = \sqrt{\frac{10^2}{12}} = 2.88 \text{ mm.}$$

$$\lambda = 325/2.88 = 112.8 < 145 \qquad \text{(Clause 7.6.6.3 of IS: 800)}$$

Maximum shear = 2.5% of total load = $\dfrac{2.5}{100}$ × 1125 = 28.125 kN (Clause 7.6.6.1 of IS: 800)

Shear due to bending = 9.34 kN.

Total shear = 28.125 + 9.34 = 37.5 kN.

Compressive force in each lacing bar = $\dfrac{V}{N}$ cosec θ = $\dfrac{37.5}{2}$ × cosec 45° = 26.52 kN.

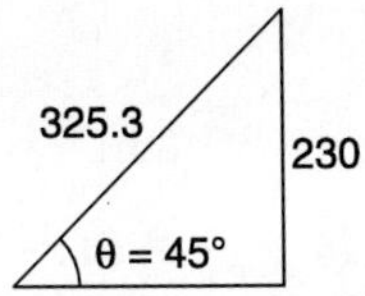

Connection

It is proposed to have bolted connection. Thread is not in shear plane.

Single shear capacity of M16 class 4.6 bolt = 37.16 kN (Table 8.8b of this book) > 26.52 kN. (Clause 10.3.3 of IS: 800)

Bearing Capacity of Bolt

Bearing capacity = $V_{dpb} = \left(\frac{2.5 k_b d t f_u}{\gamma_{mb}}\right)$ (Clause 10.3.4 of IS: 800)

The value of k_b is the least of

$$k_b = \frac{e}{3_{do}};\ \frac{p}{3_{do}} - 0.25;\ \frac{f_{ub}}{f_u};\ 1.0$$

$$\frac{e}{3_{do}} = [(60/2)/(3 \times 18)] = 0.56 \quad (1)$$

$$\frac{p}{3_{do}} - 0.25 \quad (2)$$

(Single bolt is proposed and hence pitch is not there).

$$\frac{f_{ub}}{f_u} = \frac{400}{410} = 0.9756 \quad (3)$$

$$1.0 \quad (4)$$

$$k_b = 0.39$$

Bearing capacity = V_{dpb} = 2.5 × 0.56 × 16 × 10 × 410/(1.25 × 1000) = 73.5 kN > 26.52 kN

Compressive Strength

Section is solid. Hence, buckling class *c*. (Table 10 of IS:800)

For λ = 112.8 and f_y = 250 MPa; from Table 9c; f_{cd} = 91.55 MPa

Load carrying capacity = 60 × 10 × 91.55/1000 = 54.93 kN > 26.52 kN

Tensile Strength

$T_{dn} = 0.9\ A_\eta\ f_u/\gamma_{m1}$ = 0.9 (60 – 18) × 10 × 410/1.25 × 1000 = 124 kN (Clause 6.3.1 of IS: 800)

T_{dg} = 60 × 10 × 250/(1.1 × 1000) = 136.4 kN (Clause 6..3 of IS:800)

Strength = 124 kN > 26.52 kN

Block Shear Strength

End distance = 30 mm and edge distance = 30 mm.

$$T_{db} = (A_{vg}\, f_y/\sqrt{3}\ \gamma_{mo}) + (0.9\ A_{tn} f_u/\gamma_{m1})$$

(or)

$$T_{db} = (0.9 A_{vn}\, f_u/\sqrt{3}\ \gamma_{m1}) + (A_{tg}\, f_y/\gamma_{mo}) \qquad \text{(Clause 6.4 of IS: 800)}$$

$$A_{vg} = 30 \times 10 = 300\ \text{mm}^2$$

$$A_{vn} = [30 - (18/2)] \times 10 = 210\ \text{mm}^2$$

$$A_{tg} = 30 \times 10 = 300\ \text{mm}^2$$

$$A_{vn} = [30 - (18/2)] \times 10 = 210\ \text{mm}^2$$

$T_{db} = (A_{vg}\, f_y/\sqrt{3}\ \gamma_{mo}) + (0.9\ A_{tn}\, f_u/\gamma_{m1}) = [(300 \times 250/\sqrt{3} \times 1.1)) + (0.9 \times 210 \times 410/1.25)]$ = 101.4 kN.

$T_{db} = (0.9\ A_{vn} f_u/\sqrt{3}\ \gamma_{m1}) + (A_{tg}\, f_y/\gamma_{mo}) = [(0.9 \times 210 \times 410/(\sqrt{3} \times 1.1)) + (300 \times 250/1.1)]$ = 108.85 kN.

Design block shear strength is the least of the two = 101.4 kN > 26.52 kN.

Hence, satisfactory.

Tie Plates at Ends

Required effective depth = Distance between centroids and the same should not be less than twice the width of flange. (Clause 7.7.2.3 of IS:800)

Distance between centroids = $350 - 2 \times c_{yy} = 350 - 2 \times 24.4 = 301.2$ mm.

Twice the width of flange = 2 × 100 = 200 mm.

Let the edge distance be 2 × dia of bolt hole = 2 × 18 = 36 mm

Required overall depth of tie plate = 301.2 + 36 = 373.2 mm. Provide 380 mm.

Width = 230 + 2 × 36 = 302 mm

Thickness = 301.2/50 = 5.02 mm

Let the thickness be 8 mm

Provide tie plate of 380 depth × 302 width and 8 thick with three nos. of M16 class 4.6 bolts.

4.9 EFFECTIVE LENGTH OF COLUMN OF WORKSHOP SHED

Spacing of the frame shown in Fig. 4.28 is 6 m in the longitudinal direction. Eave girder is provided in the longitudinal direction.

Effective length = L_e

Unsupported length = L

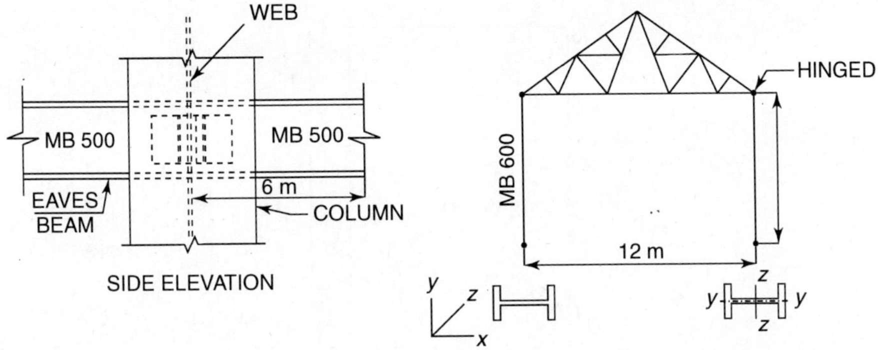

Fig. 4.28

Since both the ends of the columns are hinged,

$$L_{ez} = L = 5000 \text{ mm}$$

$$k_c = \text{stiffness of column}$$

$$k_b = \text{stiffness of beam}$$

$$\lambda_z = L_{ez}/r_z = 5000/242.4 = 20.63$$

To find L_{ey}

There are two ISMB 500 as eaves beams one on either side of column

$$k_c = (I/L)_c = 2651 \times 10^4/5000 = 5302 \text{ mm}^3$$

$$k_b = 45218.3 \times 10^4/6000 = 75363.8 \text{ mm}^3$$

Since the base and top of column is pinned,

$\beta_1 = 1$ (Fig. 28 of Annex-D of IS:800-Sway frame)

$\beta_2 = k_c/(k_b + k_b + k_c) = 0.03$ which is almost equal to zero

Strong beam weak column

From Fig. 28 of Annex-D of IS:800, for $\beta_1 = 1$ and $\beta_2 = 0.03$, $L_{ey}/L = 2$

$$L_{ey} = 2L = 2 \times 5000 = 10000 \text{ mm}$$

$$\lambda_y = L_{ey}/r_{min} = 10000/41.2 = 243$$

4.10 EFFECTIVE LENGTH OF COLUMN IN SECOND FLOOR

A single bay-two floor frame is restrained in the plane perpendicular to the frame by beams connected to it at nodes. The beams in the plane of the frame are MB 600 and the columns are ISMB 500. Find the effective length of columns. Column bases are fixed.

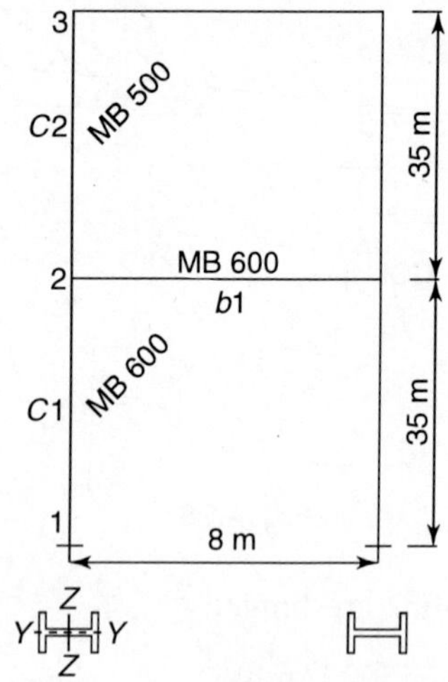

Fig. 4.29

Joint 2

$$k_{b1} = I_z/L = 91813 \times 10^4/8000 = 114766.25 \text{ mm}^3$$

$$k_{c2} = k_{c1} = I_z/L = 45218.3\ 10^4/3500 = 129195.14 \text{ mm}^3$$

$$\beta_1 = kc_2/(k_{c1} + k_{c2} + k_{b1}) = 0.35$$

Joint 3

$$\beta_2 = k_{c2}/(k_{c2} + k_{b2}) = 0.53.$$

The frame is not braced. Hence, refer to Fig. 4.28 of Annex-D of IS: 800-sway frame.

For $\beta_1 = 0.35$ and $\beta_2 = 0.53$ the value of $L_{ey}/L = 1.33$; $L_{ey} = 1.38\ L = 3.5 \times 1.33 = 4.655$ m.

Example 4.8 *Portal frame*

A portal frame of span 18 m has a column of height 7 m above the pedestal up to the eave level. Brick wall is built on either side of the column web up to the eave level and the analysis of the portal frame gives the following reactions at the base of the column. Factored axial force is 100 kN and horizontal reaction is 40 kN. Design the column using steel for which $f_y = 250$ MPa.

Solution: Force analysis

Since there is no moment at the base, the base is hinged.

The critical section is at the knee '*B*'.

The axial force at this point is (axial force at base) – (self weight of member).

Ignoring self weight, the axial load at the knee can be considered 100 kN.

Moment at $B = M = Hh = 40 \times 7 = 280$ kNm.

The effective length about major axis *ZZ* of the column is to be computed.

The top of column is restrained in *ZZ* axis by rafter

The frame is a moment resisting frame and not restrained against sway.

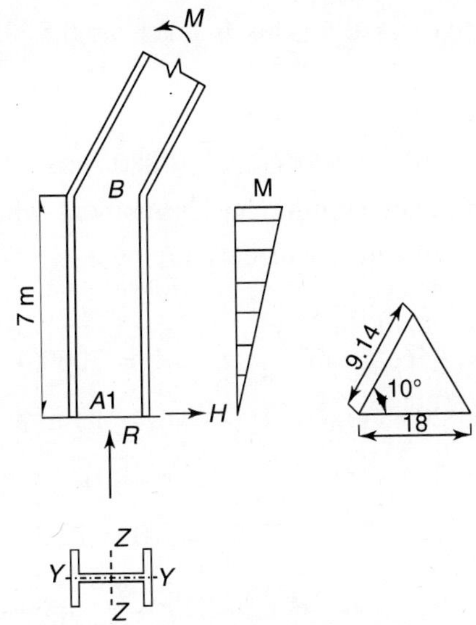

Fig. 4.30

Assume equal moment of inertia for column and rafter. $I_b = I_c = I$

$$k_c = (3/4) \times (I_c/L_c), \text{ Since the base is hinged.}$$

$$= 0.75 \times (I_c/7) = 0.107 \times I_c$$

$$k_b = (I_b/L_b) = I_b/9.14 = 0.109 \times I_b$$

$$\beta_2 = k_c/(k_b + k_c) = (0.107 \times I_c)/[(0.107 \times I_c) + (0.109 \times I_b)] = 0.5$$

Base of column is pinned and top is fixed. Referring to Fig. 28 of Annex-D of IS: 800, For $\beta_1 = 1.0$ and $\beta_2 = 0.5$, the value of $L_{ez} = 2.5 \times L = 2.5 \times 7 = 17.5$ M.
The effective length about Y axis = 0 as the column is sandwiched between brick walls.

$$\lambda_r = 0;$$

Design axial force P = 100 kN and moment $M_u = 40 \times 7 = 280$ kNm.

$$L_{ex} = 17.5 \text{ m } L_{ey} = 0$$

$$e = M/p = \frac{280}{100} = 2.8 \text{ m}$$

This is too high and hence moment will dominate.

$$M_d = 1.2\, Z_e\, f_y/1.1 \qquad \text{(Clause 8.2.1.2 of IS:800)}$$

Z_e required = $M_u/1.2\ (f_y/1/1) = 280 \times 10^6/1.2\ (250/1.1) = 1.027 \times 10^6$ mm^3.
Based on axial force

Assuming stress reduction factor as 0.5. This is taken as 0.5 since the column may fail by buckling before yielding.

$$f_{cd} = 0.5\, f_y/\gamma_{mo} = 0.5 \times 250/1.1 = 113.6 \text{ MPa.}$$

Area required = Load/stress = 100 × 1000/113.6 = 880 mm^2.

This is only a rough estimate since column carries moment which dominates.

The required area and section modulus will be more.

Try ISMB 600

$$A = 156.21 \text{ cm}^2 \qquad Z_z = 3060.4 \text{ cm}^3$$

$$t_f = 20.8 \text{ mm} \qquad t_w = 12 \text{ mm}$$

$$r_z = 242.4 \text{ mm} \qquad r_y = 41.2 \text{ mm}$$

$$L_{ez}/r_z = 17.5 \times 1000/242.4 \times 10 = 72.2$$

Section Strength

$$\varepsilon = \sqrt{(250/f_y)} = \sqrt{(250/250)} = 1.0$$

$$N_d = A_g\, f_y/\gamma_{mo} = 15621 \times 250/1.1 = 3550227 \text{ N} = 3550.227 \text{ kN}$$

(Clause 9.3.1 of IS: 800)

$$b/t_f = 10^5/20.8 = 5.05 < 9.4\,\varepsilon$$

Depth of web = 600 – 2 × (20.8 + 20) = 518.4 mm

$$d/t_w = 518.4/12 = 43.2 < 84\,\varepsilon$$

Hence, section is plastic. (Table 2 of IS: 800)

Since minor axis is completely restrained by brick wall, effective length on *ZZ* axis will govern.

To find buckling class,

$$h/b_f = (600/210) > 1.2$$

$$t_f < 40 \text{ mm}$$

Buckling class of cross section on *ZZ* axis is a (Table 10 of IS:800)

For $KL/r = 72.2$, $f_{cd} = 178.7$ MPa (Table 9a of IS:800)

Section strength in compression = P_{dz} = 15621 × 178.7/1000 = 2791.5 kN

$$M_{dz} = \beta_p Z_p\, f_y/\gamma_{mo} \qquad \text{(Clause 8.2.1.2 \& Table 46 Annex-H)}$$

$$= 1.0 \times 3510.63 \times 10^3 \times 250/(1.1 \times 10^6) = 797870 \times 10^3 \text{ Nmm} = 797.87 \text{ kNm}$$

$$M_{dy} = \beta_p Z_{py}\, f_y/\gamma_{mo}$$

Depth of web = 600 – 2 × 20.8 = 558.4 mm.

$$Z_{py} = (4 \times 105 \times 20.8 \times 52.5) + \left(2 \times \frac{558.4}{2} \times \frac{12}{2} \times \frac{12}{4}\right) = 468691.2 \text{ mm}^3$$

$$M_{dy} = 1.0 \times 468691.2 \times 250/(1.1 \times 10^6) = 106.52 \text{ kNm}$$

Section Strength (Clause 9.3.1 of IS:800)

$$N_d = A_g f_y/\gamma_{mo} \qquad \text{(Clause 9.3 of IS:800)}$$

$$= 15621 \times 250/1.1 \times 1000 = 3550.23 \text{ kN.}$$

$$n = N/N_d = 100/3550 = 0.028$$

Hence,

$$M_{ndy} = M_{dy} = 106.52 \text{ kNm (Clause 9.3.1.2 of IS: 800)}$$

$$M_{ndz} = 1.11 \times 797.87 \times (1 - 0.028) = 860.7 \text{ kNm}$$

Should be less than Md_z

Hence, $M_{ndz} = 797.87$ kNm

$$\left(\frac{M_y}{M_{ndy}}\right)^{\alpha_1} + \left(\frac{M_z}{M_{ndz}}\right)^{\alpha_1} \leq 1.0 \qquad \text{(Clause 9.3.1.1 of IS:800)}$$

$$\alpha_1 = 5n \geq 1 = 5 \times 0.028 = 0.14$$

This should be equal to or more than 1.0. Hence, $\alpha_1 = 1.0$ (Table 17 of IS:800)

$$\alpha_2 = 2$$

$$\left(\frac{0}{106.52}\right)^1 + \left(\frac{280}{797.87}\right)^2 = 0.123 < 1.0$$

$$\frac{N}{N_d} + \frac{M_y}{M_{dy}} + \frac{M_z}{M_{dz}} = 0.028 + 0 + 0.351 = 0.379 < 1.0$$

Overall Member Strength (Clause 9.3.2 of IS:800)

Bending and axial compression

$$\frac{P}{P_{dy}} + K_y \frac{C_{my} M_y}{M_{dy}} + K_{LT} \frac{M_z}{M_{dz}} \leq 1.0$$

$$KL_y/r_y = 0.65 \times 7000 = 41.2 = 110.4$$

$$f_{cd} = 103.51 \text{ MPa. (Table 9b of IS: 800)}$$

$$P_{dy} = 15621 \times 103.5/1000 = 1617 \text{ kN.}$$

$$P/P_{dy} = n_y = 100/1617 = 0.062$$

Second term = 0 since there is no moment on minor axis.

$$\lambda_z = \sqrt{f_y (KL/r)^2/\pi^2 E} \qquad \text{(Clause 7.1.2.1 of IS: 800)}$$

$$= \sqrt{250 \times 72.2^2/\pi^2 \times 2 \times 10^5} = 0.812$$

C_{my} (Table 18 of IS:800)

Column is fixed at one end and free at other end.

$$\psi = 0$$

$$0.6 + 0.4 \times 0 = 0.6$$

$$C_{my} = 0.6 = C_{mz}$$

K_{LT} (Clause 9.3.2.2 of IS:800)

$$= 1 - \frac{0.1\,\lambda_{LT}\,\eta_y}{(C_{mLT} - 0.25)} \geq 1 - \frac{0.1\,\eta_y}{(C_{mLT} - 0.25)}$$

$$\lambda_{LT} = \sqrt{\beta_p Z_p f_y/M_{cr}} \leq \sqrt{1.2\, Z_e f_y/M_{cr}} \qquad \text{(Clause 8.2.2 of IS:800)}$$

$$M_{cr} = \beta_p Z_p f_{crb} \qquad \text{(Clause 8.2.2.1 of IS:800)}$$

$$f_{crb} = \frac{1.1 \times \pi^2 E}{(L_{LT}/r_y)^2}\left(1 + \frac{1}{20}\left(\frac{L_{LT}/r_y}{h_f/t_f}\right)^{0.2}\right)^{0.5}$$

To find L_{LTZ}

Both the ends are fully restrained against torsion. Since brickwork is there on either side of web of column; we can consider both the flanges are fully restrained.

Hence, $L_{LTZ} = 0.7 \times L = 0.7 \times 7 = 4.9$ m (Table 15 of IS:800)

$$KL_{LTZ}/r_z = 4900/242.4 = 20.2$$

$$h_f = 600 - 20.8 = 579.2 \text{ mm}; \; t_f = 20.8 \text{ mm}$$

$$f_{crb} = \frac{1.1 \times \pi^2 E}{(20.2)^2} \times \left(1 + 0.05\left(\frac{20.2}{579.2/20.8}\right)^2\right)^{0.5} = 5395.2 \text{ MPa}$$

From Table 14, for $K_L/r_y = 20.2$ and $h_f/t_f = (579.2/20.8) = 27.85$., $f_{crb} = 5440$ MPa

Calculated value = 5395.2 MPa

$$M_{cr} = \beta_p Z_p f_{crb} = 1.0 \times 3510.63 \times 10^3 \times 5440/10^6 = 19098 \text{ kNm}$$

$$\lambda_{LT} = \sqrt{\frac{797.87}{19098}} = 0.204$$

$$K_{LT} = 1 - \frac{0.1 \times 0.204 \times 0.061}{(0.6 - 0.25)} = 1 - 0.0036 = 0.9964$$

$$\frac{P}{P_{dy}} + K_y \times \frac{C_{my} M_y}{M_{dy}} + K_{LT} \times \frac{Mz}{M_{dz}} = 0.061 + 0 + 0.9964 \times 0.351 = 0.411 << 1.0$$

The factor is far less than 1.0, we can try with lower sections such as ISMB 400.

Example 4.9 *Find the moment at the base of the single column for gantry girder for the following design data.*

Load from roof = 400 kN

Load from gantry = 400 kN at 250 mm from face of column.

Crane leg = 6 m. Roof leg = 2 m

Columns are braced perpendicular to minor axis against wind.

Boundary conditions.

Base is fixed. The column roof system is braced.

Solution: Refer Fig. 4.31

Effective length of the column in *ZZ* axis = l_{ez} = 1.5 × *L* = 1.5 × 8 = 12 m.

Effective length of the column in *YY* axis = l_{ey} = 1.5 × *L* = 0.65 × 8 = 5.2 m.

In *YY* axis, the column is fixed at the base and laterally supported by gantry girder at 6 m above base. Hence, the column can be assumed as fixed-fixed on *YY* axis.

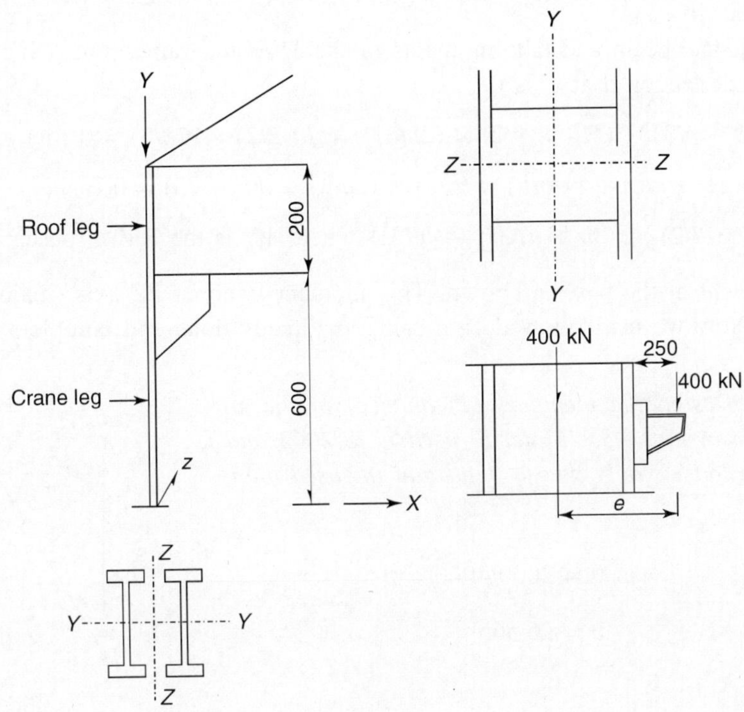

Fig. 4.31

Force Analysis

The column is a propped cantilever. The proped cantilever is analysed by a conjugate beam.

$$M = P \times e$$

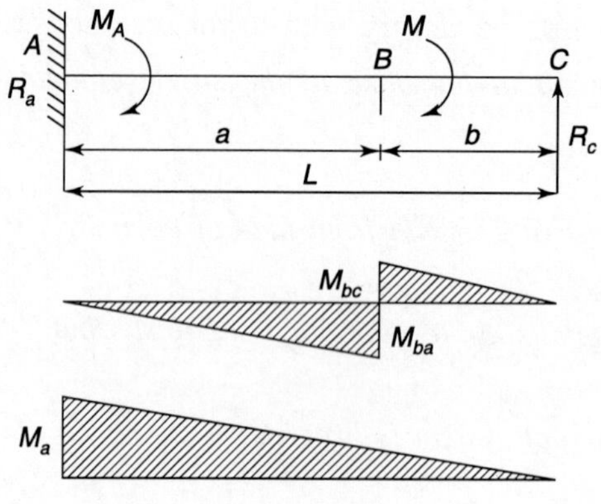

Fig. 4.32

Form hinged-hinged beam

The *BM* on either side of the brackets are

$$M_{ba} = R_A \times a = M_a/L;\ M_{bc} = R_c \times b = M\ (1 - a)/L$$

Consider conjugate beam and take moments of the *M*/*EI* diagram about point *C* and equate it to zero as the deflection d at *C* is 0.

$(M_{ba} \times a/2) \times \{b + a/3\} + (M_{bc} \times b/2) \times (2b/3) + (Ma\ L/2) \times (2L/3) = 0$ (or)

$(Ma/L) \times (a/2) \times (b + a/3) + (Mb/L) \times (b/2) \times (2b/3) + (M_A\ L^2/3) = 0$ (or)

$M_A = (3M/L^3) \times \{(a^2/2) \times ((3b + a)/3)) - (b^3/3)\}$ where M_A is the BM at base.

Now the moment at the base is known. This moment is about *ZZ* axis (major axis). Axial load is known. Now we can follow design steps as already done and complete the design of the column.

Example 4.10 *Design the members (1) and (2) of the knee bracket as shown in Fig. 4.33. Material used is IS:2062 and f_y = 250 MPa. The load is only due to dead and imposed loads.*

To find θ

$$\tan \theta = 2000/4000 = 0.5$$

$$\therefore \qquad \theta = 26.565°$$

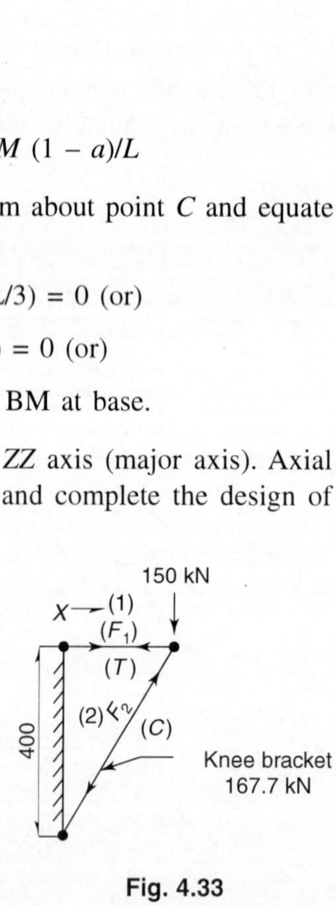

Fig. 4.33

Force in knee bracket (member)

$$\cos\theta = \cos 26.565° = 150/F_2$$

$$F_2 = 150/\cos\theta = 167.7 \text{ kN}$$

Member (2) is a compression member.

Force in Tension Member (1)

Taking moment about the point z to find, tension in this member

$$150 \times 2 = F_1 \times 4$$

Therefore, $F_1 = 75$ kN

Length of (2) = $\sqrt{(2^2 + 4^2)}$ = 20 = 4.47 m

Design of Tension Member (1)

Length of (1) = 2000 mm

Factored load = 1.5 × 75 = 112.5 kN

Area required = (112.5 × 1000)/(250/1.1) = 495 mm^2

Try ISMC 75

Tensile strength based on the following are to be calculated.

(1) As governed by yielding of gross section (Clause 6.2 of IS:800)

$T_{dg} = A_g f_y/\gamma_{mo}$ = 867 × 250/(1.1 × 1000) = 197 kN

(2) As governed by rupture of net section (Clause 6.3 of IS:800)

The joints at end of member is proposed with weld. Hence, there is no reduction in area of cross section. Therefore, $A_n = A_g$.

$T_{dn} = 0.9\, A_n f_u/\gamma_{m1}$ = 0.9 × 867 × 410/(1.25 × 1000) = 256 kN (Considering weld joint)

Least of the two is the strength which is = 197 kN > 112.5 kN

Hence, ISMC 75 is adequate for member (1).

Design of Compression Member (2)

Member (2) is to be designed as a compression member for 167.7 kN compressive force.

Let f_{cd} = 0.5 × 250 = 125 MPa

Factored load = 1.5 × 167.7 = 251.55 kN

Area required = 251.55 × 1000/125 = 2012.4 mm^2

Let us try box ISMC 150. Area = 4176 mm^2

$$I_z = 1558.8 \times 10^4 \text{ mm}^4$$

$$I_y = 1368.8 \times 10^4 \text{ mm}^4$$

$$Z_z = 207.8 \times 10^3 \text{ mm}^3$$

$$Z_y = 182.5 \times 10^3 \text{ mm}^3$$

$$r_z = 61.1 \text{ mm}$$

$$r_y = 57.3 \text{ mm}$$

Since the member is hinged at both ends; effective length = unsupported length (Table 11 of IS: 800)

Therefore, $L_e = L = 4470$ mm

$$KL/r_y = 4470/57.3 = 78$$

The section is a box section built up of channel sections.

Hence, buckling class is c. (Table 10 of IS:800)

From Table 9c of IS:800,

For $KL/r_y = 70$ and $f_y = 250$ MPa, $f_{cd} = 152$ MPa

For $KL/r_y = 80$ and $f_y = 250$ MPa, $f_{cd} = 136$ MPa

For $KL/r_y = 78$ and $f_y = 250$ MPa, $f_{cd} = 152 - [(152 - 136) \cdot 10] \times 8 = 139.2$ MPa

Strength of section in compression = 139.2 × 4176/1000 = 581.3 kN > 251.55 kN

We can even try lesser section.

Let us try box ISMC 125.

Area = 3238 mm^2

$$r_z = 50.7 \text{ mm}$$

$$r_y = 49.5 \text{ mm}$$

$$KL/r_y = 4470/49.5 = 90.3$$

From Table 9c of IS:800,

For $KL/r_y = 90$ and $f_y = 250$ MPa, $f_{cd} = 121$ MPa

For $KL/r_y = 100$ and $f_y = 250$ MPa, $f_{cd} = 107$ MPa

For $KL/r_y = 90.3$ and $f_y = 250$ MPa, $f_{cd} = 121 - [(121 - 107)\ 10] \times 0.3 = 120.58$ MPa

Strength of section in compression = 120.58 × 3238/1000 = 390.4 kN > 251.55 kN

EXERCISE

1. What is meant by effective length of a compression member?
2. Explain slenderness ratio of a compression member.
3. Write down the step-by-step design procedure for the design of column subjected to axial load with biaxial bending.
4. Write down the general rules in the design of (a) battens and (b) lacing.
5. Find the axial load carrying capacity of a ISMB 250 column of 4.5 m unsupported length between fixed end supports. The yield stress of the material is 250 MPa.
6. Design a column using two ISMB members kept at a suitable distance between centre to centre of web to carry a load of 1950 kN due to dead and imposed loads. The unsup-

ported length is 7 m and its minor axis is restrained at middle of its length. Both the ends are having pinned connection. Yield stress is 250 MPa.

7. Design batten plate connection for the column for which details are given below. The column is made of two ISMC 350 facing each other and having the outer to outer dimension of 350 mm in both the axes. The axial load is 1200 kN due to dead and imposed loads. The unsupported length of the column is 6500 mm and the ends are free to rotate. Yield stress of the material is 250 MPa.

Chapter 5

Design of Flexural Members

5.1 STRUCTURAL ANALYSIS

Reactions, shear force, bending moment and deflection are required for the design of beams.

For simply supported beam of effective span l

1. Bending moment due to uniformly distributed load $w = wl^2/8$
2. Bending moment due to concentrated load W at centre $= Wl/4$
3. Bending moment due to concentrated load W at a distance a from one support $= (Wab/l)$, where $b = l - a$.

w = uniformly distributed load in kN/m

l = effective span in m

W = concentrated load in kN

Moment is in kNm

Note: For other end conditions and type of load, the formulas for bending moments, shear force and deflections, refer relevant standard books.

5.2 DESIGN OF BEAMS

Beams are structural members loaded perpendicular to their longitudinal axis. If the load is applied perpendicular to its minor axis, it causes lateral bending. In simple words, design of beam can be called sizing of beam.

The selection of section of required size for any structure involves two stages:

(1) The analysis of the structure to determine the forces, the shears and moments present at different points.

(2) The detail design of each section to take with these forces, shears and moments.

In the permissible stress design or allowable stress design or working stress design, we have to ensure that the calculated stresses in the structure at all points are less than the permissible stresses.

In the limit state designs, the design load effect shall not exceed the design strength of the structure.

5.3 DEFLECTION

Table 6 of IS:800 gives the deflection limits for members under various loading conditions.

The maximum deflection of a beam is given by the following general expression:

Maximum Deflection = $(C \times WL^3/EI)$

where C = a constant

W = total unfactored load in N

L = span in mm

E = Young's modulus of elasticity of the material of the beam in N/mm^2

I = moment of inertia of the cross section of the beam in mm^4.

The constant C depends upon the support condition and also upon the pattern of loads that are applied to it. Table 5.1 shows the value of this constant C for a number of commonly encountered conditions of support and load. For a cantilever of length L carrying a point load of W at its free end, the maximum deflection will occur at the free-end, and will have a value of:

$$\frac{1}{3}\frac{WL^3}{EI}$$

W is in N, L is in mm, E is in N/mm^2, I is in mm^4 and deflection will be in mm.

For a simply supported beam of span L carrying a total uniformly-distributed load of W, the maximum deflection will be at midspan, and will have a value of:

$$\frac{5}{384}\frac{WL^3}{EI}$$

It should be noted that W in this formula ('capital' or 'upper case' W) is the total load shown in the diagram of Table 5.1. It will be expressed in N or kN. For a simply-supported beam carrying a uniformly-distributed load, the deflection may be given as:

$$\frac{5}{384}\frac{wL^4}{EI}$$

Please note the difference. Total load $W = w \times L$. In this formula 'lower case 'w' is used and hence the formula has L^4.

In this case, w is the intensity of a distributed load, and will be expressed in N/mm (or) kNm. Obviously, the two forms of expression represent the same function, because the total load W is equal to the intensity of load w multiplied by the length L.

Note: The maximum deflection occurs at the point marked "A" on each diagram, and has the value

$$C \times \frac{WL^3}{EI}$$

Table 5.1 Deflection and bending moment for some common loading conditions

<table>
<tr><th colspan="3">Simply Supported</th></tr>
<tr><th>Condition of Loading</th><th>Value of C</th><th>B.M. max</th></tr>
<tr><td>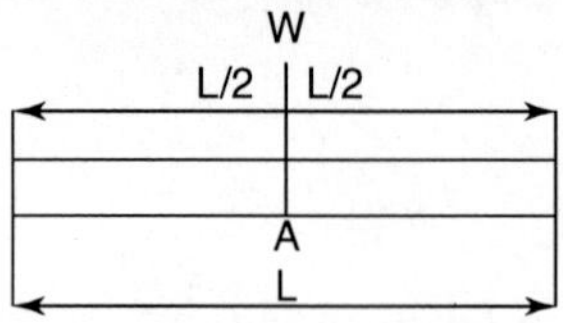</td><td>$\frac{1}{48}$</td><td>W.L/4</td></tr>
<tr><td>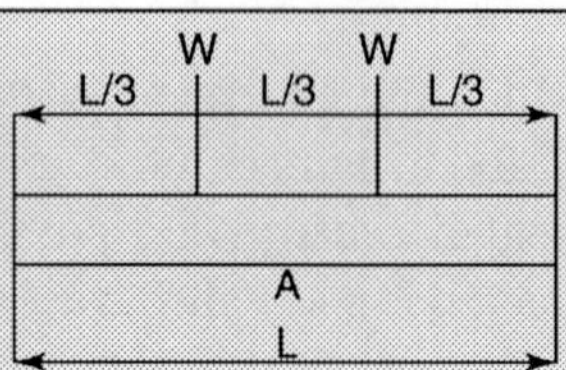</td><td>$\frac{23}{648}$</td><td>W.L/3</td></tr>
<tr><td>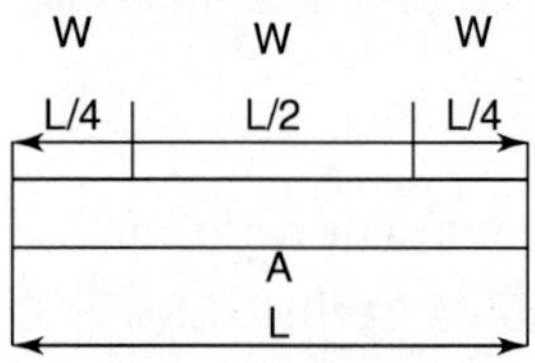</td><td>$\frac{11}{384}$</td><td>WL/4</td></tr>
<tr><td>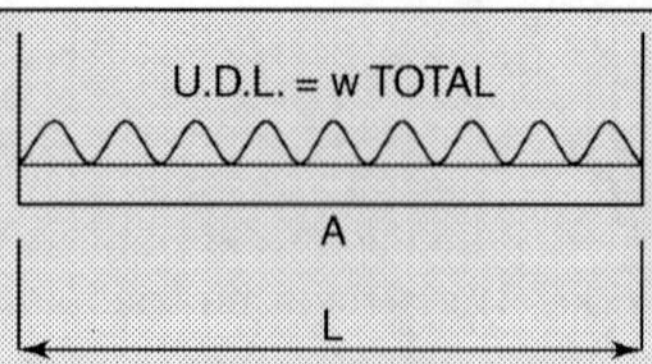</td><td>$\frac{5}{384}$</td><td>W.L/8</td></tr>
</table>

<table>
<tr><th colspan="3">Fixed Supported</th></tr>
<tr><th>Condition of Loading</th><th>Value of C</th><th>B.M. max</th></tr>
<tr><td>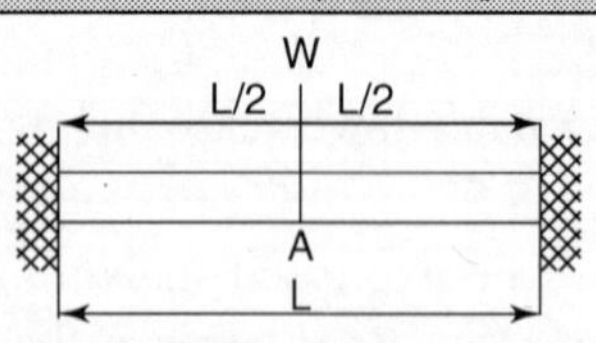</td><td>$\frac{1}{192}$</td><td>At ends = W.L/8
At A = W.L/8</td></tr>
<tr><td>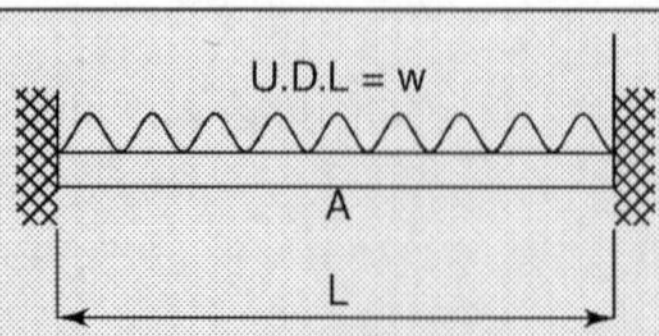</td><td>$\frac{1}{384}$</td><td>At ends = W.L/12
At A = W.L/24</td></tr>
</table>

Contd...

Contd...

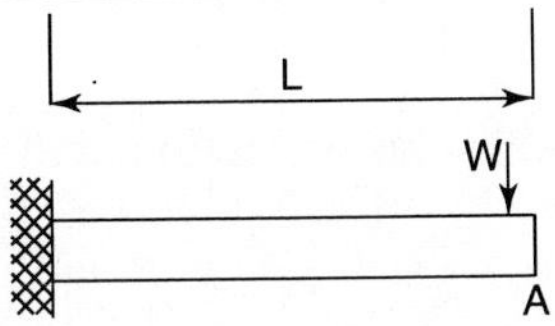	$\frac{1}{3}$	W.L
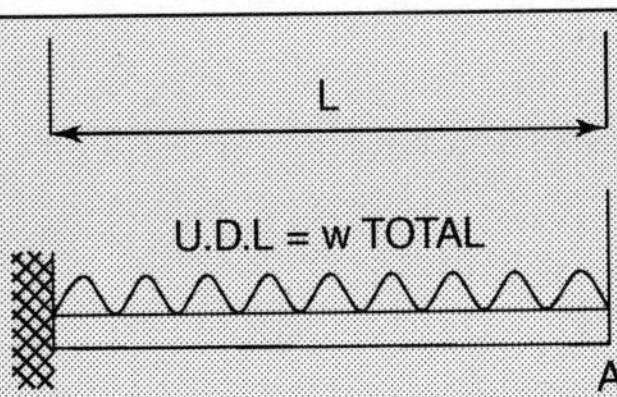	$\frac{1}{8}$	W.L/2

where E = Young's modulus and L is the span.

With respect to uniformly distributed load, total load $W = w \times L$ where w is the load per metre run of the beam.

The maximum value of BM also occurs at point A. For the fixed supports, except for the four fixed-ended examples, the maximum BM occurs at the fixed ends.

Simply Supported Beams

(a) Simply supported beam with central concentrated load

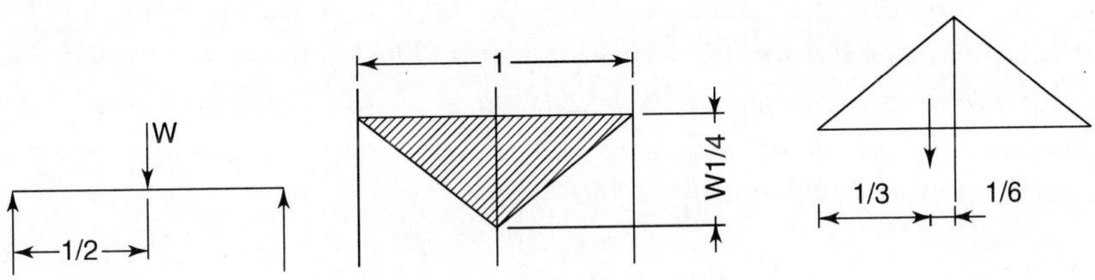

(a) (b) (c)

Fig. 5.1

$$\text{Half the area of moment diagram} = \frac{1}{2} \times \frac{L}{2} \times \frac{WL}{4} = \frac{WL^2}{16}$$

$$\text{Moment about centre} = \frac{2 \times WL^2}{16} \times \frac{1}{3} \times \frac{1}{2} \frac{WL^3}{48}$$

(Moment of area of moment diagram)

$$\text{Deflection} = \frac{WL^3}{48EI}$$

(b) Simply supported beam with uniformly distributed load

Deflection = $(5wl^4/384EI)$

If w is in kN/m, span is in m, E is in kN/sqm and I is in m^4; the deflection will be in metres.

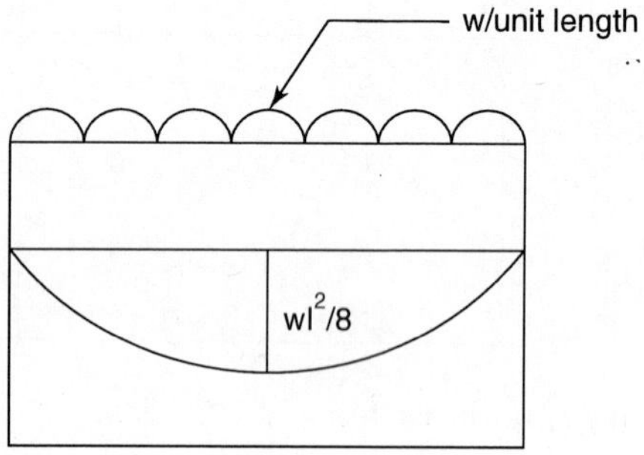

BMD

Fig. 5.2

5.4 DESIGN EXAMPLES

Example 5.1 *Design of platform beams-I*

There are four columns A1, B1, A2 and B2 supporting a platform. Floor grill having self weight of 350 N/m² is provided for the entire platform. The platform is to carry imposed load of 5 kN/m². All columns are connected by beams. Size of floor grill is 1 m × 1.2 m. Provide supporting members for floor grills and design them. Also design the column connecting beams. Material specification is IS:2062 and f_y = 250 MPa.

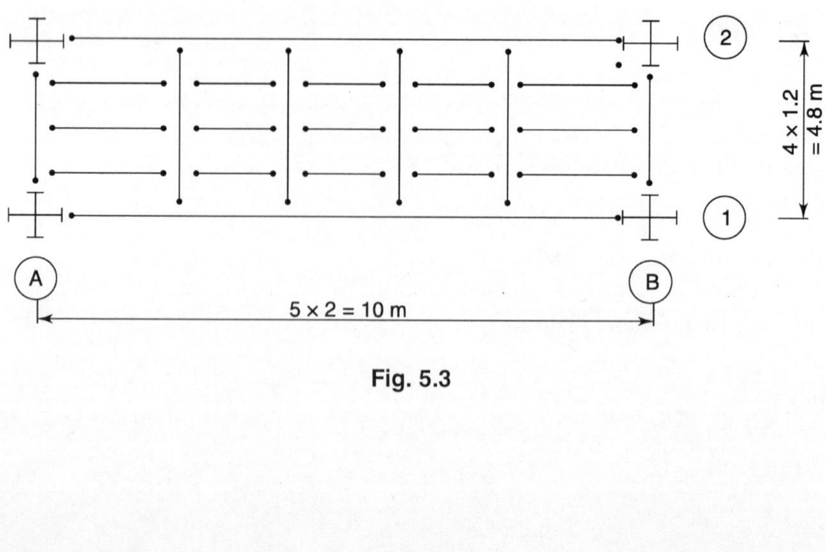

Fig. 5.3

Solution: If the imposed load is not given, one can take it from IS:875 (Part 2) depending on the functional requirements. Refer to Fig. 5.3 which gives the layout of beams and supporting members for the grills. All are simply supported at ends.

The size of grill is 1 m × 1.2 m. The supporting members are kept at 2 m between column connecting beams ($A1 - B1$) and ($A2 - B2$) since this is the shorter span compared to 10 m span.

Dead Load

Weight of grill	= 350 N/m^2
Imposed load	= 5 kN/m^2

Load on tie beam *PQ* (Figs. 5.4 & 5.5 of this book). Span of the beam $P - Q$ is 2 m. The adjuacent beams are spaced at 1.2 m on either side. This beam will carry the load per metre of span from hatched area as shown in Fig. 5.4.

Governing area	= 1 × 1.2 = 1.2 m^2
Due to grill 350 N/m^2 × 1.2	= 420 N/m
Due to self weight	= 50 N/m (Assumed)
Due to I.L 5 kN/m^2 × 1.2 m^2	= 6000 N/m
Total	= 6470 N/m

Total load on beam $P - Q$ = 6470 × 2 = 12940 N

Shear force = 12940/2 = 6470 N

Moment = (6470 × 2^2/8) = 3235 Nm

Factored moment = 1.5 × 3235 = 4852.5 Nm and Factored shear = 1.5 × 6470 = 9705 N

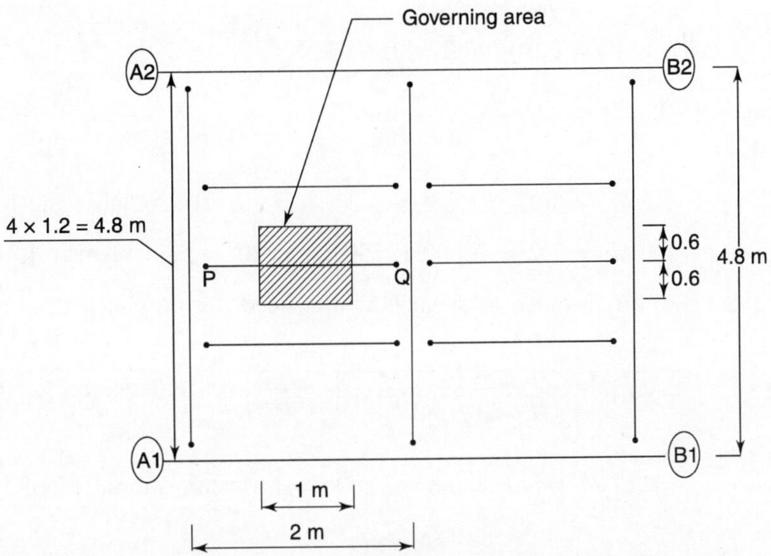

Fig. 5.4

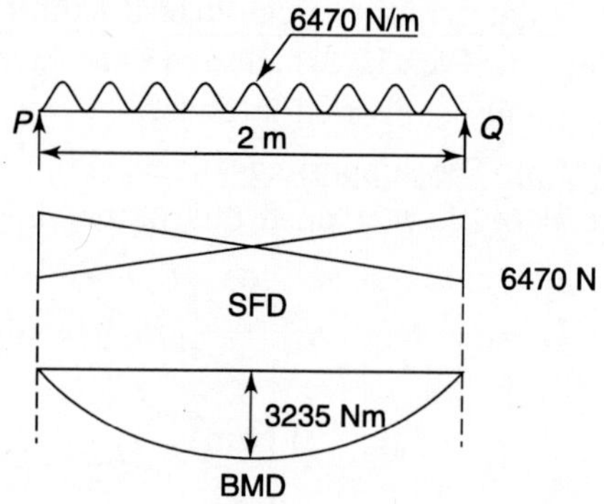

Fig. 5.5

$$M_d = 1.2 Z_e f_y/\gamma_{mo} \text{ (Clause 8.2.1.2 of IS:800)}$$

$$Z_{e\,(\text{required})} = M_d \times \gamma_{mo}/1.2 \times f_y = 4852.5 \times 1.1/1.2 \times 250 = 17792.5 \text{ mm}^3$$

Try ISMC 75

$$Z_{zz} = 20300 \text{ mm}^3$$

$$b/t_w = (40/7.3) = 5.48 < 9.4\,\varepsilon$$

$$d/t_w = [75 - 2 \times (7.3 + 6.5)]/4.4 = 9.86 < 42\,\varepsilon$$

$$\varepsilon = \sqrt{(250/f_y)} = \sqrt{(250/250)} = 1.0$$

Section is plastic (Table 2 of IS:800).

Shear capacity

$$V_d = \frac{A_v f_{yw}}{\gamma_{mo} \times \sqrt{3}} = (75 \times 4.4 \times 250/1.1 \times \sqrt{3} \times 1000) = 43.3 \text{ kN.}$$

$$0.6 \times V_d = 0.6 \times 43.3 = 25.98 \text{ kN.}$$

Factored shear = 9.705 kN < $0.6 \times V_d$

Hence, low shear.

$$M_d = 1.2 Z_e f_y/\gamma_{mo} \qquad \text{(Clause 8.2.1.2 of IS:800)}$$

$$M_d = 1.2 \times 20300 \times 250/1.1 \times 10^6 = 5.54 \text{ kNm} > 4.8525 \text{ kNm}$$

Since $d/t_w = 9.86 < 67\varepsilon$ resistance to shear buckling need not be checked.

$$f_{crb} = \frac{1.1 \times \pi^2 E}{(L_{LT}/r_y)^2}\left(1 + \frac{1}{20}\left(\frac{L_{LT}/ry}{h_f/t_f}\right)^2\right)^{0.5} \qquad \text{(Clause 8.2.2.1 of IS:800)}$$

$$L_{LT} = 0.85L = 0.85 \times 2000 = 1700 \text{ mm (Table 15 of IS:800)}$$

This channel will be fully welded with the beam spanning 4.8 m (or) provided with web cleats for connecting to the above beam. Hence, we can consider fully restrained against torsion. Top flange is partially restrained by the grills with clip (or) weld.

$$r_y = 12.1 \text{ mm}$$

$$\frac{h_f}{t_f} = \frac{75 - 73}{7.3} = 67.7/7.3 = 9.3$$

$$(L_{LT}/ry) = (1700/12.1) = 140.5.$$

$$f_{crb} = \frac{1.1 \times \pi^2 \times 2 \times 10^5}{140.5^2}\left(1 + \frac{1}{20}\left(\frac{140.5}{9.3}\right)^2\right)^{0.5}$$

(Clause 8.2.2.1 of IS:800)

$$= 110.08 \times 3.52 = 387.82 \text{ MPa}$$

$$M_{cr} = \beta_p \, z_p / f_{crb} = 1.0 \times 24.17 \times 387.82/10^3 = 9.37 \text{ kNm}$$

(Clause 8.2.2 of IS 800)

$$\lambda_{LT} = \sqrt{\beta_p z_p / M_{cr}} = \sqrt{1.0 \times 24.17 \times 10^3 / 9.37 \times 10^6} = 0.05 < 0.4$$

(Clause 8.2.2 (c) of IS 800).

Hence, torsional buckling need not be checked.

Design of Tie Beam of Span 4.8 m

Consider beam *R-S* (Refer Fig. 5.6)

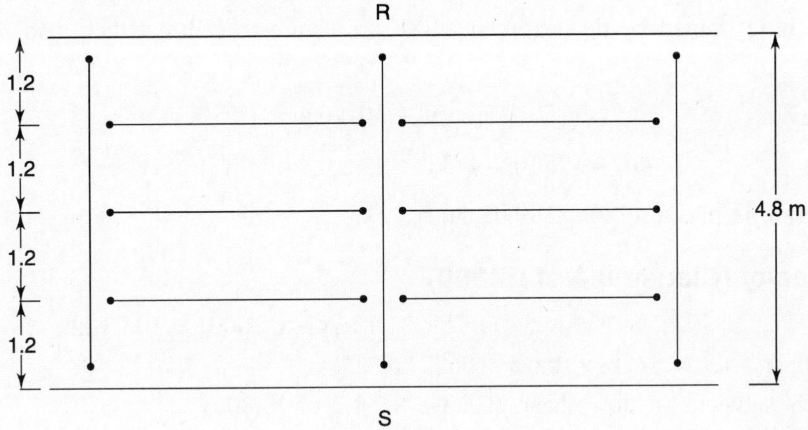

Fig. 5.6

Loads are from tie beam @ 1.2 m interval = Reaction of beam $P - Q$ = 6470 N = 6.47 kN

Self weight = 450 N. (Assumed)

$$\text{Shear force} = \frac{6.4 \times 3}{2} + \frac{0.45 \times 4.8}{2} = 10.785 \text{ kN}$$

Maximum moment = $10.785 \times 2.4 - 6.47 \times 1.2 - 0.45 \times 2.4^2/2 = 16.824$ kNm.

Factored moment = $1.5 \times 16.824 = 25.236$ kNm.

Factored shear = $1.5 \times 10.785 = 16.18$ kN

$$M_d = 1.2 Z_e \frac{f_y}{\gamma_{mo}}$$

$$Z_e \text{ required} = \frac{25.236 \times 10^6 \times 1.1}{1.2 \times 250} = 92532 \text{ mm}^3$$

Try ISMC 150 for which $Z_e = 103900$ mm^3

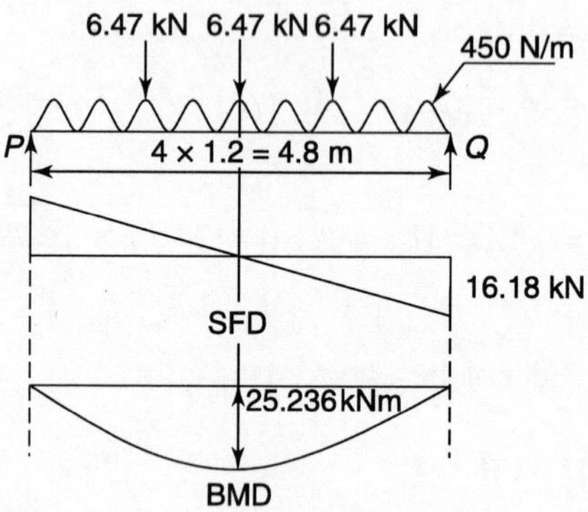

Fig. 5.7

This beam is restrained by tie beams at 1200 mm centre to centre which also supports the grills

$$d/t_w = (150 - 2 \times (9 + 10))/6.4 = 17.5 < 42\varepsilon$$

$$b/t_f = (75/9) = 8.33 < 9.4\varepsilon$$

Hence, section is plastic (Table 2 of IS:800)

Shear Capacity (Clause 8.4 of IS:800)

$V_d = (A_v \times f_{yw})/(\gamma_{mo} \times \sqrt{3}) = (150 \times 5.4) \times 250/(1.1 \times \sqrt{3} \times 1000) = 106.3$ kN

Factored shear force = $16.18 < 0.6 \times 106.3$ kN

The beam is subjected to low shear. (Clause 8.2.1.2 of IS:800)

$$\beta_b = 1.0$$

$$Z_p = 119.82 \times 10^3 \text{ mm}^3 \qquad \text{(Table 46 of IS:800)}$$

$M_d = \beta_b \times Z_p \times f_y/\gamma_{mo} = 1.0 \times 119.82 \times 10^3 \times 250/(1.1 \times 10^6) = 27.23 \text{ kNm} > 25.236$ kN.

Resistance to shear buckling need not be checked since d/t_w is less than 67ε (Clause 8.4.2.1 of IS:800)

Check for Deflection

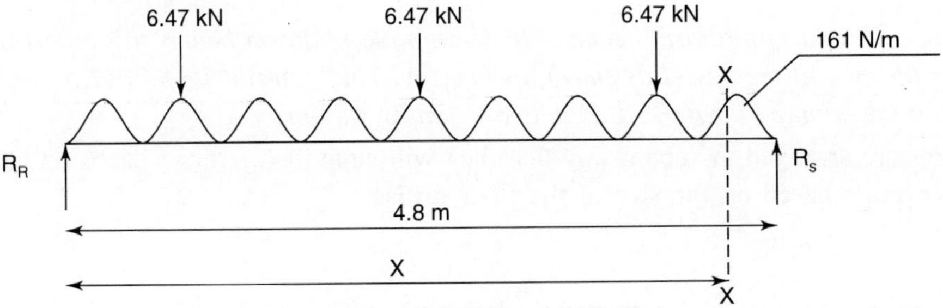

Fig. 5.8

Unfactored load is considered.

$$EI\,\frac{\partial^2 y}{\partial x^2} = -M$$

Considering section xx at x from R_R

$$M = R_R \times X - (0.161 \times x^2/2) - 6.47(x - 1.2) - 6.47 \times (x - 2.4) - 6.47 \times (x - 3.6)$$

$$EI\,\frac{\partial^2 y}{\partial x^2} = -\ 10.785 \times x + \left(0.161 \times \frac{x^2}{2}\right) + 6.47(x - 1.2) + 6.47 \times (x - 2.4) + 6.47 \times (x - 3.6)$$

$$EI\,\frac{\partial y}{\partial x} = -\,\frac{10.785X^2}{2} + (0.161 \times X^3/6) + (6.47(x - 1.2)^2/2) + (6.47 \times (x - 2.4)^2/2) + (6.47 \times (x - 3.6)^2/2) + C_1$$

$$Ely = -\ (10.785 \times x^3/6) + (0.016 \times x^4/24) + (6.47 \times (x - 1.2)^3/6) + (6.47 \times (x - 2.4)^3/6) + (6.47 \times (x - 3.6)^3/6) + C_1 \times x + C_2$$

When $x = 0$; $y = 0$ at $RC_2 = 0$

When $x = 4.8$ also $y = 0$.

$El_y = 0 = -(10.785 \times 4.8^3/6) + (0.161 \times 4.8^4/24) + (6.47 \times 3.6^3/6) + (6.47 \times 2.4^3/6) + (6.47 \times 1.2^3/6) + 4.8C_1$

$$C1 = 26.7$$

Therefore, δ at centre where x is equal to 2.4

$$= [-(10.785 \times 2.4^3/6) + (0.161 \times 2.4^4/24) + (6.47 \times 1.2^3/6) + 26.7]/EI$$

$$= 22.76/EI = 22.76/(2 \times 10^8 \times 779.4 \times 10^4)$$

$= 1.46 \times 10^{-14}$ m $= 1.46 \times 10^{-11}$mm which is far less.

Deflection limit as per Table 6 of IS:800 = span/240 = 4800/240 =20 mm

Hence, adequate.

Note: Similarly, the column connecting beams can be designed.

Example 5.2 *Design of platform beams – II. Design the platform beams to support floor grill for entire platform and the vessel as shown in Fig. 5.9. The material is IS:2062, f_y = 250 MPa. Imposed load on floor 2.5 kN/m². All beams are simply supported.*

The beams are arranged in such a way that they will support (a) vessel based on its location and (b) floor grills based on the size of the floor grills.

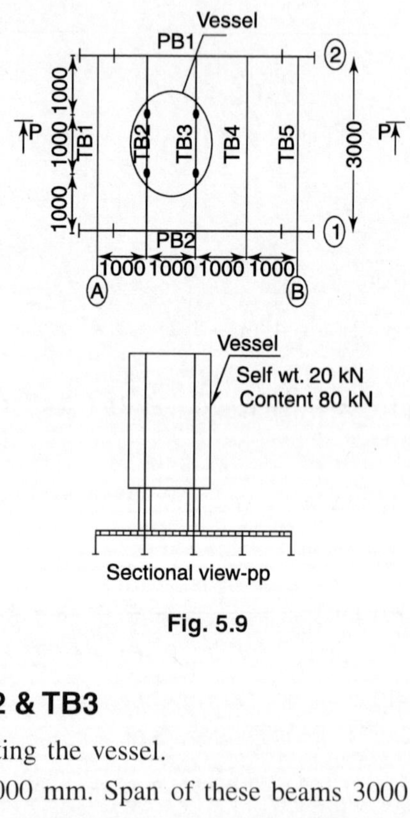

Fig. 5.9

Design of Tie Beam TB2 & TB3

These two beams are supporting the vessel.

Spacing of these beams 1000 mm. Span of these beams 3000 mm.

Imposed load/m = 1 × 1 × 2500	= 2500 N/m
Weight of floor grill = 1 × 1 × 400	= 400 N/m
Self weight of beam (assumed)	= 250 N/m
Total	= 3150 N/m = 3.15 kN/m

To be more precise, the floor grills will not be provided below the vessel and imposed load need not be considered for the plan area occupied by the vessel. However, as an approximation, we have considered the self weight of grills and the imposed load on this area also. The vessel is supported by 4 legs at a spacing of 1000 mm.

Load from Vessel

Total load = 80 + 20 = 100 kN

There are four supports.

Load on each support = 100/4 = 25 kN

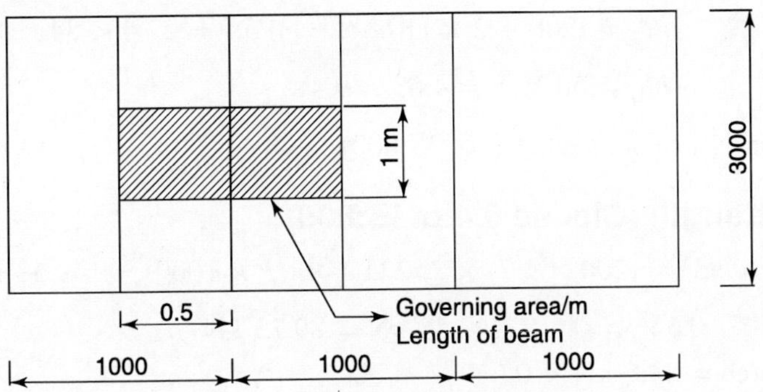

Fig. 5.10

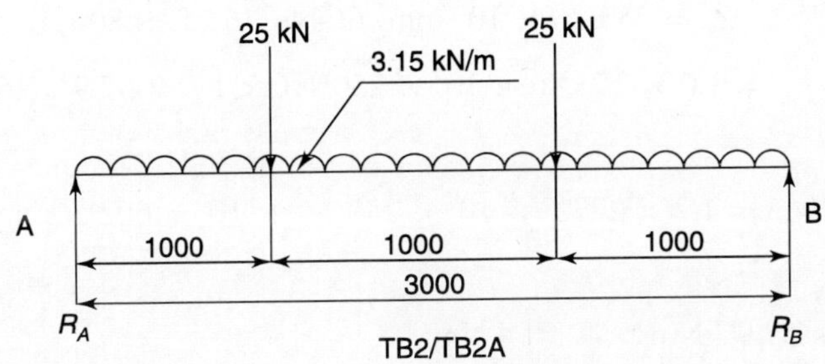

Fig. 5.11

To find reaction R_A and R_B

The load on beam is as shown in Fig. 5.11.

Taking moment about B,

$$R_A \times 3 = 25 \times 2 + 25 \times 1 + 3.15 \times 3^2/2$$

$$R_A = 29.725 \text{ kN}$$

R_B = total load – R_A = (2 × 25 + 3.15 × 3) – 29.725 = 29.725 kN

Factored shear force = 1.5 × 29.725 = 44.6 kN

Moment at load point = 29.725 × 1 – 3.15 × 1^2/2 = 28.15 kNm

Moment at centre = (29.725 × 1.5) – (3.15 × 1.5^2/2) – 25 × 0.5 = 28.54 kNm

$$M_{max} = 28.54 \text{ kNm} = 28.54 \times 10^6 \text{ Nmm}$$

Factored moment = 1.5 × 28.54 × 10^6 = 42.81 × 10^6 Nmm.

Z_e required = $M/(1.2 \times (f_y/1.1))$ = 42.81 × 10^6 × 1.1/(1.2 × 250) = 156.9 × 10^3 mm³

Try ISMB 200

$$Z_e = 223.5 \times 10^3 \text{ mm}^3$$

$$d/t_w = (200 - 2 \times (10.8 + 11))/5.7 = 27.4 < 84\varepsilon$$

$$b/t_f = 50/10.8 = 4.63 < 9.4\varepsilon$$

Hence, the section is plastic. (Table 2 of IS:800)

Design Shear Strength (Clause 8.4 of IS:800)

$V_d = (A_v \times f_{yw})/(\gamma_{mo} \times \sqrt{3}) = (200 \times 5.7) \times 250/(1.1 \times \sqrt{3} \times 1000) = 149.58$ kN

$$0.6\,V_d = 0.6 \times 149.58 \text{ kN} = 89.75 \text{ kN}$$

Factored shear force = 44.6 kN < 0.6 V_d

The beam is subjected to low shear (Clause 8.2.1.2 of IS:800)

Since the section is plastic β_b = 1.0

$$Z_p = 253.86 \times 10^3 \text{ mm}^3 \text{ (Table 46 of IS:800)}$$

$M_d = \beta_b \times Z_p \times f_y/\gamma_{mo} = 1.0 \times 253.86 \times 10^3 \times 250/1.1 \times 10^6 = 57.9$ kNm (Clause 9.2.1 of IS:800).

$M_d = 1.2 \times Z_e \times f_y/\gamma_{mo} = 1.2 \times 223.5 \times 10^3 \times 250/1.1 \times 10^6 = 60.95$ kNm (Clause 9.2.1 of IS:800).

Moment capacity is 57.9 kNm > 42.81 kNm.

The beam is laterally supported by ties to carry floor grill.

Resistance to shear buckling need not be checked since d/t_w is less than 67ε (Clause 8.4.2.1 of IS:800)

Design of Tie Beams TB4

This beam carries floor grill only and it will be designed for its own self weight, self weight of grill and imposed load.

Load/*m* = 3.15 kN/m

$$R_A = R_B = 3.15 \times 3/2 = 4.725 \text{ kN}$$

Maximum bending moment at centre = $wl^2/8$ = 3.15 × 3^2/8 = 3.54 kNm.

Factored moment = 1.5 × 3.54 = 5.31 kNm.

Z_e required = 5.31 × 10^6 × 1.1/1.2 × 250 = 19470 mm³

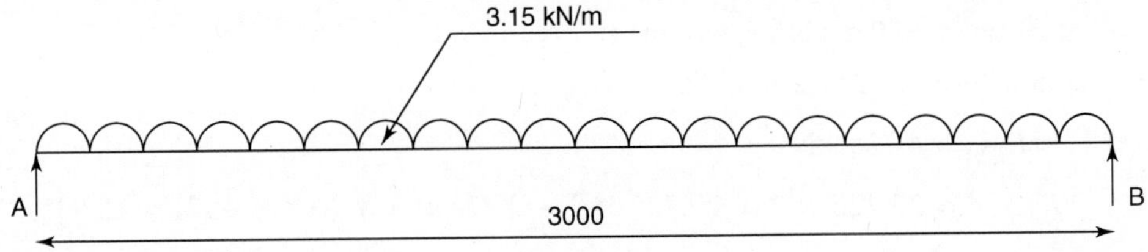

Fig. 5.12

Try ISMC 100; Z_e = 37300 mm³

$$b/t_f = 50/7.5 = 6.67 < 9.4\varepsilon$$

$$d/t_w = (100 - 2 \times (7.5 + 9))/4.7 = 14.25 < 84\varepsilon$$

Hence, section is plastic. (Table 2 of IS:800)

Factored shear force = 1.5 × 4.725 = 7.09 kN = V

$$V_d = (A_v \times f_{yw})/(\gamma_{mo} \times \sqrt{3}) \quad \text{(Clause 8.4 of IS:800)}$$

$$= (100 \times 4.7) \times 250/(1.1 \times \sqrt{3} \times 1000) = 61.67 \text{ kN}$$

$$0.6\,V_d = 0.6 \times 61.67 \text{ kN} = 37 \text{ kN} > \text{Factored shear force of } 7.09 \text{ kN}$$

The beam is subjected to low shear. (Clause 8.2.1.2 of IS:800)

Since the section is plastic β_b = 1.0

$$Z_p = 43.83 \times 10^3 \text{ mm}^3 \quad \text{(Table 46 of IS:800)}$$

$$M_d = \beta_b \times Z_p \times f_y/\gamma_{mo} = 1.0 \times 43.83 \times 10^3 \times 250/1.1 \times 10^6 = 9.96 \text{ kNm.}$$

$$M_d = 1.2 \times Z_e \times f_y/\gamma_{mo} = 1.2 \times 37.3 \times 10^3 \times 250/1.1 \times 10^6 = 10.2 \text{ kNm.}$$

Moment capacity is 9.96 kNm > 5.31 kNm

Hence for TB4; provide ISMC 100

TB1 & TB5 are subjected to half the load of TB4.

However, for uniformity, provide ISMC 100

Provide ISMC 100 for TB1, TB4, & TB5

Design of Primary Beam PB1 & PB2

Both the beams are symmetrically loaded.

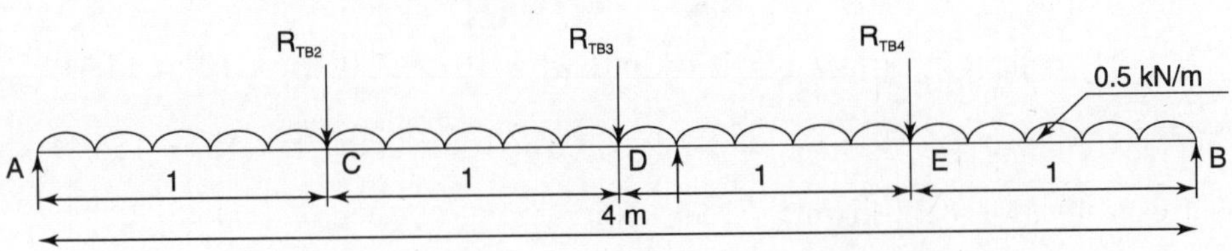

Fig. 5.13

Self weight of beam = 0.5 kN/m (assumed)

R_{TB2} = reaction from TB2 = 29.725 kN = R_{TB3}

$$R_{TB4} = 4.725 \text{ kN}$$

$$R_A \times 4 = 29.725 \times 3 + 29.725 \times 2 + 4.725 \times 1 + (0.5 \times 4^2/2) = 157.35 \text{ kNm}$$

$$R_A = 157.35/4 = 39.34 \text{ kN}$$

Factored shear force = 1.5 × 39.34 = 59.00 kN at *A*.

$$R_B = (2 \times 29.725 + 4.725 + (0.5 \times 4) - 39.34 = 24.84 \text{ kN}$$

Factored shear force at *B* = 37.25 kN.

Bending moment at $C = M_c = 39.34 \times 1 - 0.5 \times 1^2/2 = 39.09$ kNm

Bending moment at $D = M_d = 39.34 \times 2 - 29.725 \times 1 - 0.5 \times 2^2/2 = 47.96$ kNm

Bending moment at $E = M_e = 39.34 \times 3 - 29.725 \times 2 - 29.725 \times 1 - 0.5 \times 3^2/2 = 26.595$ kNm

$$M_{max} = 47.96 \text{ kNm.}$$

Factored moment = $1.5 \times 47.96 \times 10^6 = 71.94 \times 10^6$ kNm.

$$M_d = 1.2 \times Z_e \times f_y/\gamma_{mo}$$

Z_e required = Moment × 1.1/1.2 × f_y = $71.94 \times 10^6 \times 1.1/(1.2 \times 250) = 263.78 \times 10^3$ mm³

Try ISMB 300

$$Z_e = 573.6 \times 10^3 \text{ mm}^3$$

$$V_d = (A_v \times f_{yw})/(\gamma_{mo} \times \sqrt{3}) \qquad \text{(Clause 8.4 of IS:800)}$$

$$= (300 \times 7.5) \times 250/(1.1 \times \sqrt{3} \times 1000) = 295.24 \text{ kN}$$

$0.6\,V_d = 0.6 \times 295.24 = 177.14$ kN > Factored shear force of 59.00 kN.

The beam is subjected to low shear. (Clause 8.2.1.2 of IS:800)

$$b/t_f = 70/12.4 = 5.645 < 9.4\varepsilon$$

$$d/t_w = (300 - 2 \times (12.4 + 14))/7.5 = 32.96 < 84\varepsilon$$

Hence, the section is plastic. (Table 2 of IS:800)

Since the section is plastic $\beta_b = 1.0$

$$Z_p = 651.74 \times 10^3 \times 10^3 \text{ mm}^3 \qquad \text{(Table 46 of IS:800)}$$

$$M_d = \beta_b \times Z_p \times f_y/\gamma_{mo} = 1.0 \times 651.74 \times 10^3 \times 250/1.1 \times 10^6 = 148.12 \text{ kNm.}$$

$$M_d = 1.2 \times Z_e \times f_y/\gamma_{mo} = 1.2 \times 573.6 \times 10^3 \times 250/1.1 \times 10^6 = 156.44 \text{ kNm.}$$

Moment capacity is 148.12 kNm > 71.94 kNm

Hence, satisfactory.

Check for Deflection

Unfactored load is considered.

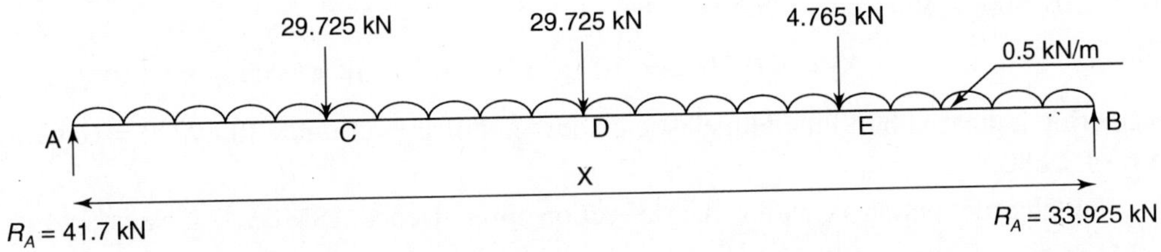

Fig. 5.14

$$EI\ d_y^2/d_x^2 = (-)\ M$$

Considering section x at a distance x from one of the supports

$$(-)\ M_X = 39.34 \times X - 29.725(X - 1) - 29.725(X - 2) - 4.765(X - 3) - 0.5x^2/2 = EI\ d_y^2/d_x^2$$

$$EI\ d_y/d_x = (-)\ 39.34\ x^2/2 + C_1 + 29.725(x - 1)^2/2 + 29.725(x - 2)^2/2 + 4.765\ (x - 3)^2/2 + 0.5\ x^3/6$$

$$EI_y = (-)\ 39.34\ {}_X{}^3/6 + C_{1X} + C_2 + 29.725(x - 1)^3/6 + 29.725(x - 2)^3/6 + 4.765(x - 3)^3/6 + 0.5 \times 4^4/24$$

at $\quad x = 0 \quad y = 0 \quad C_2 = 0$

at $\quad x = 4 \quad y = 0$

$$0 = (-)\ (39.34/6) \times 4^3 + 4C_1 + 29.725 \times 3^3/6 + 29.725 \times 2^3/6 + 4.765 \times 1^3/6 + 0.5 \times 4^4/24 = (-)\ 245.44 + 4C_1$$

$$C_1 = 245.4/4 = 61.36$$

$$EI_y = (-)\ 39.34\ {}_X{}^3/6 + 61.36\ {}_X + 29.725({}_X - 1)^3/6 + 29.725({}_X - 2)^3/6 + 4.765\ (x - 3)^3/6 + 0.5 \times 4^4/24$$

at $\quad x = 1 \quad y_c = ?$

$$EI_y = (-)\ 41.7/6 + 61.36 \times 1 + 0.5/24 = +\ 54.82$$

at $\quad x = 2 \quad y_d = ?$

$$EI_y = (-)\ 39.34 \times 2^3/6 + 61.36 \times 2 + 29.725 \times 1^3/6 + 0.5 \times 16/24 = 80.51 \text{ (maximum)}$$

at $\quad x = 3 \quad y_e = ?$

$EI_y = (-)\ 39.34 \times 3^3/6 + 61.36 \times 3 + 29.725 \times 2^3/6 + 29.725 \times 1^3/6 + 0.5 \times 3^4/24 = +\ 53.33$

E for steel = 2×10^5 N/mm^2 = 2×10^8 kN/m^2

I for ISMB 300 = 8604 cm^4 = 8.604×10^{-5} m^4

$$Y_d = [80.51/(2 \times 10^8 \times 8.604 \times 10^{-5})] \times 1000 = 4.68 \text{ mm}$$

Considering industrial building with elastic cladding, deflection limit = 4000/240 = 16.67 mm. (Table 6 of IS:800)

The calculated deflection is within the deflection limit. Hence, ISMB300 is satisfactory.

Note Moment capacity is 148.12 kNm. Actual bending moment = 79 kNm. Actual deflection is 4.75 mm which is within the deflection limit 16.67 mm. We can even try ISMB 250.

5.5 DESIGN OF CRANE GANTRY

1. Introduction

In the industries, crane gantry girders spanning between columns are used to support cranes (bridge girders). These crane gantry girders are supported by columns. The load on these crane gantry girders are from wheels running over these and hence they are subjected to moving loads with impact.

The hoisting capacity, weight of trolley, wheel base (spacing of wheels), spacing of column and minimum hook approach are the factors deciding the size of these crane gantry girders.

These crane gantry girders are subjected to lateral load which is taken care by the top flange. Hence, top flange is provided with channels with web kept on flange (or) angle (or) plate on the edge of flanges.

Columns below the crane gantry girders are called crane leg and the column supporting the roof above the support level of crane gantry is called the roof leg. The crane gantry girders are supported laterally at column locations.

Very high capacity cranes at heavy industries has laterally restrained crane gantry plate girders with web stiffeners.

Example 5.3 *Design of crane gantry.*

Design the crane gantry using rolled shapes of material specification IS:2062 having the yield stress f_y = 250 MPa for the following design data.

Top of crane leg: 8.5 *m*

Crane capacity: 2 × 200 *kN. Lifting capacity of each crane 200 kN.*

Column spacing = Crane gantry span = 6 m

Span of bridge girder = Centre to centre of rail = 20 M.

Wheel base = 3.4 m

Hook approach = 1.2 m

Weight of crane bridge = 200 kN

Vertical clearance = 3.4 m

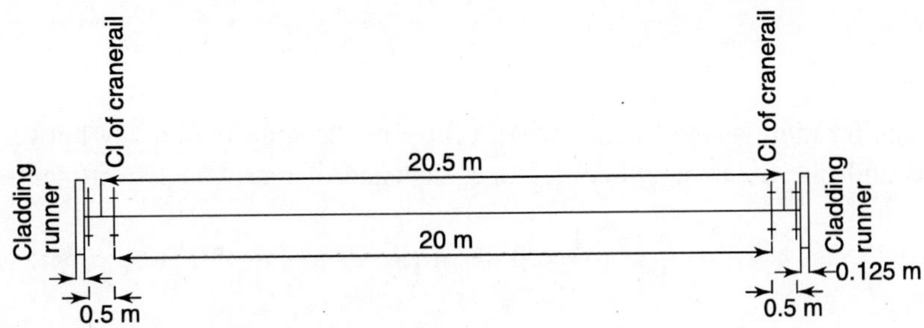

Fig. 5.15 *Plan showing the span of crane gantry*

Weight of trolley	= 60 kN (Refer Fig. 5.16)
Weight of rail	= 290 N/m
Height of rail	= 75 mm
W_t (moving load)	= Capacity + Weight of trolley = 200 + 60 = 260 kN
W_c	= Weight of Crane bridge girder = 200 kN.
Hook approach = a	= 1.2 m
Wheel base = b	= 3.4 m

Note: Values of a, b, weight of trolley and self weight of crane bridge girder can be obtained from manufacturer's mannual.

Wheel Load

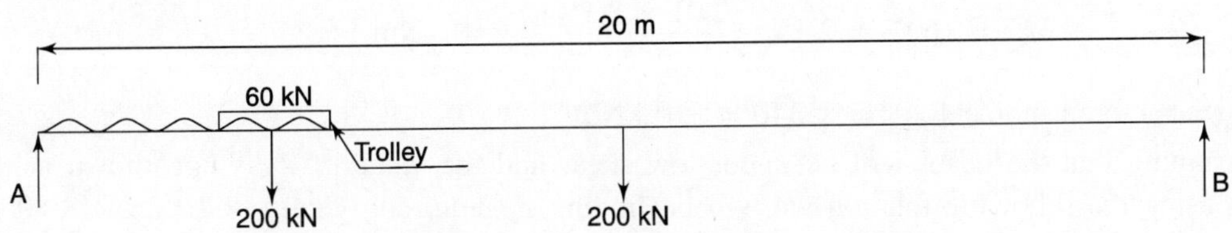

Fig. 5.16 *Trolley location for maximum wheel load*

Maximum static wheel load at A will occur, when the crane hook is at a minimum distance of 1.2 m (minimum hook approach)

$$R_A = \frac{W_c}{2} + W_t(L - a)/L = \frac{200}{2} + 260(20 - 1.2)/20 = 344.4 \text{ kN}$$

Load per wheel $= \dfrac{344.4}{2} = 172.2$ kN

Impact allowance 25%	= 43.0 kN
Design wheel load	= 172.2 + 43 = 215.2 kN

Wheel base = 3.4 m

Span of crane gantry = 6 m

Maximum bending moment will occur when centre of the span is midway between centre of gravity of loads and one (nearest) wheel. Hence, distance of one wheel from centre of span

$$= \frac{b}{4} = \frac{3.4}{4} = 0.85 \text{ m (Refer to Fig. 5.17)}$$

Let the self weight of girder $= \frac{2 \times 215.2}{250} \left(\text{From empirical relation } \frac{W}{250}\right) = 1.72$ kN/m

Weight of rail = 0.29 kN/m

Total = 2.01 kN/m

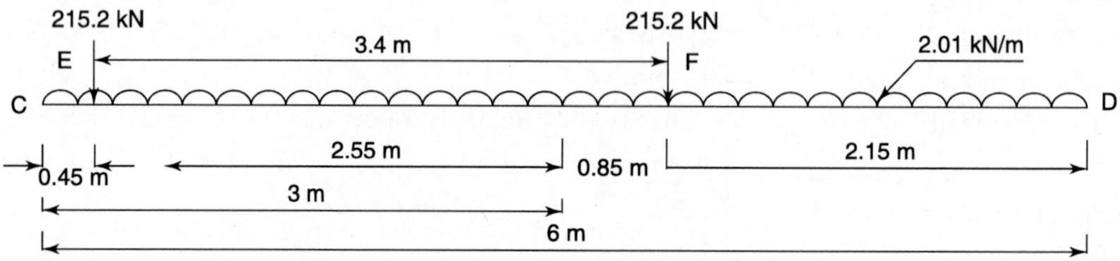

Fig. 5.17 *Wheel location for maximum bending moment*

$$R_D = \frac{215.2 \times (3.85 + 0.45)}{6} + \frac{2.01 \times 6}{2} = 160.3 \text{ kN and } R_C = 282.2 \text{ kN}$$

$$M_F = 160.3 \times 2.15 - \frac{2.01 \times 2.15^2}{2} = 340 \text{ kNm}$$

M_u = Factored moment = 1.5 × 340 = 510 kNm

Assuming that the beam will be under low shear and the value of Z_p is not known at this stage, using the following relation we will be finding Z_e required. (Clause 8.2.1.2 of IS:800).

$$M_d = 1.2 \times Z_e \times f_y/\gamma_{mo},$$

$$Z_e \text{ required} = \frac{M_d}{1.2 \times (f_y/\gamma_{mo})} = \frac{510 \times 10^6}{1.2 \times (250/1.1)} = 18{,}70{,}000 \text{ mm}^3$$

Choosing a trail section involves a number of iterations. With the help of standard steel tables, iterations can be reduced. Depth can be kept arround (span/12) and the width of top flange can be kept arround (span/20).

Approximate depth = 6000/12 = 500 mm.

Approximate width = 6000/20 = 300 mm.

Try ISMB 600 with ISMC 350 on flange

Properties of Section

ISMB 600	ISMC 350
A (mm^2) 15621	5366
t_f (mm) 20.8	13.5
t_w(mm) 12.0	8.1
W_f (mm) 210	100
I_{zz} (mm^4) 91813 × 10^4	10008 × 10^4
I_{yy} (mm^4) 2651 × 10^4	430.6 × 10^4
C_{yy} (mm)	24.4

To find C.G. taking moment of area about base

$$Y_t = \frac{15621 \times 300 + 5366 \times (600 + 8.1 - 24.4)}{15621 + 5366} = 372.54 \text{ mm}$$

$$Y_c = 600 + 8.1 - 372.54 = 235.56 \text{ mm}$$

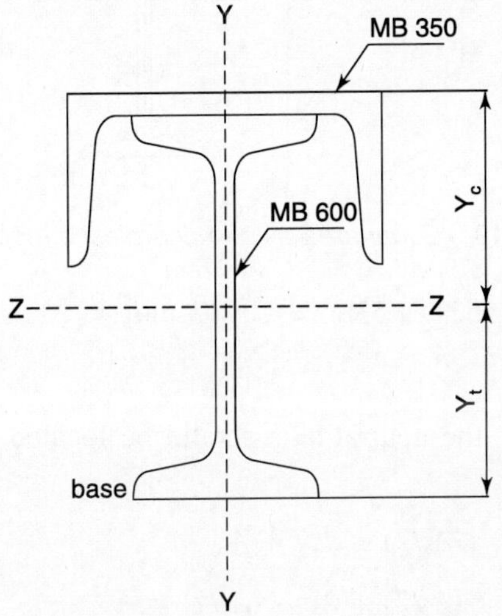

Fig. 5.18 *Neutral axis of crane gantry*

$$Y_t = 372.54 \text{ mm}$$

$$Y_c = 235.56 \text{ mm}$$

Distance between C.G. of built-up section and that of I section alone

$$= 372.54 - 300 = 72.54 \text{ mm} = 7.254 \text{ cm.}$$

Distance between C.G. of built-up section and that of channel alone = $\overline{Y_c}$

$$\overline{Y_c} = 235.56 - 24.40 = 211.16 \text{ mm} = 21.116 \text{ cm.}$$

$$I_{zz} = [(91813 + 156.21 \times 7.254^2) + (430.6 + 53.66 \times 21.116^2)] = 124390 \text{ cm}^4$$

$$I_{yy} = 2651 + 10008 = 12659 \text{ cm}^4$$

$$Z_{zzc} = \frac{124390}{23.556} = 5280.6 \text{ cm}^3 = 5.28 \times 10^6 \text{ mm}^3$$

$$I_{tf} \text{ of tension flange} = \frac{20.8 \times 210^3}{12} = 1605 \times 10^4 \text{ mm}^4$$

$$I_{cf} \text{ for compression flange} = \frac{2651}{2} + 10008 = 11333.5 \times 10^4 \text{ mm}^4$$

$$Z_{y1} \text{ for top flange only} = 11333.5 \times 10^4/175 = 647.63 \times 10^3 \text{ mm}^3$$

Plastic Modulus

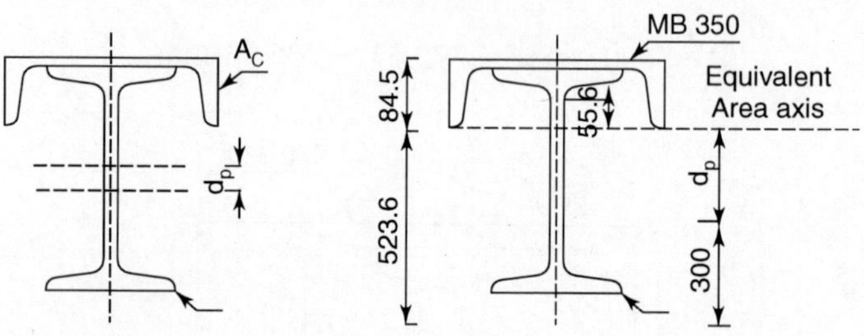

Fig. 5.19 *Equal area axis to find plastic modulus*

Total Area $A = A_I + A_C = 15621 + 5366 = 20987 \text{ mm}^2$

$A/2 = 10493.5 \text{ mm}^2$

For plastic section modulus, the neutral axis should be located at a point which divides the total area into two equal parts

$$(A_I/2) + d_p\, t = (A_I/2) - d_p\, t + A_c$$

$$d_p\, t + d_p\, t = (A_I/2) + Ac - (A_I/2) = A_c$$

$$d_p = \frac{A_c}{2t} = \frac{5366}{2 \times 12} = 223.6 \text{ mm}$$

Plastic section modulus below the axis dividing area in two equal parts.

$$\Sigma A\bar{y} = 20.8 \times 210(523.6 - 20.8/2) + (523.6 - 20.8) \times 12 \times (523.6 - 20.8)/2 = 3758.5 \times 10^3 \text{ mm}^3$$

Above this axis.

$$\Sigma A\bar{y} = 5366(84.5 - 24.4) + 210 \times 20.8(84.5 - 8.1 - 10.4) + 55.6 \times 12 \times 55.6/2 = 629.33 \times 10^3 \text{ mm}^3$$

$$Z_{pz} = 3758.5 \times 10^3 + 629.33 \times 10^3 = 4388 \times 10^3 \text{ mm}^3$$

For the top flange only

$$Z_{py} = 20.8 \times 210^2/4 + (350 - 2 \times 13.5)^2 \times 8.1/4 + 2 \times 100 \times 13.5\ (175 - 13.5/2)$$

$$= 894.86 \times 10^3 \text{ mm}^3$$

Moment Capacity

$$\varepsilon = \sqrt{\frac{250}{250}} = 1.0$$

$$\frac{b}{t_f} \text{ of } I \text{ section} = \left[\frac{210}{2}\right]/20.8 = 5.04 < 9.4\varepsilon$$

$$\frac{b}{t_f} \text{ of channel} = [100]/13.5 = 7.4 < 9.4\ \varepsilon$$

$$\frac{d}{t_w} \text{ of } I \text{ section} = [600 - 2 \times (20.8 + 20)]/12 = (518.4/12) = 43.2 < 84\ \varepsilon$$

Section is plastic. (Table 2 of IS:800)

Local Moment Capacity

Assuming that the beam will be in low shear,

$$M_d = 1.2 \times Z_{zzc} \times f_y/1.1 = \frac{1.2 \times 5.28 \times 10^6 \times 250}{1.1 \times 10^6} = 1440 \text{ kNm}$$

(Clause 8.2.1.2 of IS:800).

Since the section is plastic $\beta_b = 1.0$

$$Z_p = 4388 \times 10^3 \text{ mm}^3$$

$$M_d = \beta_b \times Z_p \times f_y/\gamma_{mo} = 1.0 \times 4388 \times 10^3 \times 250/1.1 \times 10^6 = 997.3 \text{ kNm.}$$

Value of M_{dz} is lesser of the two values 1440 kNm and 997.3 kNm.

Hence, M_{dz} = 997.3 kNm.

To calculate M_{dy}

$$M_{dy} = \frac{f_y}{1.1} \times Z_{py} \text{ (top flange)} = \frac{250}{1.1} \times \frac{894.86 \times 10^3}{10^6} = 203.38 \text{ kNm}$$

$$1.2 \times Z_{yt}\frac{f_y}{1.1} = 1.2 \times \frac{647.63 \times 10^3}{10^6} \times \frac{250}{1.1} = 176.63 \text{ kNm}$$

Vaue of M_{dy} is lesser of the two values 203.38 kNm and 176.63 kNm.

Hence, M_{dy} = 176.63 kNm and as already calculated, M_{dz} = 997.3 kNm.

Moment Due to Horizontal Load

$$F_H = 10\% \text{ of } (60 + 200) = 26 \text{ kN}$$

$$W_H = \frac{26}{2} = 13 \text{ kN per wheel}$$

$$R_{HD} = 13(3.85 + 0.45)/6 = 9.32 \text{ kN}$$

$$M_y \text{ at } F = 9.32 \times 2.15 = 20 \text{ kNm}$$

$$M_{uy} = 1.5 \times 20 = 30 \text{ kNm}$$

Combined Capacity

$$\frac{510}{997.3} + \frac{30}{176.63} = 0.68 < 1.0$$

The section is adequate. Since it is far less than 1.0; we can try ISMB 500 in place of ISMB 600 ISMB 550 is not available in the market.

Check for Buckling Resistance

The design bending strength of laterally unsupported beam as governed by lateral torsional buckling = design buckling strength = $M_d = \beta_b Z_p f_{bd}$ (Clause 8.2.2 of IS:800)

$$\beta_b = 1.0 \text{ since section is plastic}$$

$$h_f = c/c \text{ dist. between flanges} = 600 + 8.1 - \frac{20.8 + 8.1 + 20.8}{2} = 583.25 \text{ mm.}$$

$L_{LT} = 1.0\ L = 6000$ mm ((Table 15 of IS:800 – For the condition, warping not restrained in both flanges).

$$t_f = 20.8 + 8.1 = 28.9 \text{ mm}$$

$$r_y = \sqrt{\frac{I_{yy}}{A}} = \sqrt{\frac{12659 \times 10^4}{(15621 + 5366)}} = \sqrt{\frac{12659 \times 10^4}{20987}} = 77.66 \text{ mm}$$

Elastic lateral torsional buckling moment (Clause 8.2.2.1 of IS:800).

The crane gantry girder is simply supported, symmetrical about minor axis and bending about major axis. Hence, elastic critical moment for lateral torsional buckling is to be calculated as per Annex-E of IS:800/2007. Here the following simplified equation is used as in Clause 8.2.2.1 of IS:800.

$$M_{cr} = \frac{\pi^2 EI_y h_f}{2 \times L_{LT}^2}\left(1 + \frac{1}{20}\left(\frac{L_{LT}/r_y}{h_f/t_f}\right)^2\right)^{0.5}$$

$$M_{cr} = \frac{\pi^2 \times 2 \times 10^5 \times 12659 \times 10^4 \times 583.25}{2 \times 6000^2}\left(1 + \frac{1}{20}\left(\frac{6000/77.66}{583.25/28.9}\right)^2\right)^{0.5}$$

$$= 2025.82 \times [1.32] = 2666.7 \text{ kNm}$$

Slenderness ratio

$$\lambda_{LTZ} = \sqrt{\frac{\beta_b Z_{pz} f_y}{M_{cr}}} = \sqrt{\frac{1 \times 4388 \times 10^3 \times 250}{2666.7 \times 10^6}} = 0.64 < 0.4$$

Hence, as per Clause 8.2.2 (c) of IS:800, lateral torsional buckling is to be checked.

Refer Fig. 2 of IS:800. Such section comes under built-up section. Buckling class of cross section is c (Table 10 of IS:800).

$\alpha_{LT} = 0.21$ for rolled steel sections. (Clause 8.2.2 of IS:800)

$$\Phi_{LTZ} = 0.5[1 + \alpha_{LT}(\gamma_{LTZ} - 0.2) + \gamma_{LTZ}^2]$$

$$= 0.5[1 + 0.21(0.64 - 0.2) + 0.64^2] = 0.751$$

$$\chi_{LTZ} = \frac{1}{\varnothing_{LTZ} + (\varnothing_{LTZ}^2 - \lambda_{LTZ}^2)^{0.5}} \leq 1.0$$

$$= \frac{1}{0.751 + (0.751^2 - 0.64^2)^{0.5}} = 0.874 < 1.0$$

Design bending compressive stress

$$f_{bd} = \frac{f_y \chi_{LT}}{\gamma_{mo}} = \frac{250 \times 0.874}{1.1} = 198.6 \text{ N/mm}^2$$

$$M_{dz} = \beta_b Z_{pz} f_{bd} = \frac{1.0 \times 198.6 \times 4388 \times 10^3}{10^6}$$

$$= 871.6 \text{ kNm} > (510 + 30) \text{ kNm}$$

Hence, satisfactory.

For Top Flange Only

$$M_{dy} = \frac{f_y}{1.1} \times Z_{yt} = \frac{250}{1.1} \times 647.63 \times 10^3/10^6 = 147.2 \text{ kNm}$$

$$\frac{510}{871.6} + \frac{30}{147.2} = 0.789 < 1.0$$

Hence, satisfactory.

Check for Shear

Maximum shear (reaction) will be there when one of the two wheels are on the support. Refer Fig. 5.20. At this condition, the factored shear force.

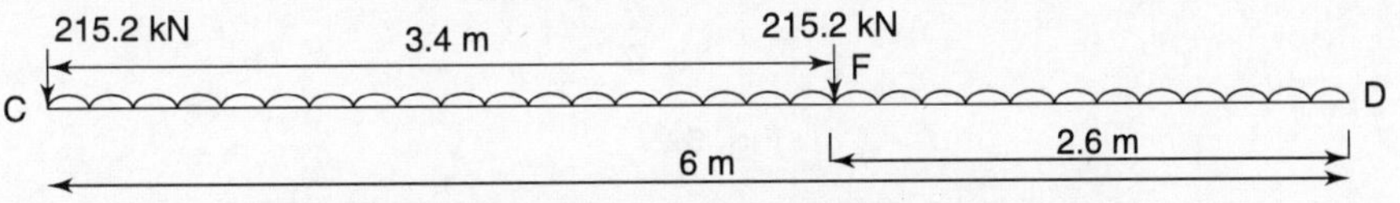

Fig. 5.20 *Location of wheels for maximum shear*

$$V = 1.5\left[215.2 + \frac{215.2 \times 2.6}{6} + \frac{2.01 \times 6}{2}\right] = 471.725 \text{ kN}$$

Design strength = $V_d = \dfrac{V_n}{\gamma_{mo}} = \dfrac{A_v f_{yw}}{\gamma_{mo}\sqrt{3}}$ (Clause 8.4.1 of IS:800)

$$= [600 \times 12 \times 250/1.1 \times \sqrt{3}] = 944.8 \text{ kN} > 471.725 \text{ kN}$$

Hence, satisfactory.

$$0.6V_d = 0.6 \times 944.8 = 566.8 \text{ kN} > 471.725 \text{ kN.}$$

The member is subjected to low shear. (Clause 8.2.1.2 of IS:800)

Weld Design (Between Flange of *I* Section and Web of Channel)

Required capacity of weld = $\dfrac{VAY}{I_z}$

$$Y = 235.56 - 24.4 = 211.16 \text{ mm}$$

$$A = 5366 \text{ mm}^2 \quad V = 471.725 \text{ kN}$$

$$I_z = 124390 \times 10^4 \text{ mm}^4$$

$$q = \frac{471.725 \times 10^3 \times 5366 \times 211.16}{124390 \times 10^4} = 430 \text{ N/mm}$$

q per side = 430/2 = 215 N/mm

This shear is taken by weld between the channel and *I* section.

Providing 4 mm continuous weld; f_{wd} = 189 MPa (Clause 10.5.7.1.1of IS:800).

Throat thickness of weld = (0.707 × 4) mm.

Strength/mm = (0.707 × 4 × 189) N = 534.5 N/mm > 215 N/mm

Provide 4 mm weld on both sides continuously which is less than the thickness of thinner part connected, (web of the channel).

Lateral Shear Force

$$\text{SF} = 13 + 13 \times \frac{2.6}{6} = 18.63 \text{ kN} \times 1.5 = 27.95 \text{ kN}$$

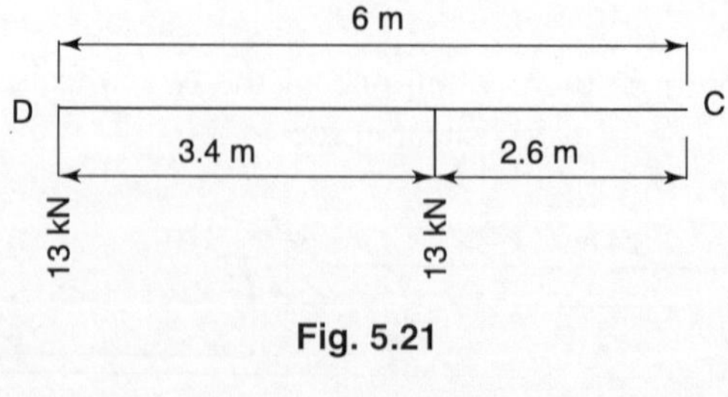

Fig. 5.21

Design Strength

$$V_d = \frac{A_v f_{vw}}{\gamma_{mo}\sqrt{3}} \qquad \text{(Clause 8.4.1 of IS:800)}$$

Shear area = Area of web of channel + Area of flange of I section.

Area of flange = A_v = 350 × 8.1 + 20.8 × 210 = 7203 mm^2

$$V_d = \frac{7203 \times 250}{1.1\sqrt{3}} = 945 \text{ kN} > 27.95 \text{ kN}$$

Hence, satisfactory.

Buckling of Web

The web of the girder shall be checked for buckling if $d/t_w > 67\varepsilon$ for web without stiffeners (Clause 8.4.2.1 of IS:800)

$$\frac{d}{t_w} = 43.2 < 67\varepsilon$$

Web buckling need not be checked. However, the calculation is given below for knowing the procedure.

Stiff bearing length (Clause 8.7.1.3 of IS:800).

Refer Fig. 5.22. The angle of dispersion is considered 45°.

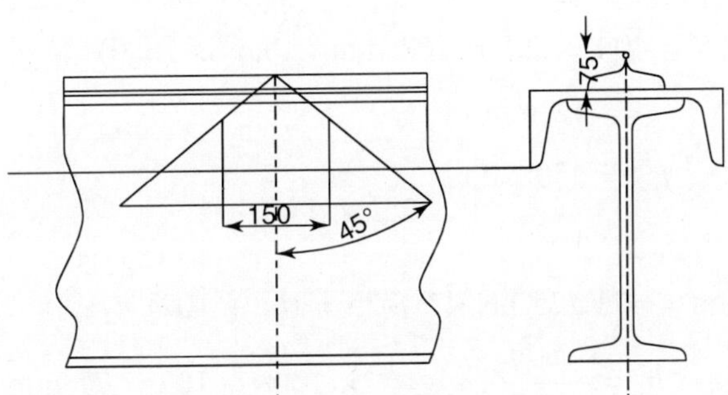

Fig. 5.22 *Buckling of web*

The wheel will have point contact at top of rail.

b_1 = width of dispersion at top of girder = twice the height of rail = 2 × 75 = 150 mm.

n_1 = width of dispersion at middle of girder = twice the height from middle of girder to top of girder.

n_1 = 2(300 + 8.1) = 616.2 mm

d = depth of web clearing root

= 600 – 2 × 20.8 – 2 × 20 = 600 – 61.6 = 518.4 mm

web slenderness

$$\lambda = \frac{2.5d}{t} = \frac{2.5 \times 518.4}{12} = 108$$

$$\frac{h}{b_f} = \frac{600 + 8.1}{350} = 1.737 \text{ (with respect to top)}$$

$$= \frac{600 + 8.1}{210} = 2.89 \text{ (with respect to bottom)}$$

In both the cases $h/b_f > 1.2$ and $t_f < 40$ mm. Hence, buckling class about *yy* axis = *b*.

For channel and built up section, buckling class is *c* (Table 10 of IS:800)

Hence, f_{cd} as per Table 9(c) for $\lambda = 108$,

$$f_{cd} = 107 - \frac{8}{10}(107 - 94.6) = 97 \text{ MPa.}$$

Strength against buckling = $(b_1 + n_1)\, t \times f_{cd}/1000 = (150 + 616.2) \times 12 \times 97/1000 = 891.85$ kN > wheel load capacity exceeds the load. Hence, satisfactory.

Web Bearing

Bearing stiffeners are to be provided when the reaction > F_w

where $F_w = (b_1 + n_2)\, t_w f_{yw}/\gamma_{mo}$ (Clause 8.7.4 of IS:800)

b_1 = stiff bearing length

$n_2 = (20.8 + 20) \times 2.5 = 40.8 \times 2.5 = 102$ mm (slope is 1:2.5)

$$t_w = 12 \text{ mm}$$

$$f_{yw} = 250 \text{ MPa}$$

$$\gamma_{mo} = 1.1$$

Equating F_w = reaction = 471.725 kN 471.725 = $(b_1 + 102) \times 12 \times 250/1.1$

$$b_1 = \frac{471.725 \times 1.1}{12 \times 250} \times 1000 - 102 = 71 \text{ mm}$$

Minimum bearing required = 71 mm

Provide bearing more than 71 mm

Check for Deflection

Wheel load (unfactored) = 172.2 kN & without impact

For maximum deflection to occur, the wheels are at equidistant from the mid span.

Deflection at mid span

$$\delta = WL^3\,[(3a/4L) - (a^3/L^3)]/(6\,EI)$$

where
$$a = \frac{L - C}{2} = \frac{6000 - 3400}{2} = 1300 \text{ mm}$$

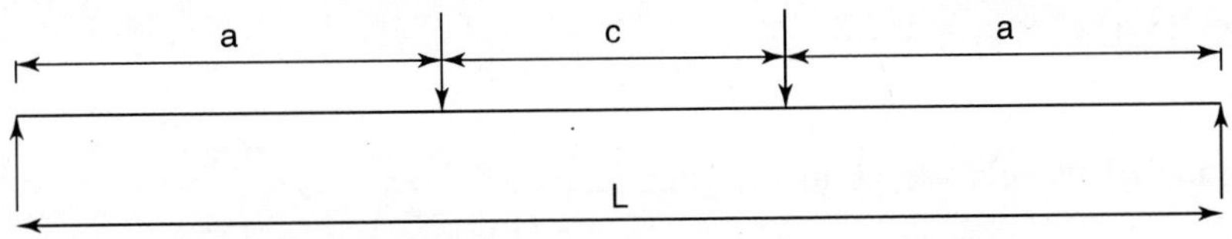

Fig. 5.23 *Location of wheels for maximum deflection*

$$I_{zz} = 124390 \times 10^4 \text{ mm}^4$$

$$\delta = \frac{172.2 \times 1000 \times 6000^3}{6 \times 2 \times 10^5 \times 124390 \times 10^4} [(3 \times 1300/4 \times 6000) - (1300^3/6000^3)]$$

$$= 24.92 \times 0.1523$$

$$= 3.8 \text{ mm} < 6000/750 = 8 \text{ mm}$$

The calculated deflection 3.8 mm is lesser than the deflection limit of 8 mm. Hence, satisfactory. (Table 6 of IS:800)

When Two Cranes are Working

Normally, the projection of end carriage from centre of wheel will be 500 mm on each side. Hence, the centre to centre of wheel of two cranes will be = 2 × 0.5 = 1 m (Refer to Fig. 5.24).

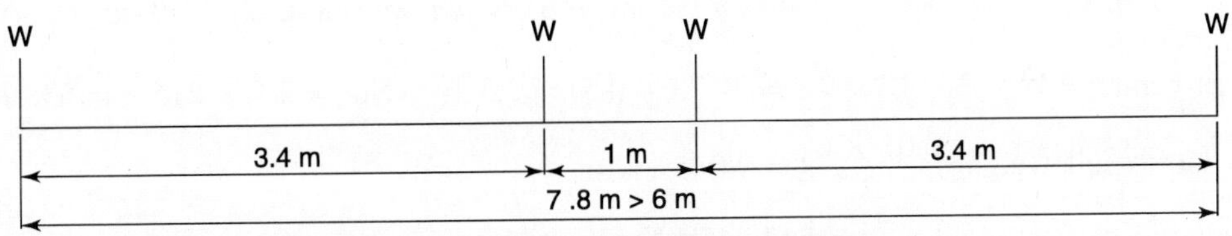

Fig. 5.24 *When two cranes are working*

All the wheels cannot be on one span since wheel base plus the gap exceed the span. Considering three wheels on one span.

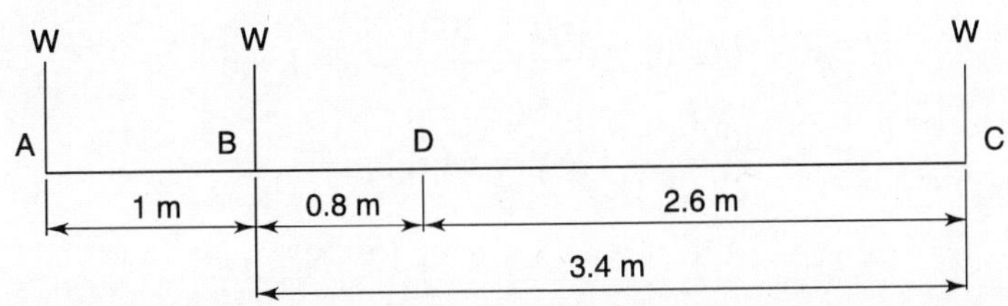

Fig. 5.25 *When three wheels of two cranes are on the crane gantry*

Taking moment about *C*

$$\frac{W \times 4.4 + W \times 3.4}{3W} = \frac{7.8}{3} = 2.6 \text{ m}$$

CG is at *D*

The disposition of wheels are as in the Fig. 5.26.

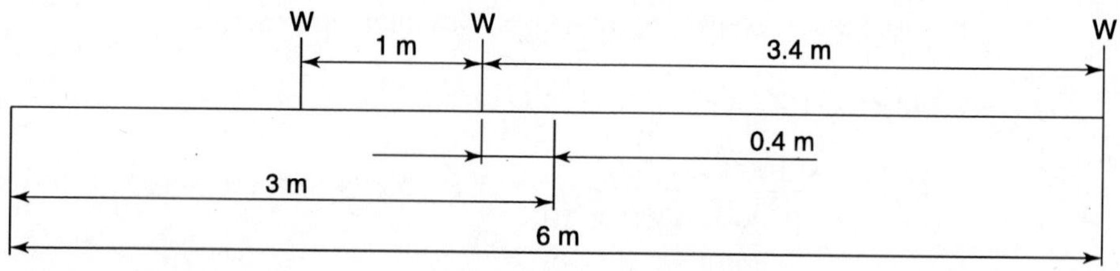

Fig. 5.26 *Maximum bending moment when two cranes are there*

In such a case the disposition as in Fig. 5.27 will give the maximum BM.

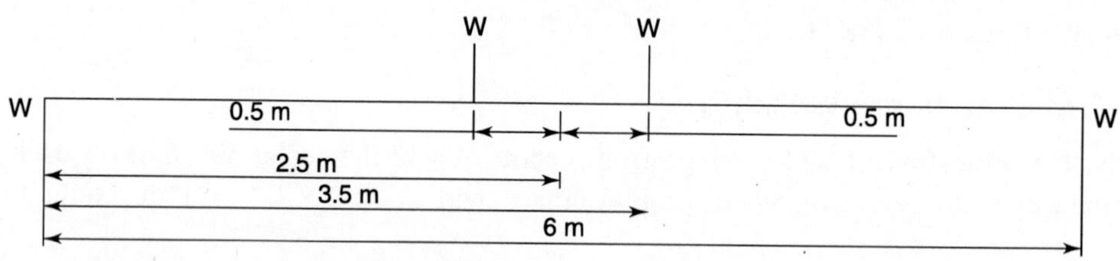

Fig. 5.27 *Location of wheels for maximum BM when two cranes are working*

BM at centre = $W \times 3 - W \times 0.5 = W(3 - 0.5) = 2.5\ W$. kNm = $2.5 \times 215.2 = 538$ kNm

BM due to self weight $\dfrac{2.01 \times 6^2}{8} = 9.045$ kNm

$$= \mathbf{547.045\ kNm}$$

Factored moment = $547.045 \times 1.5 = 820.6$ kNm < M_{dz} 997.3 kNm (local moment capacity)

Moment Due to Horizontal Load

$$F_H = 10\% \text{ of } 2\ \frac{(60 + 200)}{2} = 26 \text{ kN}$$

$$W_H = \frac{26}{2} = 13 \text{ kN per wheel}$$

$$R_{HD} = 13 \times (2.5 + 3.5)/6 = 13 \text{ kN.}$$

$$M_{dy} = 13 \times 2.5 = 32.5 \text{ kNm or } (13 \times 3) - (13 \times 0.5) = 32.5 \text{ kNm}$$

Factored $M_{dy} = 1.5 \times 32.5 = 48.75$ kNm.

Combined Capacity

$$\frac{820.6}{997.3} + \frac{48.6}{176.63} = 0.82 + 0.27 = 1.09 > 1.0$$

Since the increase is about 9% which is marginal, this can be allowed.

5.6 DESIGN OF WELDED PLATE GIRDERS

Introduction

A plate girder is a built up *I* beam using plates. These are used in highway bridges, railway bridges and for long span with heavy load. The designer has got many choices in selecting thickness of web and flange, width of flange and depth of web. Hybrid girders using different material specifications for web and flanges can also be designed. Though the connections can be bolted (or) welded, mostly welded plate girders are used in India. Deep plate girders are used for hanging boiler components in higher capacity top supported boilers.

Flange plates can be curtailed where not required and the profile can follow the BMD to make it economical. Failure may be due to yielding of tension flange, buckling of compression flange and stability of thin and deep web plate.

Due to deep and thin webs, knowledge in plate stability theory will help.

The elements of plate girders are:

(a) Web plate
(b) Flange plates
(c) Intermediate vertical stiffeners
(d) Vertical stiffeners at concentrated load points
(e) Bearing stiffeners
(f) Horizontal stiffeners
(g) Web splice
(h) Flange splice.

Elements mentioned as (a) and (b) above forms the plate girder.

Other elements are based on requirement.

Design Steps

(1) Assume self weight (w = (W/250) kN/m, where W is the total load carried)
(2) Calculate maximum BM and SF
(3) Keeping span depth ratio with in 1/8 to 1/12, find the depth. Alternatively, assuming thickness of web, we can find depth using the following relation.

$$d = (M \times K/f_y)^{0.33}$$

$$K = d/t_w \text{ (web slenderness)}$$

Also we can use

$$A_f = M/(f_y \times d) - A_W/6$$

$$A_w = \text{Area of web plate}$$

and

$$A_f = \text{Area of each flange}$$

(4) Check the thickness of web based on serviceability requirement as per Clause 8.6.1.1 and based on compression buckling requirement as per Clause 8.6.1.2 of IS:800/2007.

(5) Width of flange can be kept 1/3 depth of section. (In Indian rolled *I* sections, the ratio is around this value). The width and thickness of flanges are to be checked for the requirement of plastic/compact/semi compact section.

(6) Based on the lateral restraint, find the moment capacity as per Clause 8.2.1 (or) 8.2.2 of IS:800.

(7) Using Clause 8.4.2.2 (a) simple post-critical method or Clause 8.4.2.2 (b) by considering tension field method. Calculate the resistance to shear buckling of web.

(8) Design of connection between flange plate and web. (Clause 8.6.3.3 of IS:800)

(9) Design of bearing stiffeners and their connections. (Clause 8.7.4, Clause 8.7.6 and Clause 8.7.9 of IS:800)

(10) Design of stiffeners at concentrated load points and their connections (Clause 8.7.3 and Clause 8.7.5 of IS:800)

(11) Design of intermediate stiffeners, if required and their connections (Clause 8.7.2 of IS:800).

(12) Design of web splice and connections. (Clause 8.6.4.2)

(13) Design of flange splice and connections. (Clause 8.6.3.2)

Dimensions of plate girder

Generally, the depth to span ratio kept as 1/8 to 1/12

D - Overall depth

d_e - Effective depth (centre of compression flange to centre of tension flange)

d - Clear depth (depth of web)

t_w - Thickness of web

t_f - Thickness of flange

b_f - Width of flange

Example 5.4 *Welded plate girder*

A plate girder is simply supported over an effective span of 16 M. It carries an uniformly distributed load of 72 kN/M in addition to its self weight and two concentrated loads of 400 kN each at 4 M from either support. The load is due to dead and imposed loads only. Design the welded plate girder. It is effectively supported in lateral direction throughout. Material IS:2062, $F_y = 250$ *MPa.*

Step 1 Computation of BM & SF

Total load	$= (72 \times 16) + 2 \times 400 = 1952$ kN
Self wt (kN/m)	$= W/250 = 1952/250 = 7.8$ say 8 kN/m
Total UDL	$= 72 + 8 = 80$ kN/m
M_{max}	$= 4160$ kNm Factored $M = 1.5 \times 4160 = 6240$ kNm
V_{max}	$= 1040$ kN Factored $V = 1.5 \times 1040 = 1560$ kN

Step 2 Sizing of web

Refer to Fig. 5.28.

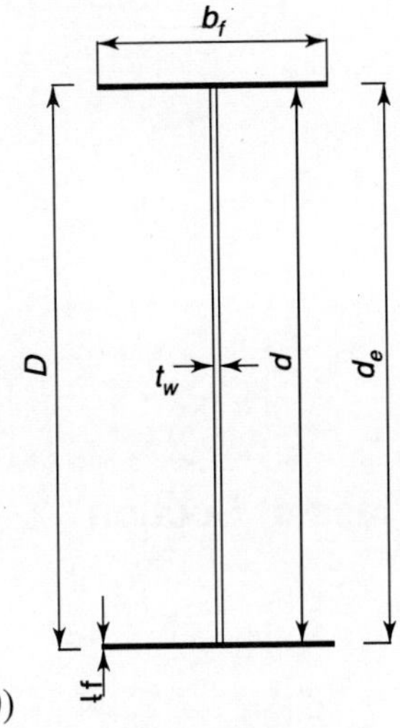

Fig. 5.28 *Section of plate girder*

To find depth of plate girder $d = (M \times K/f_y)^{0.33}$

$$= (6240 \times 10^6 \times 180/250)^{0.33} = 1532 \text{ mm}$$

$$K = d/t_w = 180 \text{ (assumed)}$$

$$t_w = 1532/180 = 8.51 \text{ mm}$$

Let $t_w = 10$ mm

For no intermediate stiffeners (Clause 8.6.1.1 of IS:800)

$\frac{d}{t_w} < 200\ \varepsilon$: $d = 2000$ mm for $t_w = 10$ mm.

To avoid buckling of compression flange (Clause 8.6.1.2 of IS:800)

$$\frac{d}{t_w} \le 345\ \varepsilon^2\text{: } \varepsilon = \sqrt{(250/250)} = 1$$

d can be 3450 mm for t_w is 10 mm

To avoid shear buckling

$\frac{d}{t_w} < 67\varepsilon$ (Clause 8.4.2.1 of IS:800): $d = 670$ mm for $t_w = 10$ mm

Provide $d = 1520$ mm & $t_w = 1520/67 = 22.69$ mm to have resistance to shear buckling.

Let $t_w = 20$ mm.

Web shear buckling need not be checked when $(d/t_w) = < 67\ \varepsilon$

In this case $d/t_w = 1520/20 = 76 > 67\varepsilon$.

Hence, shear buckling is to be checked.

Step 3 Sizing of flanges

$$A_f = \frac{M}{f_y\, d} - \frac{A_w}{6}$$

$$= \frac{6240 \times 10^6}{250 \times 1520} - \frac{1520 \times 20}{6} = 11{,}354 \text{ mm}^2$$

The width of flange can be about 30% of the depth as in the case of IS rolled sections.

Let $b_f = 0.3 \times d = 456$ mm. Provide 460 mm

$t_f = (11354/460) = 24.68$ mm. Provide 32 mm

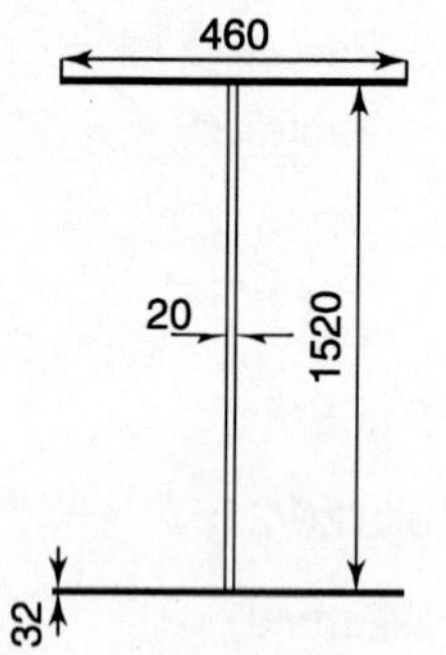

Fig. 5.29 *Proposed section of girder*

Class of Section

$$\frac{b}{t_f} = \frac{(460 - 20)/2}{32} = 6.875 < 8.4\varepsilon \text{ (Table 2 of IS:800)}$$

$$\frac{d}{t_w} = \frac{1520}{20} = 76 < 84\varepsilon$$

The section is plastic.

Shear Capacity (Clause 8.4 of IS:800)

$$V \leq V_d$$

$$V_d = \frac{V_n}{\gamma_{mo}}$$

$$V_\eta = V_p = \frac{A_v f_{yw}}{\sqrt{3}} = \frac{1520 \times 20 \times 250}{10^3 \times \sqrt{3}} = 4387.8 \text{ kN}$$

$$V_d = \frac{4387.8}{1.1} = 3988.96 \text{ kN}$$

$$0.6\,V_d = 0.6 \times 3988.96 = 2393 \text{ kN} > 1560 \text{ kN}$$

Hence, safe against shear and it is in low shear.

Moment Capacity (Clause 8.2.1.2 of IS:800)

$$Md = \beta_b Z_p \frac{f_y}{\gamma_{mo}} \leq 1.2 Z_e \frac{f_y}{\gamma_{mo}}$$

Z_p Plastic modulus $= 2b_f\, t_f\,(D - t_f)/2 + t_w d^2/4$

$$= 2 \times 460 \times 32 \times 1552/2 + 20 \times 1520^2/4 = 34{,}397{,}440 \text{ mm}^3$$

Elastic Modulus

$$I_Z = \left(\frac{460 \times 1584^3}{12} - \frac{440 \times 1520^3}{12}\right) = 2.358 \times 10^{10} \text{ mm}^4$$

$$Z_e = I_Z/(1584/2) = 29{,}777{,}256 \text{ mm}^3$$

$$\beta = \frac{Z_e}{Z_p} = \frac{29777256}{34397440} = 0.8657$$

$$\left(\text{Shape factor} = \frac{Z_p}{Z_e} = 1.155\right)$$

Section is plastic. Hence, $\beta_b = 1.0$

$$M_d = \frac{1.0 \times 34397440 \times 250}{10^6 \times 1.1} = 8599.4 \text{ kN}$$

$$M_d = 1.2\, Z_e \frac{f_y}{\gamma_{mo}} = \frac{1.2 \times 29777256 \times 250}{1.1 \times 10^6} = 8121 \text{ kNm.}$$

Hence, $M_d = 8121$ kNm which is greater than M_z which is = 6240 kNm as already calculated. Hence, satisfactory.

Since the compression flange is restrained throughout, lateral torsional buckling need not be checked.

Bearing Stiffeners (Clause 8.7.4 of IS:800)

Assuming width of support as 600 mm and minimum stiff bearing provided by support = 600/2 = 300 mm = b_1

Dispersion length (slope 1:2.5)

$$\eta_2 = 2.5 \times 32 = 80 \text{ mm}$$

Local capacity of web = F_w

$$= (b_1 + \eta_2)\, t_w \times \frac{f_{yw}}{\gamma_{mo}} = \frac{(300 + 80) \times 20 \times 250}{1000 \times 1.1}$$

$$= 1727 \text{ kN} > \text{support reaction } 1560 \text{ kN}$$

Since local capacity of the web is more than the reaction, bearing stiffener is not required.

At the Location of Concentrated Load

$$F_w = (b_1 + \eta_2)\, t_w \times \frac{f_{yw}}{\gamma_{mo}}$$

Considering that the concentrated load is imparted by ISMB 450 for which t_w = 9.4 mm and radius at root = 15 mm.

$$b_1 = 2(9.4 + 15) = 48.8 \text{ mm}$$

$$\eta_2 = 2.5 \times 2 \times 32 = 160 \text{ mm}$$

$$F_w = \frac{(48.8 + 160) \times 20 \times 250}{1000 \times 1.1} = 949 \text{ kN} > 400 \text{ kN}$$

Hence, stiffener for web is not required.

Design of Weld to Connect the Flanges and Web

Fillet weld on either side of web to connect the flanges is proposed.

Shear force/mm = $(VA\bar{Y}/I)$ = [1560 × 460 × 32 × (1532/2) × 1000]/(2.358 × 10^{10}) = 746 N/mm.

Force on each side = 372 N/mm.

Web thickness = 20 mm.

Flange thickness = 32 mm.

Minimum size of weld = 6 mm (Table 21 of IS:800).

Throat thickness = (0.707 × 6) mm.

Strength = (0.707 × 6) × 189 = 802 N/mm > 372 N/mm.

Provide 6 mm weld on either side of web.

Design of the Same Plate Girder with Thin Web

Let the web depth = 1500 mm

$$t_w = 16 \text{ mm}$$

$$A_f = M/(f_y \times d) - \frac{A_w}{6}$$

$$= \frac{6240 \times 10^6}{250 \times 1500} - \frac{1500 \times 16}{6} = 12640 \text{ mm}^2$$

Let $b_f = 0.3d = 0.3 \times 1500 = 450$ mm

Provide 500 mm

$$\therefore \quad t_f = \frac{12640}{500} = 25.28 \text{ mm}$$

Provide 40 mm

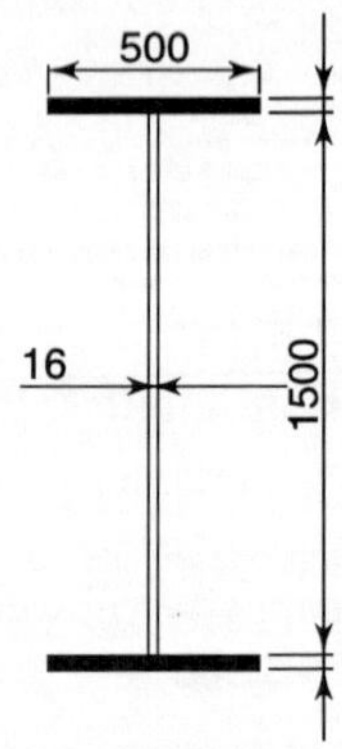

Fig. 5.30 *Proposed section of girder with thin web*

Section Class

$$\frac{b}{t_f} = \frac{(500 - 16)/2}{40} = 6.05 < 8.4\varepsilon. \text{ Section is plastic.}$$

$$\frac{d}{t_w} = \frac{1500}{16} = 93.75 < 105\varepsilon. \text{ Section is compact.}$$

Hence, section is compact.

Moment Capacity

$$M_d = \beta_b Z_p \frac{f_y}{\gamma_{mo}} \qquad \text{(Clause 8.2.1.2 of IS:800)}$$

β_b = 1.0 for compact section.

$$Z_p = [2b_f t_f (D - t_f)/2] + [t_w \times d^2/4] = [2 \times 500 \times 40 \times 1540/2] + [16 \times 1500^2/4] = 39.8 \times 10^6 \text{ mm}^3$$

$$I_{eZ} = \frac{500 \times 1580^3}{12} - \frac{484 \times 1500^3}{12} = 28218.8 \times 10^6 \text{ mm}^4$$

$$Z_e = (I_{eZ}/(1584/2)) = 35.72 \times 10^6 \text{ mm}^3$$

$$\frac{Z_e}{Z_p} = \frac{35.72 \times 10^6}{39.8 \times 10^6} = 0.897$$

Shape factor = 1.11. This is calculated only for verification.

For range of shape factor for *I* sections, refer to Table 46 of IS:800.

$$M_d = 1.0 \times \frac{Z_p \times f_y}{\gamma_{mo}}$$

$$= \frac{1.0 \times 39.8 \times 10^6 \times 250}{1.1 \times 10^6}$$

$$= 9045 \text{ kNm} > 6240 \text{ kNm}$$

To Avoid Buckling of Compression Flange

$$\frac{d}{t_w} < 345\, \varepsilon_f^2 \quad \text{(Clause 8.6.1.2 of IS:800)}$$

$$\varepsilon_f^2 = \left(\left(-\frac{250}{250}\right)^{0.5}\right)^2 = 1 \text{ and } \frac{d}{t_w} \text{ is only } 93.75$$

Compression flange buckling requirement is met.

Shear Resistance of the Web (Clause 8.4.2.1 of IS:800)

$$\frac{d}{t_w} = \frac{1500}{16} = 93.75 > 67\varepsilon \text{ for web without stiffeners.}$$

Resistance to shear buckling is to be verified.

Simple post-critical method (Clause 8.4.2.2-a of IS:800)

$$V_n = V_{cr}$$

$$V_{cr} = A_v \tau_b$$

$$\tau_{cre} = \frac{K_v \pi^2 E}{12\,(1 - \mu^2)\,(d/t_w)^2}$$

$\mu = 0.3$ Poisson's ratio

$K_v = 5.35$ when transverse stiffeners are provided only at supports

$$\tau_{cre} = \frac{5.35 \times \pi^2 \times 2 \times 10^5}{12\,(1 - 0.3^2) \times 93.75^2} = 110 \text{ N/mm}^2$$

$$\gamma_w = \sqrt{f_{yw}}/(\tau_{cre} \times \sqrt{3}) = \sqrt{(250/(\sqrt{3} \times 110))} = 1.145 < 1.2$$

When $0.8 < \gamma_w < 1.2$

$$\tau^b = 1 - 0.8\,(1.145 - 0.8)\,\frac{250}{\sqrt{3}} = 104.5 \text{ N/mm}^2$$

Shear capacity corresponding to web buckling

$$V_{cr} = dt_w\,\tau_b = \frac{1500 \times 16 \times 104.5}{1000} = 2508 \text{ kN} > 1560 \text{ kN}$$

Hence, satisfactory.

Note: Since the compression flange is restrained, there is no need to check for lateral torsional buckling.

Bearing Stiffener

Refer to Figs. 5.31 & 5.32.

$\eta_2 = 2.5 \times 40 = 100$ mm; $b_1 = 300$ mm as already calculated.

Local capacity of web = F_w

$$F_w = (b_1 + \eta_2)\,t_w \times \frac{f_y}{\gamma_{mo}} = \frac{400 \times 16 \times 250}{1000 \times 1.1} = 1454 \text{ kN} < 1560 \text{ kN (Clause 8.7.4 of IS:800)}$$

Since the local capacity of web is less than the reaction, bearing stiffener is required Design load = 1560 – 1454 = 106 kN (Clause 8.7.6 of IS:800). Refer to Figs. 5.30 and 5.31.

Area of the Stiffener in Contact with Flange

Bearing stress = $f_y/(0.8 \times 1.1) = \dfrac{250}{0.88} = 284$ N/mm^2 (Clause 8.7.5.2 of IS:800)

Area of stiffener required = $\dfrac{106 \times 1000}{284} = 373$ mm^2

Adopting 6 mm thick, width required = $\dfrac{373}{6} = 62$ mm

Required width of each stiffener = 62/2 = 31 mm.

Provide 6 × 50 mm.

$$l = 0.7 \times 1500 = 1050 \text{ mm (Clause 8.7.1.5 of IS:800)}$$

$$A_q = 2 \times 6 \times 50 = 600 \text{ mm}^2$$

$$F_{psd} = (600 \times 250/(0.8 \times 1.1))/1000 = 170 \text{ kN} > 106 \text{ kN (Clause 8.7.5.2 of IS:800)}$$

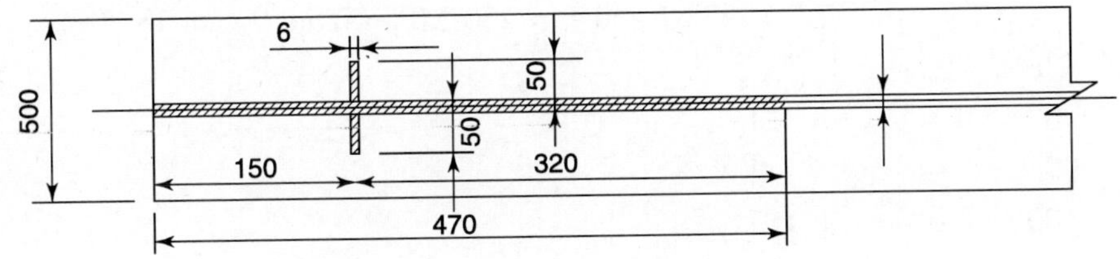

Fig. 5.31 *Bearing stiffener*

Buckling resistance of stiffeners (Clause 8.7.1.5 of IS:800)

For this, the effective section is the full area of stiffener (as in Clause 8.7.1.2 of IS:800) with an effective length of web on each side of the centre line of the stiffeners, limited to 20 times the web thickness. In this case, since the bearing length is 300 mm and the bearing stiffeners will be at midpoint, the web beyond the stiffener is only 150 mm.

Hence, effective length of web = 150 + 20 × 16 = 470 mm.

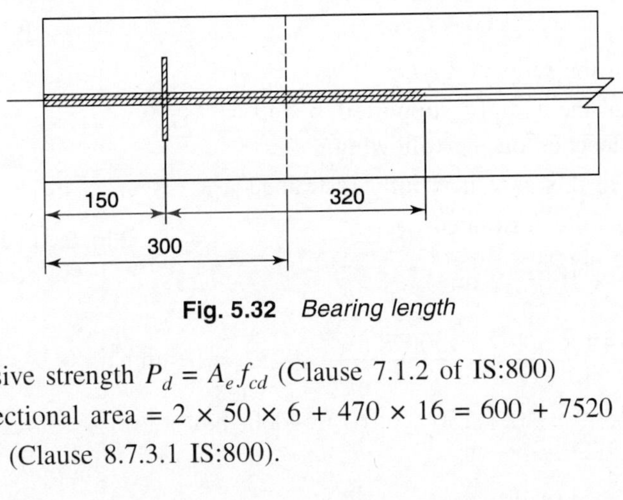

Fig. 5.32 *Bearing length*

Design compressive strength $P_d = A_e f_{cd}$ (Clause 7.1.2 of IS:800)

A_e = Effective sectional area = 2 × 50 × 6 + 470 × 16 = 600 + 7520 = 8120 mm^2

Buckling class *c* (Clause 8.7.3.1 IS:800).

To find f_{cd}

$$r = \sqrt{\frac{I_z}{A}} = \sqrt{\frac{\frac{470 \times 16^3}{12} + 2\left(\frac{6 \times 50^3}{12} + 300 \times 33^2\right)}{8120}} = 10.75 \text{ mm}$$

$$r_y^2 = \left\{2\left(\frac{50 \times 6^3}{12}\right) + \frac{16 \times 320^3}{12} + 16 \times 320 \times 160^2\right\} + [(16 \times 150^3/12)$$

$$+ (16 \times 150 \times 75^2)]/8120$$

$$r_y = \sqrt{r_y^2} = 154.97 \text{ mm}$$

Least radius of gyration = 10.75 mm.

$$KL/r = 1050/10.75 = 97.67$$

$$f_{cd} \text{ for } KL/r = 97.67 \text{ and } f_y = 250 \text{ MPa (Table 9 c of IS:800).}$$

K/r		f_{cd}
90	-----	121
97	-----	111.2
100	-----	107

$$P_d = A_e \times f_{cd} = \frac{8120 \times 111.2}{1000} = 903 \text{ kN} < 1560 \text{ kN}$$

Hence, not satisfactory. Increase the stiffener size and rework.

Torsional Stiffeners

Note: In this case, since the plate girder is laterally supported, torsional stiffeners are not required. (Clause 8.7.9 of IS:800)

However, we will work out as an example,

$$I_S \geq 0.34\ \alpha_s\ D^3 T_{cf}$$

value of α_s depends on the ratio of L_{LT}/r_y

Effective length for the simply supported beams = L_{LT} (Table 15 of IS:800) depends on the following.

Torsional restraint. In this case it is fully restrained.

Compression flange fully restrained.

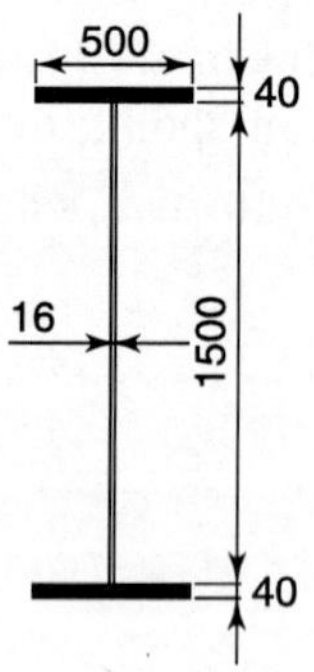

Fig. 5.33 *Designed section of girder*

Hence, $L_{LT} = 0.75 \times 16 = 12$ m.

$$I_{yy} = \frac{2 \times 40 \times 500^3}{12} + \frac{1500 \times 16^3}{12} = 8.33 \times 10^8 \text{ mm}^4$$

$$\text{Area} = 2 \times 500 \times 40 + 1500 \times 16 = 64000 \text{ mm}^2$$

$$r_y = \sqrt{\frac{L_{yy}}{A}} = 114 \text{ mm}$$

$$L_{LT}/r_y = \frac{12000}{114} = 105$$

$$\alpha = 30/105^2 \text{ (Clause 8.7.9 of IS:800)} = 2.72 \times 10^{-3}$$

$$\text{Required } I_S = 0.34\ \alpha_s\ D^3\ T_{ef} = 0.34 \times 2.72 \times 10^{-3} \times 1580^3 \times 40 = 1460 \times 10^5 \text{ mm}^4$$

Provided I_s

$$I_s = \frac{470 \times 16^3}{12} + 2\left(\frac{6 \times 50^3}{12} + 300 \times 33^2\right) = 9.38 \times 10^5 \text{ mm}^5 < \text{required.}$$

Hence, increase the length (projection) of the stiffener to 80 mm. Refer Clause 8.7.1.2. As per this. The outstand can be 14 t_p ε = 14 × 6 × 1 = 84 mm (Clause 8.7.1.2 of IS:800)

$$I_s = \frac{470 \times 16^3}{12} + 2\left(\frac{6 \times 80^3}{12} + 480 \times 48^2\right)$$

$$= 160426 + 2(256000 + 1105920) = 28.8426 \times 10^5 < 804 \times 10^5 \text{ mm}^4$$

6 mm plate cannot be projected beyond this limit 14 t_p ε. Hence, we will try with 8 mm plate. Maximum projection = 14 × 8 = 112 mm

Provide 8 × 112 stiffener.

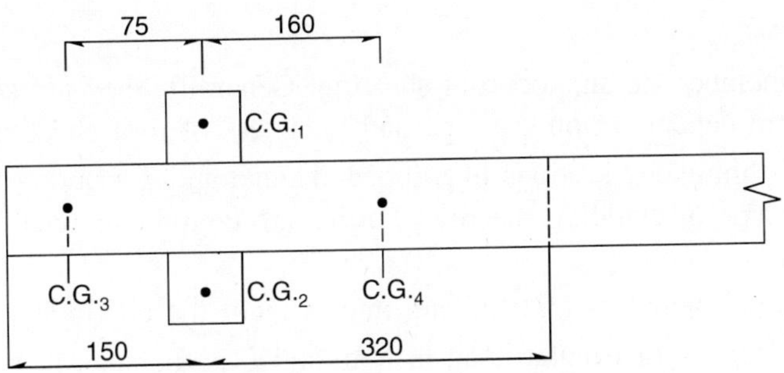

Fig. 5.34 *Intermediate stiffener*

C.G.$_1$ & C.G.$_2$ indicate C.G. of stiffener on each side of web. C.G.$_3$ & C.G.$_4$ indicates C.G. of web of plate girder on each side of stiffener.

$$I_s = \frac{470 \times 16^3}{12} + 2\left(\frac{8 \times 112^3}{12} + 896 \times 64^2\right)$$

$$= 160426 + 2\,(936618.6667 + 3670016) = 93.736 \times 10^5 \text{ mm}^4 < \text{required.}$$

Maximum projection available = $\frac{500 - 16}{2}$ = 242 mm = flange projection

Thickness possible = $\frac{242}{14}$ = 17.28 mm (Clause 8.7.1.2 of IS:800)

Providing 16 × 240 mm stiffener. Area = 16 × 240 = 3840 mm^2

$$I_s = \frac{470 \times 16^3}{12} + 2\left(\frac{16 \times 240^3}{12} + 3840 \times 128^2\right) = 1628 \times 10^5 > 1460 \times 10^5 \text{ mm}^4$$

Hence, torsional stiffeners are more than adequate.

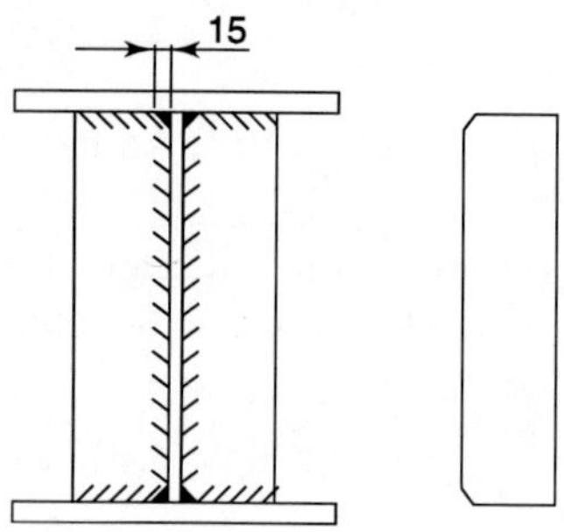

Fig. 5.35 *Stiffener detail*

5.7 DESIGN OF PURLIN

General

Purlins are flexural members to support roof sheeting. Generally, they are kept at a spacing of 1000 mm to 1400 mm depending on the type and thickness of roof sheeting.

Angle channels, *I*, thin gauge sections like lipped channel and *Z* sections are used as purlins. Spacing depends on type of cladding/sheeting. Purlins are continuous over supports below the roof covering sheets.

Maximum bending moment = $wL^2/10$ taking into account the continuity.

where w is the uniformly distributed load in N/m and L is the span in m.

Design shall be as per Clause 8.9 of IS:800.

Load Combinations

w_d = dead load, w_l = imposed load, w_w = wind load

Case 1: No wind

Load on major axis = $(w_d + w_l) \cos \theta$

Load on minor axis = $(w_d + w_l) \sin \theta$

Case 2: With wind and no imposed load

Load on major axis = $(w_w - w_d \cos \theta)$

Load on minor axis = $(w_d \sin \theta)$

Case 3: With wind and with imposed load

Note: (This case may not occur because maximum imposed load occurs when the sheets are erected. When the design wind force is existing, erection of sheets cannot be carried out).

Load on major axis = $(w_w - w_d \cos \theta - w_l \cos \theta)$

Load on minor axis = $(w_d + w_l) \sin \theta$

BM = $wl^2/10$ considering purlin as continuous beam (Clause 8.9 of IS:800).

= $Wl/10$ (where W is the total load = wl)

Example 5.5 *Design of purlin.*

The trusses for a factory building are spaced at 6 m centre to centre and the purlins are spaced at 1.35 m centre to centre. The slope of the top chord of the truss is 15° to the horizontal. The span of the truss is 10 m. The vertical load from roof sheet is equal to 180 N/m² while the wind load on roof surface normal to the roof is equal to 1200 N/m². Design the purling using (a) I section and (b) Angle section. Use material with yield stress 250 MPa.

Solution:

$$\text{slope} = 15^\circ$$
$$\sin\theta = 0.259$$
$$\cos\theta = 0.966$$

(a) Design of *I* section purlin

(i) Computation of dead load (W_d) on purlin

Self weight of roof sheeting = 180 × 1.35 = 243 N/m

Self weight of purlin (assumed) = 120 N/m

Total W_d = 363 N/m

Wind load/m run.

$$W_w = 1200 \times 1.35 = 1620 \text{ N/m}$$

Bending moment

Refer to Fig. 5.37. Wind load is along the weaker axis of purlin. Dead load and imposed load, being gravity loads acts vertically downward.

$$M_{zz} = (W_w + W_d \cos\theta)\, L^2/10 = (1620 + 363 \times 0.966) \times 6^2/10 = 7094 \times 10^3 \text{ Nmm.}$$

Since wind load is considered, the load factor is 1.2 (Table 4 of IS:800:2007)

Factored moment = $1.2 \times 7094 \times 10^3 = 8513 \times 10^3$ Nmm = M_{uz}

$$M_{yy} = W_d \sin\theta \times L^2/10$$
$$= 363 \times 0.259 \times 6^2/10$$
$$= 338 \times 10^3 \text{ Nmm}$$

Since wind load is also considered, load factor is 1.2 (Table 4 of IS:800)

Factored moment = $1.2 \times 338 \times 10^3 = 405.6 \times 10^3$ Nmm = M_{uy}

Selection of *I* section

The ratio of elastic section module about major axis and minor axis for *I* section is approximately six.

Hence, let us assume $Z_{ez}/Z_{ey} = 6$; $Z_{ey} = Z_{ez}/6$;

$$M_{uz} = 1.2 \times Z_{ez} \times f_y/\gamma_{mo} \text{ (Clause 8.2.1.2 of IS:800)}$$

$$f_y/\gamma_{mo} = \frac{M_{uz}}{1.2 \times Z_{ez}}$$

$$M_{uy} = 1.2 \times (Z_{ez}/6) \times f_y/\gamma_{mo}$$

$$f_y/\gamma_{mo} = \frac{6 \times M_{uy}}{1.2 \times Z_{ez}}$$

$$\frac{M_{uz}}{1.2 \times Z_{ez}} + \frac{6 \times M_{uy}}{1.2 \times Z_{ez}} = f_y/\gamma_{mo}$$

$$\frac{M_{uz}}{1.2 \times f_y/\gamma_{mo}} + \frac{6 \times M_{uy}}{1.2 \times f_y/\gamma_{mo}} = Z_{ez}$$

$$Z_{ez} = \frac{8513 \times 10^3 + 6 \times 405.6 \times 10^3}{1.2 \times (250/1.1)} = 40.14 \times 10^3 \text{ mm}^3$$

Try ISMB 100

Class of section

Width of flange = 75 mm

$b = 75/2 = 37.5$ mm

$t_f = 7.2$ mm

$t = 4$ mm

$b/t_f = 37.5/7.2 = 5.2 < 9.4\ \varepsilon$

$d/t_w = (100 - 2 \times (7.2 + 9)/4 = 67.6 < 84\ \varepsilon$

Section is plastic (Table 2 of IS:800)

Section strength (Clause 9.3.1.1 of IS:800)

$(M_y/M_{ndy})^{\alpha_1} + (M_z/M_{ndz})^{\alpha_2} < 1$

Since there is no axial load $n = 0 < 0.2$.

$M_{ndy} = (1.2 \times 10.9 \times 10^3 \times 250/1.1)/10^6 = 2.97$ kNm

$M_{ndz} = (1.11 \times M_{dz}\ (1 - n)) \le M_{dz}$

$M_{ndz} = (1.11 \times M_{dz}\ (1 - 0)) = 1.11\ M_{dz} \le M_{dz}$

Hence, $M_{ndz} = M_{dz}$

$M_{ndz} = (1.2 \times 51.5 \times 10^3 \times 250/1.1)/10^6 = 14.06$ kNm

$\alpha_1 = 1$ (Table 17)

$\alpha_2 = 2$ (Table 17)

$$(M_y/M_{ndy})^{\alpha_1} + (M_z/M_{ndz})^{\alpha_2} = \left(\frac{405.6 \times 10^3}{2.97 \times 10^6}\right)^1 + \left(\frac{8513 \times 10^3}{14.06 \times 10^6}\right)^2 = 0.14 + 0.37 = 0.51 < 1$$

Hence, satisfactory. But the section is not fully utilized.

(b) Design of angle section purlin

$$M_u = 1.2 \times Z_{ez} \times (f_y/\gamma_{mo})$$

$$Z_{ex} \text{ required} = \frac{8513 \times 10^3}{1.2 \times (250/1.1)} = 31214 \text{ mm}^3$$

Trial section

Minimum depth = $L/45$ = 6000/45 = 133.3 mm.

Minimum width = $L/60$ = 6000/60 = 100 mm.

Try ISA 150 × 115 at 16.2 kg/m with thickness 8 mm.

$$M_{ndz} = 1.2 \times Z_{ez} \times (f_y/\gamma_{mo}) = (1.2 \times 44200 \times 250/1.1)/10^6 = 12.05 \text{ kNm}$$

$$M_{ndy} = 1.2 \times Z_{ey} \times (f_y/\gamma_{mo}) = (1.2 \times 27200 \times 250/1.1)/10^6 = 7.42 \text{ kNm}$$

$(M_y/M_{ndy})^{\alpha_1} + (M_z/M_{ndz})^{\alpha_2} = 0.055 + 0.71 = 0.76 < 1.0$

The section is not fully utilized.

Try ISA 130 × 130 × 8

$$Z_{ez} = 34.5 \times 10^3 \text{ mm}^3 = Z_{ey}$$

$$M_{ndz} = 1.2 \times Z_{ez} \times (f_y/\gamma_{mo}) = (1.2 \times 34500 \times 250/1.1)/10^6 = 9.41 \text{ kNm} = M_{ndy}$$

$$(M_y/M_{ndy})^{\alpha_1} + (M_z/M_{ndz})^{\alpha_2} = (0.4056/9.41)^{\alpha 1} + (8.513/9.41)^{\alpha 2} = 0.86 < 1.0$$

Hence, satisfactory. Still further smaller sections can be tried.

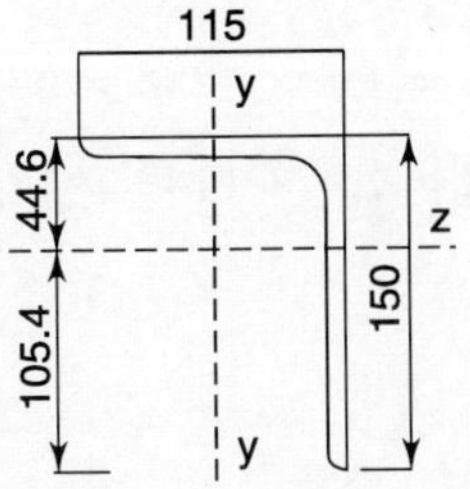

Fig. 5.36 *Angle section*

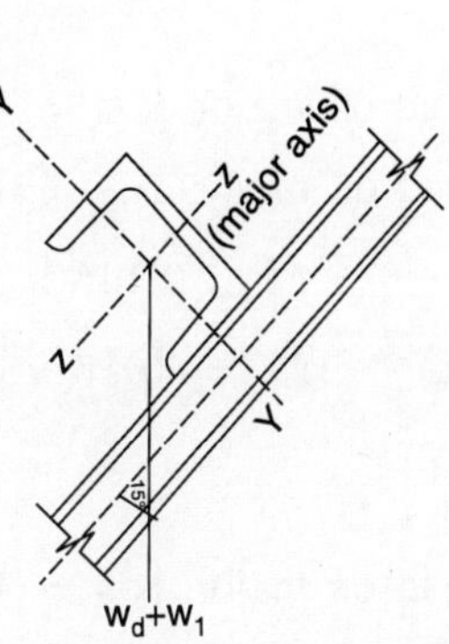

Fig. 5.37 *Purlin over rafter*

Note: ISA130 × 130 × 8 is economical but it is marginally less than the minimum width to span ratio.

Example 5.6 *ISMC Purlin*

Design a purlin using ISMC for the following data:

Wind force on roof sheet = 1300 N/m²

Dead load of A.C. sheet = 159 N/m²

Roof slope = 15° to Horizontal.

Span of purlin 6 m

Spacing of purlin 1.4 m

Span of truss = 20 m

Solution: Imposed load = 750 – [(15 – 10) × 20] = 650 N/m^2 (IS: 875/part 2)

Spacing of purlin = 1.4 m

Span of purlin = 6 m

Assuming self weight of purlin = 100 N/m

$$w_d = (159 \times 1.4 + 100) = 323 \text{ N/m}$$

$$w_l = 1.4 \times 650 = 910 \text{ N/m}$$

$$W_d = 6 \times 323 = 1938 \text{ N (load per purlin)}$$

$$W_l = 6 \times 910 = 5460 \text{ N (load per purlin)}$$

$$W_d + W_l = 7398 \text{ N}$$

Roof Slope = 15°

cos 15° = 0.966

sin 15° = 0.259

Case 1

No wind

Load on major axis = $(W_d + W_l) \cos \theta = 7398 \times \cos 15° = 7146$ N

Load on minor axis = $(W_d + W_l) \sin \theta = 7398 \times \sin 15° = 1915$ N

$M_z = (7146 \times 6/10) = 4288$ Nm. Factored = 1.5 × 4288 = 6432 Nm

$M_y = (1915 \times 6/10) = 1149$ Nm. Factored = 1.5 × 1149 = 1723.5 Nm

Case 2

Wind load + D.L.

Load on major axis = $(W_w - W_d \cos \theta)$

= (1300 × 1.4 × 6) – (1938 × cos 15°) = 9048 N

Load on minor axis = $(W_d \sin \theta) = 502$ N

$M_z = (9048 \times 6/10) = 5429$ Nm. Factored = 1.2 × 5429 = 6514.8 Nm

$M_y = (502 \times 6/10) = 301$ Nm. Factored = 1.5 × 301 = 451.5 Nm

Case 3

Wind load + D.L. + I.L.

Load on major axis = $(W_w - W_d \cos \theta - W_l \cos \theta)$

= (1300 × 1.4 × 6) – (7398 × cos 15°) = 3774 N

Load on minor axis = $(W_d + W_l) \sin \theta = 1915$ N

$M_z = (3774 \times 6/10) = 2264$ Nm. Factored = $1.2 \times 2264 = 2716.8$ Nm

$M_y = (1915 \times 6/10) = 1149$ Nm. Factored = $1.2 \times 1149 = 1379$ Nm

Assuming $Z_z / Z_y = 6$, $Z_{required} = ((M_z + 6M_y)/(1.2 \times f_y/\gamma_{mo}) \times 10^3$

Case	1	2	3
$Z_{required} \times 10^3$ mm^3	61.5	33.8	40.3

$Z_{required} = 61.5$ mm^3 Try *MC*125

$Z_z = 66.6 \times 10^3$ mm^3; $Z_y = 13.1 \times 10^3$ mm^3

$I_z = 416.4 \times 10^4$ mm^4 and $I_y = 59.9 \times 10^4$ mm^4

$M_{ndz} = 1.2 \times Z_{ez} \times (f_y/\gamma_{mo}) = (1.2 \times 66600 \times 250/1.1)/10^6 = 18.2$ kNm

$M_{ndy} = 1.2 \times Z_{ey} \times (f_y/\gamma_{mo}) = (1.2 \times 13100 \times 250/1.1)/10^6 = 3.6$ kNm

There is no axial load. Hence, $n = 0$.

$\alpha_1 = 1$ (Table 17 of IS:800)

$\alpha_2 = 2$ (Table 17 of IS:800)

Case	1	2	3
$(M_y/M_{ndy})^{\alpha 1} + (M_z/M_{ndz})^{\alpha 2} =$	0.48 + 0.12 = 0.6	0.125 + 0.12 = 0.25	0.38 + 0.02 = 0.4

In all the three cases above, $(M_y/M_{ndy})^{\alpha 1} + (M_z/M_{ndz})^{\alpha 2} < 1.0$ is satisfied. Considering the channel purlin is laterally supported throughout its length by the sheet,

Deflection = $\delta_x = (5/384) \times (WL^3/EI)$

$= (5/384) \times (6^3 \times 1000^3/(2 \times 10^5 \times 416.4 \times 10^4) = 3.38W \times 10^{-3}$ (*W* is in *N*)

$\delta_y = 23.5W \times 10^{-3}$; $\delta_{tot} = \sqrt{(\delta x^2 + \delta y^2)}$

Case	δ_x mm	δ_y mm	δ_{tot} mm
1	24.15	45	51.07
2	30.60	11.8	32.8
3	12.76	45	46.77

Note:

(1) *W* is unfactored load.

(2) Case 1 is downward deflection and cases 2 & 3 are upward deflections.

(3) The calculated deflections should not exceed the deflection limits as per Table 6 of IS:800 for the type of roof cladding used. In this case, sheeting is A.C. Hence, we can

consider the same as brittle. Hence, deflection limit = (Span/180) = 6000/180 = 33.33 mm. Case 1 and case 3 deflections exceed this limit. Either sag rods (or) higher section of purlin is needed to limit the deflections.

5.8 DESIGN OF SAG ROD

Tension in Sag Rod

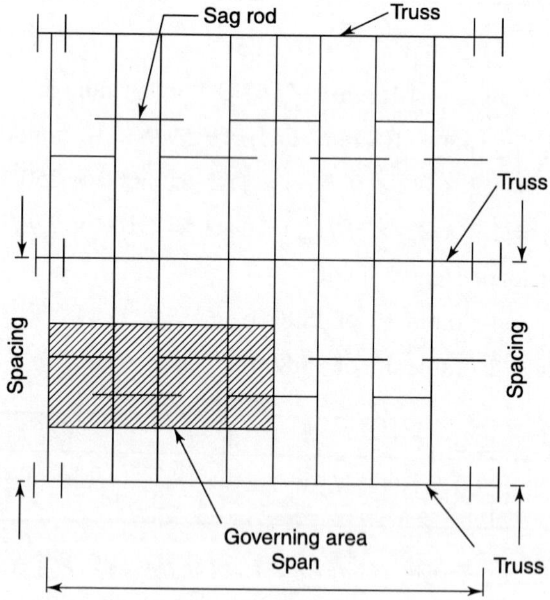

Fig. 5.38 *Plan*

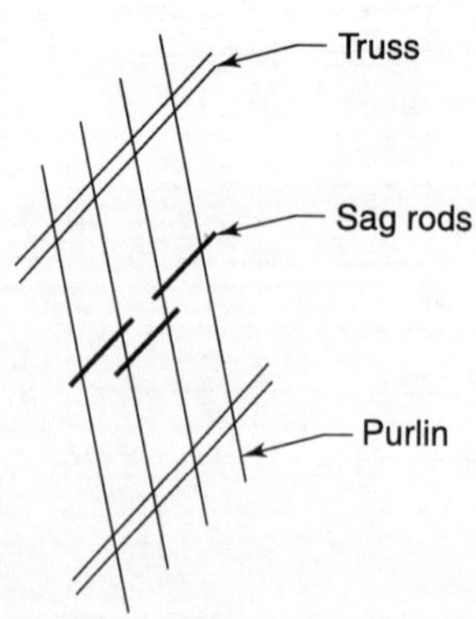

Fig. 5.39 *Truss, purlin and sag rod*

It is proposed to provide one sag rod at middle of trusses/mid span of purlin.

$$\text{Plan area} = \frac{\text{Span}}{2} \times \frac{\text{Spacing}}{2} \times \text{Span} \times \frac{\text{Spacing}}{4}$$

$$\text{Slope area} = \frac{\text{Span} \times \text{Spacing}}{4 \times \cos\theta}$$

$$\sin\theta = \frac{T}{W_g} \text{ and } T = W_g \sin\theta$$

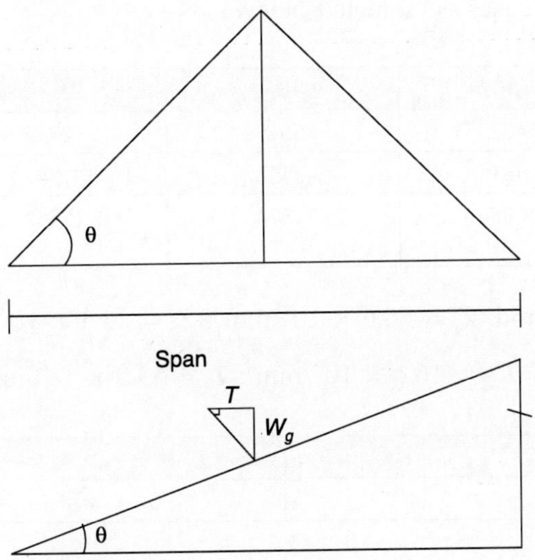

$$\text{Tension in sag rod} = \frac{\text{Span} \times \text{Spacing}}{4 \times \cos\theta} \times W_g \sin\theta$$

$$T = \frac{L_1 \times L_2}{4 \times \cos\theta} \times W_g \sin\theta$$

$$L_1 = \text{span of truss} = 20 \text{ m}, \; L_2 = \text{spacing of truss} = 6 \text{ m}$$

$$w_g = \text{gravity load} = (w_d + w_l) = (159 + 650) = 809 \text{ N/m}^2, \; \theta = 15^\circ$$

Slope distance = 10.35 m

Tension in sag rod = $(L_1 \times L_2 \times (w_d + w_l) \sin\theta)/4\cos\theta$ =
$(20 \times 6 \times 809 \times \sin 15^\circ)/(4 \times \cos 15^\circ) = 25126/4 \cos 15^\circ = 6503$ N

Number of purlins = 10.35/1.4 = 7.4 say 8, Weight of purlin = 8 × 6 × 125 = 6000 N

Tension due to this in sag rod = 6000 × sin θ/4 cos θ = 402 N

Total tension = 6503 + 402 = 6905 N,

$$T_{dg} = A_g \times f_y/\gamma_{mo}$$

$$A_g = (T_{dg} \times \gamma_{mo}/f_y) = 6905 \times 1.1/250 = 30.4 \text{ mm}^2.$$

Area of sag rod required = 30.4 mm^2

$$(\pi d^2/4) = 30.4 \text{ mm}^2. \text{ Hence, } d = \sqrt{(4 \times 30.4)/\pi)} = 6.22 \text{ mm}$$

Provide 16Φ rod as sag rod at mid span. This should be continued to the other side of slope (or) to be connected to ridge tie (or) to rafters.

Providing sag rod one set at mid span, the section strength and deflections are checked as below.

$Z_{required}$ for all the three cases is tabulated below.

	Case 1	*Case 2*	*Case 3*	*Remark*
M_x N-m	6432	6515	2717	No change
M_y N-m	431	113	345	$1/4^{th}$
$Z_{required} = (Mz + 6M_y) \times 1.1/1.2 \times 250$	33066	26374	17552	$Z_{required} = 33.1 \times 10^3 mm^3$

Try ISMC125 with sag rod. $Z_z = 66.6 \times 10^3$ mm^3; $Z_y = 13.1 \times 10^3$ mm^3

$$I_z = 416.4 \times 10^4 \text{ mm}^4;\ I_y = 59.9 \times 10^4 \text{ mm}^4$$

Case	1	2	3
Section strength	0.24	0.16	0.12

The values are low compared to the limit of 1.0.

Deflection

$$\delta z = (5/384) \times (WL^3/EI)$$

$$= (5/384) \times [W \times 6^3 \times 1000^3/(2 \times 10^5 \times 416.4 \times 10^4)] = 3.38\ W \times 10^{-3}$$

$$\delta_y = (5/384) \times [W \times (6/2)^3 \times 1000^3/(2 \times 10^5 \times 59.9 \times 10^4)] = 2.94\ W \times 10^{-3}$$

Case	δ_x *mm*	δ_y *mm*	δ_{tot} *mm*
1	24.2	5.63	24.8
2	30.6	1.5	30.6
3	12.8	5.63	14

Note:

(1) W is unfactored

(2) The maximum deflection is 30.6 mm. Span/deflection Ratio as per Table 6 of IS:800/2007, the deflection limit for brittle cladding (A.C. Sheet is 180) = 33.33 mm

Hence, satisfactory.

Provide ISMC 125 purlins with 16Φ sag rod at mid span of purlin.

EXERCISE

1. A simply supported beam of span 4.5 m is carrying an uniformly distributed load of 15 kN/m throughout its span and a concentrated load of 40 kN at mid span. Find the maximum bending moment and shear force.
2. Using the design data given in (5.1) above, design the beam. The loads are due to dead and imposed loads. Use rolled ISMB having yield stress of 250 MPa.
3. Calculate the deflection of a beam shown below, carrying static load due to dead and imposed loads.

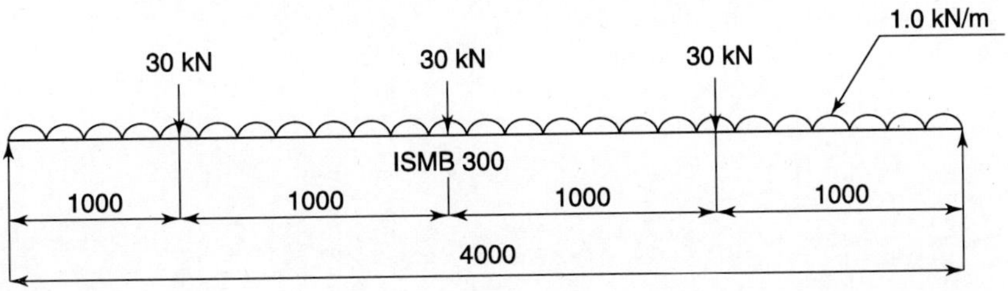

The yield stress is 250 MPa.

4. Design the crane gantry beam using ISMB and ISMC having yield stress – 250 MPa for the following design data.

   Load to be carried 200 kN.

   Column spacing 5 M

   Span of the bridge glider 18 M

   Wheel base 3.4 m

   Hook approach 1.1 m

   Weight of the bridge glider 180 kN
5. Design a plate glider using material having yield stress of 350 MPa and ultimate tensile stress of 490 MPa for the following data.

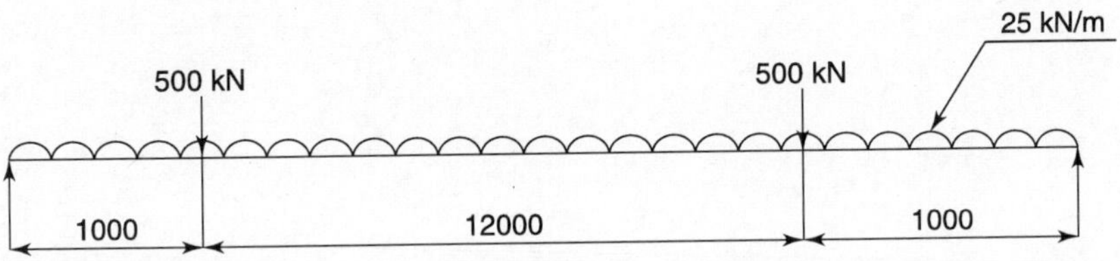

The girder is laterally restrained about minor axis at the concentrated load points.

6. What are the points to be considered to provide the following for a plate glider?
   (a) Vertical stiffeners
   (b) Vertical and horizontal stiffeners
   (c) Bearing stiffeners
   (d) Stiffeners at load points
   (e) Torsional stiffeners

Chapter 6

Column Bases

General

The steel columns transfer the load to column bases resting on reinforced cement concrete foundation. These foundations transfer the load to the soil. The steel column base will be large enough to distribute the load on reinforced cement concrete foundation so that the intensity of pressure on concrete is within the bearing strength of concrete which is 0.6 f_{ck} as per Clause 7.4.1 of IS:800. Large base plate results in larger projection beyond the steel column and thereby resulting in thicker base plate. For heavily loaded columns the base plate will be provided with stiffeners.

There are two types of bases for industrial steel columns.

(1) Slab bases (Clause 7.4.3 of IS:800).

(2) Stiffened base (or) gusseted base (Clause 7.4.2 of IS:800)

Anchor bolts are provided to take up tension. Shear keys are provided to take up shear. Shear resistance at the contact surface between the bottom of base plate and top of concrete surface can be calculated using a friction coefficient of 0.45 (Clause 7.4.1 of IS:800).

The concrete foundation shall be strong enough to carry the axial force, moment and shear.

Though the intensity of pressure on concrete may not be uniform; it is considered uniform.

The stiffened base plate has got the following two advantages.

(a) Better distribution of intensity of pressure on concrete foundation pedestal.

(b) Reduction in thickness of base plate of the column due to the effect of stiffeners.

The assumption in the design of base plate is that it is rigid enough to distribute the load (intensity of pressure) uniformly over the contact surface. The rigidity is ensured by adequate thickness of the base plate or with stiffened base plate.

6.1 PRINCIPLE OF SLAB BASES (CLAUSE 7.4.3 OF IS:800)

A base plate without stiffeners/gusset plates is called a slab base.

$$t_s = \sqrt{\frac{2.5w(a^2 - 0.3b^2)\gamma_{mo}}{f_y}} > t_f$$

a = longer projection

b = shorter projection

Bearing area required = axial load on column/bearing strength of concrete

The slab can be square or rectangular. It can be circular for circular columns.

Thickness as per Clause 7.4.3.1 of IS:800

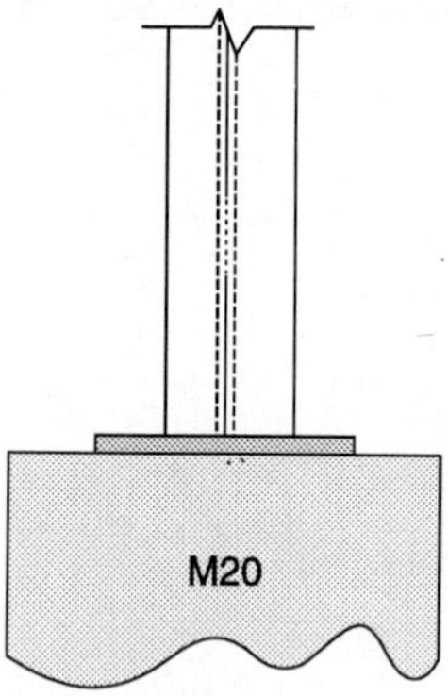

Fig. 6.1

$$t_s = \sqrt{(2.5w(a^2 - 0.3\ b^2))//(f_y/\gamma_{mo})} > t_f$$

t_s = the slab thickness in mm

w = the pressure (or) loading on the underside of the base of the base in MPa

a = the greater projection of the plate beyond the column in mm.

b = the lesser projection of the plate beyond the rectangle circumscribing the column in mm

f_y = yield stress of the material of the slab base

t_f = thickness of flange of the column.

Slab bases (base plate with out stiffeners) are adopted up to 500 kN load. For slab bases, the area will be comparitively smaller and thickness of base plate will be more.

If the size of the base plate is larger than that required to limit the bearing stress on the reinforced cement concrete pedestal or column, an equal projection c of the base plate beyond the face of the steel column and gusset may be taken as effective in column load as given in Fig. 9 of IS:800. In such case use only c^2 in place of $(a^2 - 0.3\ b^2)$ is to be used in the above equation. Round off this required thickness (t_s) to next higher thickness equal to nominal thickness of plates rolled. Also refer to Clause 7.4.3.3 of IS:800 with respect to machining of base plate. However, if machining is required add a minimum of 2 mm to the calculated thickness for machining.

6.2 GENERAL RULES

(1) The bearing strength of concrete should be assumed to be an average plastic rectangular stress block in compression with a maximum stress level of 0.6 f_{ck} (Clause 7.4.1 of IS:800), where f_{ck} is the characteristic 28-day cube strength of the concrete. (Refer to IS:456)

(2) If the contact surface between the base plate and the column, including any stiffeners used, are machined to have full continuous contact, then compressive forces may be assumed to be transmitted by direct bearing from the column to the base plate. Otherwise, the welds or the bolts connecting these elements should be designed to transmit

all compressive forces. The welds and the fasteners should be designed to transmit all shear forces and tensile forces at the connection.

(3) The thickness of the base plate should not be less than the thickness of the column flange.

6.3 SHEAR TRANSFER FROM COLUMN TO CONCRETE

General Rules

(1) Holding-down or anchor bolts cast in sleeves are not used to transfer shear force to the foundation. If anchor bolts are designed to take up tension and shear, the same shall be checked for combined shear and tension (Clause 10.3.6 of IS:800).

(2) Do not use friction to transfer shear force to the foundation in cases of high shear forces and alternating dynamic loads. Transfer of shear forces by friction must only be allowed in the cases where the coefficient of friction used can be guaranteed by tests and the contact pressure is also guaranteed. Normally, the resistance due to friction between the base plate and the reinforced cement concrete pedestal is not considered in the design.

It is advisable to transfer shear forces to the foundation through shear keys. The shear keys are plates welded to the bottom of the steel column base plate and kept in a pocket in the concrete foundation and grouted.

Anchor bolts kept in position and concrete is cast with no tolerance for erection of columns should normally be avoided since no tolerence will be there for erection. Such bolts may resist shear forces. These bolts are designed for combined tension and shear. The bearing stress on the concrete should be checked assuming the shear is transferred to the top of the concrete in a depth equal to three times the diameter of the anchor bolt. Anchor bolts can be kept in pockets provided in the concrete pedestal or the same can be provided with sleeve pipes to give tolerance for erection.

6.4 RULES TO DETERMINE TRIAL SIZE OF BASE PLATE

(a) Allow, generally, 125 mm to 150 mm projection from the column periphery to arrive at base plate size and round off the same to nearest 50 mm. This will help to provide anchor bolts with sufficient enge/end distance.

(b) Choose a trial thickness base plate without stiffeners using the following rules:

- Thickness of 30 – 35 mm for 100 – 125 mm outstand beyond column;
- Thickness of 40 – 50 mm for 150 – 200 mm outstand beyond column.

(c) The bolt spacing should be more than or equal to six times the diameter of the bolt.

(d) For stiffened base plates, the thickness may be reduced by about 10 mm from the recommended thickness.

(e) The edge distance of the base plate from the centre of the anchor (or) holding-down bolt should be between 2.5 and 3 times the diameter of the bolt.

(f) Grade 8.8 bolt should preferably be used where large bending moments are transmitted to the foundation.

(g) The stiffeners when used should extend from the column to 20 mm short of the edge of the base plate.

(h) The depth (vertical height) of the stiffener plate should be two times the outstand from the face of the column. Also note that for battened or laced columns Clause 7.7.2.2 of IS:800 should be followed.

(i) The corner of the stiffener plate may be trimmed at a ratio of 2 vertical: 1 horizontal.

(j) The thickness of the stiffener plate should not be less than 10 mm or depth/16 whichever is the greater.

(k) The stiffener should be designed to withstand compression from the compressive stress block, allowing for buckling effects.

(l) The minimum area of base plate should be the factored axial load on the column including self weight of the column divided by 0.6 f_{ck} (Clause 7.4.1 of IS:800) when there is no moment.

(m) There will be no tension in the anchor (or) holding – down bolts if the following condition is satisfied:

$$M/P = e < D/6$$

where D = depth of the base plate (dimension perpendicular to the axis of bending)

M = applied bending moment about an axis perpendicular to D

P = axial load from column

A = plan area of the base plate

(i.e.) If $M/Z < P/A$; tension will not be there.

6.5 STRESSES

(1) Bearing strength in concrete = $0.6 \times f_{ck}$ (Clause 7.4.1 of IS:800)

Example: for M20 = $0.6 \times 20 = 12$ MPa.

(2) Design bond stress in concrete: 1.2 MPa for $M20$ (IS:456 Table 21)

(3) Yield stress of M.S rod IS:432 (IS:456 Table 22) for diameter > 20 mm = 250 MPa.

(4) Shear stress = $f_y/(\gamma_{mo}\sqrt{3})$

For values of f_y, refer to Table 22 of IS:456.

6.6 DESIGN FOR THICKNESS OF BASE PLATE

w = actual bearing pressure on concrete

= P/A if axial load only is there

= $P/A + M_z/Z_z$ if uniaxial bending moment is also there.

= $P/A + M_z/Z_z + M_y/Z_y$ if biaxial bending moments are there.

P = factored axial load in N

A = area in sq. mm

M_y = factored moment about y axis (Nmm)

M_z = factored moment about z axis (Nmm)

$Z_{e\,y}$ = Elastic section modulus about y axis = $(B \times W^2/6)$
= (dimension of base plate parallel to y axis) × (dimension of base plate perpendicular to y axis)^2/6

Z_{ez} = Elastic section modulus about z axis = $(B^2 \times W/6)$ = (dimension of base plate perpendicular to z axis)^2 × (dimension of base plate parallel to z axis)/6

This bearing pressure w on concrete shall be less than the bearing strength of concrete $0.6\,f_{ck}$ (Clause 7.4.1 of IS:800).

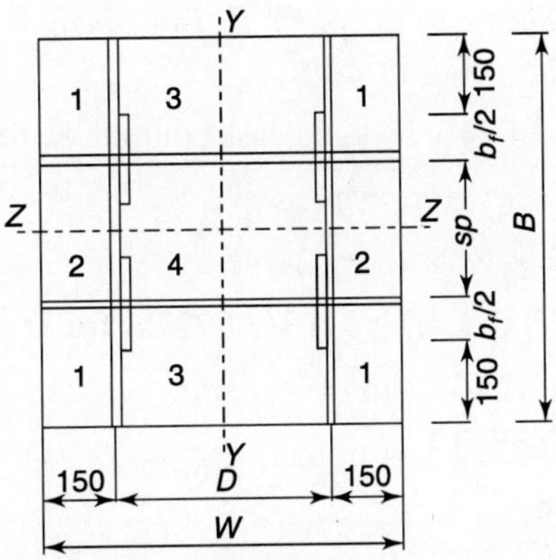

Fig. 6.2

b_f = width of flange.

6.7 COLUMN WITH GUSSETED BASE

Note:

1. Gussets help distribute the loads and moment on columns uniformly to the base.
2. Gussets being plates must be strengthened at regular intervals to prevent local buckling; hence additional plates between column face and end are required
3. The gusset should generally be thick enough so that the stresses in it are generally low, to permit efficient transfer of loads and moment.
4. The gussets are welded to the base plate. Generally, fillet weld is adopted.

6.8 DESIGN PROCEDURE FOR CALCULATING THE THICKNESS OF STIFFENED BASE PLATE

Consider the portion (1)

Two adjacent edges clamped (or) fixed. (by stiffeners)

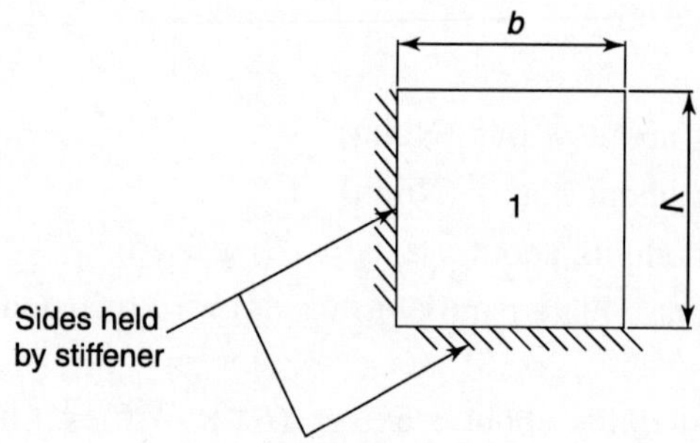

Fig. 6.3

$$a/b = (150 + (b_f/2)/150) \text{ (Refer to Fig. 6.2)}$$

Refer: *[Theory of Plates and Shells by Stephen P. Timoshenko & S. Woinowsky Krieger and find maximum coefficient for this value of a/b].*

$M = wa^2$ (or) wb^2 as in last column of the table.

Find moment factor $M = w \times (a^2 \text{ (or) } b^2)$

$M = 1.2 \times Z_e \times (f_y/\gamma_{mo})$

Z_e of plate required $= M/(1.2 \times (f_y/\gamma_{mo})) = M/227.3$ for steel with yield stress 250 MPa.

$= (M/227.3)\ \text{mm}^3 = 1 \times t^2/6$

$t = \sqrt{6M/227.3} =$ mm

Similarly, find t for part 2, 3 and 4

Adopt the greater thickness.

In case $a/b > 2$; additional stiffener can be introduced so that $a/b < 2$.

Stiffened base gusset base

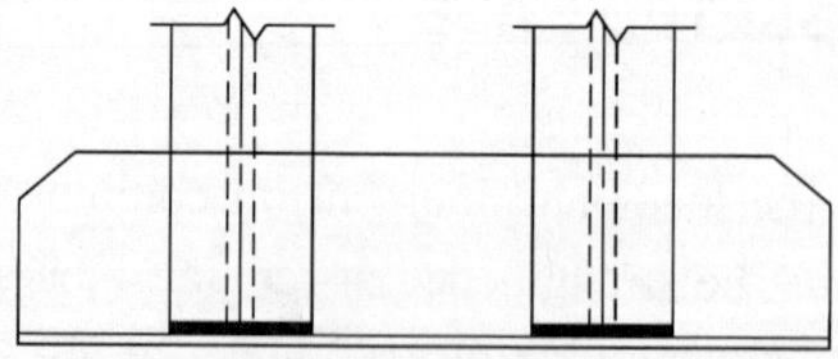

Fig. 6.4 *Stiffened base gusset base*

6.9 ANALYSIS OF COLUMN-TO-FOUNDATION CONNECTION

Fixed bases will transfer moment in addition to shear and direct loads to the foundation. Pinned bases transfer only direct load and shear force. Creating a pinned connection at the base with full rotational freedom to avoid moment transfer is difficult. Almost all bases will have a certain degree of fixity. The fixed bases are used in industrial structures for cantilever columns for columns of portal frames with horizontal loads to prevent sway. Full fixity is very difficult to achieve because the foundation is not infinitely rigid. The designer should take extreme care in the design of the connection of the structure to the foundation and reflect the assumptions made in the analysis. Since soil is a spring, isolated footings provide pinned base. When the foundation of a column rests on single pile, the column base can be considered pinned. When the column base rests on two piles; the base is fixed in one direction and pinned in the other direction. If the column base rests on more than two piles; the column base can be considered fixed on both the axes. A column base resting on a pedestal on raft foundation can be considered fixed. If the pedestals/foundations are connected by tie beams; the column base can be considered fixed in the relevant axis. An isolated footing sufficiently anchored into the soil which is hard rock, the column base can be considered fixed.

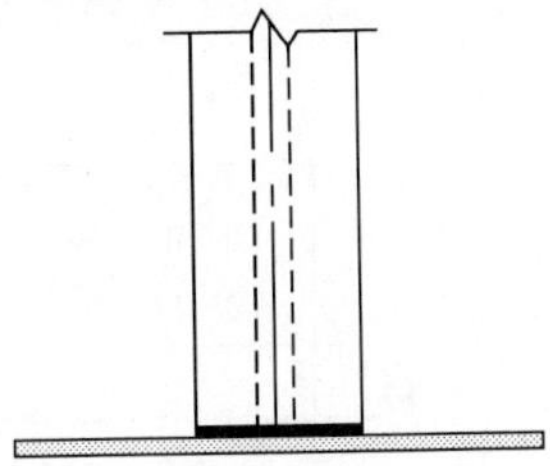

Fig. 6.5

6.10 DESIGN OF BASE PLATE AND ANCHOR BOLT (HOLDING DOWN BOLT OR FOUNDATION BOLT)

Example 6.1 *An ISMB 400 column is subjected to an axial load of 466.7 kN. The load is due to dead and live load only. The pedestal is reinforced cement concrete M20. Base plate is IS:2062 with f_y = 250 MPa. Design the slab base.*

Solution:

$$\text{Factored load} = 1.5 \times 466.7 = 700 \text{ kN}$$

$$\text{Bearing stress} = 0.6\, f_{ck} = 0.6 \times 20 = 12 \text{ MPa}$$

$$\text{Area of base plate required} = 700 \times 1000/12 = 5833 \text{ mm}^2$$

$$\text{Let us keep Length} = 500 \text{ mm}$$

$$\text{Width} = 58333/500 = 116.7 \text{ mm}$$

Let us keep the width as 250 mm to accommodate the flange of the column and the anchor bolts.

$$w = \frac{P_u}{A_{bp}} = \frac{700 \times 1000}{500 \times 250} = 5.6 \text{ N/mm}^2 < 12 \text{ N/mm}^2$$

$$t_s = \sqrt{\frac{2.5 \times 5.6 \times (55^2 - 0.3 \times 50^2) \times 1.1}{250}} = 11.8 \text{ mm}$$

Provide 12 mm thick base plate.

$$\text{Area of the base plate} = 500 \times 250 = 1{,}25{,}000 \text{ mm}^2 > 58333 \text{ mm}^2$$

Size of base plate is larger than required. Hence, we have to calculate c (Clause 7.4.1.1. of IS:800).

Area of the hatched portion in terms of

$$c = 2 \times [(2c + t_f) \times (b_f + 2c)] + (D - 2t_f - 2c) \times (2c + t_w)$$

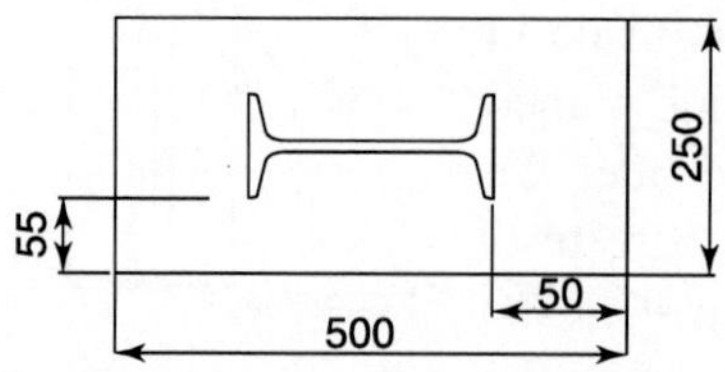

Fig. 6.6(a)

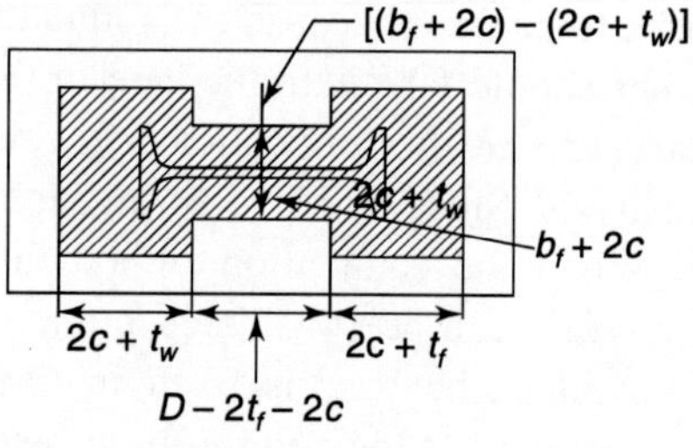

Fig. 6.6(b)

$$= 2 \times [(2c + 16) \times (140 + 2c)] + (400 - 2 \times 16 - 2c) \times (2c + 8.9)$$
$$= 2 \times [(280c + 4c^2 + 2240 + 32c)] + [736c + 3275.2 - 4c^2 - 17.8c]$$
$$= 560c + 8c^2 + 4480 + 64c + 718.2c + 3275.2 - 4c^2$$
$$= 4c^2 + 1342.2c + 7755.2$$

Equating this to the required area, we can find the value of c

$$4c^2 + 1342.2c + 7755.2 = 58333$$
$$c^2 + 335.55c - 12644.45 = 0$$

$$c = \frac{(-335.55 \pm \sqrt{335.55^2 + 50577.8})}{2}$$

$$= \frac{(-335.55 \pm 404)}{2} = 34.225\text{mm}$$

$$t_s = \sqrt{2.5 \times 12 \times 34.225^2/(250/1.1)} = 12.43 \text{ mm}$$

Next higher nominal thickness of rolled plates in India is 16 mm. Hence, provide 16 mm.

Example 6.2 *A column is subjected to an axial load of 466.7 kN and 53.3 kNm moment about zz axis as shown in Fig. 6.6 (c). The load is due to dead and live loads only. The pedestal is reinforced cement concrete M20. Base plate is IS:2062 with f_y = 250 MPa. Design the slab base and anchor bolts.*

Factored load = 466.7 × 1.5 = 700 kN

Factored moment = 1.5 × 53.3 = 80 kN

Using M20 concrete, bearing strength = 0.6 f_{ck} = 0.6 × 20 = 12 MPa. (Clause 7.4.1 of IS:800)

Considering only axial forces,

$$\text{Area required} = \frac{700 \times 1000}{12} = 58333 \text{ mm}^2$$

Since moment is also there; increase the size. Alternatively limiting the maximum intensity of pressure on concrete $(P/A) + (M/Z)$ to bearing strength 12 MPa (Clause 7.4.1 of IS:800).

$$(P/A) + (M/Z) = 12 \text{ MPa.}$$

$$\frac{700 \times 1000}{B \times L} + \frac{80 \times 10^6}{\left(\frac{B \times L^2}{6}\right)} = 12$$

$$\text{Assuming } B = 400 \text{ mm}$$

$$\frac{700 \times 1000}{400 \times L} + \frac{80 \times 10^6 \times 6}{400 \times L^2} = 12$$

$$\text{Solving } L = 397 \text{ mm.}$$

Providing 400 mm is sufficient to limit the bearing pressure on concrete to 12 MPa. (Clause 7.4.1 of IS:800) But to provide anchor bolt, we require some space for tightening to have required edge/end distance. Hence, keeping 125 mm from the outer face of column, $L = 200 + 2 \times 125 = 450$ mm.

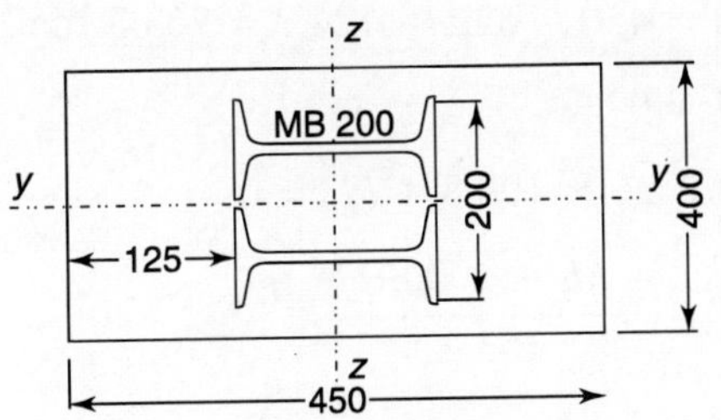

Fig. 6.6(c)

Maximum intensity of pressure =

$$w_{max} = (P/A) + (M/Z) = (700 \times 1000/400 \times 450)$$

$$+ (80 \times 10^6 \times 6/400 \times 450^2) = 9.82 \text{ MPa.}$$

$$a = 125 \text{ mm}, b = 100 \text{ mm (see Fig. 6.6 (c))}$$

$$t_s = \frac{\sqrt{2.5 \times 9.82 \times (125^2 - 0.3 \times 100^2) \times 1.1}}{250} = 36.93 \text{ mm}$$

$$\frac{P}{A} = 3.89 \text{ MPa}$$

$$\frac{M}{Z} = 5.93 \text{ MPa}$$

Provide 40 mm thick base plate.

Since M/Z is greater than P/A, minimum stress w_{min} = "$P/A - M/Z$" is negative. Hence, tension is there.

$$P/A - M/Z = 3.89 - 5.93 = (-)\ 2.04 \text{ MPa.}$$

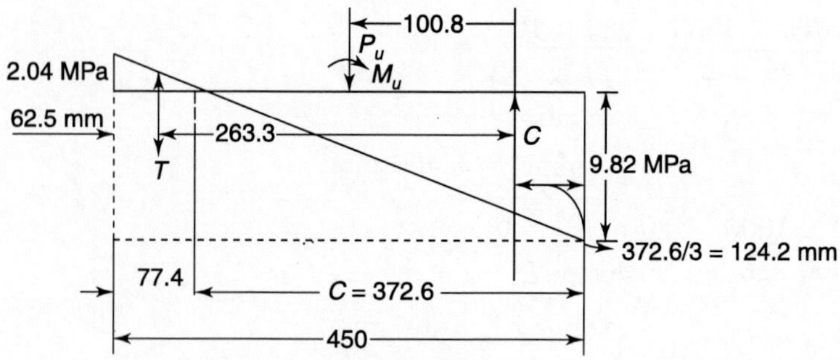

Fig. 6.7

By the similar triangle principle, we can calculate the lever arm. Keeping 62.5 mm from centre of bolt to edge of base plate,

$$\text{Lever arm} = 450 - 62.5 - 124.2 = 263.3 \text{ mm.}$$

Moment towards $C \Sigma M = 0$

$$-T \times 263.3 - P_u \times 100.8 + M_u = 0$$

$$-T = \frac{-M_u + 100.8P_u}{263.3}$$

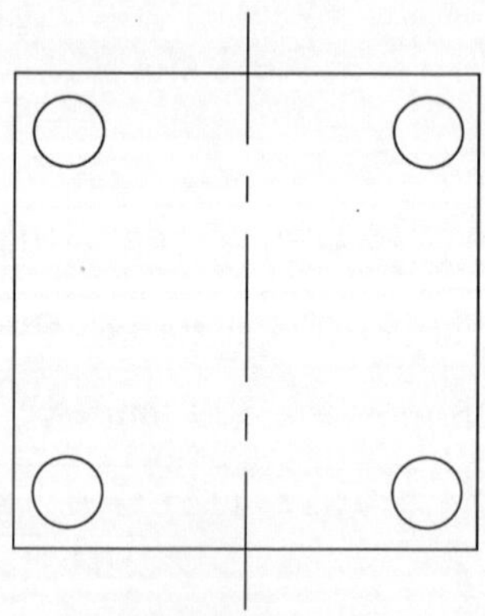

Fig. 6.8

$$T = \frac{M_u - 100.8P_u}{263.3}$$

$$T = \frac{80 \times 10^6 - 100.8 \times 700 \times 1000}{263.3 \times 1000} = 35.85 \text{ kN}$$

$$\text{Tension/bolt} = 35.85/2 = 17.93 \text{ kN}$$

$$\text{Design strength} = \frac{A_g f_y}{\gamma_{mo}} \text{ (Clause 6.2 of IS:800)}$$

M.S. rods are used for foundation bolts (IS 2062; IS 432)

$$f_y = 250 \text{ MPa}$$

$$A_g \times 250/1.1 = 17.93 \times 1000$$

$$A_g = 17930 \times 1.1/250 = 78.88 \text{ mm}^2$$

$$\frac{\pi d^2}{4} = 78.88$$

$$d = \frac{4 \times 78.88}{\pi} = 10.02 \text{ mm}$$

Minimum bolt diameter to be used in structural connections = 12 mm

Rupture strength (Clause 6.3.2 of IS:800)

$$= 0.9\, A_n\, f_u/\gamma_{m1} = 0.9 \times 84.3 \times 415/(1.25 \times 1000) = 25.2 \text{ kN} > 17.93 \text{ kN}$$

Hence, provide M12 threaded 16 dia rod with M12 washer and nut.

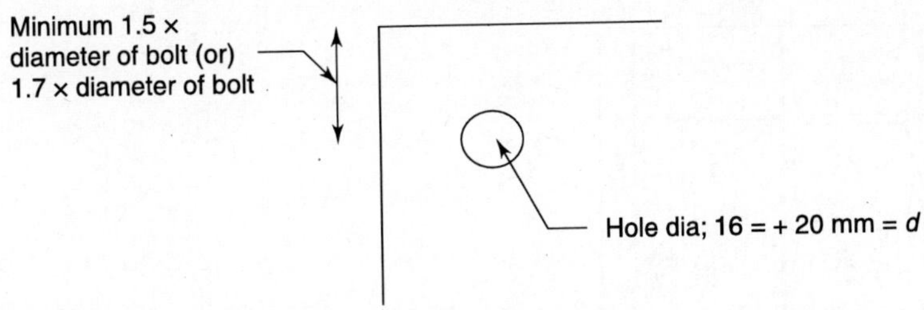

Fig. 6.9

Diameter of oversize hole for 16 mm diameter rod = 16 + 4 = 20 (Table 19 of IS:800)

$$\text{Min. edge distance } e = 1.5 \times d_h$$

$$= 1.5 \times 20 = 30 \text{ mm}$$

Provided edge distance = 62.5 mm which is more than required.

Length of anchor bolt:

Design bond stress in limit state design for plain bars in tension

= 1.2 MPa (Clause 26.2.1.1 of IS:456).

$$\pi dl \times \text{bondstress} = 17930 \text{ N}$$

$$l = \frac{17930}{\pi \times 16 \times 1.2} = 297 \text{ mm}$$

Provide 300 mm long $M12$ threaded 16 mm rod as bolts, 2 Nos. on each side with 62.5 mm edge distance

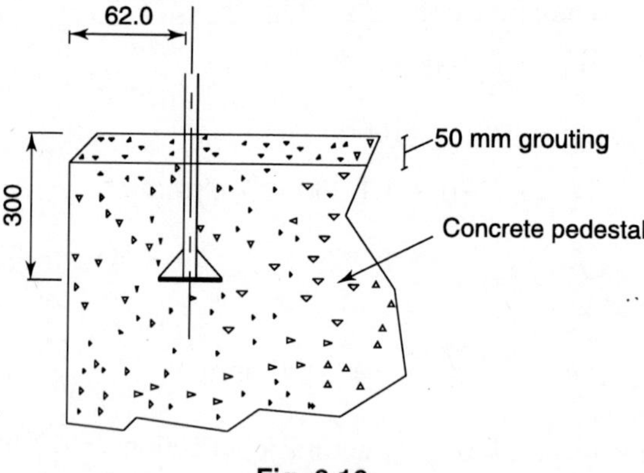

Fig. 6.10

Example 6.3 *Design the base plate with stiffened base (or) gusseted base for the design data as in Example 6.2:*

The stiffeners can be provided as shown.

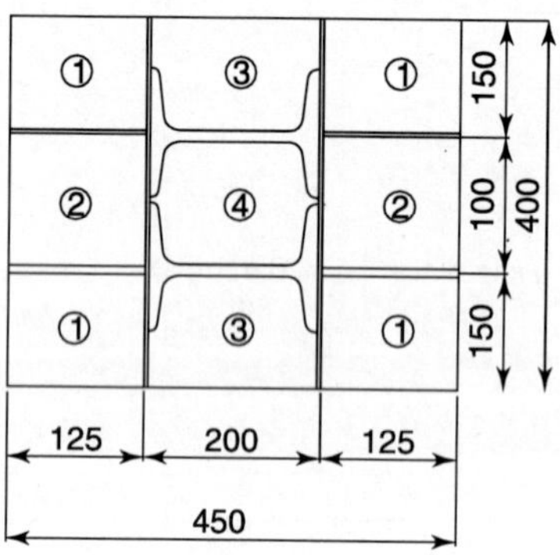

Fig. 6.11

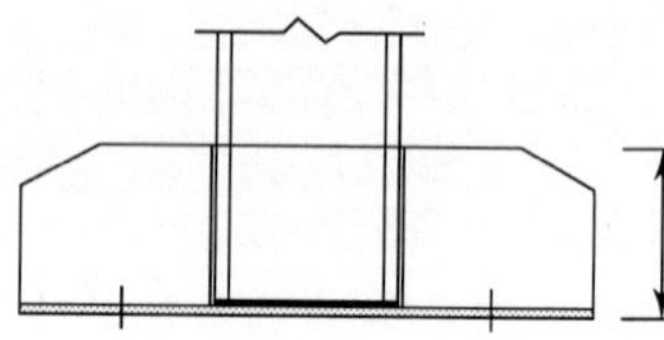

Fig. 6.12

Type 1: 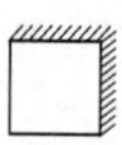 b = 125 mm; a = 150 mm, two adjacent edges clamped.

Type 2: 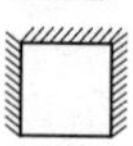 b = 100 mm, a = 125 mm, three edges clamped

Type 3: 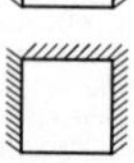 b = 150 mm, a = 200 mm, three edges clamped

Type 4: 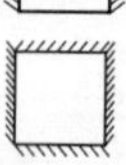 b = 100 mm, a = 200 mm, all the four edges clamped

Type	*b in mm*	*a in mm*	*b/a*	*Maximum B.M. Coeff* α	*B.M. in Nmm* $\alpha q b^2$
1	125	150	0.83	0.09	13533
2	100	125	0.8	0.08	7640
3	150	200	0.75	0.08425	18615
4	100	200	0.5	0.0829	8141

The maximum bearing pressure = 9.82 MPa

Maximum B.M. = 13533 Nmm

Considering 1 mm width,

$$\text{Bending strength} = \frac{1.2Z_e f_y}{\gamma_{mo}} = \frac{1.2 \times 1 \times t^2 \times 250}{6 \times 1.1} = 45.45t^2$$

$$\text{equating; } t = \sqrt{\frac{13533}{45.45}} = 17.25 \text{ mm}$$

(Clause 8.2.1.2 of IS:800). Provide 20 mm thick base plate.

Note that the thickness has come down to 20 mm from 40 mm for slab base due to the provision of stiffeners.

Design of Stiffener

Height of stiffener = 200 mm (Clause 7.7.2.3 and 7.4.1.1 of IS:800)

Load taken by the stiffeners are as shown below.

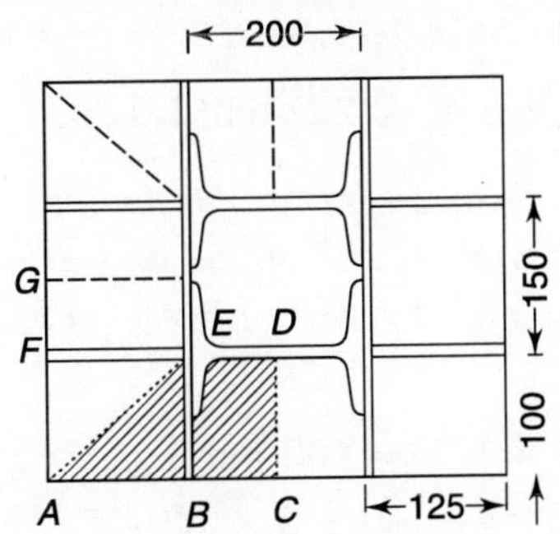

Fig. 6.13

Load on stiffener *BE*:

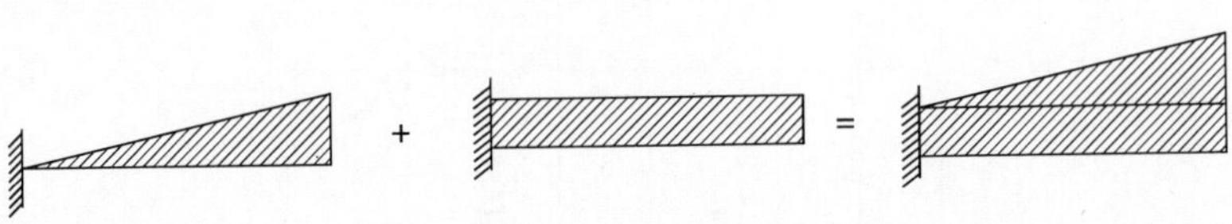

Fig. 6.14

Area enclosed by *ABCDE*

Force due to rectangular load = 100 × 150 × 9.82 = 147300 N.

Moment due to this = 147300 × (½) × 150 = 11047.5 Nmm.

$$\text{Total triangular load} = \frac{1 \times 125 \times 150 \times 9.82}{2} = 92062.5 \text{ N}$$

Moment at *E* due to triangular load = 92062.5 × (2/3) × 150 = 9.21×10^6 Nm

Total moment = 9.217×10^6 N mm (1)

$$Z_e = \frac{t \times 200^2}{6}$$

$$M = \frac{1.2 \times Z_e \times f_y}{1.1} = \frac{1.2 \times t \times 200^2 \times 250}{6 \times 1.1} = 1818181.82t \quad (2)$$

Equating 1 & 2; t = 5.07 mm

Design of stiffener *FE*:

$$\text{Total triangular load} = 92062.5 \text{ N}$$

$$\text{Moment at } E = 92062.5 \times (2/3) \times 125 = 7.672 \times 10^6 \text{ Nmm}$$

$$\text{Total rectangular load} = \frac{100 \times 125 \times 9.82}{2} = 61375 \text{ N}$$

$$\text{Moment at } E = 61375 \times 125/2 = 3.84 \times 10^6 \text{ Nmm}$$

$$\text{Total moment} = 11.51 \times 10^6 \text{ Nmm}$$

$$t = 11.51 \times 10^6/1818181 = 6.33 \text{ mm}$$

Hence, provide all stiffeners 8 mm thick 200 mm height.

Example 6.4 *Two ISMB 350 shapes are kept at 350 mm centre to centre of web as shown in Fig. 6.15. The moment about the major axis is 80 kN and axial load is 700 kN. These include the effect of wind also. Design the gusseted base to rest on M20 concrete.*

Size of base plate

Parallel to web 600 mm (350 + 2 × 125) = 600 mm.

Perpendicular to web = (350 + 140 + 2 × 130) = 750 mm

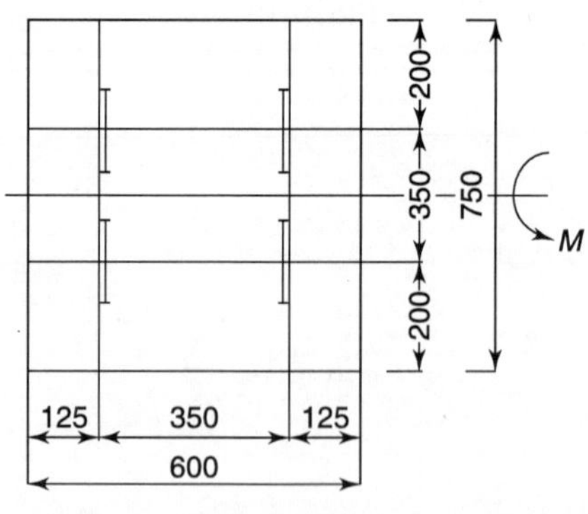

Fig. 6.15

Load and moment include wind.

The partial safety factor for loads and moment is 1.2 (Table 4 of IS:800).

Factored axial load = $P = 1.2 \times 700 = 840$ kN

Factored moment = $M = 1.2 \times 80 = 96$ kN m (due to wind)

Concrete M20 bearing stress = $0.6 f_{ck} = 0.6 \times 20 = 12$ MPa (Clause 7.4.1 of IS:800)

Yield stress in IS:432 rod = 250 MPa

Maximum actual bearing stress

$$w_{max} = P/A + M/Z$$

$$= 840/(0.6 \times 0.75) + 96 \times 6/(0.6 \times 0.75^2) = 1866 + 1707 = 3573 \text{ kN/m}^2$$

$$= 3.6 \text{ MPa} < 12 \text{ MPa}$$

Hence, size of base plate is adequate.

Minimum actual bearing stress

$$P/A - M/Z = 1866 - 1707 = 159 \text{ kN/m}^2 = 0.16 \text{ N/mm}^2.$$

Since value of M/Z is less than the value of P/A; there is no tension.

Hence, it is enough if nominal anchor bolt is provided.

Provide 4 number of M24 bolt. The anchor bolts will be made from IS:432/IS:2062 rods of diameter 28 mm.

Design strength in tension

For tension, net area (area at threaded portion) is considered which is approximately between 0.7 to 0.8 times the gross area. If the net area is known the same shall be taken.

For 3 mm pitch of thread, the net area of *M*24 bolt = 353 mm, (IS:4218 Part III & Table 8.3 of this book)

Design strength in tension due to yielding = T_{dg}

$$= A_g \times f_y/\gamma_{mo} = (\pi \times 28^2/4) \times 250/1.1 \times 1000 = 140 \text{ kN (Clause 6.2 of IS:800)}$$

Design strength in tension due to rupture = $T_{dn} = 0.9 \times A_n \times f_u/\gamma_{m1}$

$$= 0.9 \times (353 \times 415)/(1.25 \times 1000) = 105.5 \text{ kN (Clause 6.3.2 of IS:800)}$$

The least of the above two is the design strength in tension = 105.5 kN.

Strength in bond for M20 concrete = 1.2 MPa (Clause 26.2.1.1 of IS:456)

Length of anchorage required (l_a)

There is no tension. Hence, it is enough if nominal length is provided. Here the length of anchorage is calculated for design tensile strength of the bolt.

Surface area: $\pi \times 28 \times l_a$

Equating design tensile strength to bond strength,

$$= (\pi \times 28 \times l_a \times 1.2)\ N = 105.5 \times 1000$$

$$la = (105.5 \times 1000)/(\pi \times 28 \times 1.2) = 999 \text{ mm}$$

For full capacity, length required = 1000 mm. Provide 700 mm embedded length since there is no tension.

Design of Base plate using BM coefficients

[Note: Refer to Theory of Plates and Shells by Stephen P Timoshenko & S. Woinowsky Krieger

M = Timoshenko coefficient × w × (dimension of the panel a or b as in the last column of the table which gives the BM coefficients)^2 = $\rho.w.a^2$

w = 3.6 MPa

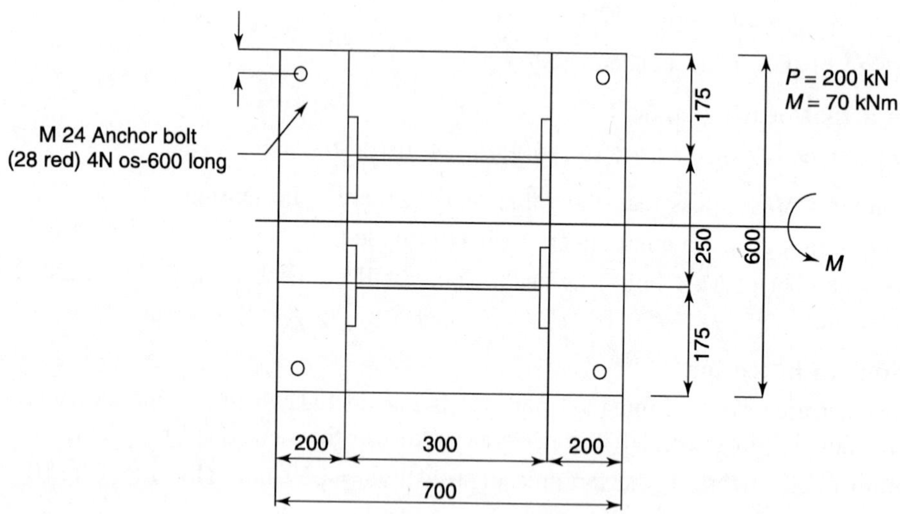

Fig. 6.16

Considering unit width of plate;

$$I = (1 \times t^3/12);\ y = t/2;\ Z = I/y = t^2/6$$

Moment capacity of plate = $1{,}2 \times f_y \times Z_e/\gamma_{mo}$

$$= 1.2 \times f_y \times (t^2/6)/\gamma_{mo} = 1.2 \times 250 \times (t^2/6)/1.1 = 45.45 \times t^2$$

Taking the maximum bending moment 22623 Nmm into consideration.

Pane l designation,	Edge condition	a mm	b mm	b/a	pp	MN mm	t mm
1.	Two adjacent sides fixed	125	200	1.6	0.1068	6007.5	-
2.	Three sides fixed	200	350	1.75	0.082	11808	-
3.	Three sides fixed	125	175	1.4	0.077	4331	-
4.	Four sides fixed	350	350	1	0.0513	22623	21.7 mm

$$t = \frac{22623}{\sqrt{45.45}} = 22.31 \text{ mm}$$

Provide 25 mm thick base plate.

Example 6.5 *A column consists of two ISMB 300 kept at 250 mm centre to centre of web is subjected to the following:*

Axial load = 175 kN.

Moment = 58 kNm

The load and moment is inclusive of the effect of wind. Design the base plate and anchor bolt. The concrete grade is M20 and the anchor bolt material is IS:432

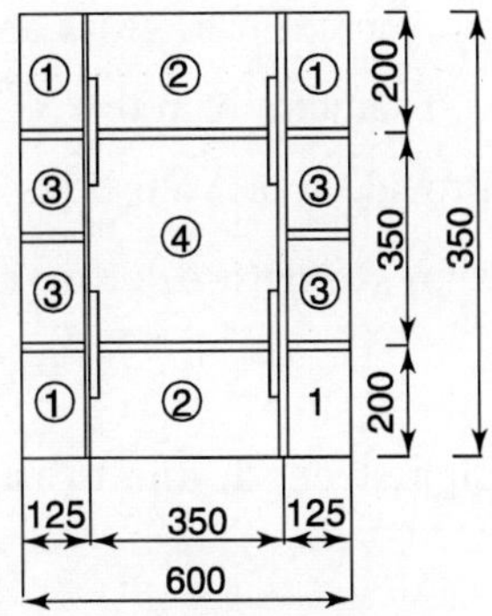

Fig. 6.17

Size of Base plate

Parallel to web = 250 + 2 × 175 = 600 mm

Parallel to flange = 300 + 2 × 200 = 700 mm

Factored load = 1.2 × 175 kN = 210 kN

Factored moment = 1.2 × 58 kNm = 69.6 kNm

$$w_{max} = P/A + M/Z = 210 \times 1000/(700 \times 600) + 69.6 \times 10^6 \times 6/(700 \times 600^2)$$

$$= 0.5 + 1.657 = 2.157 \text{ MPa} < 0.6 f_{ck} = 12 \text{ MPa}$$

The size of base plate is adequate

$$w_{min} = 0.5 - 1.657 = -1.157 \text{ MPa. (tension)}$$

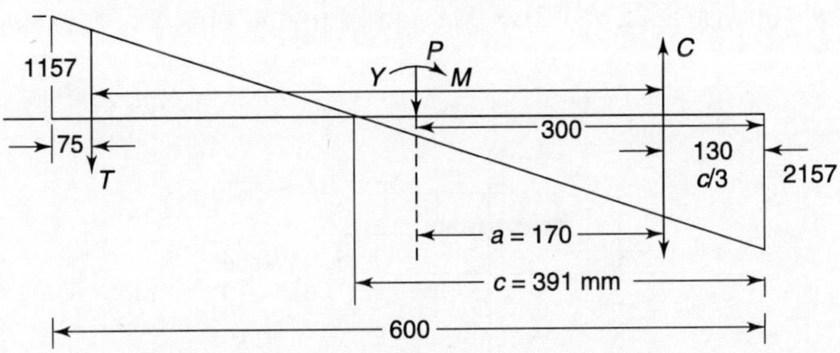

Fig. 6.18

$$c = w_{max}/[(w_{max} + w_{min})] = (2.157/(2.157 + 1.157) \times 600 = 391 \text{ mm}$$

Taking moment about C

$$M - P \times a - T \times \bar{y} = 0$$

Therefore, $T = ((M - Pa)/\bar{y}) = (69.6 - 210 \times 0.17)/0.395 = 85.82$ kN

$$\bar{y} = 600 - 75 - c/3 = 525 - 391/3 = 525 - 130 = 395 \text{ mm}$$

$$a = 300 - 130 = 170 \text{ mm}$$

Tension in each bolt, considering total as 4 (two on tension side and other two on compression side) = (85.82/2) = 42.91 kN.

Tensile strength of anchor bolt = 250/1.1 = 227.3MPa

Therefore, net area of bolt = (42.91 × 1000/227.3) = 188.78 mm^2.

Approximate gross area = 188.78/0.7 = 270 mm^2

Diametre of rod = $\sqrt{(270 \times 4/\pi)}$ = 18.52 mm

Provide four M20 bolts (rod diameter 22 mm)

Design strength due to yielding (Clause 6.2 of IS:800)

$$T_{dg} = A_g f_y/\gamma_{mo} = (\pi \times 22^2/4) \times (250/1.1) = 86.43 \text{ kN}.$$

Design strength due to rupture (Clause 6.3.2 of IS:800)

$$T_{dn} = 0.9\, A_{nc}\, f_u/\gamma_{m1} = (0.9 \times 297) \times (415/1.25) = 88.63 \text{ kN}.$$

Design strength = 86.43 kN > 42.91 kN.

Length of anchor bolts

Tension in anchor bolt 42.91 kN

Diameter of rod 22 mm.

Bond strength for M20: 1.2 MPa (Clause 26.2.1.1 of IS:456)

Force due to bond stress = Circumference × Embedded length × Bond strength

$= \pi d l_a \tau_o = \pi \times 22 \times l_a \times 1.2 = 42.91 \times 1000$

l_a = 42.91 × 1000/(π × 22 × 1.2) = 518 mm. Provide 600 mm embedded length

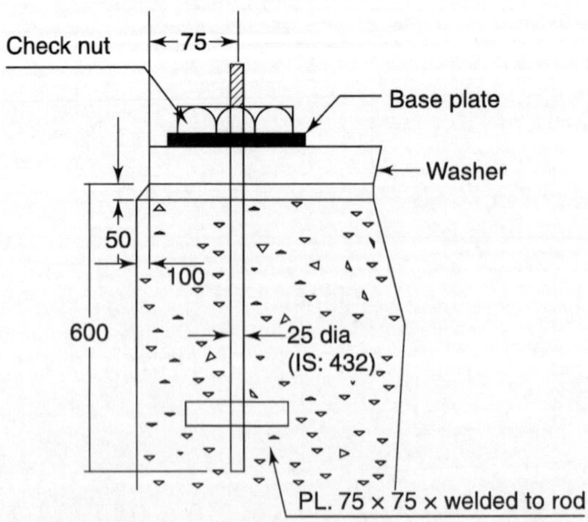

Fig. 6.19

Example 6.6 *Factored moment on column base = 10 kNm.*

Factored axial load = 100 kN.

Column size ISMB 250

Design the column base resting on M20 concrete

Flange width of ISMB 250 = 125 mm

Let us have the size of base plate in the same ratio of depth/width = 250/125 = 2

For no tension condition,

$$\frac{P}{A} - \frac{M}{Z} = 0$$

Bearing pressure $\dfrac{100 \times 10^3}{B \times 2B} - \dfrac{10 \times 10^6 \times 6}{B \times (2B)^2} = 0$

$$\frac{10^5}{2B^2} - \frac{6 \times 10^7}{4B^3} = 0$$

Multiplying both sides by $2B$ and resolving, $B = 300$

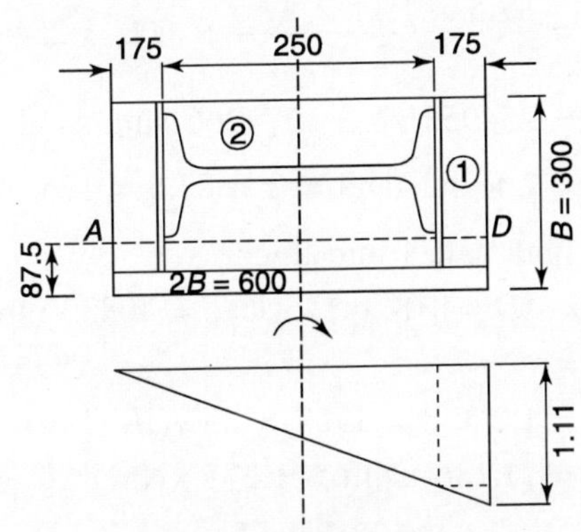

Fig. 6.20(a)

Maximum bearing pressure

$$\frac{10^5}{2 \times 300^2} - \frac{6 \times 10^7}{4 \times 300^3} = 1.11 < 12 \text{ MPa (Clause 7.4.1 of IS:800)}$$

The maximum bearing stress is too low. This indicates the size of base plate is more than that required. But the size of base plate is arrived at keeping the minimum bearing stress as zero (no tension condition). If the size of base plate is reduced, there will be tension and anchor bolts are to be designed

Thickness of base plate

Cantilever projection = 175 mm **(Part 1)**

Maximum Pressure = 1.11 MPa

$$M_u = 1.11 \times 175^2/2 = 16996.9 \text{ Nmm} = 1.2\, Z_e\, f_y/\gamma_{mo}$$

Therefore, Z_e required $= \dfrac{16996.9 \times 1.1}{1.2 \times 250} = 62.32 \text{ mm}^3 = t^2/6$

Therefore, $t = \sqrt{62.32} \times 6 = 19.33$ mm (1)

For Part 2

$$\therefore \quad b/a = 150/250 = 0.6$$

Three edges fixed/clamped/continuous. Hence, bending moment coefficient = 0.102

$$M = 0.102 \times 150^2 \times 1.11 = 2547.45 \text{ Nmm}$$

$$Z_e \text{ required} = 2547.45 \times 1.1/1.2 \times 250 = 9.34 \text{ mm}^3$$

$$t = \sqrt{6} \times 9.34/6 = 7.5 \text{ mm} \qquad (2)$$

Requirement 1 is 19.33 mm.

Hence, provide 20 mm thick base plate.

Design of stiffener:

Bending moment at face $AD = \dfrac{87.5 \times 1.11 \times 87.5}{2} \times 600 = 2549531$ Nmm

Per gusset/stiffener = 2549531/2 = 1274766 Nmm = 1.275 kNm

Let the height of stiffener = 2 × outstand = 2 × 87.5 = 175 mm

Provide a plate of 175 mm high and 8 mm thick

Section modulus of gusset at $AD = \{(8 \times 175^2/6)\} = 40833 \text{ mm}^3$

Design strength in bending =

$$1.2\, Z_e\, f_y/\gamma_{mo} = 1.2 \times 8 \times 1752 \times 250/(1000 \times 6 \times 1000 \times 1.1)$$

$$= 11.136 \text{ kNm} > 1.275 \text{ kNm}$$

Maximum shear $= \dfrac{87.5 \times 600 \times 1.11}{1000} = 58.275$ kN, Shear per plate = 29.14 kN

$$V_d = \text{Design shear strength} = A_v\, f_y/(\sqrt{3} \times \gamma_{mo})$$

$$= 175 \times 8 \times 250/(\sqrt{3} \times 1.1 \times 1000) = 183.7 \text{ kN} > 29.14 \text{ kN}$$

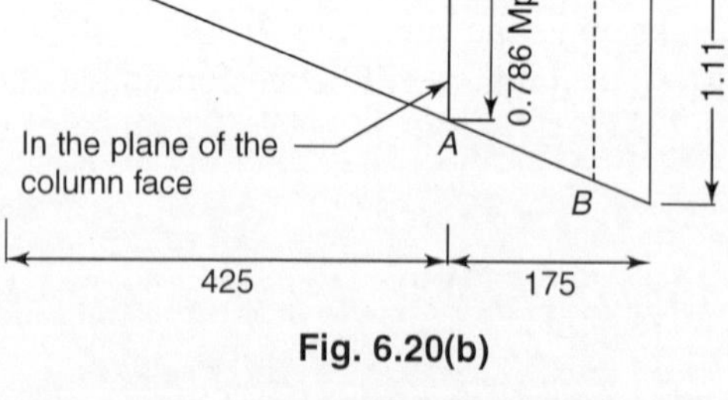

Fig. 6.20(b)

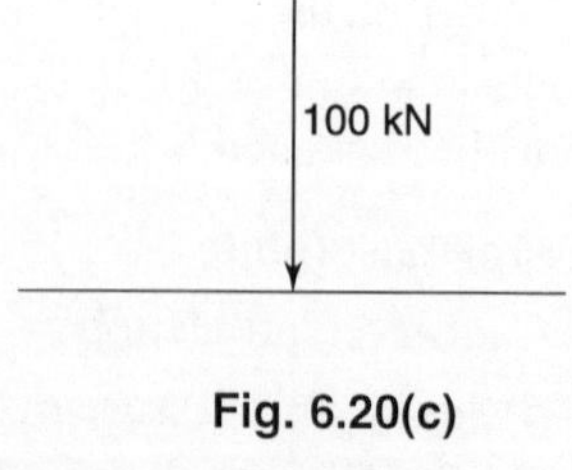

Fig. 6.20(c)

$0.6V_d = 0.6 \times 183.7 = 110$ kN > 29.14 kN. Hence, low shear.

Provide 8 thick stiffener for 175 mm height as shown in Fig. 6.20c.

Alternate arrangement

Pressure at face AD = 0.786 MPa (by Similar triangle)

$$\text{BM at face } AD = [((0.786 \times 175^2)/2) + (1/2 \times (1.11 - 0.786) \times 175 \times 2/3 \times 175)] \times 300 = 4602937.5 \text{ Nmm}$$

$$\text{Per plate} = 4602937.5/(2 \times 1000) = 2.3 \text{ kNm}$$

Design strength in bending 11.14 kNm (as already calculated)

Shear

$$0.786 \times 175 \times 300 + 1/2 \times (1.11 - 0.786) \times 175 \times 300 = 49770 \text{ N} = 49.77 \text{ kN}$$

$$\text{Per plate} = 24.885 \text{ kN}$$

$$\text{Shear capacity} = V_d = 183.7 \text{ kN (as already calculated)}$$

$$0.6\ V_d = 110 \text{ kNm} > 24.885 \text{ kN. Low shear.}$$

Hence, satisfactory.

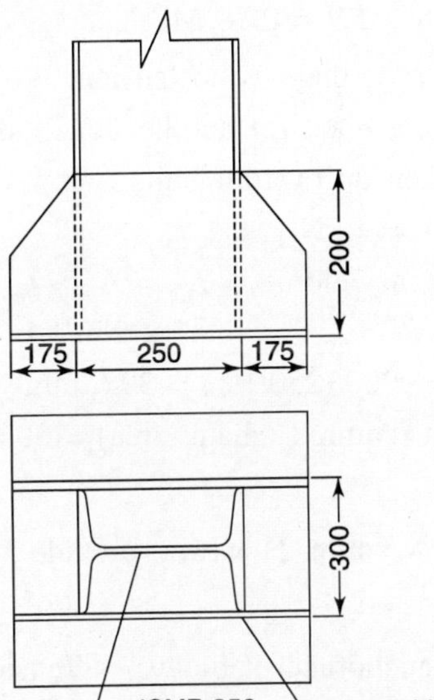

Fig. 6.21

Note: The thickness of base plate is to be calculated.

Example 6.7 *The column mark nos: F3a & F4b in an industrial structure is subjected to the following forces (Factored)*

(a) In one load combination with wind (dead load + Imposed load + wind load).

Maximum axial load = 1220 kN

Maximum bending moment (M) = 101 kNm

Maximum shear (S) = 57 kN

(b) In another load combination with wind

Maximum tension (T) = 36 kN

The size of base plate and location of bolt hole are in Fig. 6.22. Design the anchor bolts.

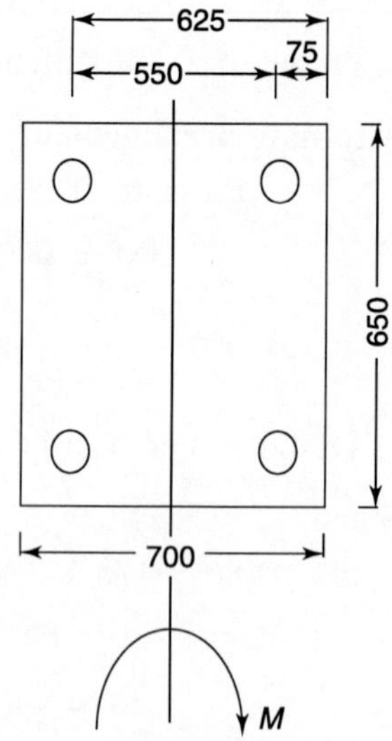

Fig. 6.22

Minimum size of base plate is decided based on the size of column, diameter of bolt, pitch of bolt and edge distance. Main criteria is to accommodate the bolts and to limit the actual maximum bearing stress within the bearing strength of concrete.

Base plate size = 700 mm × 650 mm

$(P/A) = (1220 \times 10^3/(700 \times 650)) = 2.68$ MPa

$(M/Z) = (101 \times 10^6 \times 6)/(650 \times 700^2) = 1.9$ MPa

$(P/A) + (M/Z) = 2.68 + 1.9 = 4.58$ MPa < 12 MPa for $M20$

$(P/A) - (M/Z) = 2.68 - 1.9 = 0.78$ MPa

Since M/Z is less than P/A; there is no tension.

Hence, bolts are to be designed for another load case where tension = 36 kN.

This tension will be taken up by four bolts.

Tension per bolt = 36/4 = 9 kN. (1)

Design strength of bolt in yielding = $T_{dg} = A_g f_y/\gamma_{mo}$ (Clause 6.2 of IS:800)

$= A_g \times 250/1.1$ (2)

Equating 1 and 2; $A_g = 9000/(250/1.1) = 39.6$ mm^2 = net area required

Approximate gross area required (shank area) = 39.6/ 0.7 = 56.6 mm^2

$\pi d^2/4 = 56.6$ mm^2

$d = \sqrt{(4 \times 56.6)/\pi} = 8.5$ mm. However, provide $M16$ threaded rods of 20 mm diameter.

Foundation bolt length

Bond length or length of anchorage of bolt L_d = Tension/$\pi d t_o$

$$L_d = (9000/\pi \times 20 \times 1.2) = 119 \text{ mm}$$

However, provide 300 mm anchorage.

where, L_d = development length of bolt in mm

d = nominal diameter of the bolt in mm

t_o = bond strength in N/mm^2 = 1.2 N/mm^2 for M20 (Clause 26.2.1.1 of IS:456)

Hence, satisfactory.

Note: For industrial structures the anchor bolts of less than *M*16 are not generally provided. Minimum anchorage is kept as 300 mm.

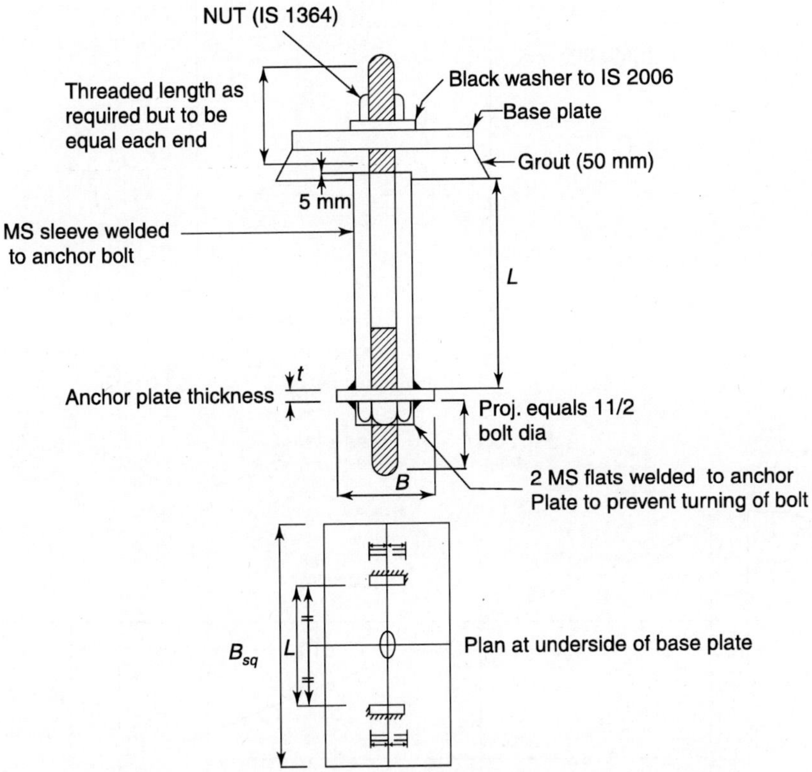

Fig. 6.23 *Details of typical holding-down bolts and anchor plates*

Example 6.8 *The column and base plate are shown in Fig. 6.24. The moment is 360 kNm and the vertical load is 220 kN. Both are factored. Ends of columns and gussets are not machined for bearing. M20 concrete is used. Design the anchor bolts.*

Bearing strength of M20 concrete = 0.6 × 20 = 12 MPa (Clause 7.4.1 of IS:800)

w_{max} = Maximum bearing stress on concrete

$= (220 \times 1000/650 \times 1000) \pm (6 \times 360 \times 10^6/650 \times 1000^2)$

$= 0.34 \pm 3.32 = 3.66$ (or) $- 2.98$ N/mm^2

w_{max} = Maximum bearing stress on concrete = 3.66 MPa < 12 MPa (Clause 7.4.1 of IS:800)

To find *NA*

$$NA = \frac{1000}{3.66 + 2.98} \times 2.98 = 449 \text{ mm}$$

Keeping holding down bolts at 100 mm from the edge of base plate.

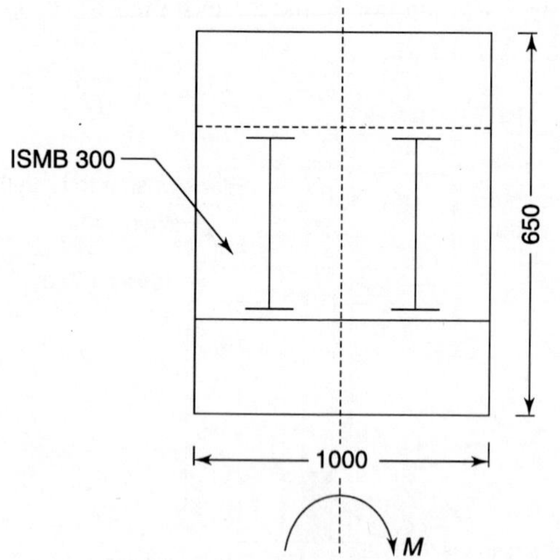

Fig. 6.24 (a) *Base plate*

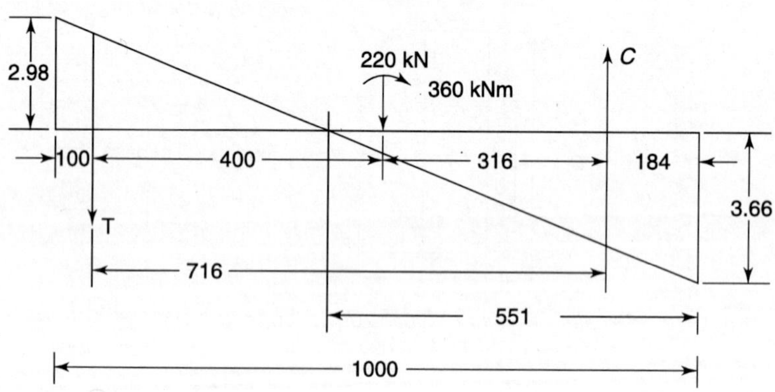

Fig. 6.24 (b) *Forces on the plate*

The compression force in concrete has a triangular distribution as shown in Fig. 6.24 (b) and centre of compression is at 551/3 = 184 mm from the compression edge.

The compression force (C) in concrete and the tensile force (T) in the bolt are to resist the vertical load and the moment.

Taking moment about C

$$T \times 716 + 360 \times 10^6 - 220 \times 10^3 \times 316 = 0$$

$$T = 360 \times 10^6 - 220 \times 10^3 \times 316/716 \times 1000 = 405.7 \text{ kN}$$

Let us provide eight bolts, four on each side.

Tension per bolt = 405.7/4 = 101.425 kN

Design tensile strength in yielding = $A_g\, f_y/\gamma_{mo} = (\pi d^2/4) \times 250/1.1 = 101.425$ kN.

Net area required = 101.425 × 1000 × 1.1/250 = 446.27 mm^2

Let us provide *M*24 threaded rod of 25 mm diameter.

Gross area = $(\pi d^2/4) = (\pi 25^2/4) = 491.07\ mm^2 = A_g$

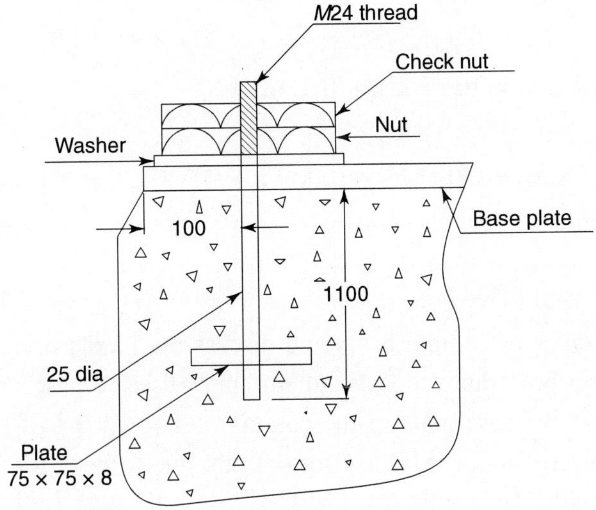

Fig. 6.25 *Typical cross section showing anchor bolt*

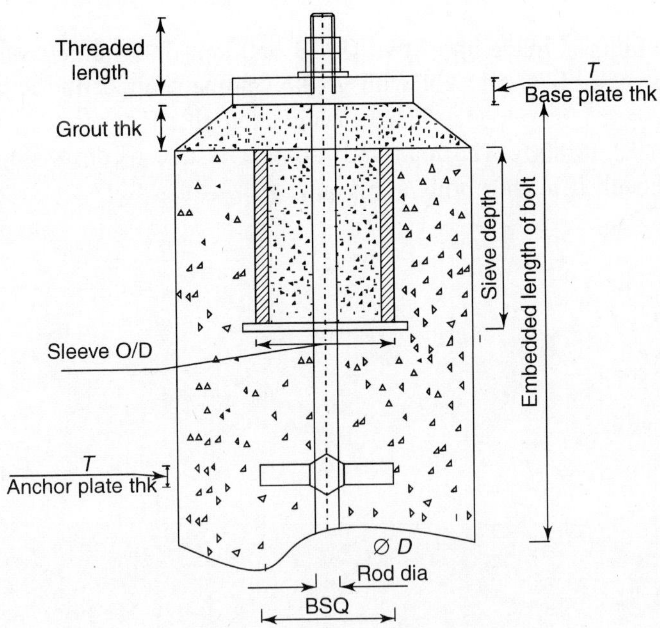

Fig. 6.26 *Typical cross section showing anchor bolt with sleeve*

Net area of $M24 = 353\ mm^2 = A_n$

Design strength due to yielding (Clause 6.2 of IS:800)

$$T_{dg} = A_g f_y/\gamma_{mo} = 491.07 \times 250/(1.1 \times 1000) = 111.61 \text{ kN}$$

Design strength due to rupture (Clause 6.3.2 of IS:800)

$$T_{dn} = 0.9\ A_{nc}\ f_u/\gamma_{m1} = 0.9 \times 353 \times 415/(1.25 \times 1000) = 105.5 \text{ kN}$$

$$\text{Design strength} = 105.5 \text{ kN} > 101.425 \text{ kN}$$

Embedding length

L_d = (Tension/bolt)/$\pi_d \times \tau_o$ = $(101.425 \times 1000)/(\pi \times 25 \times 1.2)$ = 1075.72 mm

Use 8 bolts of 25 mm dia rod with *M*24 thread and length of anchorage 1100 mm.

EXERCISE

1. Write down the types of column bases and discuss the merits of each over the other.
2. Explain the design procedure to designed stiffened/gusseted base.
3. How will you size the base plate of the column?
4. A column is subjected to an axial load of 450 kN due to dead load and imposed load. The bearing strength of concrete pedestal is 12 MPa. (Clause 7.4.1 of IS:800) The size of the column is ISMB 400. Design the slab base plate using the material having yield stress of 250 MPa. Also suggest stiffeners and calculate the thickness of stiffened base plate.
5. A built-up column is made up of two ISMB 350 kept 300 mm *c*/*c* of web. The column carries an axial load of 600 kN with 75 kN moment about the axis parallel to web. Design a gusseted base plate. The bearing strength of concrete pedestal is 12 MPa (Clause 7.4.1 of IS:800). The load and moment is due to dead, imposed and seismic loads. Also check if anchor bolts are required.

Chapter 7

Welded Structural Connections

General

Welding is done to join steel structural components.

Requirements of welds and welding shall conform to IS:816 and IS:9595. There are many types of welding. Electrodes are selected based on the properties of parent material. The common electrodes used for carbon steel structures are E6013 & E7018.

Lap Joint (Clause 10.5.1.2 of IS:800)

Minimum lap = four times the thickness of thinner part joined or 40 mm whichever is more.

Size of Weld (Clause 10.5.2.3 of IS: 800)

The size of fillet weld shall not be less than 3 mm and the minimum size of the first run of a single run fillet weld shall be as in Table 21 of IS:800/2007.

Effective Throat Thickness (Clause 10.5.3 of IS: 800)

The effective throat thickness of a fillet weld shall not be less than 3 mm and not to exceed $0.7t$ (or) $1.0t$, where t is the thickness of the thinner part connected.

The effective throat thickness = $K \times$ size of weld (Table 22 of IS:800)

Value of $K = 0.7$ if the plates/parts joined are at 60 deg to 90 deg. 0.707 is generally adopted.

Reduction in Weld Strength (Clause 10.5.7.3 of IS:800)

If the length of the welded joint, l_j of a splice or connection in a compression or tension member is greater than 150 t_t; the design capacity of weld f_{wd} shall be reduced by a factor

$$\beta_{tw} = (1.2 - (0.2\ l_j/150t_t)) < \text{or} = 1.0$$

where l_j = length of the joint in the direction of force transfer, and

t_t = throat size of the weld.

This limitation shall not apply to flange to web connection where the welds are loaded for the full length.

Intermittent Welds (Clause 10.5.5 of IS:800)

Effective length shall not be less than four times the weld size, with a minimum of 40 mm.

The clear spacing between the effective lengths of intermittent fillet weld shall not exceed 12 and 16 times the thickness of thinner plate joined for compression and tension joint respectively and in no case be more than 200 mm.

Actual compressive or tensile or shear stress of a fillet weld, should be less than or equal to design strength (f_{wd}) of weld.

$$f_{wd} = f_u/(\gamma_{mw} \times \sqrt{3}) \text{ (Clause 10.5.7.1.1 of IS:800)}$$

In the case of shop weld, γ_{mw} = 1.25 and in the case of site weld γ_{mw} = 1.5 (Table 5 of IS:800).

Butt Weld

Actual stresses in the butt welds shall be less than the design strength as governed by the parent metal welded together. (Clause 10.5.7.1.2 of IS:800)

7.1 WELDS AND WELDING

Electric arc welding is the most common method of connecting metal to metal by fusion. A low-voltage, high-current generator is used to create an electric arc between an electrode and the elements to be welded. The heat generated melts the metal, which then fuses to form a weld. The molten metal is protected either by flux or by inert gas to prevent oxidation and embrittlement

7.1.1 Welding Processes

Shielded metal arc welding (SMAW)

This is the most common method adopted for welding structural components. The consumable electrode has a core of filler metal with an outer covering of flux. The flux melts during the welding operation to protect the molten metal from atmospheric contamination due to oxidation.

Gas shielded metal arc welding (GMAW)

The electrode is a continuous spool of wire acting as the filler metal. Inert gas supplied continuously during the welding process protects the molten metal from oxidation.

Submerged arc welding (SAW)

This is a low hydrogen process. The electrode is a continuous spool of wire and the flux is a granular material fed from a hopper directly in front of the arc. Welding takes place submerged under a flow of granular flux.

Other welding processes are not discussed here.

7.2 TYPES OF WELDS

(a) Fillet welds

(b) Groove welds

(c) Slot welds

(d) Plug welds

7.2.1 Fillet Weld

Two steel faces may be joined by a fillet weld which normally has a triangular cross section. The strength of the weld depends on the throat thickness which is a multiple of the leg length. The effective throat thickness varies with the angle between the surfaces of steel joined together. When the angle between the surfaces to be joined is less than 60° or more then 120° then fillet welds are not permitted.

Generally, the angle between the surfaces is kept at 90°. For this, the throat thickness = 0.707 times the size of weld.

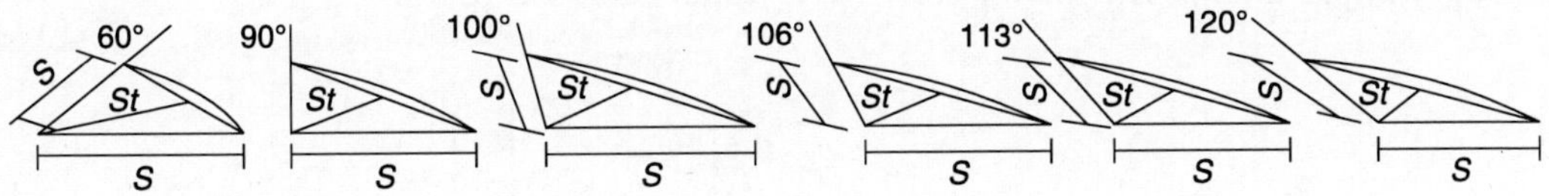

Fig. 7.1

Table 7.1 Effective throat thickness of fillet welds

Angle between the faces to be joined	*Effective throat thickness as a multiple of leg length*
60° – 90°	$S_t/S = 0.707$
91° – 100°	$S_t/S = 0.65$
101° – 106°	$S_t/S = 0.60$
107° – 113°	$S_t/S = 0.55$
114° – 120°	$S_t/S = 0.50$

S_t = throat thickness of fillet weld; S = leg length of fillet weld.

Design Rules

(1) End returns Fillet welds terminating at the ends of connected parts should be returned round the corners for a distance of at least twice the leg length. Also see Fig. 7.11 of this book. (Clause 10.5.1.1 of IS:800)

(2) Lap joints The minimum lap should not be less then $4t$ or 40 mm where t is the thickness of the thinner part joined. (Clause 10.5.1.2 of IS:800)

(3) Single fillet welds A single line of fillet weld must not be subjected to bending moment about its longitudinal axis (Clause 10.5.1.3 of IS:800)

(4) Intermittent fillet welds (Clause 10.5.5 of IS:800) Intermittent weld shall not be used where moisture ingress is possible. Intermittent welds have free space through which moisture enters and oxidation starts. Hence, wherever intermittent welds are proposed; continuous seal weld of at least 2 to 3 mm is also proposed to avoid moisture ingress. Intermittent fillet welds should not be subject to frequent stress changes (e.g., as in a gantry girder or a wind girder). Intermittent fillet weld shall have an effective length of not less than four times the weld size (or) 40 mm whichever is more. The longitudinal spacing between intermittent runs of fillet weld should not exceed $12t$ for compression elements and $16t$ for tension elements and shall not be less than 200 mm, where t is the minimum thickness of the connected part.

(5) Effective length of fillet weld (Clause 10.5.4 of IS:800) The effective length of fillet weld is its actual length less twice the length of the size of weld or leg length. The effective length should not be less than four times the leg length. When fillet weld is done at the rounded toe of a rolled section, the weld size shall not exceed 3/4 of the thickness at toe.

t_t = weld throat thickness = $0.7t$ where $t < T$ (refer Figs. 7.2 & 7.3)

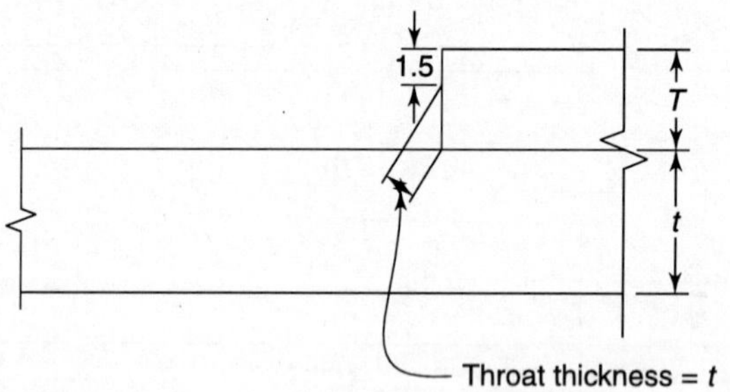

Fig. 7.2

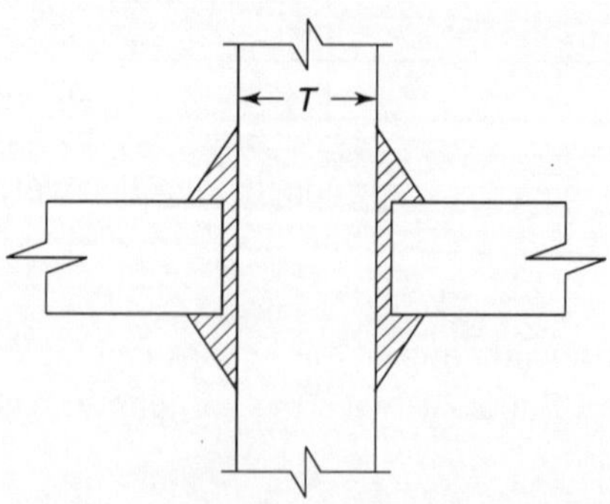

Fig. 7.3

7.2.2 Butt Welds

Partial penetration butt welds

The throat thickness of a partial penetration butt weld carried out from one side only should be taken as 5/8 of the thickness of thinner part joined. (Clause 6.1.4.2(b) of IS:816)

Partial penetration butt weld with superimposed fillet weld

The capacity of a partial penetration butt weld with superimposed fillet weld should be calculated with reference to its throat thickness as illustrated in Figs. 7.4 and 7.5 of this book.

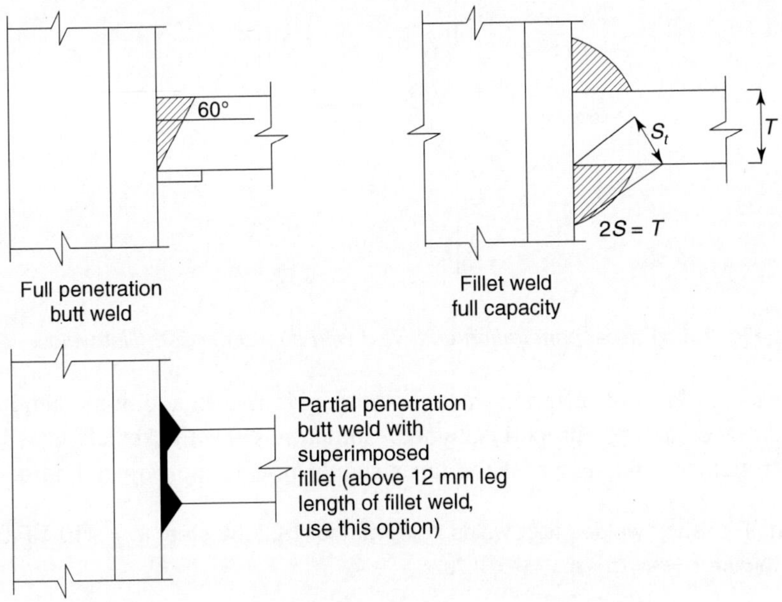

Fig. 7.4 *Full penetration and partial penetration butt weld*

The calculated force in the weld should not exceed the design strength of the weld. Assuming that the angle of bevel preparation is 45° the force equilibrium at the connection is given by:

$$T = 2(T_w \cos 45° + V_w \cos 45°)$$

where T = applied tensile force at the connection

T_w = tension on the fusion face

V_w = shear on the fusion face

From this analogy check that the shear and tensile forces on the fusion face do not exceed the design strength respectively.

Full preparation butt weld

The strength of a full preparation butt weld is equal to the strength of the parent material, provided a suitable welding consumable is used. The preparation for butt weld is very important. The shape of edge preparation depends on accessibility and the thickness of the metal parts being

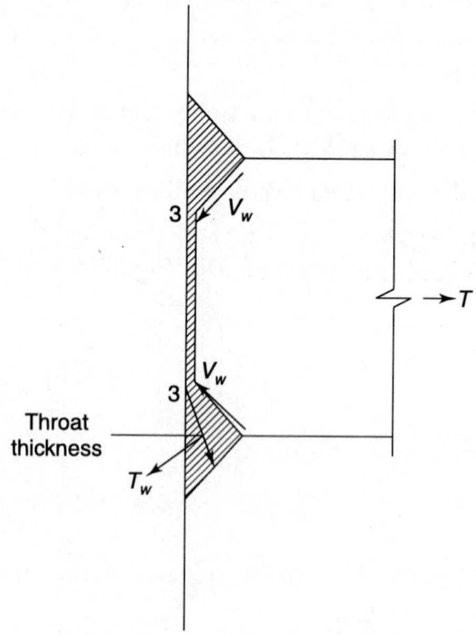

Fig. 7.5 *Partial penetration butt weld with superimposed fillet weld*

joined by butt welds. Plates of differing thickness may give rise to additional bending moments if their centroidal axes are not properly aligned. Sometimes the thicker element is tapered to meet the thinner element. The slope of the taper should not be more than 1 in 4.

Table 7.2 Strength of shop welded fillet welds in kN/mm for IS:2062 steel (f_u = 410 MPa)

Throat thickness = $t_t = t \times \cos 45° = 0.707t$

where t = size of weld.

Capacity = $f_u/(\gamma_{mw} \times \sqrt{3}) = (f_{wd} \times t_t)$ N/mm

Size of weld (or) Leg length t in mm	*Throat thickness t_t in mm*	*Design strength of shop weld per unit length in N/mm.*
3	2.12	401.7
4	2.83	535.6
5	3.54	669.5
6	4.24	803.4
8	5.66	1071.2
10	7.07	1339
12	8.48	1606.8
15	10.61	2008.6
18	12.73	2410.3
20	14.14	2678.1

Contd...

Contd...

22	15.55	2946
25	17.68	3347.6
f_u =	410 N/sq.mm	

For obtaining the strength of site weld, multiply the shop weld capacity by 0.83.

In the case of shop welded connection, design strength of weld is 189.37 N/mm^2 and that for site weld is 157.8 N/mm^2 for f_u = 410 MPa material. Design strength of shop weld per unit length in N/mm = 0.7071 × Size of weld × 189.37.

Table 7.3 Strength of full penetration shop butt welds in kN/mm run for IS:2062 steel (f_u = 410 MPa). (Clause 6.1.4.1 of IS:816)

t_e = thickness of thinner part in mm

Thickness of thinner part connected (t) in mm.	*Shop welded*	
	Shear = $(t_e \times f_y)/(\sqrt{3} \times \gamma_{mw})$ in N/mm	Tension (or) compression normal to effective area and tension (or) compression parallel to the axis of the weld = $(f_y/\gamma_{mw}) \times t_e$ in N/mm
6	1136	1200
8	1515	1600
10	1894	2000
12	2272	2400
15	2840	3000
18	3409	3600
20	3787	4000
22	4166	4400
25	4734	5000
28	5302	5600
30	5681	6000
35	6628	7000
40	7575	8060

Note: For obtaining the strength of single groove butt weld, the above values shall be multiplied by (5/8).

Table 7.4 Minimum size of fillet welds in mm

Thickness of thinner plate connected in mm	12	20	40	56	150	>150
Minimum weld size in mm	5	6	8	10	12	16

Also refer to Table 21 of IS: 800. Weld size is determined by the thinner of the two parts joined. The weld size can be 0.7 to 0.8 times thickness of the thinner part joined.

The capacity of a linear mm of fillet weld per mm length is given in Table 7.2.

Maximum size of fillet welds that is permitted along edges of connected parts shall be

(a) Material less than 6 mm thick, not greater than the thickness of the material.

(b) Material 6 mm or more in thickness not greater than the thickness of the material less 1.5 mm (Clause 6.2.5 of IS:816 and 10.5.8.1 of IS:800).

Fillet weld applied to the toe of a rolled section shall not exceed ¾ of the thickness of the section at toe.

7.3 COLUMN BASE CONNECTIONS

In welded connections for columns with gusseted bases, the gusset plates, angle cleats, stiffners, welding, etc., in combination with bearing area of the column, all fabricated flush for bearing, shall be sufficient to take the loads, bending moments and reaction to the base plate without exceeding specified stresses, but not less than 60% of the axial load shall be calculated as taken by welding.

If there is moment, sufficient welding shall be provided to transmit the full moment to the base.

Where the end of the column and the gusset plates are not machined for complete bearing, welding connecting them to the base plate shall be sufficient to transmit all the forces to which the base is subjected.

7.4 DESIGN EXAMPLES – FILLET WELDS

Example 7.1 *Design a shop fillet welded lap joint to connect two plates of 200 mm × 16 mm. Matreial is IS:2062, f_y = 250 MPa. Also design the joint with cover plates.*

Strength of plate due to yielding = $T_{dg} = \frac{f_y A_g}{\gamma_{mo}} = \frac{250 \times 200 \times 16}{1.1 \times 1000}$ = 727 kN (Clause 6.2 of IS:800)

Strength of plate due to rupture = $T_{dn} = 0.9\ A_n\ f_u/\gamma_{m1}$ (Clause 6.3 of IS:800)

$= 0.9 \times 200 \times 16 \times 410/1.25 \times 1000 = 944.64$ kN

Hence, strength of plate = least of T_{dg} and T_{dn} = 727 kN.

Minimum size of weld for 16 mm plate = 5 mm (Table 21 of IS:800)

Maximum size of weld = 16 – 1.5 = 14.5 mm (Clause 10.5.8.1 of IS:800)

Consider 8 mm weld.

Throat size = $0.707 \times t = 0.707 \times 8 = 5.656$ mm = t_t

Strength/mm = $(f_u/\sqrt{3} \times 1.25) \times 5.656 = (410/1.25 \times \sqrt{3}) \times 5.656 = 1071$ N/mm (Clause 10.5.7.1.1 of IS:800).

Required weld length = $\frac{727 \times 1000}{1071}$ = 679 mm

Length provided = $100 + 2\sqrt{200^2 + 50^2} < 679$ mm

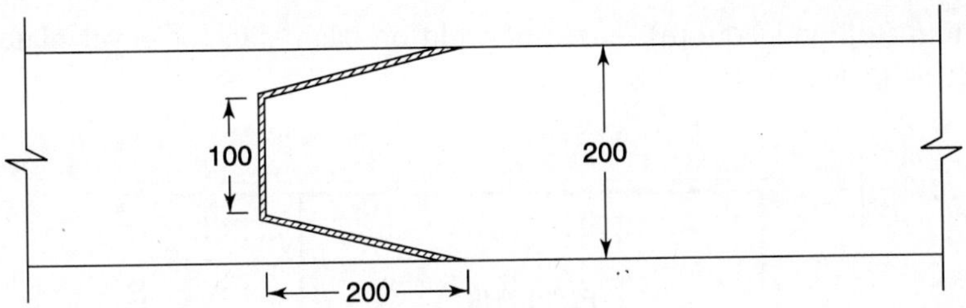

Fig. 7.6

Insufficient

Increasing the weld size to 12 mm

Throat thickness $t_t = 0.707 \times 12 = 8.424$ mm

$$\text{Strength} = \left[\frac{410}{\sqrt{3} \times 1.25}\right] \times 8.424 = 1607 \text{ N/mm}$$

$$\text{Required weld length} = \frac{727 \times 1000}{1607} = 452 \text{ mm}$$

Provided 512 mm is adequate.

Joint Using Cover Plates

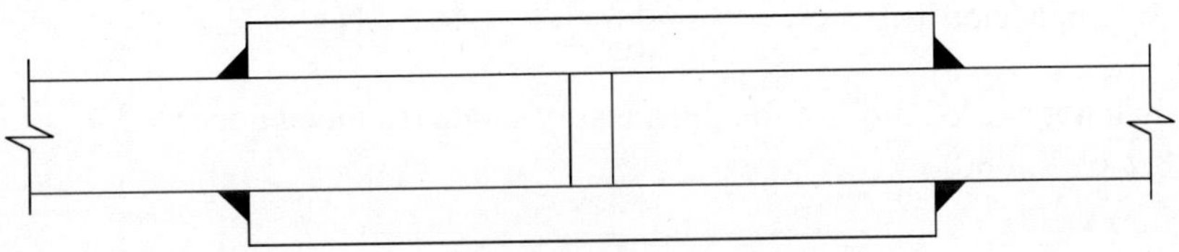

Fig. 7.7

Let the width of cover plate be = 200 – 2 × 15 = 170 mm

15 mm space is left for welding

$$\text{Required area of cover plate} = 1.05 \times 200 \times \frac{16}{2} = 1680 \text{ mm}^2$$

$$\text{Width} = 170 \text{ mm}; \qquad \text{Thickness} = \frac{1680}{70} = 9.88 \text{ mm}$$

Use 10 mm thick 170 mm wide cover plate on either side.

Use 6 mm fillet weld. Strength = 189 × 0.707 × 6 = 802 N/mm

$$\text{Length of weld required} = \frac{727000}{802} = 906 \text{ mm}$$

$$\text{Length of sides} = \frac{906 - 170 \times 2}{2 \times 2} = 141.5 \text{ mm}$$

Provide 200 mm (top and bottom) length of weld on each side for cover plate and on both sides of the joint as shown in Fig. 7.8.

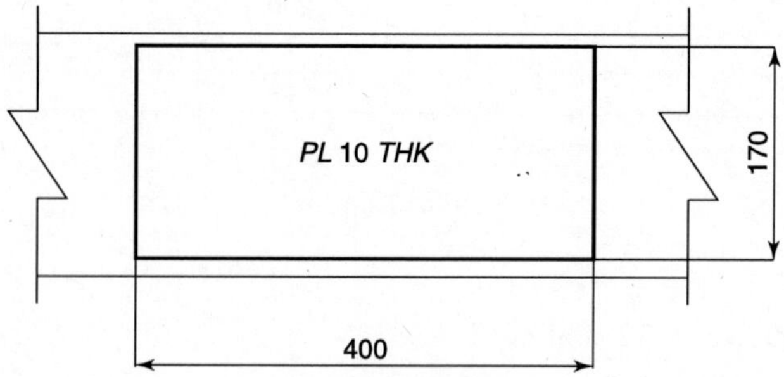

Fig. 7.8

Example 7.2 *ISA 65 × 65 × 6 is used as a tension member and connected to 8 mm thick gusset plate. Design a shop fillet welded joint for a tensile force of 160 kN (factored).*

Distance of C.G. Line of angle = C_{yy} = 18.1 mm.

Resistance by outstanding leg = $F_1 = \dfrac{160 \times (65 - 18.1)}{65} = 115.4$ kN

Resistance by connected leg = $F_2 = \dfrac{160 \times (18.1)}{65} = 44.6$ kN

Weld size shall not exceed 3/4th of the thickness of angle (rounded edge) = 3/4th of 6 = 4.5 mm (Clause 10.5.8.2 of IS:800)

Provide 4 mm weld

Strength/mm = 189.37 × 0.707 × 4 = 535.6 N/mm (Refer to Table 7.2)

$$L_1 = (F_1/535.6) = 215.5 \text{ mm}$$

$$L_2 = (F_2/535.6) = 83.3 \text{ mm.}$$

L_1 and L_2 are effective lengths of welds.

Actual length = effective length + 2 × size of weld (Clause 10.5.4.1 of IS:800)

$$L_1 = 215.5 + 2 \times 4 = 223.5 \text{ mm. Provide 225 mm.}$$

$$L_2 = 83.3 + 2 \times 4 = 91.3 \text{ mm. Provide 95 mm.}$$

End return 2 × 4 = 8 mm (Clause 10.5.1.1 of IS:800)

Check for Block Shear: (Clause 6.4.2 of IS:800)

Gusset thickness = 8 mm

Block shear strength = (225 + 95) × 8 × (250/√3 × 1.1)/1000 = 336 kN > 160 kN

Hence, satisfactory.

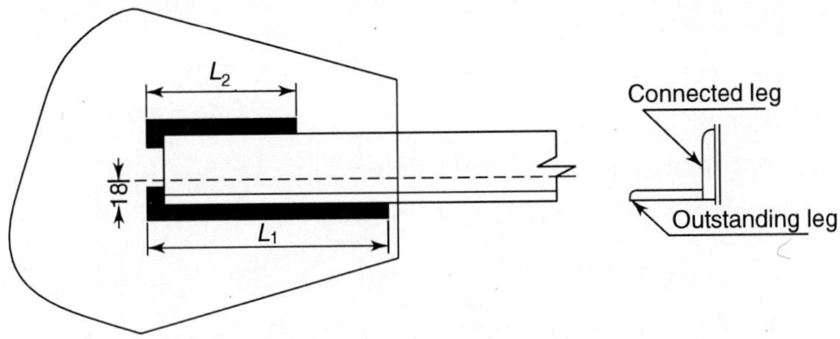

Fig. 7.9

Example 7.3 *Design the fillet welded (at shop) connection with 8 mm thick gusset for an angle tie ISA 60 × 60 × 6 subjected to a tensile force of 50 kN due to dead load and imposed load.*

Refer Fig. 7.10.

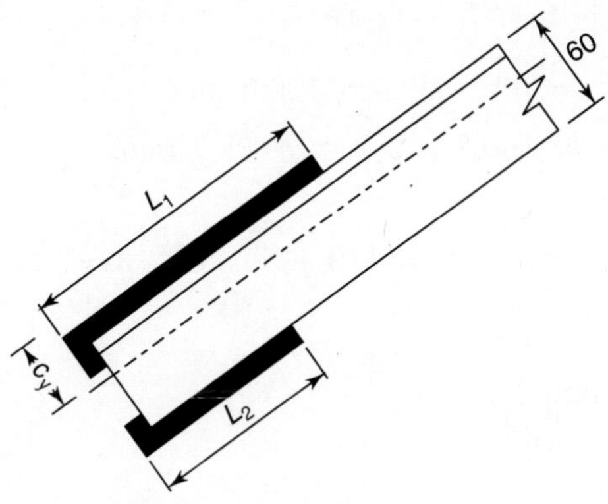

Fig. 7.10

Taking moment of force in weld about C.G.

$$L_1 \times c_y = L_2\,(60 - c_y) \tag{1}$$

$$L_1 + L_2 = \text{length of weld required} \tag{2}$$

Resolving 1 and 2, we can get L_1 and L_2

$$\text{Force} = 50 \text{ kN} = 50000 \text{ N}$$

$$\text{Thickness of angle} = 6 \text{ mm.}$$

When fillet weld is done at the rounded toe of a rolled section, the weld size shall not exceed 3/4 of the thickness at toe.

$$\text{Size of weld} = (3/4) \times 6 = 4.5 \text{ mm}$$

Provide 4 mm weld

Strength of 4 mm weld = 535.6 N/mm (Refer to Table 7.2).

L_1, L_2 are effective length of welds.

Factored load = 1.5 × 50 = 75 kN

Length of weld required = 75000/535.6 = 140 mm

$$L_1 + L_2 = 140 \text{ mm} \quad (1)$$

c_y for ISA 60 × 60 × 6 = 16.9 mm

$$L_1 \times 16.9 = L_2 (60 - 16.9) = 43.1 L_2 \quad (2)$$

$$L_1 = (43.1/16.9) L_2 = 2.55L_{-2} \quad \text{(From equation 2)}$$

Substituting in (1), we get

$$2.55 L_2 + L_2 = 140 \text{ mm}$$

$$L_2 = 140/3.55 = 39.4 \text{ mm} + 2 \times 4 = 47.4 \text{ mm}$$

Provide 50 mm

$$L_1 = 140 - 39.4 = 100.6 \text{ mm}$$

Provide $L_1 = 100.7 + 2 \times 4 = 108.7$ mm.

Provide 110 mm

Total $= L_1 + L_2 = 110 + 50 - 8 - 8 = 144 \text{ mm} > 140 \text{ mm}$

Size of weld 4 mm

Return weld = 2 × 4 = 8 mm

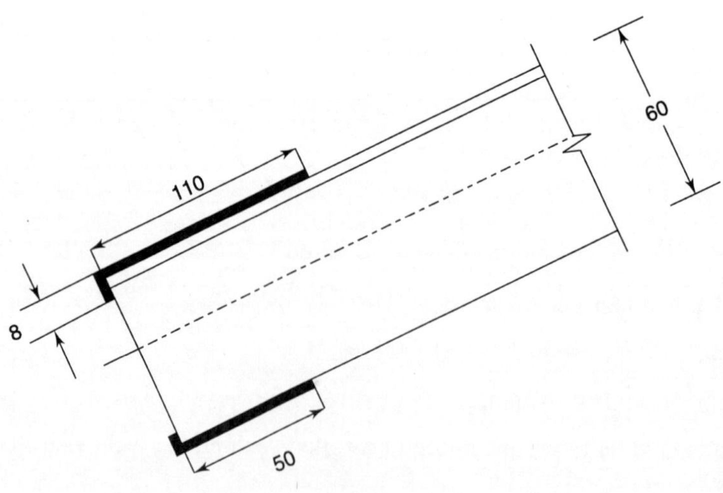

Fig. 7.11

Note:

(a) Calculated length is effective length (39.4 mm & 100.7 mm). To arrive at actual length; twice the weld size is to be added (Clause 10.5.4.1 of IS:800).

(b) Block shear shall be checked as in Example 7.2.

Example 7.4 *Gusset and rafter weld joint*

A 16.0 M span truss has a rafter consisting of 2 ISA 75 × 75 × 8 and the same is on either side of gusset subjected to a compression of 80 kN. Design the shop fillet welded connection to gusset plate. The compressive force is due to dead load + imposed load.

Solution: Section properties of one ISA 75 × 75 × 8 are, $A = 1138 \text{ mm}^2$

$$C_{yy} = 21.4 \text{ mm}, t = 8 \text{ mm}$$

Maximum weld size = 0.75 × 8 = 6 mm (Clause 10.5.8.2 of IS:800).

Provide 6 mm weld.

Strength of weld = F_{wd} = 803.4 N/mm (Refer to Table 7.2 of this book).

Factored load = 1.5 × 80 = 120 kN (If the force is due to wind load, the factor is 1.2.)

Weld length required = 120000/803.4 = 149.4 mm

Length of weld per angle = 149.4/2 = 74.7 mm = $L_1 + L_2$

Weld length ratio

$$(75 - 21.4)\ L_2 = L_1 \times 21.4$$

$$\frac{L_2}{L_1} = \frac{21.4}{75 - 21.4} = 0.4$$

$$L_1 + L_2 = L_1 + 0.4\ L_1 = 1.4\ L_1 = 74.7 \text{ mm}$$

$$L_1 = 74.7/1.4 = 53.36 \text{ mm}$$

$$L_2 = 0.4 \times 53.36 = 21.34 \text{ mm}$$

Refer to Fig. 7.10.

Weld length on the upright leg is = $L_1 + 2 \times 6$ = 65.36 mm Provide 80 mm.

Weld length on the connected leg = $L_2 + 2 \times 6$ = 33.34. Provide 60 mm

Provide 12 mm long return weld

Total effective length of weld = (80 – 12) + (60 – 12) = 116 mm > 74.7 mm

Example 7.5 *Design the shop fillet welded connection for two flats each of thickness 8 mm subjected to 40 kN pure tension due to wind force. Material specification is IS:2062 f_y = 250 MPa.*

Tensile force = 40,000 N

Minimum thickness of weld = 3 mm (Table 21 of IS:800).

Using 6 mm fillet weld

Throat thickness = 0.707 × 6 = 4.242 mm

Strength of fillet = 803.4 N/mm

Total length required = 1.2 × 40,000/803.4 = 59.75 mm (Refer to Table 7.2 of this book)

Divided equally on either side = 30 mm + 2 × 6 = 42 mm

Provide 50 mm

Welds must be returned at ends for a length = 2 × 6 = 12 mm

Note: The weld size can be 0.7 times the thickness of thinner plate joined. The maximum size of weld shall not exceed the thickness of thinner plate joined. (Clause 10.5.3.1 of IS:800).

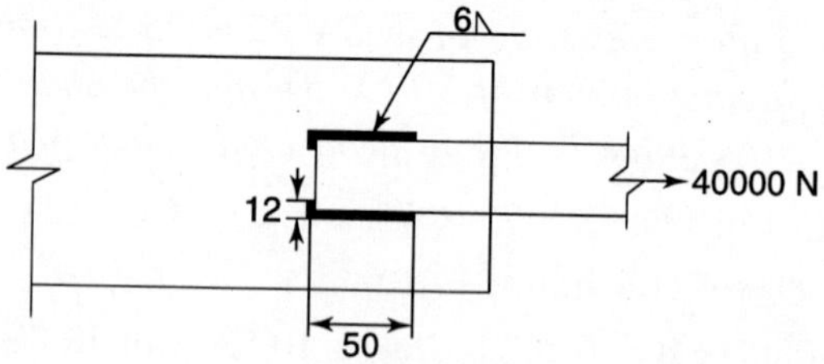

Fig. 7.12

Example 7.6 *Design a single angle in tension member subjected to a tensile force of 65 kN due to dead load and imposed load. Also design the shop fillet welded connection with 8 mm gusset plate. The tension is due to D.L + I.L.*

Tensile force in the single angle member = 65 kN

Factored force = 1.5 × 65 = 97.5 kN. (Table 4 of IS:800)

To choose an angle

$$T_{dg} = A_g f_y/\gamma_{mo} \text{ (Clause 6.2 of IS:800)}$$

$$A_g = T_{dg}/(f_y/\gamma_{mo})$$

$$\text{Net area required} = \frac{97.5 \times 1000}{(250/1.1)} = 429 \text{ mm}^2$$

Gross area required = net area required = 429 mm^2 (since it has welded connection)

Try ISA 50 × 50 × 6

Gross area = 568 mm^2 C_{yy} = 14.5 mm

Size of weld = 0.75 × thickness of rolled section at toe = 0.75 × 6 = 4 mm

Provide 4 mm fillet weld,

Strength of 4 mm fillet weld = 535.6 N/mm (Refer to Table 7.2 of this book).

To avoid distortion, moment about face C.G.

$$L_1 \times 14.5 = L_2\ (50 - 14.5)$$

$$L_1 = 2.45\ L_2$$

Length of weld required = 1.5 × 65,000/535.6 = 182 mm = $L_1 + L_2$ = 2.45 $L2$ + $L2$ = 3.45 $L2$

$$L_2 = 182/3.45 = 52.76 \text{ mm.}$$

Adding ends = 52.76 + 2 × 4 = 60.76 mm.

Provide 65 mm

$$L_1 = 2.45 \times 52.76 + 2 \times 4 = 137.3 \text{ mm.}$$

Provide 140 mm

However, in practice, the weld is being done equal to 140 mm length on either leg which will create minor eccentricity.

Provide return weld = $2t = 2 \times 4 = 8$ mm

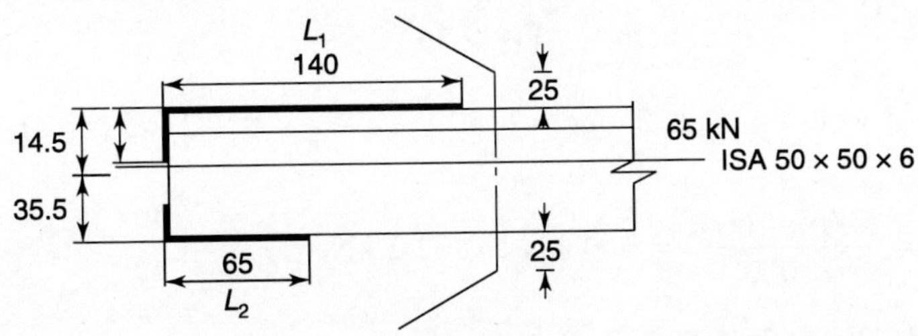

Fig. 7.13

Example 7.7 *Connection between column base and column*

A plate formed I section of total depth 500 mm and having flange width of 350 mm, thickness of flange 20 mm and that of web 12 mm is subjected to an axial load of 800 kN and moment of 152 kNm. The size of weld between the web and flange is 8 mm. Calculate the fillet weld size for connecting this column to column base at the shop. The load is purely due to dead load and imposed load. Material of the plate is IS:2061 having f_y = 250 MPa. Thickness of base plate is 25 mm.

Solution: The forces are assumed to be imposed by the column through its flanges.

Then the force at one of the flanges due to the direct load, assuming flush bearing. 60% is assumed to be transfered through the weld.

Flush bearing is achieved by machining the bottom of column, top of base plate and blue matching to 80%.

Factored force per flange = F_s = (1.5 × 800/2) × 0.6 = 360 kN

Force due to bending moment = Moment/column depth = 1.5 × 152 × 10^3/454 = 502.2 kN

Total load in each flange = 360 + 502.2 = 862.2 kN

The maximum length of weld which can be done on the inside face of flange is 350 – 12 – 2 × 8 = 322 mm and 350 mm outside as shown in Fig. 7.14 and deducting 50 mm for end runs, the effective length available = 622 mm.

Therefore, force/mm = 862.2/622 = 1.386 kN = 1386 N.

Using 12 mm shop fillet weld, strength = 1606.8 N/mm run (Refer to Table 7.2 of this book).

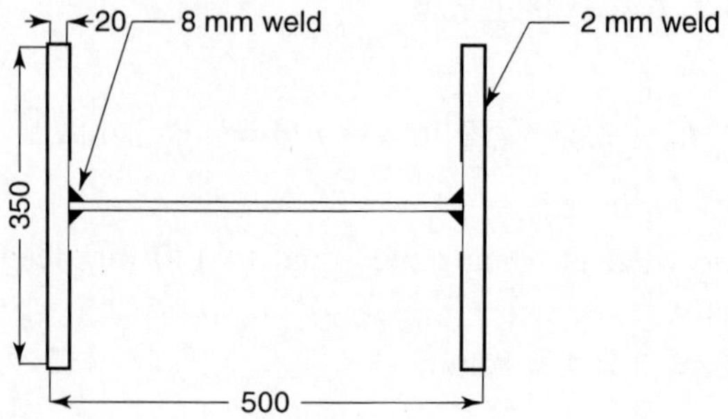

Fig. 7.14

Effective length of weld on each flange = (350 – 12 – 2 × 8 – 2 × 2 × 12) + (350 – 2 × 12) = 600 mm

$$\text{Design strength} = \frac{600 \times 160.8}{1000} = 964 \text{ kN} > 862.2 \text{ kN}$$

Provide 12 mm fillet weld as shown in Fig. 7.14.

Example 7.8 *Connection between column base and column*

Design the shop fillet welded connection between column and column base for a column of size ISMB 400 carrying an axial load of 750 kN due to dead and imposed loads. Size of base plate 700 × 450 × 25 mm.

Solution: Factored load = 1.5 × 750 = 1125 kN

Thickness of web of ISMB 400 = 8.9 mm

Size of weld 0.7 × 8.9 = 6.23 mm. Provide 6 mm weld.

Thickness of flange of ISMB 400 = 16 mm

Size of weld 0.7 × 16 = 11.2 mm. Provide 12 mm weld.

Design Strength of Shop Weld

Strength of 6 mm weld per mm = 803.4 N/mm and that for 12 mm weld = 1606.8 N/mm (Table 7.2 of this book).

Clear depth of web = h_1 = 334.4 mm.

Effective length of weld on web = 2 × (334.4 – 2 × size of weld) = 2 × (334.4 – 2 × 6) = 644.8 mm

$$\text{Design strength} = \frac{644.8 \times 803.4}{1000} = 518 \text{ kN}$$

Effective Length of Weld on Flange

Inside = 2 × (140 – 8.9 – 2 × 2 × 12) = 2 × 83.1 = 166.2 mm.

Outside = 2 × (140 – 2 × 12) = 232 mm.

Total effective length = 166.2 + 232 = 398.2 mm.

$$\text{Design strength} = \frac{398.2 \times 1606.8}{1000} = 639.22 \text{ kN}$$

Total design strength of weld = 518 + 639.8 = 1157.8 kN > 1125 kN.

Hence, satisfactory.

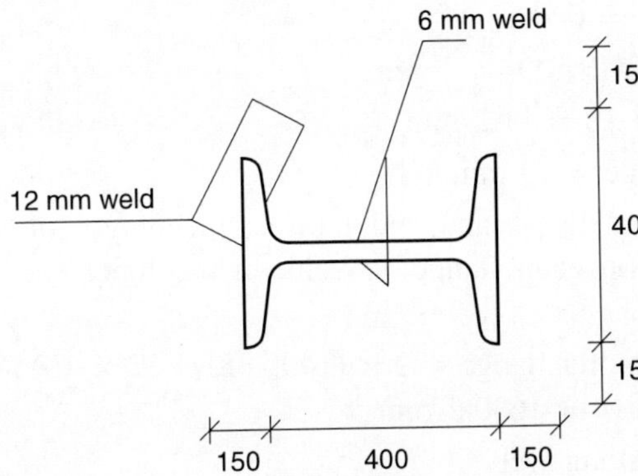

Fig. 7.15

Note: In practice the bottom of column and top of base plate are machined to achieve at least 80% contact. Only 20% of load may be transferred through weld. In this solution, we have considered 100% of axial load is transferred through the weld.

Example 7.9 *Connection between stiffened base plate and column*

A column is subjected to axial load and bending moment of 500 kN and 120 kNm during seismic load condition. The column is provided with a base plate of 450 × 700 × 25 mm along with stiffener plates of 16 mm thick as shown in Fig. 7.16. Design the shop fillet weld to carry the load. Size of column ISMB 400

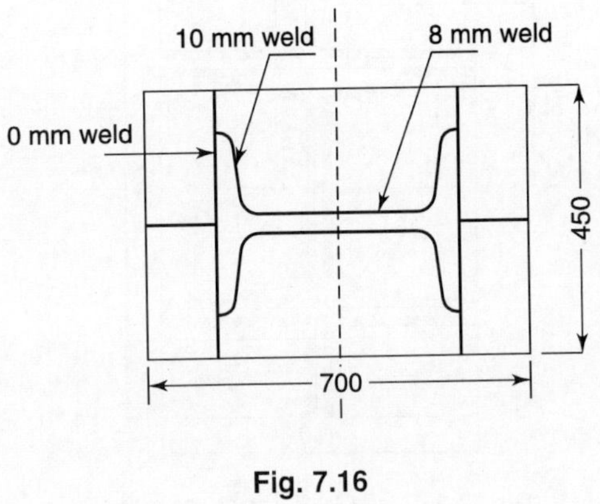

Fig. 7.16

The load and moment is inclusive of seismic condition.

Hence, the partial safety factor for load is 1.2. (Table 4 of IS:800)

Factored load = P = 1.2 × 500 = 600 kN

Factored moment = M = 1.2 × 120 kNm = 144 kNm

Thickness of web of ISMB 400 = 8.9 mm (steel property table)

Weld size can be = 0.7 × 8.9 = 6.23 mm

Let the weld size on web = 6 mm

Thickness of flange of ISMB 400 = 16 mm

Weld size can be = 0.7 × 16 = 11.2 mm

Let the weld size on flange = 10 mm

After allowing for web thickness and end returns, the length of weld on each flange is = Effective length of weld on each flange = Weld on the inner side of flange + weld with gussets.

Effective length of weld along flange = 2 × (140 – 8.9 – 2 × 10) + 2 × (450 – 140 – 2 × 10) + 2 × (450 – 16 – 2 × 10) = 1630.2 mm.

Effective length of weld along web

= effective length on web + effective length of gusset along web (on both sides).

Effective length of weld along web = 2 × [400 – 2 × 16 – 2 × 6] + 2 × (700 – 400 – 2 × 16 – 2 × 10) = 1208 mm.

The sectional properties of the weld, namely, the area of cross section and moment of inertia of the weld are:

$$A_w = 0.707 \times 6 \times 1208 + 0.707 \times 10 \times 1630.2 = 5124 + 11525.5 = 16649.5 \text{ mm}^2$$

I_w = Moment of inertia of the web weld + Moment of inertia of the flange welds all taken about the centre of gravity of the column major axis (axis about which bending moment is acting).

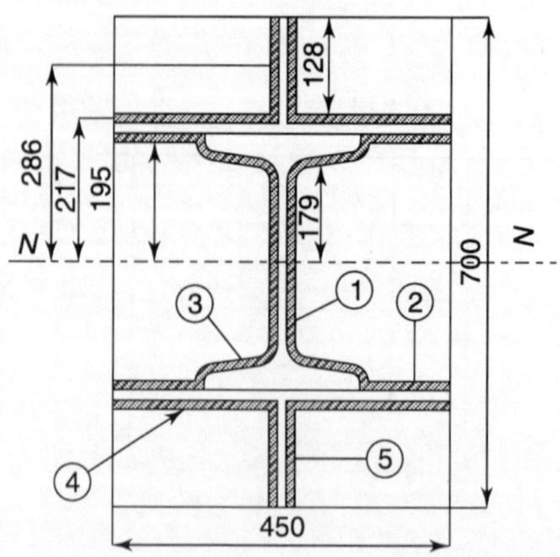

Fig. 7.17 *C.G. distance of weld*

Moment of Inertia

- Length of 6 mm weld = (400 – 2 × 16 – 2 × 6) = 356 mm

  $Iz_1 = (0.707 \times 6 \times 356^3/12) \times 2 = 31.9 \times 10^6 \text{ mm}^4$
- Length of 10 mm weld = (450 – 140 – 2 × 2 × 10)/2 = 135 mm

  $Iz_2 = 4 \times [(135 \times (0.707 \times 10)^3/12) + 0.707 \times 10 \times 135 \times 195^2] = 145.2 \times 10^6 \text{ mm}^4$
- Length of 10 mm weld = (140 – 8.9 – 2 × 2 × 10)/2 = 45.55 mm

  $Iz_3 = 4[(45.55 \times (0.707 \times 10)^3/12) + 0.707 \times 10 \times 45.55 \times 179^2] = 41.28 \times 10^6 \text{ mm}^4$
- Length of 10 mm weld = (450 – 16 – 2 × 2 × 10)/2 = 197 mm

  $Iz_4 = 4[(197 \times (0.707 \times 10)^3/12) + 0707 \times 10 \times 197 \times 217^2] = 262.6 \times 10^6 \text{ mm}^4$
- Length of 10 mm weld = (7000 – 400 – 2 × 16 – 2 × 10)/2 = 124 mm

Area = 0.707 × 10 × 124 = 876.68 mm^2

- $Iz_5 = 4[(0.707 \times 10 \times 124^3/12) + 876.68 \times 286^2] = 291.3 \times 10^6 \text{ mm}^4$

  $Izz = Iz_1 + Iz_2 + Iz_3 + Iz_4 + Iz_5 = 772.28 \times 10^6 \text{ mm}^4$

Distance of the outermost weld of the flange is y = 700/2 = 350 mm.

The stresses due to axial force and bending = $(P/A_w) + (M_y/z)$

$= [600 \times 1000/16649.5] + [(144 \times 10^6)/(772.28 \times 10^6/350)]$

$= [36.03 + 65.3] = 101.33 \text{ N/mm}^2 < 189 \text{ N/mm}^2$

Note: 1. End returns 2 × size of weld is deducted everywhere. It is enough if the same is deducted only where the weld terminates.

2. It is sufficient to consider 20% of axial load if 80% contact between bottom of column and top of base plate is achieved by machining. However, 100% moment is to be considered.

Example 7.10 *Design the fillet welded joint at shop between web and flange of the plate girder in Fig. 7.18 (the portion above NA is shown). The factored shear force = 700 kN. The material is IS:2062 f_y = 250 MPa. Thickness of web = 20 mm. Thickness of flange = 25 mm.*

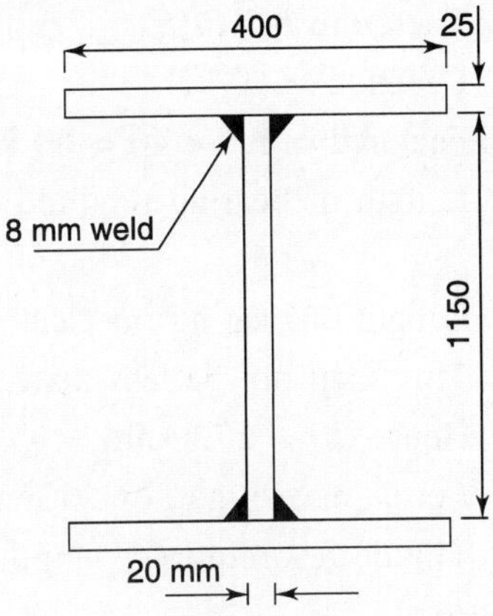

Fig. 7.18

Minimum size of weld for thicker part 25 mm = 6 mm (Table 21 of IS:800).

Providing 8 mm weld

$$\text{Horizontal shear} = \tau_{vf}\,\text{cal} = \frac{VA_f\,\overline{y_f}}{I_{zz}\,\Sigma t}$$

Throat thickness = t = 0.7071 × 8 = 5.656 mm

$$\Sigma t = 2 \times 5.656 = 11.312 \text{ mm}$$

$$A_f\,y_f = 400 \times 25 \times [(1150/2) + (25/2)] = 587.5 \times 10^4 \text{ mm}^3$$

I_{zz} = [(400 × 25^3/12) + (400 × 25 × 587.5^2)] = 520833 + 3451562 = 3452 × 10^3 = 3.452 × 10^9 mm^3

τ_{vf} cal = [(700 × 1000 × 587.5 × 10^4)/(3.452 × 10^9 × 11.312)] = 105.32 N/mm^2 < 189.37 N/mm^2

Since the stress (105.32 N/mm^2) is far less than the design strength (189.37 N/mm^2) weld size can be reduced.

Example 7.11 *ISMB 300 beam transmits 40 kN reaction on one of the flanges of the column ISMB 500. Design a fillet welded (shop) seat angle connection.*

Properties of ISMB 300

$$\tau_f = 12.4 \text{ mm}$$

$$r_b = 14 \text{ mm}$$

$$t_w = 7.5 \text{ mm}$$

Width of flange of beam = 140 mm.

Seat angle is kept below the bottom flange of beam to support the beam. The length of angle can be kept more than the width of the flange of beam provided the flange of the column is wider. Let us keep 10 mm projection of seat angle beyond the flange of beam. Therefore, length of seat angle = 140 + 2 × 10 = 160 mm.

Width of flange of ISMB 500 = 180 mm > 160 mm.

(Length of bearing × t_w) × (f_{yw}/γ_{mo}) > Reaction

$$b \times t_w\ (f_{yw}/\gamma_{mo}) > R$$

Therefore b $R/(t_w(f_{yw}/\gamma_{mo}))$

Reaction = 40 kN

Factored = 1.5 × 40 = 60 kN

Length of bearing required at the root line = $b = R/(t_w(f_{yw}/\gamma_{mo}))$

= 60 × 1000 × 1.1/7.5 × 250 = 35.2 mm

Length of bearing on cleat = $b_1 = b - (t_f - \gamma_b)$ = 35.2 – (12.4 + 14) = 8.8 mm

This shall not be less than ½ × b = 17.6 mm

Hence, b_1 = 17.6 mm

Let us have a gap of 10 mm between the face of flange of column and the beam.

Therefore, required length of outstanding length of angle = 35.2 + 10 = 45.2 mm = $(b_1 + g)$

Providing ISA 75 × 75 × 10 angle, radius at root of angle = 7 mm.

$$b_2 = (b_1 + g) - (t_a + \gamma_a) = 27.6 - 10 - 7 = 10.6 \text{ mm}$$

Considering the load is acting over an area of 140 × 10.6; uniformly distributed load = 60000/(140 × 10.6) = 40.4 N/mm^2 = 40.4 N/mm

$$M_u = 40.4 \times 10.6^2/2 = 2271 \text{ Nmm}$$

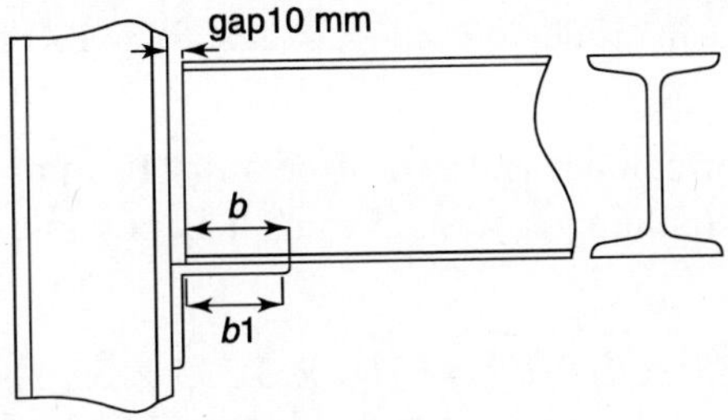

Fig. 7.19

Moment capacity of the leg of the angle = $1.2 \times \left(\frac{250}{1.1}\right) \times \left(\frac{1 \times t^2}{6}\right)$ = 2271 Nmm.

Required thickness of angle = t = 7.07 mm

Provide ISA 75 × 75 × 10 which is satisfactory

Design of Welded Connection

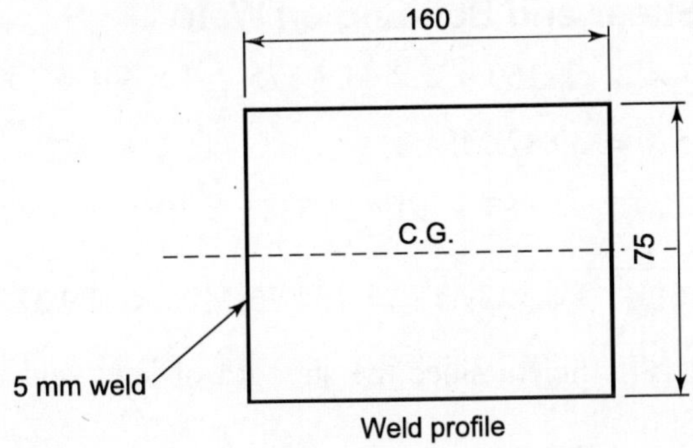

Fig. 7.20

Thickness of angle = 10 mm

Thickness of flange of beam = 12.4 mm

Thickness of flange of column = 17.2 mm

Thinner part is angle for which thickness = 10 mm.

Maximum size of weld = 3/4 × 10 = 7.5 mm (Clause 10.5.8.2 of IS:800).

Thickness of thicker part is the flange of the column = 17.2 mm for which minimum size of weld = 5 mm (Table 21 of IS:800).

Hence, minimum weld size = 5 mm and maximum size of weld = 7.5 mm.

Using 5 mm size of weld, strength of shop weld

= 189.37 × 0.7071 × 5 = 669.5 N/mm (Refer to Table 7.2 of this book).

$$\text{Required length of weld/side} = \frac{60 \times 1000}{2 \times 669.5} = 44.81 \text{ mm}$$

Take the length of the weld per side = 2 × (75 × 2 × 5) = 130 mm

Throat size = 0.707 × 5 = 3.535 mm.

The gap between the face of column and end of beam is 10 mm. The weld size between the seat angle and column flange (column face) is 5 mm. It is possible to weld before erection of beam.

$$MI = [(3.535^3 \times 160/12) + (160 \times 3.535 \times 37.5^2)] + 2[(3.535 \times 75^3/12)]$$

$$= 589 + 795375 + 124277 = 920.2 \times 10^3 \text{ mm}^4$$

$$Z = 920.2 \times 10^3/37.5 = 24.5387 \times 10^3 \text{mm}^3$$

Moment at Face of Column

Distance = 10 + (75/2) = 47.5 mm

M = (60 × 47.5/1000) = 2.85 kNm

Combined Effect of Shear and Bending on Weld

Area of weld = 0.707 × 5 × 2 × [(160 – 2 × 5) + (75 – 2 × 5)] = 1520 mm^2

Shear stress = 60000/1520 = 39.47 MPa

Stress due to bending = $M/Z_e = 2.85 \times 10^6/(24.5387 \times 10^3) = 116.14$ MPa

Combined stress = $\sqrt{116.14^2 + 3 \times 39.47^2} = 134.76$ MPa < 189.37 MPa.

If it is site weld also it is sufficient since the strength of site weld is 157.8 MPa. Hence, satisfactory.

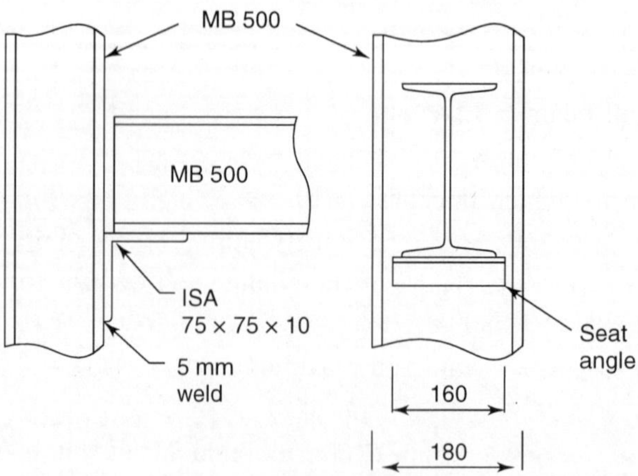

Fig. 7.21

Example 7.12 *Flange plate connection to flange of beam*

When the moment capacity of the available rolled sections is marginally inadequate to take up the bending moment; the same can be strengthened by providing additional plates on the flanges. These plates should be made to act along with the rolled sections in meeting the requirements. Hence, it is necessary to design the connections between them. The same way, plate girders with more than one flange plate can also be designed.

Design the shop fillet welded connection between the flange plate and flange of the rolled I section for a maximum shear of V = 250 kN in a beam made up of ISMB 600 with 180 × 6 mm plate on each flange as shown in Fig. 7.22. Yield stress for the beam and plates is 250 MPa.

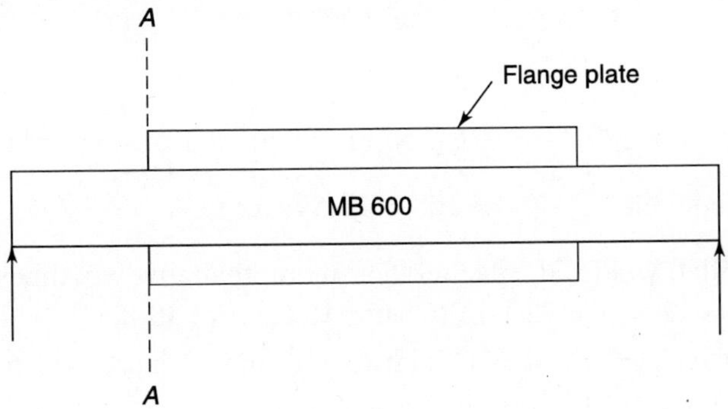

Fig. 7.22

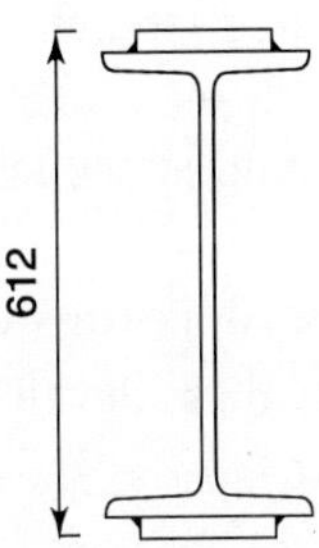

Fig. 7.23

$$I_z = I_z \text{ of } MB\ 600 + I_z \text{ of plate about } xx \text{ axis of beam}$$
$$= 91813 + (18 \times 0.6^3/12 + 18 \times 0.6 \times 30.3^2) \times 2 = 111644.4 \text{ cm}^4$$

Factored shear = 1.5 × 250 = 375 kN

Horizontal shear/linear centimetre = VAY/I_z

$$= \frac{(375 \times 1000 \times 180 \times 6 \times 303)}{111644.4 \times 10^4} = 110 \text{ N/mm}$$

Minimum weld size for 6 mm plate = 3 mm (Table 21 of IS:800).

Maximum size of weld = 6 – 1.5 = 4.5 mm.

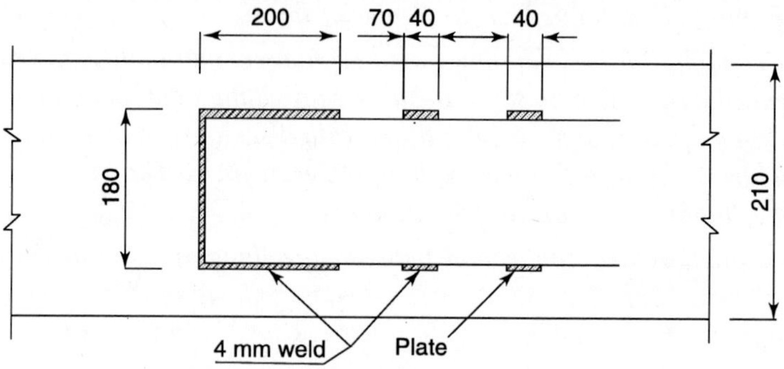

Fig. 7.24

Provide 4 mm weld.

Strength = 0.7071 × 4 × 189.37 = 535.6 N/mm (Table 7.2 of this book).

$$\% \text{ length of weld required} = \frac{110}{535.6} \times 100 = 20.5\%$$

Use intermittent fillet welds at spacing not more than twelve times the thickness for compression member = 12 × 6 = 72 mm (Clause 10.5.5.2 of IS:800)

Minimum length of weld = 4 × 4 = 16 mm (or) 40 mm (Clause 10.5.5.1 of IS:800)

Use 40 mm intermittent weld at 70 mm clear spacing.

$$\text{Pitch} = 40 + 70 = 110 \text{ mm}$$

$$\text{Strength} = \frac{(40 \times 535.6)}{189.37} = 113 > \text{Required strength of 110 N/mm}$$

Hence, provide 4 mm weld of 40 mm long at a pitch of clear spacing of 70 mm. Pitch will be 110 mm.

(a) Flange plate connection, (b) built-up section, (c) weld detail

At the end of the cover plate the welds must develop strength of plate.

Plate strength = $T_{dn} = (0.9\, A_n\, f_u/\gamma_{m1})$ (Clause 6.3.1 of IS:800)

$$= 0.9 \times 180 \times 6 \times 410/1.25 = 318816 \text{ N}$$

$$T_{dg} = A_g \times f_y/\gamma_{mo} = (180 \times 6 \times 250)/(1000 \times 1.1) = 245.45 \text{ kN}$$

Hence, design strength of plate in tension = Least of T_{dg} and T_{dn} = 245.45 kN

$$\text{Required length of 4 mm weld} = \frac{245450}{535.6} = 458.3 \text{ mm}$$

Effective length of weld at the end = 180 – 2 × 4 = 172 mm.

Required effective length of weld along the plate = (458.3 – 172)/2 = 143.15 mm.

Required length on sides = 143.15 + 2 × 4 = 151.1 mm

Provide 200 mm length.

Note: If the beam is open to the sky, seal weld of 2 mm in the clear spacing is to be provided to avoid water ingress and thereby corrosion.

7.5 ECCENTRIC CONNECTIONS (FILLET WELDS)

Example 7.13 *Connection between gantry bracket to column*

A gantry girder column ISMB 400 is to be provided with a bracket of ISMB 300 to carry the load of 150 kN from gantry girder at a distance of 300 mm from the face of the column. Design a fillet welded connection between the bracket and column at shop.

Solution: $e = 300$ mm

$V = 150 \times 1.5$ kN = 225 kN (Clause 3.5.1 and Table 4 of IS:800)

$$M = 225 \times 0.3 = 67.5 \text{ kNm.}$$

Fillet weld is proposed.

Properties of MB300 are, $A = 5626$ mm, $b_f = 140$ mm, $t_f = 12.4$ mm, $t_w = 7.5$ mm

$$z = 573.6 \text{ cm}^3$$

The flange weld be designed to resist the total bending moment through equal and opposite (couple) of forces through the angle cleats. Proposed arrangement is shown in Fig. 7.25 and the weld profile is shown in Fig. 7.26. ISA50 × 50 × *t* angle cleats are proposed.

Weld between the column and cleats.

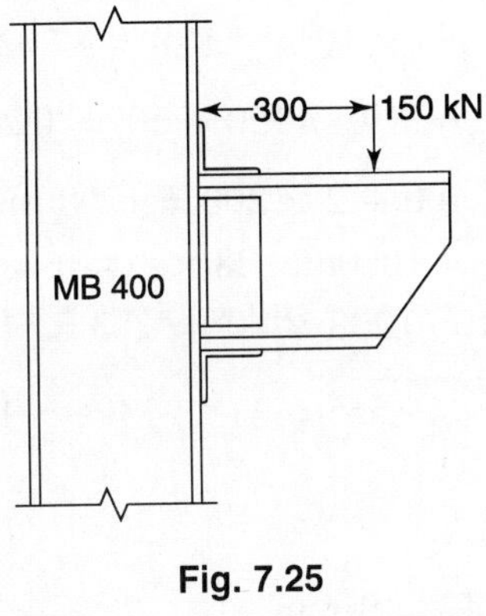

Fig. 7.25

$$Iw_x = 2\left(\frac{100 \times 1^3}{12} + 100 \times 1 \times 200^2\right) + 4\left(\frac{1 \times 50^3}{12} + 50 \times 1 \times 175^2\right)$$

$$+ 4\,(50 \times 1^3/12 + 50 \times 1 \times 100^2) + (1 \times 200^3/12)$$

$$= 8 \times 10^6 + 6.17 \times 10^6 + 2 \times 10^6 + 0.67 \times 10^6 = 16.84 \times 10^6 \text{ mm}^4$$

Try ISA 50 × 50 × 6 angles for flange cleat.

Weld size 0.75 × 6 = 4 mm. (Clause 10.5.8.2 of IS: 800 & Clause 6.2.5.2 of IS:816)

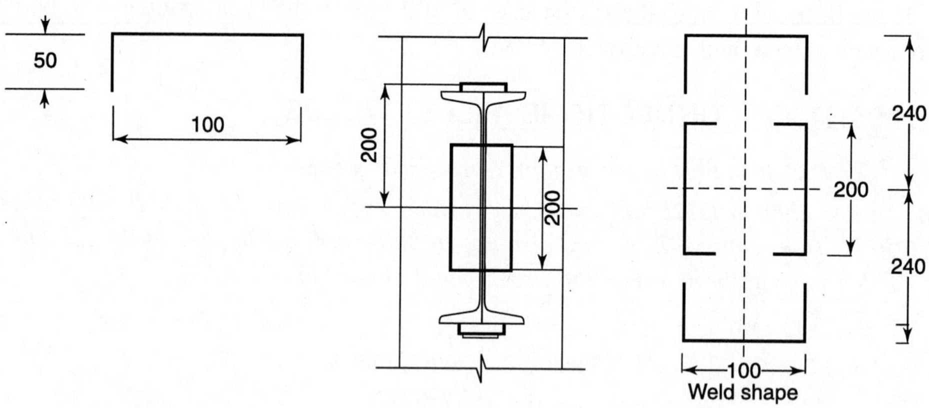

Fig. 7.26

Provide 4 mm weld.

Design strength of weld = 0.7071 × 4 × 189.37 = 535.6 N/mm (Table 7.2 of this book).

Length of cleat = 100 mm.

Second moment of area of this weld shape = $I_w = (16.84 \times 10^6 \times a_e)$ mm^4

a_e throat thickness of weld of 4 mm.

$$a_e = 0.7071 \times a = 0.7071 \times 4 = 2.828 \text{ mm.}$$

$$I_{wx} = 16.84 \times 10^6 \times 2.828 = 47.62 \times 10^6 \text{ mm}^4$$

$$Z_w = I_{wx}/y = I_{wx}/(150 + 50) = 0.238 \times 10^6 \text{ mm}^3$$

Length of weld = 8 × 50 + 2 × 100 + 2 × 200 = 1000 mm

Stress due to direct shear = 225 × 1000/(2.828 × 1000) = 79.56 MPa.

Stess due to bending = 67.5 × 10^6/0.238 × 10^6 = 283.6 MPa.

Combined stress = $\sqrt{283.6^2 + 3 \times 79.56^2}$ = 315.3 MPa > 157.8 MPa (site weld)

Inadequate.

Try ISA 90 × 90 × 12 angle.

Weld size not to exceed 0.75 × 12 = 9 mm.

Use 8 mm weld.

$$Iw_x = 2\left(\frac{100 \times 1^3}{12} + 100 \times 1 \times 240^2\right) + 4\left(\frac{1^3 \times 90}{12} + 90 \times 1 \times 195^2\right)$$

$$+ 2\left(\frac{1 \times 200^3}{12}\right) + 4\left(\frac{1^3 \times 90}{12} + 90 \times 1 \times 100^2\right)$$

$$= 11520016.7 + 13689030 + 1333333 + 3600030$$

$$= 30142410 \text{ mm}^4 \times a_e$$

$$Z_e = \frac{30142410}{240} = 125593 \times a_e = 125593 \times 0.707 \times 8 = 710356 \text{ mm}^3$$

$$= 710.356 \times 10^3 \text{ mm}^3.$$

Shear Strength

Total length of weld = 8 × 90 + 2 × 100 + 2 × 200 = 1320 mm

Area = 1320 × 0.707 × 8 = 7466 mm^2

Stress due to direct shear = 225 × 1000/7466 = 30.14 MPa.

Stress due to bending = 67.5 × 10^6/710.356 × 10^3 = 95.02 MPa.

Combined stress = $\sqrt{95.02^2 + 3 \times 30.14^2}$ = 108.4 MPa < 189.37 MPa.

Weld Between Bracket Cleat on the Flange of Beam

Fig. 7.27 *Weld from plan (View A)*

Due to moment, there will be tension in one of the flanges and compression in another flange.

Length of weld on each flange = 2 × 90 + 100 = 280 mm (as in Fig. 7.25)

Area = 280 × 0.707 × 8 = 1584 mm^2

Tensile force due to bending moment = (67.5 × 10^6)/300 = 225000 N

Stress = τ = 225000/1584 = 142 MPa < 189.37 MPa

Web will take the shear.

Length of weld on web = 4 × 90 + 2 × 200 = 760 mm

Area = 760 × 0.707 × 8 = 4298.56 mm^2

Stress = τ = (225 × 10^3)/4298.56 = 52.3 MPa < 189.37 MPa

Hence, the connection is satisfactory.

Note: The connection between fixed beam and column flange (moment connection perpendicular to the plane of moment) can be done in the same way.

Example 7.14 *Welded bracket (force in the plane of welds)*

The bracket is as shown in Fig. 7.28. It is welded on all four sides and the load is as shown. Design the shop fillet welded (at shop) connection and the bracket plate. Material shall be IS:2062 f_y = 250 MPa. The load is due to dead load and imposed load.

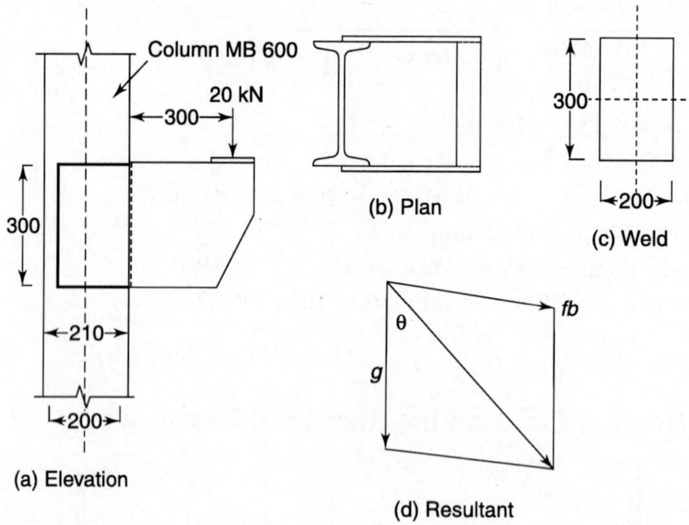

Fig. 7.28 *Welded bracket – (a) and (b) Bracket, (c) Welds, (d) Resultant forces*

Solution: Shear = 20 kN.

Partial safety factor for load = 1.5 (Table 4 of IS:800)

Factored shear = 1.5 × 20 × 1000/2 = 15000 N per bracket.

Factored moment = 1.5 × 20 × 0.3 = 9 kNm

Moment per bracket plate = 4.5 kNm

First we have to arrive at the size of the bracket plate. The flange width of the column in this case is 210 mm. Let us provide the plate for 200 mm leaving 10 mm for welding. The depth of plate can be anywhere between 250 mm and 500 mm. Let us provide 300 mm.

Assume throat thickness of weld of 1 mm

Length of weld = 2(300 + 200) = 1000 mm

Shear due to direct load, q = 15000/1000 = 15 N/mm run

Torsion due to bending moment

$b = M \times r/I_p$, where I_p = polar moment of inertia of the welds = $I_x + I_y$

$$I_p = 2(200 \times 150^2 + 300^3/12) \times 1 + 2(300 \times 100^2 + 200^3/12) \times 1$$
$$= 2083.3 \times 10^4 \text{ mm}^4$$

$$r = \sqrt{100^2 + 150^2} = 180.28 \text{ mm}$$

Due to torsion = $4.5 \times 10^6 \times 180.28/2083.3 \times 10^4$ = 39 N/mm run

The angle between shear and flexural stress for the weld farthest away,

$$\cos\theta = 100/180.28 = 0.5546$$

Resultant force = $\sqrt{q^2 + b^2 + 2qb\cos\theta}$

$$= \sqrt{15^2 + 39^2 + 2 \times 15 \times 39 \times 0.5546} = 48.94 \text{ N/mm}$$

Required size of weld = t = 48.94/ (0.707 × 189.37) = 0.365 mm

The size of weld required is too small.

Design of Plate Thickness

Eccentricity to edge of flange of column = 300 – 100 = 200 mm.

Moment at the edge of flange = 15 × 0.2 = 3 kNm

$$Z_e = t \times 300^2/6$$

M_d = Bending Strength = $1.2 \times Z_e \times f_y/\gamma_{mo}$ = 1.2 × ((t × 300²)/6) × (250/1.1) = 4.09 × 10⁶ t

Providing 12 mm thick plate

$$M_d = (4.09 \times 10^6 \times 12/10^6) = 49.1 \text{ kNm}$$

$$M_d = 49.1 \text{ kNm} > 3 \text{ kNm}.$$

Shear Strength = $A_v f_{yw}/\gamma_{mo}$ = 300 × 12 × 250/(1.1 × $\sqrt{3}$) = 472 kN >> 15 kN.

It is general engineering practice to provide plate of thickness not less than 12 mm for such brackets. Alternatively angle sections (or) channel section can be provided.

Provide 12 mm thick plate with 8 mm fillet weld.

Example 7.15 *Welded bracket (force not in the plane of welds)*

Design the fillet weld (at shop) for the bracket as shown in Fig. 7.29. The material of construction is IS:2062 f_y = 250 MPa. The load is only due to dead load.

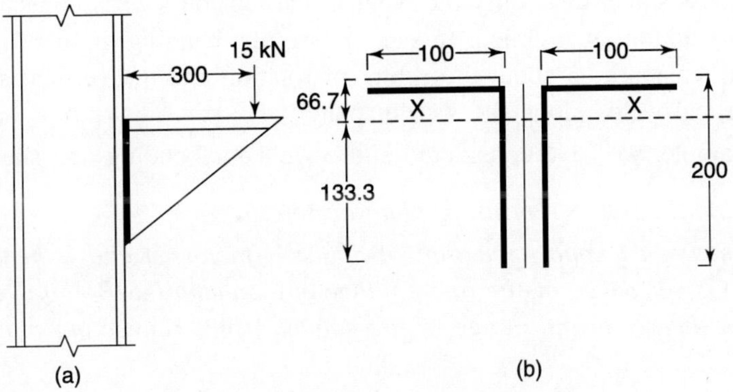

Fig. 7.29 *(a) Bracket, (b) Weld detail*

Solution: The bending moment is perpendicular to the plane of welds. The bracket is proposed with a rolled section welded as shown in Fig. 7.29 with horizontal fillets underside the seat and vertical along the web. The load is due to dead load. Material of construction IS:2062 and f_y = 250 MPa.

Factored load = 1.5 × 15 = 22.5 kN

Moment = 22.5 × 0.3 = 6.75 kNm

Assume throat thickness of 1 mm of fillet weld.

Centre of gravity of welds from top is = (2 × 200 × 1 × 100)/(400 + 200)

= 67 mm from the top of horizontal welds

$$I_{xx} = 2 \times \left(\frac{1 \times 200^3}{12} + 200 \times 33^2 + \frac{100 \times 1^3}{12} + 100 \times 67^2\right) = 2.670 \times 10^6 \text{ mm}^4$$

Bending strength considering horizontal weld at top

$$= 1.2 \times (2.67 \times 10^6/67) \times 250/(1.1 \times 10^6) = 10.87 \text{ kNm} > 6.75 \text{ kNm.}$$

Stress = $6.75 \times 10^6/(2.67 \times 10^6/67) = 169.4$ MPa.

Shear = $(P/l_t\, l_w) = 22500/1 \times 600 = 37.5$ N/mm (Clause 10.5.9 of IS:800)

Top is in tension. Hence, resultant at top is to be checked.

Resultant = $\sqrt{169.4^2 + 3 \times 37.5^2} = 181$ N/mm < $f_u/(\sqrt{3} \times \gamma_{mw}) = 189.37$ MPa. (Clause 10.5.10.1.1 of IS:800)

Though 1 mm size of throat is sufficient as per calculation, provide 4 mm fillet weld = 0.7071 × 4 × 189.37 = 535.6 N/mm > 181 N/mm

Note: The stress in the horizontal weld at the top is considered for bending stress calculations, since failures generally occur in the tension side for fillet welds in such joints.

For design purposes sometimes, only the weld in the tension side is considered and a point of rotation about the bottom is assumed. In such a case, the bending moment is resisted by the moment of weld in tension area about the centre of rotation and the vertical shear is assumed to be taken the vertical welds along the web or plate.

In the above example, we have considered entire weld for bending and shear.

Example 7.16 *Moment connection using plate on top flange*

ISMB 500 beam having a support reaction of 300 kN (factored) and a bending moment of 100 kNm (factored) is connected at site to the flange of a column ISMB 450. Design the welded connection between the beam and flange of the column using connecting plate at top flange. F_y *= 250 MPa.*

Provide a connecting plate at the top flange of the beam

$$\text{Force in top plate due to moment} = \frac{\text{moment}}{\text{lever arm}} = \frac{100 \times 10^3}{500} = 200 \text{ kN}$$

To find the gross area of cross section of plate, equating $T_{dg} = A_g f_y/\gamma_{mo}$ (Clause 6.2 of IS:800) and the force in top plate due to moment.

$$T_{dg} = A_g f_y/\gamma_{mo} = 200 \times 1000 \text{ N.}$$

Required gross area of plate = $A_g = 200 \times 1000 \times \gamma_{mo}/f_y = 200 \times 1000 \times 1.1/250 = 880 \text{ mm}^2$

Parameter	*ISMB 500*	*ISMB 450*
Depth D	500 mm	450 mm
Width of flange B	180 mm	150 mm
Thickness of flange T	17.2 mm	17.4 mm
Thickness of web t	10.2 mm	9.4 mm
Clear height of web h_1	424.1 mm	379.2 mm

Width of flange of column b_f = 150 mm. Let us provide a 150 mm wide plate.

This plate will be tapered to 130 mm to enable welding between the plate and top flange of beam.

Required thickness of the top plate = 880/130 = 6.76 mm.

Provide a plate 12 mm thick, 150 mm wide tapered to 130 mm at the end.

Thickness of thicker part among top plate, flange of beam and flange of column = 17.4 mm

Minimum fillet weld size = 5 mm (Table 21 of IS:800)

Let us provide 6 mm fillet weld

Strength of site weld per mm = $(f_u/\gamma_{mo} \times \sqrt{3})$ = $(410/1.5 \times \sqrt{3})$ = 157.8 N (Clause 10.5.7.1.1 of IS:800)

Strength of 6 mm site weld/mm = 0.707 × 6 × 157.8 = 669 N/mm

Length of the weld required = $\dfrac{200 \times 10^3}{669}$ = 298.8 mm.

At the end weld will be done for the width of plate 130 mm.

Effective length of this weld = 130 – 2 × 6 = 118 mm (Clause 10.5.4.1 of IS:800).

Weld required on sides = $\dfrac{298.8 - 118}{2}$ = 90.4 mm. However, provide 250 mm.

Total length of the weld = 2 × 250 + 130 = 630 mm > 298.8 mm.

The length of the plate between the column flange and the starting point of weld shall be equal to or little more than the width of plate = 150 mm.

Hence, required length of the top plate = 150 + 250 = 400 mm.

This plate shall be connected to the flange of the column by single V butt weld.

Incomplete penetration butt welds welded from one side only, should have a throat thickness of at least 7/8 of the thickness of the thinner of the parts joined. 5/8 of the thickness of thinner part shall be considered for design. (Clause 6.1.4.2 of IS:816)

Strength of single bevel butt weld between the plate and column flange

$$= (5/8) \times 12 \times 150 \times 250/1.5 = 187.5 \text{ kN} < 200 \text{ kN}.$$

Since this weld will be done at site, partial safety factor of 1.5 is taken.

Now the weld is in adequate.

Hence, let us increase the thickness of plate to 16 mm and rework.

Strength = (5/8) × 16 × 150 × 250/1.5 = 250 kN > 200 kN

Hence, satisfactory.

Design of Stiffeners on the Web of Column

Force in the two stiffeners $= \dfrac{M}{D} = \dfrac{100 \times 10^6}{500} = 200$ kN

Stiffeners will be provided on both sides of web of column.

Force per stiffener = 100 kN.

Thickness of web of column = 9.4 mm.

Let us assume the size of weld between the stiffener and the web of column as 8 mm.

To find required length = $100 \times 10^{3} = L \times 0.7071 \times 8 \times 189.37$ (this can be done at shop.)

$$\text{Hence, } L = \frac{100 \times 10^3}{0.7071 \times 8 \times 189.37} = 93.35 \text{ mm on each side of web of column.}$$

Let us provide stiffener plates of length to accommodate this weld.

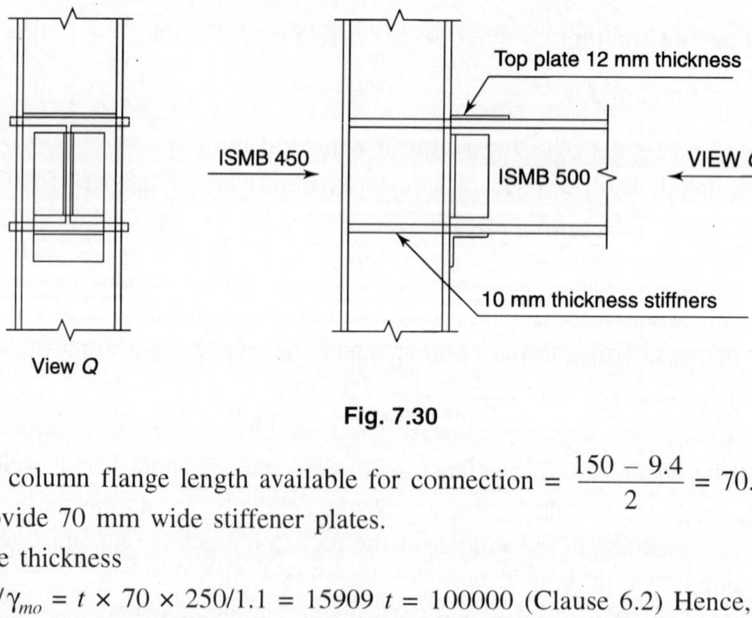

Fig. 7.30

Maximum column flange length available for connection $= \dfrac{150 - 9.4}{2} = 70.3$ mm

Let us provide 70 mm wide stiffener plates.

To find the thickness

$T_{dg} = A_g f_y/\gamma_{mo} = t \times 70 \times 250/1.1 = 15909\, t = 100000$ (Clause 6.2) Hence, $t = 6.3$ mm and

$T_{dn} = 0.9\, A_n\, f_u/\gamma_{m1} = 0.9 \times t \times 70 \times 410/1.25 = 100000$ (Clause 6.3.1 of IS:800).

Hence, $t = 4.84$ mm

Provide 10 mm thick stiffener plates.

Total length of the weld that can be accommodated to connect the stiffener with the column flange = 4 × 70 = 280 mm.

Required length = 4 × 93.5 = 374 mm.

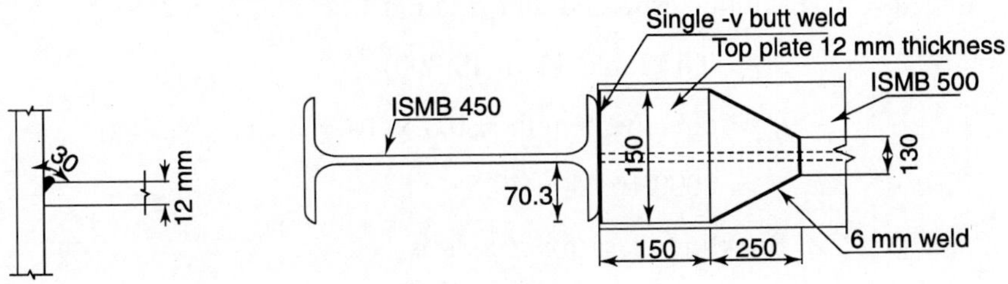

Fig. 7.31

The remaining length of the weld can be adjusted on the stiffener plate connecting the web of column.

Let us provide 40 mm of weld length on each side of the stiffener plate joining it with column web.

Total length of weld provided = 280 + 4 × 40 = 440 mm > 374 mm. Hence, satisfactory.

Note: Design of cleat and seat angles to take up shear are to be done similar to other worked examples of this book.

Example 7.17 *Strap joint (plate splice) for columns*

An ISMB 400 column of 10 m length is hinged at base and fixed at top. Since 10 m long rolled ISMB400 is not readily available, a splice is to be provided. The column is restrained against local buckling on weaker axis. Axial load is 600 kN and the moment about major axis is 112.5 kNm due to dead, imposed and wind loads.

Design the splice using splice plates and fillet weld at shop.

Solution: Factored axial load = 1.2 × 600 = 720 kN

Factored moment = 1.2 × 112.5 = 135 kN

Sectional properties of ISMB 400 are

$$A = 7846 \text{ mm}^2$$

$$r_x = 161.5 \text{ mm}, \ r_y = 28.2 \text{ mm}$$

$$z_x = 1022.9 \text{ cm}^3$$

Width of flange 140 mm

Flange thickness 16 mm.

Thickness of web = 8.9 mm

Buckling class $h/b_f = 400/140 = 2.85 > 1.2$

$$t_f = 16 \text{ mm} < 40 \text{ mm}$$

Hence, buckling class on *ZZ* axis is a and *YY* axis is *b*. (Table 10 of IS:800)

End conditions:

One end fixed = Translation and rotation restrained.

Other end hinged = Translation restrained and rotation free.

$$K = 0.8 \text{ (Table 11 of IS:800)}$$

$$L_e = \text{effective length} = 0.8 \times 10 = 8 \text{ m.}$$

$$KL_z/r_z = 8000/161.5 = 49.54$$

This is a major axis *ZZ*. Buckling class for *Z* axis is '*a*'.

$$f_{cd} = 205 \text{ MPa (Table 9(a) of IS:800).}$$

$$KL_y/r_y = 0$$

As the column is restrained in weaker axis, only r_z controls the design.

Design compressive strength = $A_e \times f_{cd}$ (Clause 6.3.1 of IS:800)

$$= 7846 \times 205/1000 = 1608.4 > 600 \text{ kN}$$

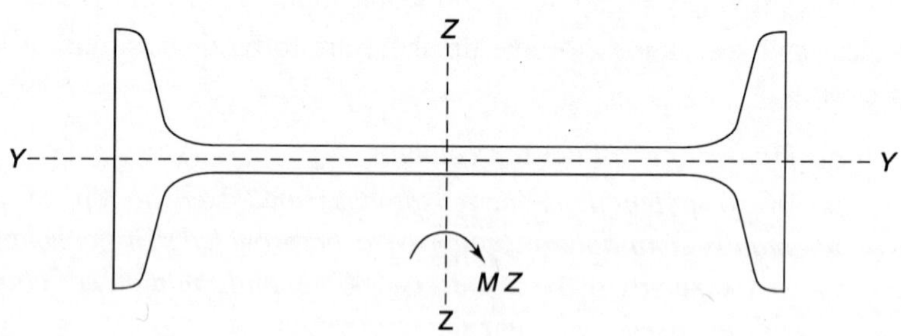

Fig. 7.32

Section Strength (Clause 9.3.1 of IS:800)

$$N_d = A_g\, f_y/\gamma_{mo} = 7846 \times 250/(1.1 \times 1000) = 1783.2 \text{ kN}$$

$$\varepsilon = \sqrt{(250/f_y)} = \sqrt{(250/250)} = 1.0$$

$$b/t_f = 70/16 = 4.4 < 9.4\,\varepsilon$$

$$d = 400 - 2 \times (16 + 14) = 340 \text{ mm.}$$

$$d/t_w = 340/8.9 = 38.2 < 84\,\varepsilon$$

Hence, section is plastic. (Table 2 of IS:800)

$$N/N_d = 720/1783.2 = 0.4$$

$$Z_p = 1176.18 \times 10^3 \text{ mm}^3 \text{ (Table 46 of IS:800)}$$

$$M_{dz} = \frac{\beta_p Z_p f_y}{\gamma_{mo}} = 1.0 \times 1176.18 \times 10^3 \times 250/(1.1 \times 10^6) = 267.31 \text{ kNm.}$$

$$M_{dz} = 1.2\ Z_e f_y/\gamma_{mo} \text{ (Clause 8.2.1.2 of IS:800)}$$

$$= 1.2 \times 1020 \times 10^3 \times 250/1.1 \times 10^6 = 278.2 \text{ kNm.}$$

The least value is M_{dz} = 267.31 kNm

$(N/N_d) + (M_z/M_{dz}) < 1.0$ (Clause 9.3.1.1 of IS:800)

$0.4 + 135/267.31 = 0.4 + 0.505 = 0.905 < 1.0$

Hence, column section is adequate.

Note: Overall member strength as in Clause 9.3.2 of IS:800 shall be checked.

Splice (Strap) Plate Design

The splice is designed for the load and moment carried by the column. If the load and moment at the exact location of splice is known from the analysis result, the splice can be designed for the same.

Flange Splice

Cross-sectional area of each flange = 140 × 16 = 2240 mm^2

Flange splice plate width shall be less than the width of flange to enable fillet welding

Required area of flange splice plate = 1.05 × Area of flange of the column

= 1.05 × 2240 = 2352 mm^2.

Let the width of flange splice plate be 120 mm leaving 10 mm on each side.

Required thickness of flange splice plate = 2352/120 = 19.6 mm.

Provide 20 mm thick flange splice plate.

Provided area of flange splice plate = 120 × 20 = 2400 mm^2 > 2352 mm^2

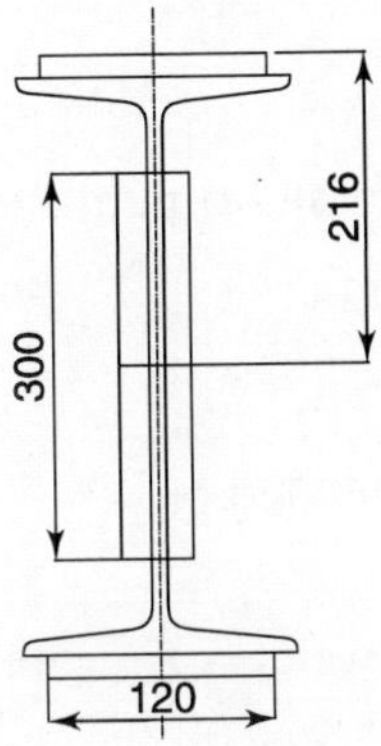

Fig. 7.33

Web Splice Plate

Clear depth of web = d = 400 – 2 × (16 + 14) = 340 mm.

Keep 300 mm width of web splice plate.

Thickness of Web = 8.9 mm

Provide 8 mm thick splice plate.

$$A = 300 \times 8 = 2400 \text{ mm}^2$$

$$A_p = 2[2400 + 2400] = 9600 \text{ mm}^2$$

Moment of inertia of splice plates = I_p

$= 2[(120 \times 20^3/12) + 120 \times 20 \times (200 + 10)^2 + (8 \times 300^3/12)] = 24784 \times 10^4 \text{ mm}^4$

Section modulus of splice plates = $Z_p = 24784 \times 10^4/(200 + 20) = 112.65 \times 10^4 \text{ mm}^3$

Intensity due to axial load = P/A_p = (720 × 1000)/9600 = 75 MPa.

Intensity due to bending = $M/Z_p = 135 \times 10^6/112.65 \times 10^4$ = 119.8 MPa.

Compression = 75 + 119.8 = 194.8 MPa < (250/1.1)

Tension = 112.65 – 75 = 37.65 MPa.

$$N = 720 \text{ kN.}$$

$$N_d = A_g f_y/\gamma_{mo} = (9600 \times 250)/(1.1 \times 1000) = 2181.8 \text{ kN.}$$

$$M_z = 135 \text{ kNm}$$

$$M_{dz} = 1.2 \times Z_e \times f_y/\gamma_{mo} = 1.2 \times (\ 112.65 \times 10^4 \times 250/1.1 \times 10^6)$$

$$= 307.2 \text{ kNm.}$$

$(N/N_d) + (M_z/M_{dz}) < 1.0$

(720/2181.8) + (135/307.2) = 0.33 + 0.44 = 0.77 < 1.0

Hence, web splice plates of size 8 × 300 and that of flange 20 × 120 is adequate.

Weld Size

Provide 8 mm fillet weld on flange strap and 6 mm fillet weld on web strap plate.

Flange Splice Plates

Total compressive stress = 194.8 MPa.

Force/Strap plate = 194.8 × 2400 = 467520 N.

Weld is in shear.

Shear strength of 8 mm shop fillet weld = F_w = 0.7071 × 8 × 189.37 = 1071.2 N/mm (Table 7.2 of this book).

Minimum length of weld required = F/F_w = 467520/1071.2 = 436.4 mm.

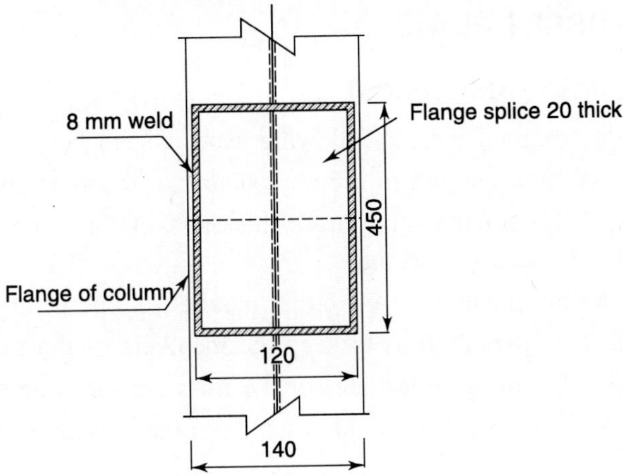

Fig. 7.34

Let the full width of plate be welded at the ends, then the effective length of the edge weld on flange splice = 120 – 2 × 8 = 104 mm.

Effective length of weld required on each face = [(436.4 – 104)/2] + 2 × 8 = 182.2 mm.

Length of strap plate required = 2 × 182.5 = 365 mm.

Provide 20 × 450 × 120 mm flange splice plate on each flange with 8 mm weld all round.

Design of Web Splice Plate

Area of flange of column = 2 × 140 × 16 = 4480 mm^2

Total area of ISMB 400 = 7846 mm^2

Area of web of column = 7846 – 4480 = 3366 mm^2

Axial load shared by web = (720/7846) × 3366 = 308.9 kN

Compressive stress on splice = (308.9 × 1000)/(300 × 8 × 2) = 64.35 MPa < (250/1.1)

6 mm fillet weld at the shop is proposed.

Strength of 6 mm weld = 0.7071 × 6 × 189.37 = 803.4 N/mm (Table 7.2 of this book).

Length of weld required to transfer 308.9 kN through the joint = 308.9 × 1000/803.4 = 384.5 mm.

Effective length of weld at end = 300 – 2 × 6 = 288 mm

Balance required = 384.5 – 288 = 96.5 mm

Effective length of weld required per side: 96.5/2 = 48.24 mm

Therefore, length required = 48.24 + 2 × 6 = 60.24 mm

Provide 8 × 300 × 200 mm web splice plates on either side of web with 6 mm fillet weld at the shop.

7.6 DESIGN OF GUSSET PLATE

7.6.1 Design Principle (Guidelines)

The gusset plates are not designed but the following aspects shall be considered:

(1) Size of the plate is large enough to accommodate the members and the welds/bolts.

(2) The thickness shall be equal to the larger thickness of the connecting members. The thickness shall not be less than 6 mm.

(3) The profile follows the members axes, either parallel (or) perpendicular to the member.

(4) The members shall be joined so as to keep the members of the truss in one plane.

(5) It shall be ensured that the gusset plates of the truss are on same side of the members.

(6) The gusset plates at supports be provided with extra stiffeners where required.

Example 7.18 *Support gusset plates*

In a 18 m span truss, bottom chord and rafter are meeting at the support and are subjected to the following forces.

Force in bottom chord = 100 kN tension.

Force in rafter = 120 kN compression

Forces include those due to wind also

Size of bottom chord is 2ISA 75 × 75 × 8.

Same size of angle is proposed for the rafter.

Calculate the length of shop fillet weld required and draw sketch to find the size of gusset plate.

Solution:

Generally, the size of gusset is arrived manually. Now with the help of AutoCAD, the same can be arrived more precisely.

Note that the centre of gravity line of angles and column are meeting at one point (*wp* = working point). In practice achieving this is not possible. The centre of gravity line of rafter will meet the centre line of column at a higher elevation to provide roof sheeting.

Factored forces (Table 2 of IS:800)

(a) bottom chord = 1.2 × 100 = 120 kN.

(b) rafter = 1.2 × 120 = 144 kN

Bottom Chord to Gusset Plate Connection

Force in each angle of the bottom chord = 120/2 = 60 kN

Thickness of angle = 8 mm.

Approximate size of weld = 0.75 × 8 = 6 mm (Clause 10.5.8.2 of IS:800).

Provide 6 mm weld.

Strength of 6 mm shop weld per mm length = 803.4 N/mm (Refer to Table 7.2 of this book).

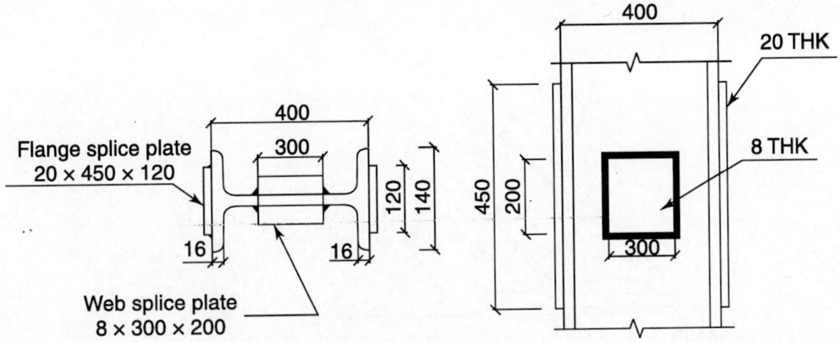

Fig. 7.35

Length of weld required = 60 × 1000/803.4 = 74.68 mm

$$L_1 + L_2 = 74.68 \text{ mm} \qquad (1)$$

Taking moment about C.G. line

$$L_1 \times C_{yy} = L_2\,(75 - C_{yy}) \qquad (2)$$

$$L_1 \times 21.4 = L_2\,(75 - 21.4) = 53.6\,L_2$$

$$L_1 = 2.5\,L_2$$

Substituting in (1); 3.5 L_2 = 74.68 mm

Effective length of weld L_2 = 21.3 mm

Actual length of weld L_2 = 21.3 + 2 × 6 = 33.3 mm

Provide 40 mm for L_2.

$$L_1 = 2.5 \times 33.3 + 2 \times 6 = 95.2 \text{ mm}$$

Provide 100 mm for L_1.

Rafter to Gusset plate connection:

Force in each angle = 144/2 = 72 kN

$$L_1 + L_2 = (72 \times 1000)/803.4 = 89.62 \text{ mm}$$

$$L_1 \times C_{yy} = L_2\,(75 - C_{yy})$$

$$L_1 = 2.5\,L_2$$

$$L_2 = 89.62/3.5 = 25.6 + 2 \times 2 \times 6 = 37.6 \text{ mm}$$

Provide 40 mm for L_2.

$$L_1 = 2.5 \times 25.6 + 2 \times 6 = 76 \text{ mm}$$

Provide 80 mm for L_1.

Based on these and leaving 25 mm projection of gusset plate from edge of angle as shown, find the size of gusset (*x*, *y*, etc) as in Fig. 7.36.

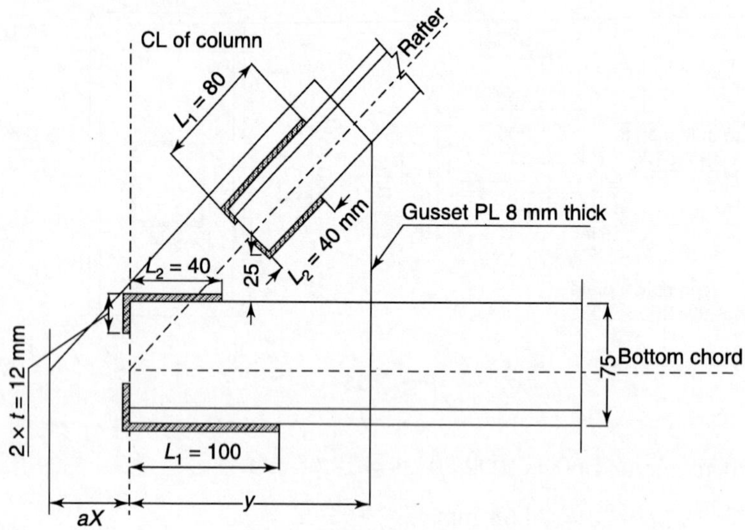

Fig. 7.36

7.7 SPLICE JOINT

Splice joint in members becomes necessary when the length of available raw material is less than that required for the member. This helps utilizing, the available material, thus saving the cost.

It is the practice to have the length of the flange cover plates each side of the splice, equal to the flange width or 230 mm whichever is greater, and the thickness, half the thickness of the upper column flange or 10 mm whichever is greater. The splice in a column has to take the axial force, moment and shear. The splice in a beam has to take shear and moment.

Column Splices

Splice of a compression member when there is no moment and full bearing is ensured are designed as an erection device.

Building columns are provided with splice approximately at every 12 M depending on the available length of rolled sections and storey height. It is advisable to splice at about 1000 mm above the platform member to enable connection of platform member or vertical bracing member to the column. The connectors and splice material shall be designed for shear, axial force and moment in the column at the location of splice.

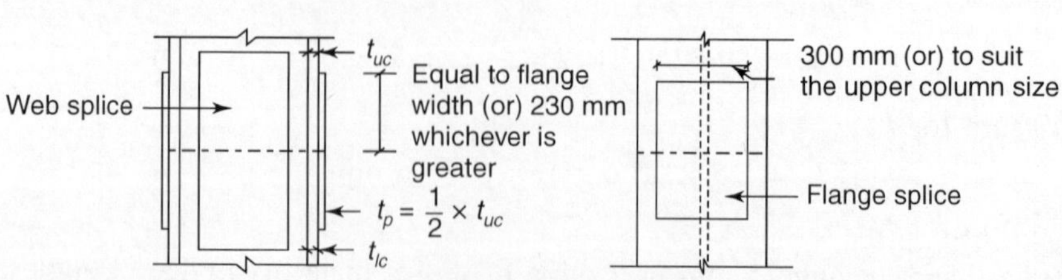

Fig. 7.37

t_{uc} = thickness of flange of upper column

t_{lc} = thickness of flange of lower column

Design of Splice for Flexural Member Rolled Section

When the span, loading, support conditions and the location of splice are known; then we can calculate the exact bending moment, shear force at the location of splice and design the web and flange splice. In the absence of all the above data, the splice is designed for full capacity. Also refer to Clause 10.7(6) of IS:800 for minimum design action on connections for splicing of flexural members.

The splicing of flexural members can be done by providing flange and web splice plates using fillet weld. The flange splice is designed to take tensile force in the flange. The length of splice plate depends on the length of weld required to take the tesile force in the flange. Splice on the web is designed to take up the shear and moment due to eccentricity in web splice connection.

These splice plates can be connected using bolts, taking into account the holes while designing the member.

Yet another method is splicing the flanges with double *V* butt weld with out splice plates and splicing the web with splice plates. This method of splice connection will help positioning floor grills or chequred plates for platform on top flange.

Example 7.19 *Design a splice connection for ISMB 400 beam having f_y = 250 MPa for its full capacity using splice plates fillet welded at site to flanges and web. Draw a neat sketch showing the connection.*

Solution: Yield stress ratio = $\varepsilon = (250/f_y)^{0.5} = (250/250)^{0.5} = 1$

Shear capacity

V_d = Design strength = $(A_v f_{yw})/(\gamma_{mo} \sqrt{3})$ (Clause 8.4 of IS:800)

A_v = Area of web = $400 \times 8.9 = 3560$ mm^2

f_{yw} = 250 MPa

γ_{mo} = 1.1 (Table 5 of IS:800)

$V_d = 3560 \times 250/(1.1 \times 1000 \times \sqrt{3}) = 467.12$ kN.

When the factored shear force is less than 0.6 V_d, the member will be under low shear. (Clause 8.2.1.2 of IS: 800).

$0.6\ V_d = 0.6 \times 467.12 = 280.3$ kN

Moment capacity

$M_d = \beta_b Z_p f_y/\gamma_{mo}$

$b/t_f = (140/2)/16 = 70/16 = 4.37 < 9.4\ \varepsilon$

Clear depth of web = $d = 400 - 2 \times (16 + 14) = 340$ mm.

$d/t_w = 340/8.9 = 38.2 < 84\ \varepsilon$

Section is plastic (Table 2 of IS:800)

β_b = 1.0 for plastic section. (Clause 8.2.1.2 of IS:800)

Plastic section modulus = Z_p = 1176.18 × 10^3 mm^3 (Table 46 of IS:800)

$M_d = \beta_b Z_p f_y/\gamma_{mo}$ = (1.0 × 1176.18 × 10^3 × 250)/(1.1 × 10^6) = 267.3 kNm

This shall be less than $1.2 Z_e f_y/\gamma_{mo}$ for simply supported beam. (Clause 8.2.1.2 of IS:800)

Elastic section modulus = Z_y = 1020 × 10^3 mm^3 (Table 46 of IS:800).

$1.2\ Z_e f_y/\gamma_{mo}$ = (1.2 × 1020 × 10^3 × 250)/(1.1 × 10^6) = 278.2 kNm

Moment capacity is 267.3 kNm.

Design data

Moment capacity or design bending strength M_d = 267.3 kNm

Design shear strength = 280.3 kN

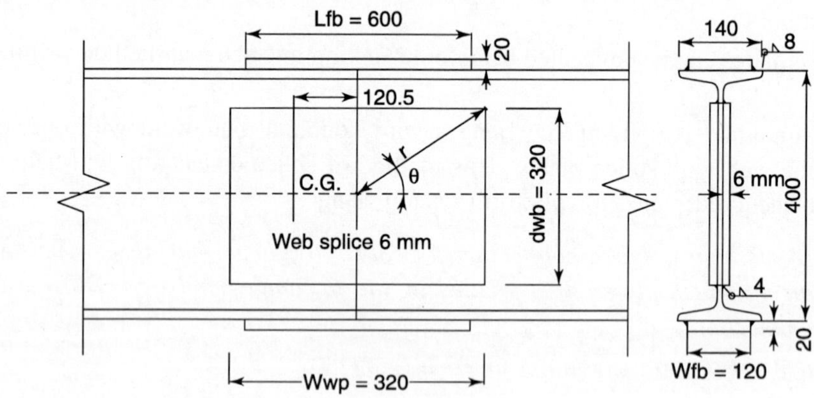

Fig. 7.38

Design of web splice

Shear area required = A_v = Design shear strength × $(\gamma_{mo}\sqrt{3})/(f_{yw})$

$$= 280.3 \times 10^3 \times 1.1 \times \sqrt{3}/250 = 2136\ mm^2$$

Clear depth of web = d = 400 – 2 × (16 + 14) = 340 mm. The depth of web splice plate shall not be more than this. Depth higher than this will hit against the root of flange.

Let us provide 320 mm depth of web splice. Web splice plate will be provided on both sides of web.

Thickness of each plate required = 2136/(2 × 320) = 3.34 mm.

This shall not be less than 50% of thickness of web and 6 mm whichever is more.

Hence, provide 320 × 6 mm splice plates.

Approximate size of weld = 0.7 × 6 = 4.2 mm

Provide 4 mm weld

Strength of site weld of 4 mm size = $0.7071 \times 4 \times f_u/(\gamma_{mw} \times \sqrt{3})$ (Clause 10.5.7.1.1 of IS:800)

$$= 0.7071 \times 4 \times 410/(1.5 \times \sqrt{3}) = 446.35 \text{ N/mm}$$

Length of weld required per splice plate on one side = (280.3 × 1000)/(2 × 446.35) = 314 mm

Provided effective length parallel to the splice joint

$$= 320 - 2 \times 4 = 312 \text{ mm (Clause 10.5.4.1 of IS:800).}$$

This itself is adequate as per calculation for taking shear. But there is an eccentricity caused by centre of gravity of welds. Due to this there will be moment.

Providing 160 mm perpendicular to the splice joint, the size of splice plate = 320 × 320 × 6 mm

C.G. of weld from the centre of joint

$$[2 \times (160 - 2 \times 4) \times (160/2)] + [(320 - 2 \times 4) \times 160]/[2 \times 152 + 312] = 74240/616$$
$$= 120.5 \text{ mm.}$$

$$e = 2 \times 120.5 = 241 \text{ mm}$$

Moment due to this eccentricity

$$= (280.3/2) \times (241/1000) = 33.78 \text{ kNm}$$

Bending shear stress = $M \times r/I_p$

r is the distance of extreme fibre from centre of gravity of web splice weld.

$$r = \sqrt{(160^2 + (320/2)^2} = 226.27 \text{ mm}$$

Throat thickness = 0.707 × 4 = 2.828 mm

Effective length of weld perpendicular to the splice joint = 320 – 2 × 4 = 312 mm.

Effective length of weld parallel to the splice joint = 320 – 2 × 4 = 312 mm.

$$I_x = 2 \times [(312 \times 2.828^3/12) + (2.828 \times 312 \times 160^2) + (2.828 \times 312^3/12)]$$
$$= 59.49 \times 10^6 \text{ mm}^4$$
$$I_y = 2 \times [(312 \times 2.828^3/12) + (312 \times 2.828 \times 160^2) + (2.828 \times 312^3/12)] =$$
$$= 59.49 \times 10^6 \text{ mm}^4$$
$$I_p = (59.49 + 59.49) \times 10^6 \text{ mm}^4 = 119 \times 10^6 \text{ mm}^4$$

Shear stress due to bending = $M \times r/I_p = (33.78 \times 10^6 \times 226.27)/(119 \times 10^6)$

$$= 64.23 \text{ MPa}$$

Shear stress = (280.3 × 1000)/[2 × (312 + 312) × 2.828] = 79.42 MPa

$$\theta = \tan^{-1}(160/160) = 45°$$

Combined stress = $\sqrt{(64.23^2 + 79.42^2 + 2 \times 64.23 \times 79.42 \times \cos 45°)}$

$$= 132.8 \text{ MPa} < 157.5 \text{ MPa}$$

Strength per sq mm of site weld = 157.5 MPa.

Hence, satisfactory.

Design of flange splice

Moment capacity = 267.3 kNm

Lever arm = centre to centre of flange = 400 – 16 = 384 mm = 0.384 m.

Tensile/compressive force in flange = F = (267.3/0.384) = 696.1 kN

Width of flange = 140 mm

Providing the width of splice plate as 120 mm leaving 10 mm on each side for welding.

Thickness required = (1.05 × 140 × 16)/120 = 19.6 mm

Provide 20 mm thick splice plate.

Area of each flange splice plate = 120 × 20 = 2400 mm^2 > Area of flange (140 × 16 = 2240 mm^2).

Approximate weld size = 0.7 × 16 = 11.2 mm

Provide 10 mm weld.

Strength of 10 mm size site weld/mm = 0.7071 × 10 × 157.8 = 1115.86 N/mm

Length of weld required = (Tension/strength of weld) = (696.1 × 1000)/1115.86 = 623.8 mm

Width of flange splice = 120 mm.

Effective length of weld at end = 120 – 2 × 10 = 100 mm.

Balance on two sides = (623.8 – 100)/2 = 262 mm.

This will be the effective length of weld.

Actual length of weld = Effective length + 2 × weld size = 262 + 2 × 10 = 282 mm.

Provide a length of 300 mm on either side of joint.

Provide splice plate of size 600 × 120 × 20 mm on each flange with 10 mm weld.

7.8 WELD AND SPLICE TABLES

Table 7.5 gives the length of weld required for connecting equal angles to gusset plates for 100% design strength in tension. This will help arriving at the size of gusset required quickly. Table 7.6 gives the details of fillet welded splice plates for Indian rolled ISMB sections used as beams.

Working sheet for Table 7.5

Let us take ISA90 × 90 × 8 as an example.

$$A_g = 1380 \text{ mm}^2 \quad C_x = 25.1 \text{ mm.}$$

Strength due to yielding = $T_{dg} = A_g f_y / \gamma_{mo}$

= (1380 × 250/1.1)/1000 = 313.64 kN (Clause 6.2 of IS:800)

Strength in rupture = $T_{dn} = (0.9A_{nc}f_u/\gamma_{m1}) + \beta A_{go}f_y/\gamma_{mo}$ (Clause 6.3.3 of IS:800)

$$A_{nc} = 1380/2 = 690 \text{ mm}^2$$

$$f_u = 410 \text{ MPa}$$

$$A_{go} = 1380/2 = 690 \text{ mm}^2$$

$$\beta = 1.4 - 0.076(w/t)\ (f_y/f_u)\ (b_s/L_c) \le (f_u \times \gamma_{mo})/(f_y \times \gamma_{m1}) \ge 0.7$$

$w = 90$ mm, $b_s = 90$ mm (Fig. 6 of IS:800)

$$t = 8 \text{ mm}$$

L_c is not known.

$$(f_u \times \gamma_{mo})/(f_y \times \gamma_{m1}) = 410 \times 1.1/250 \times 1.25 = 1.4432$$

Keeping $L_c = 2 \times$ size of angle leg (as an approximation) $= 2 \times 90 = 180$ mm.

$\beta = 1.4 - 0.076(90/8)\ (250/410)\ (90/180) = 1.4 - 0.076 \times 3.43 = 1.14 < 1.4432\ \&\ > 0.7$

Therefore, $T_{dn} = [(0.9 \times 690 \times 410/1.25) + (1.14 \times 690 \times 250/1.1)]/1000 = 382.46\text{ kN} < T_{dg}$

100% tension capacity = Least of T_{dg} and T_{dn} = 313.64 kN

Thickness of member (angle) = 8 mm

Thickness of gusset plate shall not be less than 8 mm.

Weld size = (3/4) × 8 = 6 mm (Clause 10.5.8.2 of IS: 800)

Provide 6 mm weld.

Strength of shop weld = 803.4 N/mm (Table 7.2 of this book).

Length of weld required = (313.64 × 1000)/803.4 = 390.4 mm

$$S_a = \text{size of angle}$$

$$L_1 + L_2 = 390.6 \text{ mm} \qquad (1)$$

$$L_1 \times C_x = L_2 \times (S_a - C_x)$$

$$L_1 \times 25.1 = L_2 \times (90 - 25.1) = 64.9 \times L_2$$

Therefore, $$L_1 = 2.5856\ L_2 \qquad (2)$$

Substituting (2) in (1)

$$2.5856\ L_2 + L_2 = 390.4 \text{ mm}$$

$$L_2 = (390.4/3.5856) = 108.9 \text{ mm}$$

$$L_1 = 2.5856 \times 108.9 = 281.52 \text{ mm}$$

These are effective lengths required.

Actual length required for $L_2 = 108.9 + 2 \times 6 = 120.9$ mm

Provide 140 mm

Actual length required for $L_1 = 281.52 + 2 \times 6 = 293.52$ mm

Provide 310 mm

Now L_1 = 310 mm

Therefore, β = 1.4 – 0.076 × 2 = 1.248 < 1.4432 but > 0.7

Therefore, T_{dn} = [(0.9 × 690 × 410/1.25) + (1.248 × 690 × 250/1.1)]/1000

= 399.4 kN > T_{dg}

Hence, since T_{dg} is not changing, there is no change in the weld length calculation.

Design strength of shop fillet weld of various sizes are given in Table 7.2 of this book. If design strength of site fillet weld is required, multiply these values by (1.25/1.5) = 0.833.

Table 7.5 Weld length for equal angles. (IS:2062 F_y = 250 MPa material)

SL. No.	*Section*	*Area (mm²) (IS:808)*	*Cx(mm)*	*100% Tension Capacity (kN)*	*Total length of weld required in mm for weld size*			
					6 mm	*8 mm*	*9 mm*	*4 mm*
(1)	*(2)*	*(3)*	*(4)*	*(5)*	*(6)*	*(7)*	*(8)*	
1	50 × 50 × 6	568	14.5	129	-	-	-	242
2	65 ×65 × 6	744	18.1	169	-	-	-	316
3	75 × 75 × 6	866	20.6	196.8	-	-	-	368
4	90 × 90 × 6	1050	24.2	238.6	-	-	-	446
5	65 × 65 × 8	976	18.9	199	248	-	-	-
6	75 × 75 × 8	1140	21.4	259	323	-	-	-
7	90 × 90 × 8	1380	25.1	313.6	391	-	-	-
8	100 × 100 × 8	1540	27.6	360	437	-	-	-
9	110 × 110 × 8	1710	30	388.6	485	-	-	-
10	75 × 75 × 10	1400	22.2	318.2	397	-	-	-
11	90 × 90 × 10	1700	25.9	386.4	482	-	-	-
12	100 × 100 ×10	1900	28.4	431.8	539	-	-	-
13	110 ×110 × 10	2110	30.9	479.5	598	-	-	-
14	130 ×130 × 10	2510	35.9	570.4	712	-	-	-
15	150 × 150 × 10	2920	40.8	663.6	828	-	-	-
16	90 × 90 × 12	2020	26.6	459	573	430	382	-
17	100 × 100 × 12	2260	29.2	514	641	481	427	-
18	110 ×110 × 12	2510	31.7	570	712	534	475	-
19	130 × 130 ×12	2990	36.7	679.5	848	636	565	-
20	150 × 150 × 12	3480	41.6	791	987	740	658	-
21	200 × 200 × 12	4690	53.9	1066	1330	997	886	-

Note: (1) The effective length given in the columns 6, 7, and 8 in the above table are (L_1 + L_2).

(2) $L_1 \times C_{yy} = L_2 \times$ (leg size of equal angle – C_{yy})

(3) From the above two (note 1 & 2) find L_1 and L_2

(4) Add 2 × size of weld on L_1 to get the actual length of weld required. Simillarly, add 2 × size of weld on L_2 to get the actual length.

(5) Return weld = 2 × size of weld

6) The above tension capacity is factored. The actual force that can be applied will depend on the partial safety factor on load as per Table 2 of IS:800.

Table 7.6 Fillet welded splice plate detail for rolled beams

Beam designation	Web splice (all dimensions are in mm)				Flange splice (all dimensions are in mm)			
	d_{wp}	t_{wp}	W_{wp}	t_{w1}	L_{fp}	W_{fp}	t_{fp}	t_{w2}
ISMB 150	90	6	50	4	240	60	12	6
ISMB 200	130	6	80	4	320	80	16	8
ISMB 250	160	6	120	4	450	105	16	8
ISMB 300	200	6	160	4	520	120	16	8
ISMB 350	270	6	200	4	600	120	20	8
ISMB 400	320	6	320	4	600	120	20	8
ISMB 450	350	6	340	4	660	130	25	10
ISMB 500	410	8	320	4	780	160	25	10
ISMB 600	480	8	400	6	1120	190	25	10

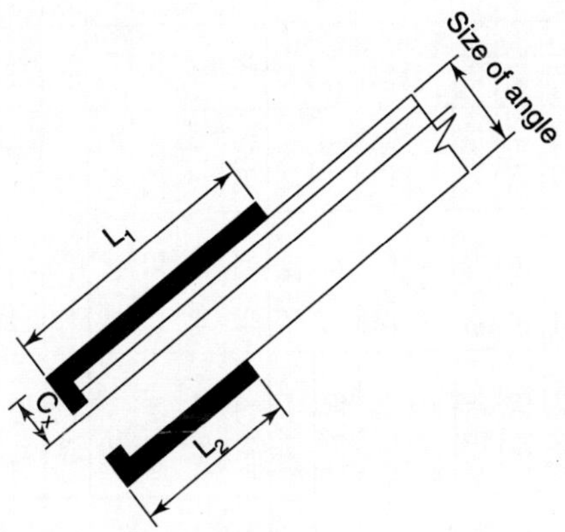

Fig. 7.39

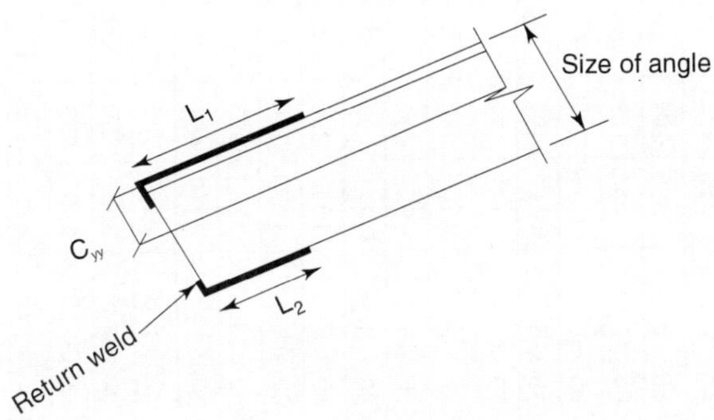

Fig. 7.40

Table 7.7 Design tensile strength of single angle fillet welded to gusset plate at site

Size	A_g in mm^2	T_{dg} in kN	W mm	t mm	Size of weld in mm	Total length of site weld required in mm	Total length of shop weld required in mm	C_{yy} in mm	L_c mm	Welded						
										w/t	b_s/L_c	β	A_{nc} mm^2	A_{go} mm^2	T_{dn} in kN	T_d in kN
ISA 50 × 50 × 6	568	129	50	6	4	289.23	241.01	14.5	205	8.3	0.24	1.31	284.0	284	168	129
ISA 60 × 60 × 6	684	155	60	6	4	348.3	290.24	16.9	250	10.0	0.24	1.29	342.0	342	201	155
ISA 60 × 60 × 8	896	204	60	8	6	304.17	253.46	17.7	214	7.5	0.28	1.3	448.0	448	265	204
ISA 65 × 65 × 6	744	169	65	6	4	378.85	315.7	18.1	273	10.8	0.24	1.28	372.0	372	218	169
ISA 65 × 65 × 8	976	222	65	8	6	331.33	276.09	18.9	235	8.1	0.28	1.3	488.0	488	288	222
ISA 75 × 75 × 6	866	197	75	6	4	440.98	367.46	20.6	320	12.5	0.23	1.26	433.0	433	252	197
ISA 75 × 75 ×8	1138	259	75	8	6	386.32	321.92	21.4	276	9.4	0.27	1.28	569.0	569	334	259
ISA 75 × 75 ×10	1402	319	75	10	7	407.95	339.94	22.2	287	7.5	0.26	1.31	701.0	701	416	319
ISA 80 × 80 × 6	929	211	80	6	4	473.06	394.19	21.8	344	13.3	0.23	1.26	464.5	464.5	270	211
ISA 80 × 80 × 8	1221	278	80	8	6	414.5	345.4	22.7	297	10.0	0.27	1.28	610.5	610.5	357	278
ISA 90 × 90 × 6	1047	238	90	6	4	533.15	444.26	24.2	390	15.0	0.23	1.24	523.5	523.5	302	238
ISA 90 × 90 × 8	1379	313	90	8	6	468.14	390.09	25.1	338	11.3	0.27	1.26	689.5	689.5	401	313
ISA 90 × 90 ×10	1703	387	90	10	7	495.54	412.93	25.9	353	9.0	0.26	1.29	851.5	851.5	502	387
ISA 90 × 90 × 12	2019	459	90	12	9	456.93	380.76	26.6	322	7.5	0.28	1.3	1009.5	1009.5	597	459
ISA 100 × 100 × 6	1167	265	100	6	4	594.25	495.18	26.7	436	16.7	0.23	1.22	583.5	583.5	334	265
ISA 100 × 100 ×8	1539	350	100	8	6	522.45	435.35	27.6	378	12.5	0.26	1.25	769.5	769.5	445	350
ISA 100 × 100 ×10	1903	433	100	10	7	533.73	461.42	28.4	396	10.0	0.25	1.28	951.5	951.5	558	433
ISA 100 × 100 ×12	2259	513	100	12	9	511.25	426.02	29.2	362	8.3	0.28	1.29	1129.5	1129.5	665	513
ISA 110 × 110 × 8	1702	387	110	8	6	577.79	481.46	30	420	13.8	0.26	1.23	851.0	851	490	387
ISA 110 × 110 × 10	2106	479	110	10	7	612.8	510.64	30.8	441	11.0	0.25	1.27	1053.0	1053	615	479
ISA 110 × 110 × 12	2502	569	110	12	9	566.24	471.85	31.6	404	9.2	0.27	1.28	1251.0	1251	734	569

$T_{dg} = A_g f_y/\gamma_{mo}$ (Clause 6.2 of IS:800)

F_y = 250 MPa, f_u = 410 MPa

$T_{dn} = 0.9\ A_{nc} f_u/\gamma_{m1} + \beta A_{go} f_y/\gamma_{mo}$ (Clause 6.33 Of IS:800)

$\beta = 1.4 - 0.76(w/t)\ (f_y/f_u)\ (b_s/L_c) \geq (f_u \times \gamma_{mo})/(f_y \times \gamma_{m1}) \geq 0.7$

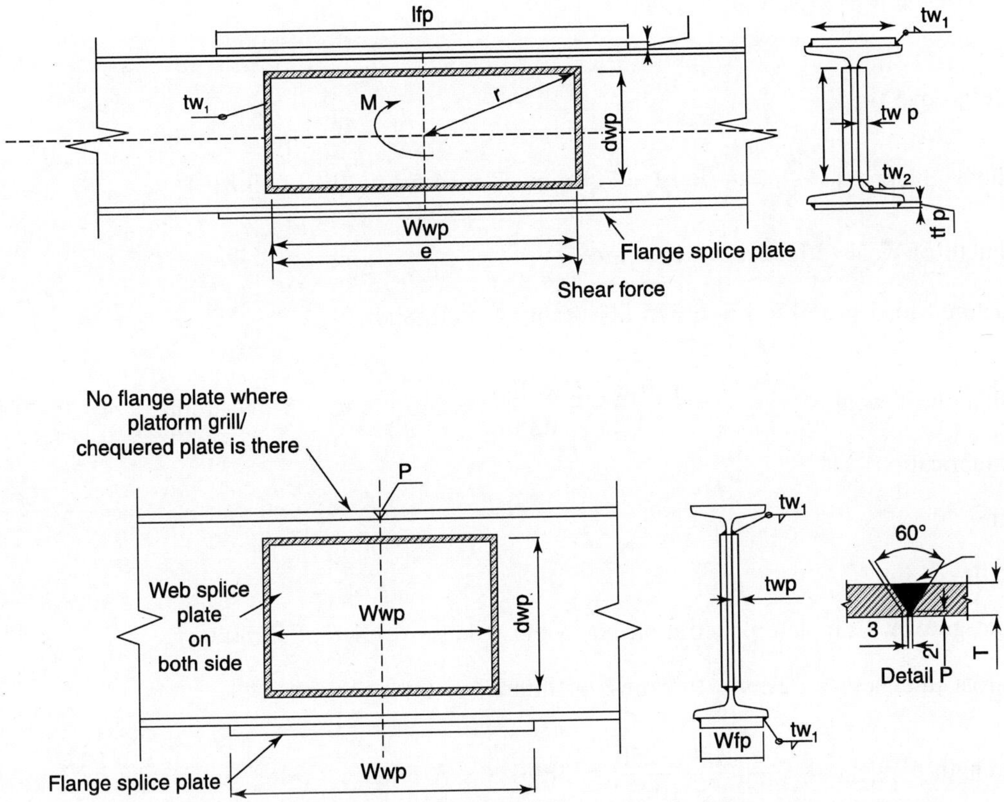

Fig. 7.41

7.9 DESIGN OF BUTT WELDED JOINTS

Example 7.20 *Two plates having different thickneses*

Two plates of thickness 16 mm and 12 mm are to be connected by a groove weld as shown in Fig. 7.42. The joint is subjected to a tensile force of 250 kN due to dead and imposed loads.

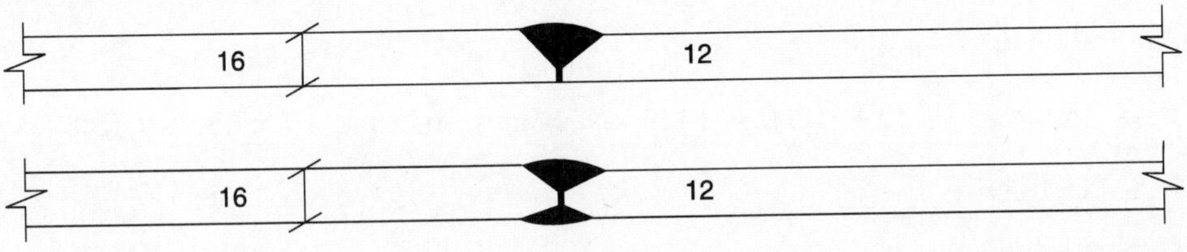

Fig. 7.42

Effective length of weld is 160 mm. Check the adequacy of shop weld (a) Single V groove weld joint (b) Double V groove weld joint

Case a:

Single groove

Throat thickness = $\frac{5}{8}t = \frac{5}{8} \times 12 = 7.5$ mm (Clause 6.1.4.2(b) of IS:816)

Length of weld (effective) = 160 mm

Factored load = 250 × 1.5 = 375 kN (Table 4 of IS:800).

Strength of weld = $\frac{L_w t_e f_y}{\gamma_{mw}} = \frac{160 \times 7.5 \times 250}{1.25 \times 1000}$ = 240 kN < 375 kN (Table 5 of IS:800 for shop fabrication)

Unsafe.

Case b:

Double groove. Complete penetration butt weld (Clause 6.1.4.1 of IS:800).

Throat thickness = 12 mm (thickness of thinner plate)

Strength of weld = $\frac{12 \times 160 \times 250}{1.25 \times 1000}$ = 384 kN < 375 kN.

Adequate.

Example 7.21 *Design of butt weld joint of flange plate*

A plate girder has flange plate of 450 × 25 mm. From the beam analysis it was found that it is subjected to a tension of 1150 kN at a location where a joint is to be provided. Design the butt weld. Steel f_y = 250 MPa.

Effective thickness of single *V* butt is T_e = (5/8) T = (5/8) × (25) = 15.63 mm (Clause 6.1.4.2(b) of IS:816)

Strength = $A_g \times f_y / \gamma_{mo}$

= 15.63 × 450 × 250/1.25 × 1000 = 1406 kN. (Clause 6.2 and 10.5.7.3.2 of IS:800).

This is > 1150 kN.

Hence, satisfactory.

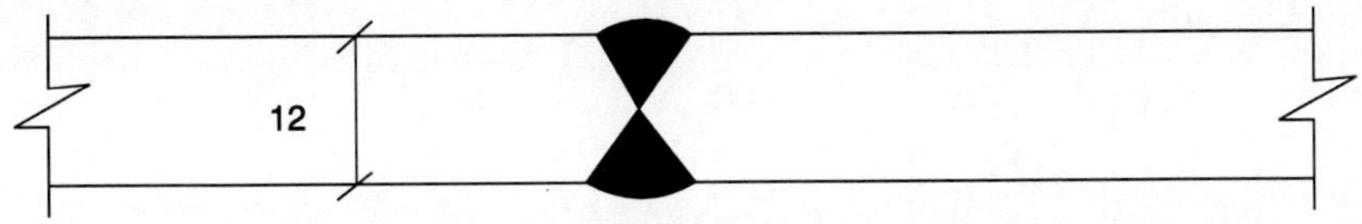

Fig. 7.43

Example 7.22 *Butt weld under bending and shear*

Web of a plate girder consists of 1250 × 12 mm plate and it has to be provided with double V bolt weld splice at a section where the shear and bending moment are V = 900 kN and M = 260 kNm. The flange plates are 420 × 25 mm size. Design web splice. Material is IS:2062 f_y 250 MPa. The shear and bending are due to dead load only.

Solution: Factored shear = $V = 1.5 \times 900 = 1350$ kN (Table 4 of IS:800).

Factored moment = $1.5 \times 260 = 390$ kNm (Table 4 of IS:800).

Size of web: 1250 × 12 mm

$$D = 1250 + 2 \times 25 = 1300 \text{ mm}$$

Providing full penetration double V butt joint; effective throat thickness = T_e = 12 mm.

Design shear strength = $V_d = A_v f_{yw}/(\gamma_{mo} \times \sqrt{3})$ (Clause 8.4 of IS:800).

$$= (1300 \times 12 \times 250)/(1.25 \times \sqrt{3}) = 1801.3 \text{ kN} > 1350 \text{ kN}.$$

$$0.6V_d = 0.6 \times 1801.3 = 1080.8 \text{ N}.$$

$$V > 06\ V_d.$$

Hence, high shear.

Bending Strength of Section (Clause 9.2.1 and 9.2.2 of IS:800)

Neutral axis is at mid depth.

$$b = (420 - 12)/2 = 204 \text{ mm. (Fig. 2 of IS:800).}$$

$$b/t_f = 204/25 = 8.16 < 8.4\ \varepsilon$$

$$d/t_w = 1250/12 = 104.16 > 84\ \varepsilon \text{ but less than } 105\ \varepsilon \text{ (Table 2 of IS:800)}$$

Section is compact.

$$M_{dv} = M_d - \beta(M_d - M_{fd}) \leq 1.2\ Z_e f_y/\gamma_{mo} \text{ (Clause 9.2.2(a) of IS:800)}$$

Plastic Design Strength of the Area of the Cross Section Excluding the Shear Area

Butt weld shall be treated as parent metal (Clause 10.5.7.1.2 of IS:800).

Z_{pf} excluding web = $2 \times 420 \times 25 \times (1250 + 25) = 26.775 \times 10^6$ mm^3

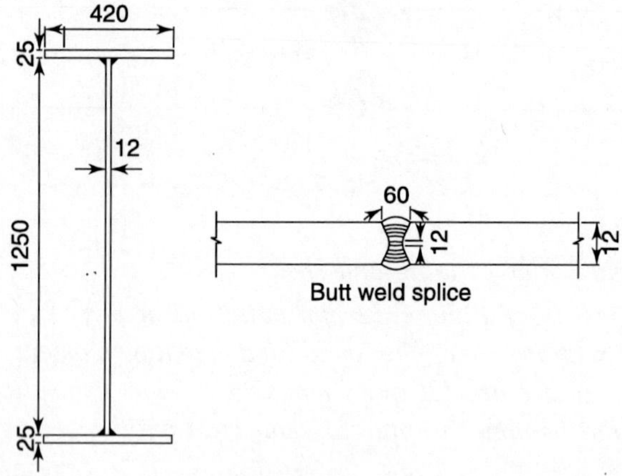

Fig. 7.44

$$M_{fd} = \beta_p Z_{pf} f_y/\gamma_{mo} = 1.0 \times 26.775 \times 10^6 \times 250/(1.25 \times 10^6) = 5355 \text{ kNm}$$

$$Z_{pz} = 2 \times 420 \times 25 \times (1250 + 25) + (1300 - 25)^2 \times 12/4 = 31.4625 \times 10^6 \text{ mm}^3$$

$$M_d = \beta_p Z_p f_y/\gamma_{mo} = 1.0 \times 31.4625 \times 250/1.25 = 6292.5 \text{ kNm}$$

$$\beta = ((2\ V/V_d) - 1)^2 = ((2 \times 1350/1801.3) - 1)^2 = 0.2489$$

$$M_{dv} = 6292.5 - 0.2489 \times (6292.5 - 5355) = 6059.2 \text{ kNm} > 390 \text{ kNm}.$$

This shall be less than $M_d = 1.2\ Z_e f_y/\gamma_{mo}$ (Clause 8.2.1.2 of IS:800)

$$Z_e = \{(420 \times 1300^3/12) - (408 \times 1250^3/12)\}/(1300/2) = 16.14 \times 10^6 \text{ mm}^3$$

$$M_d = 1.2 \times 16.14 \times 10^6 \times 250/1.25 \times 10^6 = 3873.6 \text{ kNm} > 390 \text{ kNm}.$$

Value of γ_{mo} is taken as 1.25 since we are checking the shop weld

Hence, satisfactory.

Example 7.23 *Butt weld in angle section under tension*

A member of a truss is made of ISA 75 × 75 × 6 and subjected to a tension of 60 kN. Butt welded splice has to be provided at shop to get the required length. Design a butt welded splice joint.

Solution: Section ISA 75 × 75 × 6

$$A = 866 \text{ mm}^2.$$

Factored load = 1.5 × 60 = 90 kN

Providing single *V* butt joint. $T_e = (5/8) \times 6 = 3.75$ mm (Clause 6.1.4.2 (b) of IS:816).

Strength = $(5/8) \times A_g \times f_y/1.25 = ((5/8) \times 866 \times 250)/(1.25 \times 1000) = 108.25$ kN > 90 kN.

Hence, satisfactory.

Example 7.24 *Butt joint of ISMB column*

Design a splice joint for ISMB 600 column with unsupported length = 7 m, with its base fixed and the top hinged. P = 400 kN. The force includes those due to wind also. Material IS:2062 f_y 250.

Since the axial force includes wind, partial safety factor is taken as 1.2 (Table 4 of IS:800).

$$\text{Factored load} = 1.2 \times 400 = 480 \text{ kN}.$$

$$A = 15621 \text{ mm}^2.\ Z_z = 3060.4 \text{ cm}^3.\ r_z = 242.4 \text{ mm}.\ r_y = 41.2 \text{ mm}$$

Providing single bevel butt joint, the effective thickness is 5/8 times the thickness of the plates (Clause 6.1.4.2(b) of IS:816). Hence, the properties of the weld are 5/8 times same as that of the the properties of the section.

Strength in compression = $A_e \times f_{cd}$

To find buckling class

$$h/b_f = 600/210 = 2.86 > 1.2 \text{ and } t_f = 12 < 40 \text{ mm}$$

Hence, buckling class on *ZZ* axis is *a* and buckling class on *YY* axis is *b*. (Table 10 of IS:800)

Effective length for the given end condition = 0.8 × actual length. (Table 11 of IS:800)

$$L_e = 0.8 \times 7 = 5.6 \text{ m},\ KL/r = 5600/41.2 = 136$$

$$f_{cd} \text{ for } KL/r = 136 \text{ is } 75.48 \text{ MPa. (Table 9b of IS:800)}$$

Design strength at weld joint = (5/8) $A_e \times f_{cd}$ = (5/8) × 15621 × 75.48 × (1.1/1.25)/1000 = 648.5 kN.

While calculating f_{cd} for members, γ_{mo} = 1.1 is considered. For shop welding γ_{m1} = 1.25 is to be considered.

Design strength at weld joint = 648.5 > 600 kN.

Hence, satisfactory.

Example 7.25 *Section of a plate girder is shown in Fig. 7.45*

Splice is to be provided for web at a section where shear force is 700 kN and moment = 250 kNm. The moment and shear are due to dead and imposed loads only. Yield stress of the material is (f_y = 355 MPa. Design the splice joint by butt welding the web.

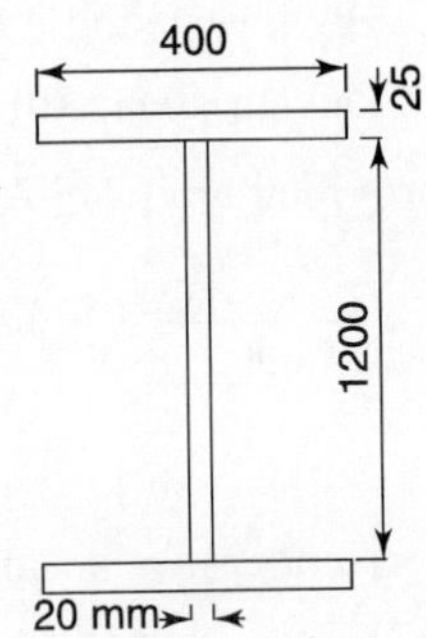

Fig. 7.45

$$d = \text{depth of web} = 1200 - 2 \times 25 = 1150 \text{ mm}.$$

$$t_w = 20 \text{ mm}$$

Depth of girder = 1200

The partial safety factor for load is 1.5 (Table 4 of IS:800).

Factored shear = 700 × 1.5 = 1050 kN.

Factored moment = 250 × 1.5 = 375 kNm.

Proposing double *V* groove.

Effective thickness of weld = $t_e = t_w$ = 20 mm (Clause 6.1.4.1 of IS:816).

Design strength of shop weld/mm length

$$= f_v = \frac{20 \times 355}{\sqrt{3} \times 1.25} = 3279 \text{ N/mm (Clause 8.4 of IS:800).}$$

$$\text{Length of weld required to resist the shear} = \frac{1050 \times 1000}{3279} = 320 \text{ mm} < d$$

However, complete penetration double V butt welding is done for the entire depth of web. (Clause 6.1.7 of IS:816).

Design shear strength = design shear strength of web (Since double V groove weld is provided) = $V_d = A_v f_{yw} / (\gamma_{mo} \sqrt{3})$ = 1150 × 20 × 355/1.1 × $\sqrt{3}$ = 4285.5 kN (Clause 6.1.7 of IS:816).

Correcting for γ_{mw}, design shear strength of weld = 4285.5 × (1.1/1.25) = 3771.24 kN.

$$0.6\, V_d = 0.6 \times 3771.24 = 2262.74 \text{ kN} > 1050 \text{ kN.}$$

Hence, low shear.

Hence, there is no reduction in bending strength. (Clause 9.2.2 of IS:800)

Alternate

Providing single groove weld

Effective thickness of weld = (5/8) × 20 = 12.5 mm (Clause 6.1.4.2 (b) of IS:816)

Design shear strength of single V butt weld/mm length

$$= [(12.5 \times 355)/(\sqrt{3} \times 1.25)] = 2049.6 \text{ kN}$$

Length of weld required to resist the shear = 1050 × 1000/2049.6 = 512.3 mm < 1150 mm

However, single V butt welding is done for the entire depth of web.

Design shear strength = (5/8)th of design shear strength of web (Since single V butt weld is provided) = $V_d = A_v f_{yw} / (\gamma_{mo} \sqrt{3})$ = 12.5 × 1150 × 355/1.1 × $\sqrt{3}$ = 2678.4 kN.

Correcting for γ_{mw}, design shear strength = 2678.4 × (1.1/1.25) = 2357 kN.

$$0.6V_d = 0.6 \times 2357 = 1414.2 \text{ kN} > 1050 \text{ kN.}$$

Hence, low shear.

Hence, there is no reduction in bending strength. (Clause 9.2.2 of IS:800)

In this case, single V butt joint is sufficient to take the shear.

7.10 ECCENTRIC CONNECTIONS (BUTT WELDED)

Example 7.26 *Butt weld joint of a bracket*

A bracket is made of ISMB 300 attached to flange of ISMB 400 column. The bracket is subjected to a factored load of 50 kN at an eccentricity of 300 mm. Check the adequacy of single bevel butt weld between the web & flange of the bracket and the flange of column.

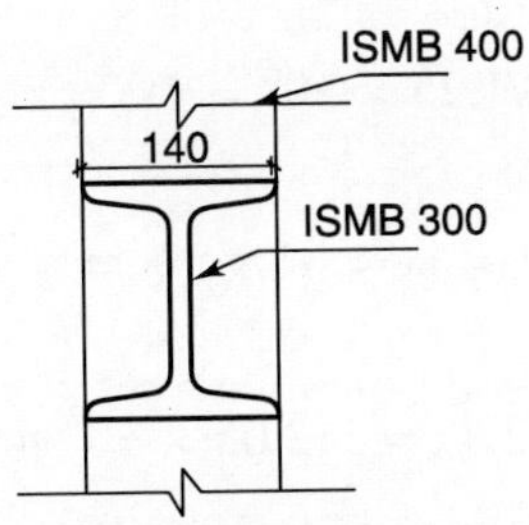

Fig. 7.46

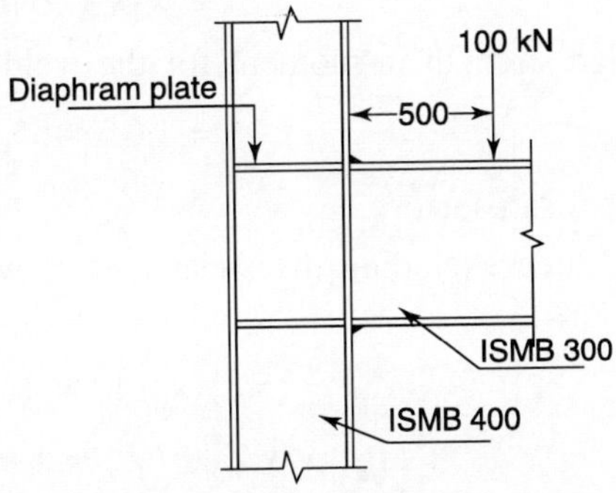

Fig. 7.47

Single level butt weld. $T_e = 5/8\,t$. (Clause 6.1.4.2(b) IS:816)

Properties of ISMB 300

A = 5625 mm^2, b_f = 140 mm, Z_{ez} = 573.6 cm^2,

t_w = 7.5 mm.

Factored shear force on the bracket = V = 50 kN.

Moment due to eccentricity of factored force = M = 50 × 0.3 = 15 kNm

Clear depth of web on which single bevel butt weld can be done

$= D - 2 \times (t_f + r_1) = 300 - 2 \times (12.4 + 14) = 247.2$ mm.

V_d = Design shear strength of the welded portion of web = V_d

$= A_v f_{yw}/(\gamma_{mo}\sqrt{3}) = t_w \times d \times f_{yw}/(\gamma_{mo}\sqrt{3}) = (7.5 \times 247.2 \times 250)/(1000 \times 1.1 \times \sqrt{3}) = 243.27$ kN.

Design shear strength of weld = (5/8) × (1.1/1.25) × 243.27 = 133.8 kN.

$$0.6V_d = 0.6 \times 133.8 = 80.3 \text{ kN} > 50 \text{ kN}.$$

Hence, low shear.

If the value of 0.6 V_d is less than 50 kN, it is in high shear and bending strength will be less. (Clause 9.2.2 of IS:800). The calculation for the same is given below.

Bending Strength of Section

$$b/t_f = 70/12.4 = 5.65 < 8.4\,\varepsilon$$

$$d/t_w = 247.2/7.5 = 32.96 < 84\,\varepsilon$$

Section is plastic. (Table 2 of IS:800)

$$Z_p = 651.74 \times 10^3 \text{ mm}^3 \text{ (Table 46 of IS:800)}$$

Design strength in moment for the section $M_d = \beta_p Z_p f_y / \gamma_{mo}$

$$= 1.0 \times 651.74 \times 10^3 \times 250/1.1 \times 10^6 = 148.12 \text{ kNm}$$

Z_p of weld = [2 × (5/8) × 12.4 × 140 × {(300 – 12.4)/2}] + [4 × (247.2/2) × (5/8) × 7.5 × (247.2/4)]

$$= 455.3 \times 10^3 \text{ mm}^3$$

Design strength in moment for the weld $M_d = \beta_p Z_p f_y / \gamma_{mo}$

$$= 1.0 \times 455.3 \times 10^3 \times 250/1.25 \times 10^6 = 91.06 \text{ kNm.} > 15 \text{ kNm}$$

Hence, satisfactory.

Weld area excluding the shear area = weld area on flanges = (5/8) × (2 × 140 × 12.4) = 2170 mm^2

$$Z_{pf} = (5/8) \times 140 \times 12.4 \times (300 - 12.4) = 312.05 \times 10^3 \text{ mm}^3$$

$$M_{fd} = \beta_p Z_{pf} f_y / \gamma_{mo} = 1.0 \times 312.05 \times 10^3 \times 250/1.25 \times 10^6 = 62.41 \text{ kNm.}$$

$$M_{dv} = M_d - \beta(M_d - M_{fd}) \text{ (Clause 9.2.2 of IS:800)}$$

$$\beta = ((2V/V_d) - 1)^2 = ((2 \times 100/133.8) - 1)^2 = 0.245$$

$$M_{dv} = M_d - \beta(M_d - M_{fd}) = 91.06 - 0.245(91.06 - 62.41) = 84.04 \text{ kNm.}$$

Design strength in moment of weld = 84.04 kNm > 15 kNm

Hence, satisfactory.

To Find the Average Shear Stress on Weld

$(5/8) \times t_w \times d \times f_{yw}/(\gamma_{mw}\sqrt{3}) = 50 \times 1000$

$$f_{vw} = (50 \times 1000 \times 1.25 \times \sqrt{3})/(5/8) \times 7.5 \times 247.2 = 92.42 \text{ MPa}$$

This exceeds $0.5 \times f_{vw}/\sqrt{3} = 0.5 \times 250/1.25 \times \sqrt{3} = 57.74$ MPa (Clause 10.5.10.2.1(b) of IS:800)

Hence, combined stress is to be checked.

To find bending stress on weld

$$M_d = \beta_p Z_p f_y / \gamma_{mo} = 15 \times 10^6$$

$$f_a = f_y = 15 \times 10^6 \times 1.25/1.0 \times (5/8) \times 455.3 \times 10^3 = 65.9 \text{ MPa}$$

(Clause 10.5.10.1.1 and 10.5.10.2.2 of IS:800)

$$\sqrt{f_a^2 + 3q^2} = \sqrt{65.9^2 + 3 \times 92.42^2} = 173.1 \text{ Mpa}$$

This should be less than $f_{yw}/(\gamma_{mw}\sqrt{3}) = 410/1.25 \times \sqrt{3} = 189$ MPa

Adequate.

Note: In this case, single *V* or double *V* cannot be provided. We can provide single bevel button fillet weld.

EXERCISE

1. Explain with sketch the effective throat thickness of a fillet weld.
2. What is meant by intermittent fillet weld? Draw a sketch to show the same.
3. Explain (a) shielded metal arc welding, (b) end returns and (c) effective length of fillet weld.
4. Explain full penetration butt weld.
5. ISA 75 × 75 × 8 is used as a tension member and connected to 8 mm thick gusset plate. Design the site fillet weld connection for a tensile force of 150 kN due to dead, imposed and seismic loads.
6. An ISMB 500 column transfers 600 kN of axial load and 100 kNm of moment about the major axis due to dead and imposed loads. Design the fillet welded shop connection between the column and base plate to transfer 100% load and moment.
7. Using the design data in (7.6) design the single *V* butt weld connection.
8. Design the fixed connection between beam ISMB 300 and flange of a column ISMB 400 to transfer shear of 100 kN and moment of 45 kNm due to dead and imposed loads.
   (a) Use cleat angles on flanges and web with fillet welds at site.
   (OR)
   (b) plate on top flange, cleat angles on bottom flange and web.
9. Check the adequacy of single *V* butt weld connection between a bracket of ISMB 250 connected to the flange of ISMB 450 to transfer shear of 80 kN and moment of 40 kNm.
10. Design the splice plate connection for web and flange of ISMB 350 beam for its full capacity using the site fillet weld.

Chapter 8

Bolted Joint – Structural Connections

8.1 GENERAL

The connections in a structure shall be designed to transmit the forces from the member connected and comply with the requirements specified in Section 10 of IS: 800.

When members are connected to the web or the flange of a section, the capacity of the web or the flange to transfer the forces locally should be checked and where necessary, local stiffening is provided.

Bolts may be used for structures not subjected to vibrations. The cost of bolts is more but it takes less time to erect structures with bolted connections. The erection work with bolts is noiseless and less skilled workers can also handle it. The stress concentration at the bolts can also be reduced by using high strength friction grip bolts.

Types of Bolts

The following types of bolts are used in structures:

(1) **Black bolts** Hexagonal black bolts are commonly used in steel works. They are made from low or medium carbon steels and should conform to IS:1363 – 1984. In metric sizes, black bolts are produced in sizes of 5 mm to 36 mm diameter and of lengths 3 to 10 times the diameter. They are designated as black bolts $M_d \times l$–IS: 1363, where d = diameter and l = length of the bolt.

(2) **Precision and semi-precision bolts** They are also known as close tolerance bolts conforming to IS: 1364–1967. Sometimes to prevent excessive slip, close tolerance bolts are provided in holes of 0.15 to 0.2 mm oversize. This may cause difficulty in alignment and delay in the progress of work.

(3) **High-strength friction grip bolts** HSFG bolts have a high yield strength. They should conform to IS: 3757 – 1985. They are tightened by torque wrenches and require hardened washers to include initial tension, which causes friction between the plate surfaces. Due to friction, there is no slip in the joint and therefore the joints with HSFG bolts are called friction type or non-slip type joints.

Definitions

The following definitions are used for bolted joints.

Nominal diameter For a bolt, the diameter of the unthreaded portion of the shank is called its nominal diameter.

Effective diameter or gross diameter For a bolt, the nominal diameter is same as the gross diameter.

Gross area The gross area of a bolt is given by its gross diameter.

Net area The net area of a bolt is the area at the root of the thread.

Pitch The distance between centres of any two adjacent bolts is called the pitch.

Diagonal pitch The distance between centres of any two adjacent bolts in the diagonal direction is called diagonal pitch [= p_2 in Fig. 8.1].

Staggered pitch The distance between centres of any two consecutive bolts in a zig-zag bolting, measured parallel to the direction of stress in the member is called staggered pitch [= p_1 in Fig. 8.1].

Gauge A row of bolts parallel to the direction of force is called a gauge line. The normal distance between two adjacent gauge lines is called the gauge distance.

Edge distance It is the distance between the edge of a member or cover plate and the centre of the nearest bolt hole.

Proof load Initial tension in HSFG bolts is known as proof load of the bolt.

Slip factor Coefficient of friction in friction type joint is known as slip factor.

Property class Bolts are grouped under different grades or property classes depending upon their strength. For example, property class 4.6 indicates that the nominal ultimate tensile strength is 400 MPa and the nominal yield stress is $4 \times 6 \times 10$ (or) $0.6 \times 400 = 240$ MPa.

Assumptions in the Theory of Bolted Joints

The analysis of joints is very complex and it is simplified by making a few assumptions as given below:

The tensile stress is uniformly distributed on the portions of the plate between the bolts.

The friction between the contact surface of the plates is neglected.

The shearing stress is uniformly distributed on the cross section of the bolts.

The bolts in a group share the direct load equally.

Bending stress in bolts is neglected.

Bearing stress distribution is uniform and the contact area is $d \times t$ where d is the diameter and t is the thickness of the plate.

Standard Clearance Hole (Clause 10.2.1(a) of IS:800)

For structural connections, the diameter of standard clearance holes for fasteners shall be as given in Table 19 of IS:800. See Fig. 8.2 of this book.

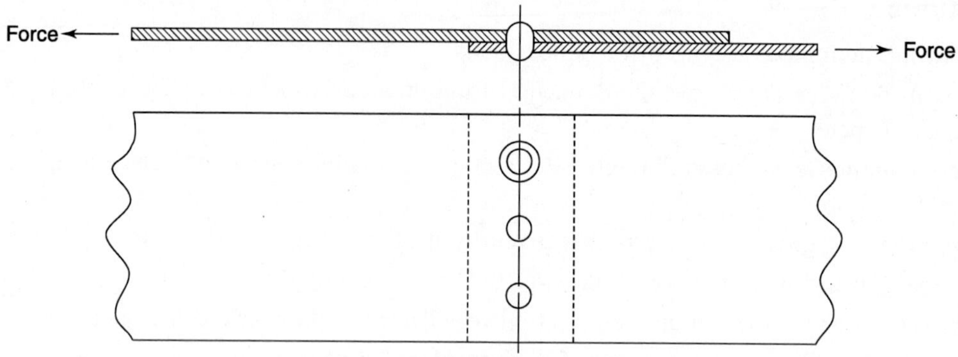

(a) Single bolted lap joint

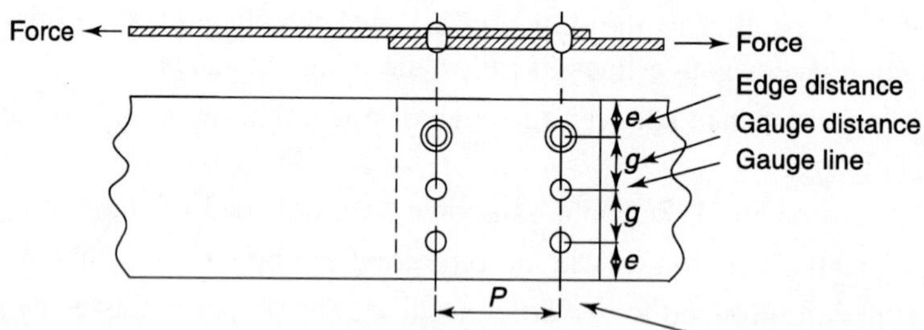

(b) Chain riveting

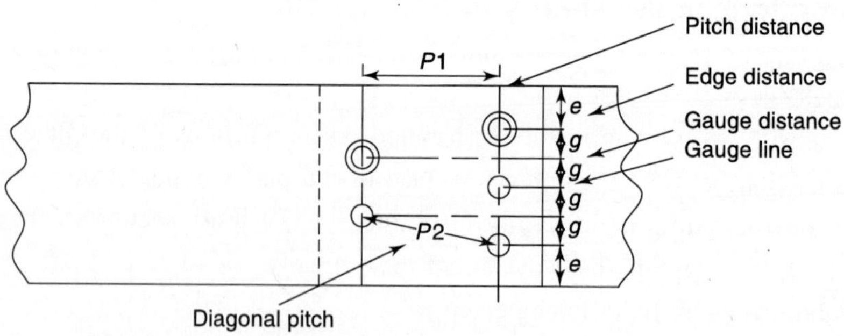

Fig. 8.1 *Pitch, edge and gauge distances*

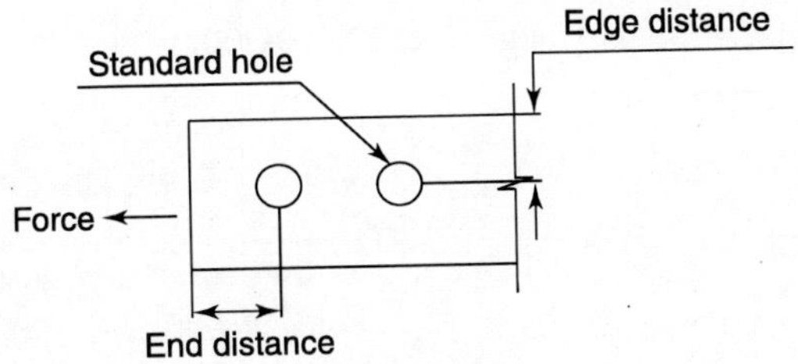

Fig. 8.2

Oversize Hole (Clause 10.2.1(b) of IS: 800)

Oversize holes, as given in the Table 19 of IS: 800/2007 may be used in slip resistant connections and holding down bolted connections. The oversize hole in the outer ply is covered by a cover plate of sufficiently large size and thickness and having a hole not larger than the standard clearance hole. In case of slip resistant connection, hardened washer is used. In foundation/ base plate, oversize holes helps the erection of columns particularly when the holding down bolts are embeded in the foundation concrete. See Fig. 8.4 of this book.

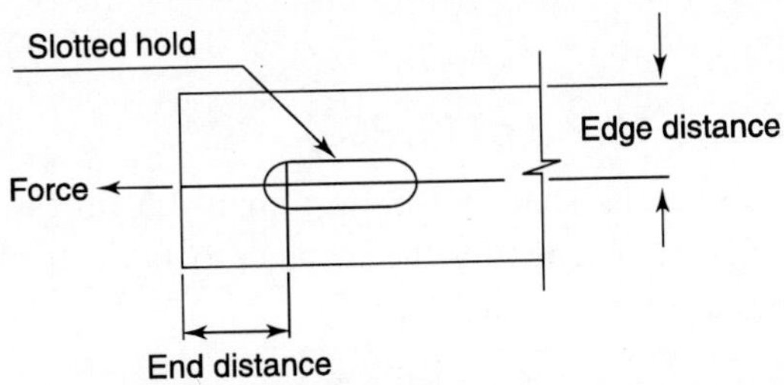

Fig. 8.3

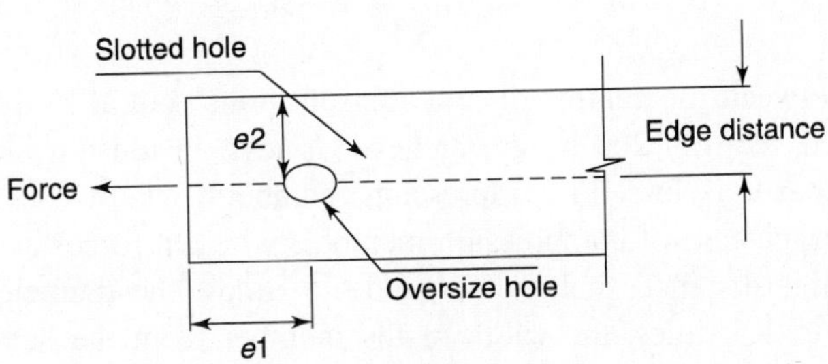

Fig. 8.4

Short and Long Slots (Clause 10.2.1(c) of IS: 800)

Slotted holes of size larger than the standard clearance hole, as given in Table 19 of IS: 800 may be used where sliding of the member connected is to be allowed. When sliding due to force (or) thermal expansion is allowed, shear force will not be transferred to the supporting member except that due to friction. See Fig. 8.3 of this book.

Table 8.1 Clearances for fastener holes (Table 19 of IS:800/2007)

Serial Number	Nominal size of bolt "d in mm"	Size of hole = Nominal dia of the bolt + Clearances in mm			
		Standard Clearance in dia & width of slot in mm	Oversize clearance in diameter in mm	Clearance in the length of slot in mm	
				short slot	long slot
1	Between 12-14	1.0	3.0	4.0	2.5d
2	Between 16-22	2.0	4.0	6.0	2.5d
3	24	2.0	6.0	8.0	2.5d
4	>24	3.0	8.0	10.0	2.5d

Spacing

Minimum spacing (Clause 10.2.2 of IS:800)

The distance between centre of bolts shall not be less than 2.5 times the nominal diameter of the bolt (Clause 10.2.2 of IS: 800). Generally the distance between centre to centre of bolts is kept at three times the nominal diameter of bolts.

Maximum spacing (Clause 10.2.3 of IS:800)

(a) The distance between the centres of any two adjacent bolts.

Shall not exceed $32t$ (or) 300 mm, whichever is less, where t is the thickness of the thinner plate.

(b) The distance between the centres of two adjacent bolts (pitch) in the direction of stress, shall not exceed $16t$ (or) 200 mm, whichever is less, in tension members and $12t$ (or) 200 mm, whichever is less, in compression members, where t is the thickness of the thinner plate. In the case of compression members wherein forces are transferred through butting faces, this distance shall not exceed 4.5 times, the diameter of the bolts for a distance equal to 1.5 times, the width of the member from the butting faces.

(c) The distance between the centres of any two consecutive bolts in a line adjacent and parallel to an edge of an outside plate shall not exceed 100 mm + $4t$ (or) 200 mm, whichever is less, in compression and tension members, where t is the thickness of the thinner outside plate.

(d) Where bolts are staggered at equal intervals and the gauge does not exceed 75 mm, the spacing specified in (b) and (c) between centre of bolts may be increased by 50% subject to the maximum spacing specified in (a).

Edge and End Distance (Clause 10.2.4 of IS:800)

The distance from the edge of the member to the centre of bolt hole measured perpendicular to the direction of load is called edge distance. The distance from the edge of member to the centre of the bolt hole measured along the direction of load is called end distance. For oversize holes, the end (or) edge distance = e_2 + Half the diameter of standard clearance hole relevant to the bolt – nominal diameter of the oversize hole (do).

For example, for 20 mm diameter bolts, the end or edge distance

$$= e_2 + (20 + (2/2)) - 24 = e_2 + 11 - 24 = e_2 - 13 \text{ mm}$$

(a) The minimum edge distance from the centre of bolt hole to the nearest edge of a plate or member shall not be less than 1.7 times the hole diameter in case of sheared or hand flame cut edges; and 1.5 times the hole diameter in case of rolled, machine flame cut, sawn and planed edges. The same is applicable for end distance also. Generally, the distance between the centre of bolt hole to nearest edge is kept at two times the hole diameter.

(b) The maximum edge distance to the nearest line of bolts from an edge of any unstiffened part should not exceed $12t\,\varepsilon$; where $\varepsilon = \sqrt{(250/f_y)}$ t is the thickness of the thinner outer plate. This is not applicable for bolts interconnecting the components of back to back tension members. Where the members are exposed to corrosive environment, the maximum edge distance shall not exceed 40 mm + $4t$ where t is the thickness of thinner connected plate.

Tacking Bolts (Clause 10.2.5 of IS:800)

(a) In case of members covered under Clause 10.2.4.3 of IS:800 when the maximum distance between centres of two adjacent bolts is exceeded, tacking bolts shall be used. But this shall not be considered for calculation.

(b) Tacking bolts shall have spacing in a line not exceeding 32 times the thickness of the thinner outside plate or 300 mm, whichever is less. Where the plates are exposed, the spacing in line shall not exceed 16 times, the thickness of the thinner outside plate or 200 mm, whichever is less. In these cases, the distance between the lines of bolts shall not be greater than the respective pitches.

(c) The requirements specified in Clause 10.2.5.2 of IS:800 shall generally apply to compression members subject to the stipulations in section 7 of IS:800 affecting the design of compression members.

(d) In tension members (section 6 of IS:800/2007) composed of two flats, angle and channels in contact back to back or separated back to back by a distance not exceeding the aggregate thickness of the connected parts, tacking bolts with spacers shall be provided at a spacing in line not exceeding 1000 mm.

For compression members, tacking bolts in a line shall be spaced at a distance not exceeding 600 mm.

8.2 CONNECTION DESIGN

The raw material cost of the finished structural component is only about 60%. The rest of the cost is in fabrication and handling in the shop. Fabrication (production) includes, cutting, edge preparation, shearing, machining, drilling, welding, surface preparation and surface finish. The bolted connections can influence up to about 60% of the cost of the production. In many cases, member sizes, should be chosen on the basis of connection requirements rather than the member design requirement. If by increasing the size of member, the connection can be simplified, one can increase the size of the member.

To avoid claims and delays in the event of complications during fabrication and erection, the drawings issued at the time of tender should show the proposed types of connections also.

Simple connections and moment connections are shown in Fig. 8.5. Only some types of connections are shown here.

8.3 SIMPLE CONNECTIONS

These depend on the shear load they transfer and are easy to fabricate and erect. In Fig. 8.5(a) to 8.5(d) the types of connections are shown.

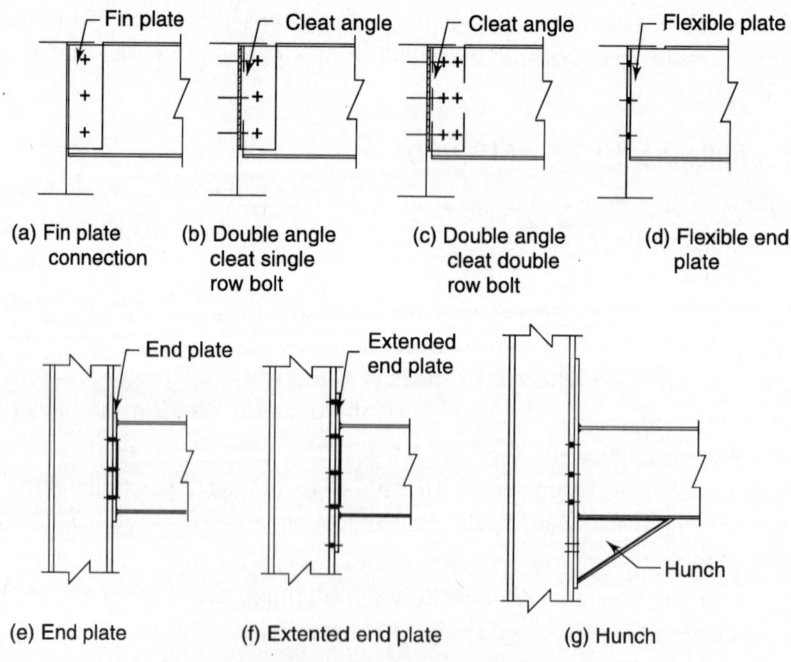

Fig. 8.5 *Beam to column connections*

Secondary beams will have notched ends and holes will be drilled in the web for bolted connection to fin plates or cleat angles. The columns will have bolt holes on the flange or web for bolted connection. The column connecting beams are called primary beams.

8.4 MOMENT CONNECTIONS

These connections are complex, depending on the moment and shear force they transfer. The fabrication cost is more compared to welded connections due to drilling and pre-assembling at shop before despatch to avoid mismatch at site. In Fig. 8.5, the standard types of connections are shown. Connections as shown in Fig. 8.5(e) and 8.5(g) are used for moment connection. The connection as shown in Fig. 8.5(g) is used for connecction with very high moment.

In moment connections, primary beams with end plates or extended end plates butt welded or fillet welded to the primary beams are used. Sometimes cleat angles for web of beam and angles for top and bottom flanges of beam are used to transfer the moment. The primary beam may also have bolt holes in its web for the connection of secondary beams. For the columns, tensile stiffening of the flanges may be required at the primary beam connection location/ elevation. Beam-to-web of column connection will be by means of angle cleats on either side of the web of the beam and with plates on top and bottom of flanges.

In a connection carrying higher moment, beams with hunched end plates are used as shown in Fig. 8.5(k) and (l). The secondary beams connected to primary (or) main beams will have flexible end plates or angle cleats on either side of web and the secondary beam will be notched. The beam connected to the web of the column shall have plates at both the flanges connected to the web and flanges of column to transfer moment. If the beam is there on only one side of the column; stiffeners shall be provided on the other side of the column.

Note: The cost of fabrication of a bolted moment connections could be six to eight times more than that of a bolted simple connection.

8.5 TRUSSES AND OPEN-WEB GIRDERS

Low-cost Connections

Trusses with diagonal members, continuous bottom and top chords where the joints are assumed to be 'pinned' will have the chords made of members connected to gusset plates bolted or welded at the intersections with the diagonal members. The diagonal members are made of double/single angles/channels/*I* sections using bolts for connections to the gusset plates. The welded connections require more handling in the shop. Whereas the bolted connections require more of drilling and matching by pre-assembling at the shop. When bolted connections are used, it is advisable to pre-assemble and check at shop to avoid site modifications when there are some missmatches.

Medium-cost Connection

Trusses may require stiffening of the chords at the connections when the forces are high. In such case, the fabrication cost increases considerably. Any moment connection at the diagonal members, will increase the cost and complexity of the connection. In the analysis of structure, the diagonal members are considered to take up axial forces only and hence they are generally connected with gusset plates which are thin and flexible. The choice of sections for the chords and the diagonals remain the same as in a low-cost connection option. Welded connections at the intersections requiring stiffening which will add to the cost and time required for fabrication.

High-cost Connections

Trusses using circular, rectangular (or) square hollow sections with welded intersections require special care since some connections may require the thickening of walls and circular sections require profile cutting which is costly and time consuming.

If the trusses are in more than one plane, then the connections become more complicated. Welding, handling and transportation processes should be considered by the designers at the conceptual design stage.

In Fig. 8.6 different types of connections of the truss are given. Low raw material weight is not necessarily the lowest cost option since welding, bolting, preassembly and transportation are also to be included. To enable the designer to account for fabrication, shipping and erection; list of points are given in para 8.06.

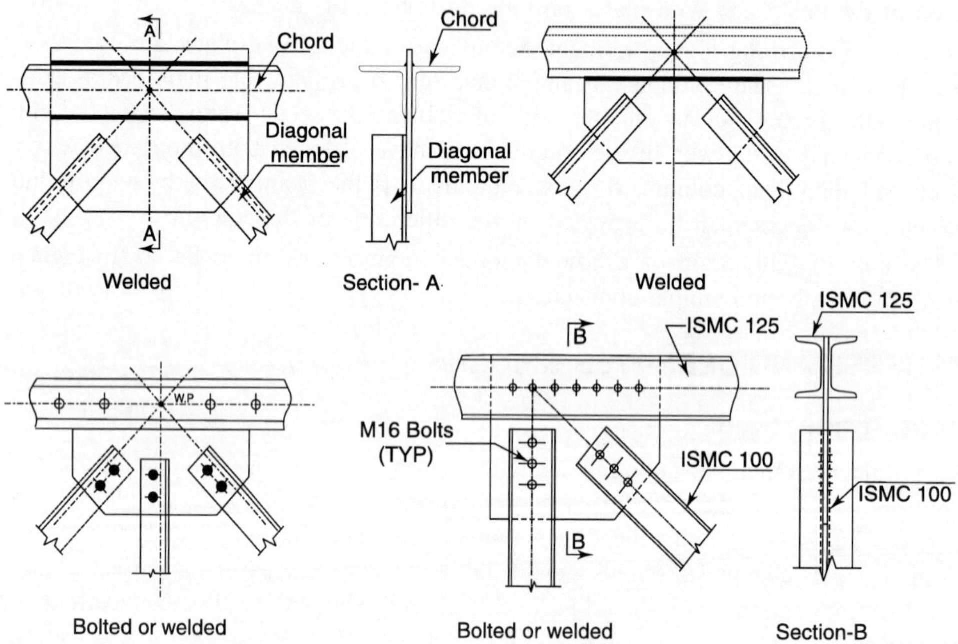

Fig. 8.6 *Connections in trusses*

8.6 LIST OF ACTIVITIES FOR DESIGN

1. Conceptual design stage

(a) Use only few sections and sizes in one structure. (Variety reduction)

(b) As far as possible try to use one grade of steel.

(c) High strength friction grip (HSFG) bolts should be used where it is absolutely essential.

(d) Welding and bolting in the same piece to be fabricated, shall not be mixed. However, it is the practice to adopt welded joints at the shop and bolted joints at the site for bolted structure.

(e) Weldability of materials is to be checked before specifying welding.

(f) To the maximum possible extent use rolled steel sections. Only when the requirement calls for, beams/columns are provided with plate formed sections.

(g) Using multi-planar trusses is to be avoided unless warranted.

2. Detailed design stage

(a) For some structures, small quantities of high grade steel may be required. Providing small quantity of high grade steel will be nearly impossible. Hence, try to use one grade of steel.

(b) Batten to increase member size to avoid stiffening at connections.

(c) Select sizes of members of a truss such that local stiffening at the intersections is avoided.

(d) To the possible extent try to avoid hunched beam-end connections though it is best to transfer moment.

(e) Avoid using *I* sections for diagonals of a truss or bracing members because flange stripping may be necessary at the intersections.

(f) Intermittent fillet welds are not advisable where there is a possibility of moisture ingress. Continuous smaller size fillet weld is cheaper than an intermittent fillet weld since intermittent weld requires more manpower.

(g) Specify the required weld size. Do not increase the size of the weld when not required.

(h) Try to use one or two grades of bolts (variety reduction).

(i) M20 Grade 8.8 for shear connection and M24 Grade 8.8 for moment connections may be proposed as a measure of standardisation with respect to a structure.

(j) Fillet weld is preferred to butt welds of equivalent strength.

(k) Try to avoid site welding. Bolted connection at site is preferred.

(l) Try to use one hole diameter on any one piece.

(m) Avoid staggered holes.

(n) Access for welding shall be checked before releasing the drawing for fabrication.

(o) Specify weld inspection and acceptance criteria clearly in the drawing.

(p) Specify surface preparation and surface finish in the drawing.

(q) Check the transportation limits before designing the components and parts. The local law is to be taken into account for export job. Generally, the width is about 3 m and length is about 22 m.

(r) That dimensional limit shall apply not only to transport but also to handling in a fabrication shop, off-site painting yards and galvanizing facilities. For galvanized structures, the size of galvanizing bath is to be considered before deciding the size and length of single piece of structural member. This may warrant more joints and this becomes a challenge to the designer.

(s) Stability of part of structure during fabrication and transport should be considered at the design stage. If required, necessary support frame (or) crates are to be proposed.

(t) Specify the lifting points in the structural components. If lugs are provided specify the removal procedure.

3. At manufacturing stage

(a) Follow the drawing and specifications.

(b) Follow safety instructions

(c) Offer for quality inspection at right stages of fabrication.

(d) Use the right machinery, and equipment for welding, drilling, machining, handling at the shop and other fabrication processes.

(e) Effects of distortion due to welding shall be considered reduced.

(f) Carry out stress relieving (heat treatment) where required.

4. At erection stage

(a) Use electrodes for welding as in welding specifications.

(b) Use right grade, size and length of bolt, nut and washer.

(c) Use right grade, size of nut and washer for bolted connections.

(d) Bolts, nuts and washers should have corrosion protection before delivery to the site.

(e) Since it is impossible to paint the top flanges after the metal decking has been installed and shear connectors have been welded in case of composite construction, provide the surface protection well before.

(f) Take care that the insulation, refractory and decorative finishes are not damaged while handling during transport and erection.

(g) Provide temporary structure during fabrication and transportation where required.

8.7 BOLT SPECIFICATIONS

Refer IS: 3757 Specification for high strength structural bolts. This code covers the requirements for large series hexagon high strength structural steel bolts of Property class 8.8 and 10.9 and in the size range of M18 to M36 (both friction type and bearing type).

(1) Bolts of property class 10.9 and 12.9 should not be hot dip galvanized.

(2) Recommended length size combinations for high strength structural bolts are given in Table 2 of this standard.

(3) A high strength structural nut IS: 6623 are to be used with these bolts.

Bolt property classes & finish	**Nut property classes & finish**
8.8 s dull black	8 s or 10 s dull black

8.8s hot dip galvanized — 10s hot dip galvanized

10.9s dull black — 10s dull black.

(4) Hardened and tempered washers to be used with these bolts shall conform to the requirement IS: 6649

(5) 8.8s, 10.9s – s denotes high strength structural bolts with a large series hexagon head. In case of hot dip galvanized bolts, the word galvanized shall be added to the designation.

IS:1367 (part 3) mechanical properties of fasteners made of carbon steel and alloy steel-bolts screws and studs.

Property class symbol consists of two figures.

Example: Class 4.6

1. (4) indicates 1/100 of the nominal tensile strength in N/mm²
2. (6) indicates 10 times the ratio between lower stress (or proof stress) and nominal tensile strength (yield stress ratio).

The multiplication of these two will give 1/10 of the yield stresses in N/mm²

Property classes 4.6. Tensile strength 400 MPa.

10 × lower yield stress/tensile strength = 2400/400 = 6

Lower yield stress = 4 × 6 × 10 = 240 MPa.

For structural bolting the lower limit of size is M12.

8.8 THICKNESS OF MEMBER (VS) MINIMUM DIA OF BOLT

Thickness of member	*Minimum diameter of bolts*
Up to 10 mm	16 mm
Over 10 mm to 16 mm	20 mm
Over 16 mm	22 mm

8.9 DESIGNATION OF BOLT

High strength structural bolt M24 × 140 – IS: 1357 – 8.8s galvanized. (Add *N* after 140 when hot dip galvanized nut (IS: 6623) is required.)

Shorter thread lengths have been provided to ensure no thread protrusion in the shear plane so that higher shear area can be adopted for bearing type connections.

Specify hex bolt M12 × 30 IS: 6639 4.6 (add *N* after 30 for nut)

(1) The hexagon nuts used with hexagon bolts shall conform to requirements given in IS: 1363 (Part-3).

(2) Approximate weight of hexagon bolts "are given in appendix A. IS:4000-Code of practice for high strength bolts in steel structures — code of practice. This standard covers the requirements for the design fabrication assembly and inspection in all types of steel structures of structural joints using high strength structural bolts conforming to IS: 3757.

(3) This standard applies to high strength bolts used in both friction type and bearing type shear joints and for tension joints.

IS:1367 gives technical supply conditions for threaded steel fasteners.

(4) IS:6639-specification for hexagon bolts for steel structures.

Scope: Requirements for hexagon bolts in the size range 12 to 39 for steel structures.

1. Table 1 dimensions for hexagon bolts for steel structures.
2. Table 2 preferred length dia combinations and clamping lengths for hexagon bolts for steel structures.
3. Mechanical properties of the bolts shall conform to the property class 4.6 or 6.6 as specified IS:1367.
4. Grade black (b)

8.10 DESIGNATION OF NUT

IS:6623 – High strength structural nuts—specification.

Covers the requirements for large series hexagon high strength structural steel nuts in property class 8 and 10 and in the size range of M16 to M36 suitable for use in both friction type and bearing type structural steel joints.

Nut specification: High strength structural nut M24 IS 6623-8s galvanized.

8.11 DESIGNATION OF WASHER (FIG. 8.7)

IS:6649-specification for hardened and tempered washers for high strength structural bolts and nuts covers the requirements for through hardened and tempered steel washers intended for assembly with large bolts and nuts in the size range M16 to M36.

Type *A* plain hole circular washer.

Type *B* sq taper for channel (6° taper); Type *C* sq taper for channel (8° taper)

Washer M24 *A* IS:6649 galvanized.

8.12 TERMINOLOGY RELATING TO BOLTED CONNECTIONS

8.12.1 Grip

The total thickness of steel structural parts to be held together, including packing plate but excluding washers is called grip (or) grip length.

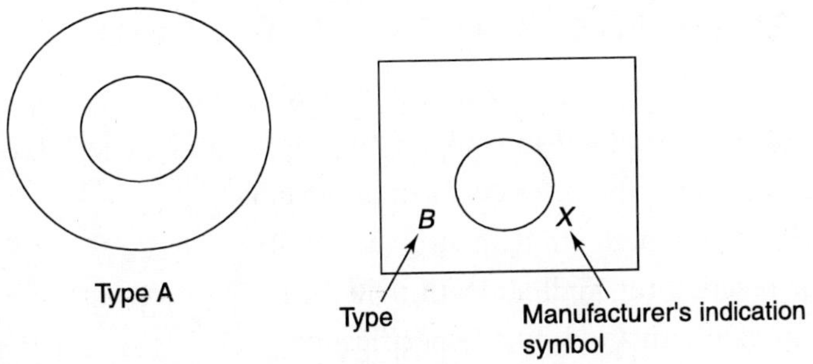

Fig. 8.7

8.12.2 Length of Bolt

The distance from the bottom of the bolt head to the extreme end of the shank as shown in Fig. 8.8.

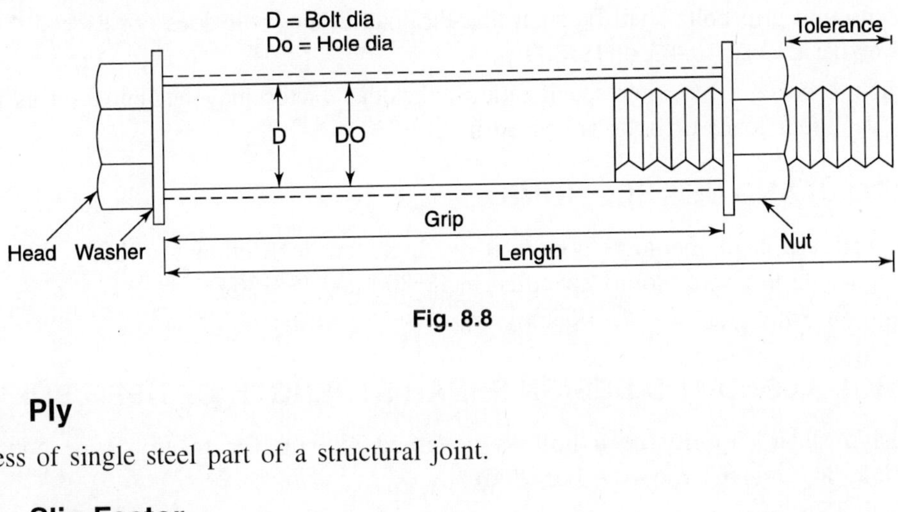

Fig. 8.8

8.12.3 Ply

Thickness of single steel part of a structural joint.

8.12.4 Slip Factor

The ratio of the shear force required to produce slip between two surfaces of a structural joint to the force clamping the two parts together.

8.13 BOLT CLASS AND STRENGTH

The most frequently used bolts are class 4.6 and 8.8 for bearing type joints. Bolt class 3.6, 5.6, 5.8 and 6.8 are also there but they are not frequently used.

Class 4.6 bolts have a minimum ultimate tensile strength of 400 N/mm^2 and a minimum stress at yield of $4 \times 6 \times 10 = 240$ N/mm^2.

Similarly, class 8.8 bolts have a minimum ultimate strength of 800 N/mm^2 and a minimum stress at yield of $8 \times 8 \times 10 = 640$ N/mm^2.

8.14 HIGH STRENGTH FRICTION GRIP (HSFG) BOLTS

Bolt class 8.8, class 10.9 and class 12.9 are used as HSFG bolts.

Generally, class 8.8 bolts are used for this purpose up to a diameter of 24 mm.

The following IS codes may be referred for more details.

IS: 3757—specification for high strength structural bolts.

IS: 4000—code of practice for high strength bolts in steel structures.

IS: 6623—High strength structural nuts—specification.

IS: 6649—Specification for hardened and tempered washers for high strength structural bolts and nuts.

ISSP: 6(4)—Use of high strength friction grip bolts.

8.15 SHEAR CONNECTIONS

In connections subjected to shear without any moment, in the plane of the friction faces, the number of friction grip bolts shall be such that the load at any bolt does not exceed the value calculated as per Clause 10.4.3 of IS: 800.

For structure covered by other specifications, the load factor may be increased as per the relevant code. Proof loads of bolts are given in IS: 3757

8.16 GROSS AND NET SECTIONS

Gross and net section of members connected by high tensile friction grip bolts shall be the same as applicable to riveted joints specified in IS: 800 and IS: 1364.

Minimum edge distance is as in IS: 800.

8.17 PROOF LOAD AND DESIGN SHEAR STRENGTH OF HSFG BOLTS

V_{dsf} = Design shear capacity of a bolt as governed by slip for friction type connection $= \mu_f \times n_e \times k_h \times F_o/\gamma_{mf}$ (Clause 10.4.1 IS: 800)

8.18 COEFFICIENT OF FRICTION (OR) SLIP FACTOR

Where the surfaces in contact comply with "the edge distance for any hole and the pitch shall conform to the requirement of IS: 800 and the surfaces are clean "as rolled" slip factor shall be assumed as 0.35.

8.19 BEARING TYPE JOINTS

8.19.1 Introduction

A joint connected with fully tensioned high strength bolts where joint slip may occur so that the applied force is transferred by shear in the bolts and bearing on the connected points. In bearing type joints with packing plates are required to carry shear, the number of bolts required

Table 8.2 HSFG bolts-proof load, design shear and tensile strength as per IS:800

S. No.	Physical property				Proof Load in KN.		Design shear strength in kN.			
							At service load		At ultimate load	
	Nominal diameter of bolt in mm (Shank)	Gross area in Sq mm A_{sb}	Net Area in Sq mm A_{nb} (Course Thread)	Ratio A_{nb}/A_{sb}	Property class 8.8	Property class 10.9	Property class 8.8	Property class 10.9	Property class 8.8	Property class 10.9
1.	12	113.14	84.3	0.75	47.21	61.37	42.92	55.79	37.77	49.1
2.	16	201.14	157	0.78	87.92	114.3	79.93	103.91	70.34	91.44
3.	20	314.29	245	0.78	137.2	178.36	124.73	162.15	109.76	142.69
4.	24	452.57	353	0.78	197.68	256.98	179.71	233.62	158.14	205.59
5.	30	707.14	561	0.79	314.16	408.41	285.6	371.28	251.33	326.73
6.	33	855.64	694	0.81	388.64	505.23	353.31	459.3	310.91	404.19
7.	36	1018.29	817	0.8	457.52	594.78	415.93	540.71	366.02	475.82

Note: (1) Design shear strength is based on μ_f = 1.0 and n_c = 1.0 For other values of μ_f and n_c multiply the design strength values in the table by the values of μ_f and n_c (2) Proff stress for property class 8.8 = 800 MPa and for property class 10.9 = 1040 MPa. (IS:1367).

Table 8.3 Treatment of surface (vs) slip factor

Treatment of surface	Slip factor
Surface not treated	0.2
Surface blasted with shot or grit with any loose rust removed no pitting	0.50
Surface blasted with shot or grit and hot dip galvanized	1.1
Surfaces blasted with shot or grit and spray metallized with zinc (thickness 50–70)	0.25
Surfaces blasted with shot or grit and pointed with ethyl zinc silicate coat (thickness 30–60 μ)	0.30
Surface blasted with shot or grit and painted with ethyl zinc silicate coat (thickness 60–80 μ)	0.30
Surfaces blasted with shot or grit and painted with alkali zinc silicate coat (thickness 60–80 μm)	0.30
Surfaces blasted with shot or grit and spray metallized with aluminium (thickness > 50 μm)	0.50

Note: The contact surfaces shall not be sand/shot blasted.
Joints subjected to shear force only.
Also refer to Table 20 of IS:800.

determined by calculation shall be increased above the numbers required by normal calculation by a factor β_{pk} = (1 – 0.0125t_{pk} where t_{pk} is the thickness of thicker packing plate. This is applicable only when the thickness of packing plate exceeds 6 mm. (Clause 10.3.3.3 of IS: 800).

8.19.2 Limitations on Bearing Type Joints

Friction type joints are preferred when shear and moment connection where slip is not acceptable.

8.20 JOINTS SUBJECTED TO SHEAR FORCE ONLY

Bearing type joints subject to shear force only, so that the shear force on any bolt does not exceed the design shear strength of bolt.

For long joints, the shear strength of any bolt shall be reduced by a factor (Clause 10.3.3.1 of IS: 800).

The shank area and net stress area for various diameter bolts are given in Table 8.4.

Table 8.4 Bolt diameter and stress area

Nominal size of bolt	*Stress area of bolt shank in mm²*	*Stress area of bolt thread in mm²*
M16	201	157
M20	314	245
M24	452	353
M30	706	561
M36	1017	817

Note: Bearing type joints subjected to shear and tensile forces shall satisfy the following condition. $(V_{sb}/V_{db})^2 + (T_b/T_{db})^2 < 1.0$ (Clause 10.3.6 of IS: 800).

8.21 FRICTION TYPE JOINTS (CLAUSE 10.4 OF IS: 800)

A joint connected with high strength pre-tensioned bolts to such an extent, that the clamping action transfers all the applied forces acting in the plane of the contact surfaces by the friction between the contact surfaces is named as friction type joints. This depends on the slip factor which in turn depends on the roughness of the contact surface and pre-tension applied.

8.22 JOINTS SUBJECTED TO SHEAR AND TENSILE FORCE

An applied pre-tension in the direction of the axis of the bolt creates friction between contact surfaces which takes the shear forces.

No slipping occurs in such joints. The joints are rigid. Number of bolts in a joint will be less and thereby the size of the gusset plate will be less. The material cost is high and special attention is required during erection.

8.23 HOLES IN MEMBERS

Unless specified otherwise in the drawings or specifications the diameter of a hole shall be 1 mm and 3 mm more than the nominal bolt diameter for bearing type joints and frictions type joints respectively. Also refer to Table 19 of IS: 800 for clearances for fastener holes.

8.24 STRESS AT ELEVATED TEMPERATURE

Yield stress will be less at elevated temperature and the same is given in Table 8.5.

Table 8.5 Bolt property class and lower yield stress (MPa)

Property Class	*Temperature + 20*	*Temperature + 100*	*Temperature + 200*	*Temperature + 250*	*Temperature + 300*
5.6	300	270	230	215	195
8.8	640	590	540	510	480
10.9	940	875	790	745	705
12.9	1100	1020	925	875	825

Young's modulus and coefficient of expansion will also change due to increase in temperature.

8.25 BOLT EDGE DISTANCES AND PITCHES

The distance between centre to centre of adjacent bolt holes in the direction of stress is called pitch. Usually, the pitch refers to the distance measured along the length of the joints which is the direction parallel to the direction of force (Clause 10.2.3 and 10.2.4 of IS: 800).

Table 8.6 Bolt edge distances (Clause 10.2.4.2 & Table 19 of IS: 800)

	Diameter of bolt (mm)	12	14	16	18	20	22	24	27	30
	Dia. of the hole	13	15	18	20	22	24	26	30	33
Edge distance	Shear or flame cut	22.1	25.5	30.6	34	37.4	40.8	44.2	51	56.1
	Rolled or machines cut	19.5	22.5	27	30	33	36	39	45	49.5

Note: (1) The end/edge distance is 1.7 times the diameter of the hole in the case of threaded/hand flame cut edges and 1.5 times the diameter of hole in the case of rolled, machined, flame cut, sawn and planed edges.

(2) It is advisable to have the edge (or) end distance twice the diameter of hole.

8.26 BOLT STRENGTSH AND CAPACITIES

Let us consider bearing type joints. As the slip between the plates takes place, shear is introduced in the bolt and at the same time bearing force is introduced on the plate and cover plates. Failure of a bearing type bolted joint can be due to

(a) Shearing of bolt.

(b) Bearing of bolt.

(c) Tension in bolt.

(d) Tearing of plate.
(e) Tension failure of plate.

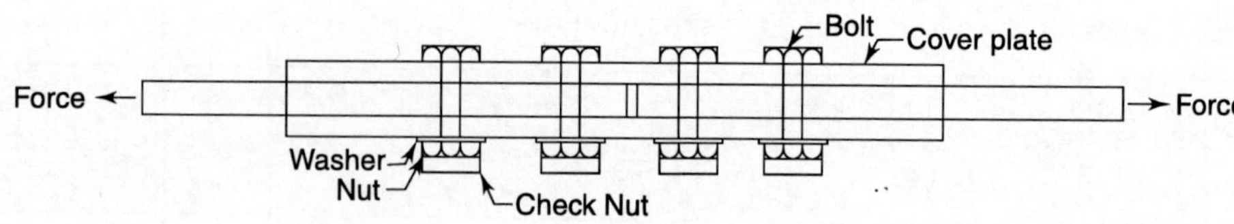

Fig. 8.9

Approximately about 40 to 70 per cent of the load slip occurs and the load is transferred to the bolt and plates. The strength of a bolt means the strength at failure as indicated in Fig. 8.10.

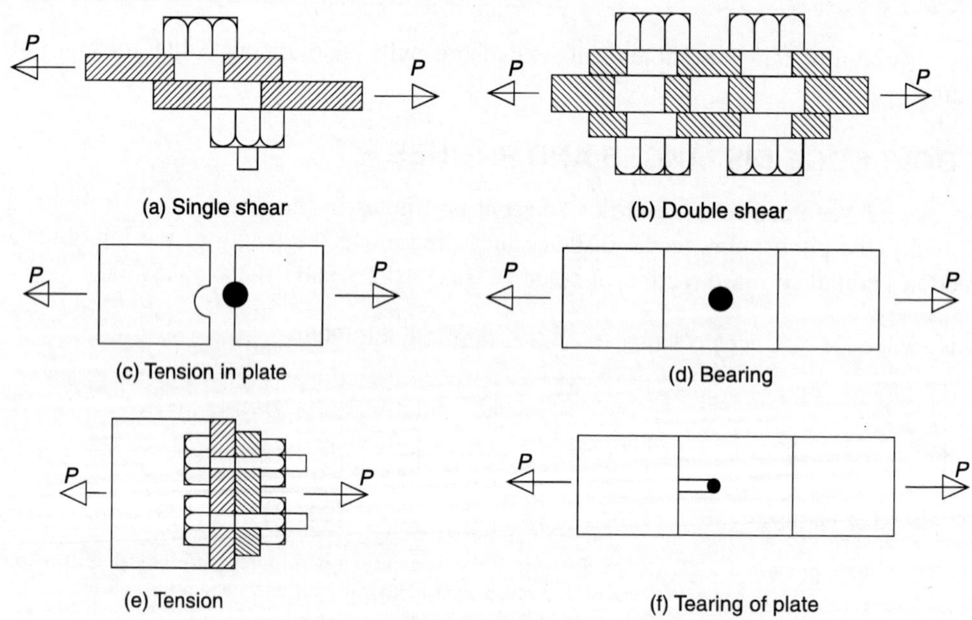

Fig. 8.10

Failure of plate, namely, tension and tearing of the plates can be avoided by providing enough edge distances and pitch distances in the joint. The design of joints should ensure no failure by shear or tension or bearing of the bolt occurs.

To avoid failure of the plates by tension or by tearing the following are the minimum requirements. (Clause 10.2.2. of IS: 800)

(a) Minimum pitch = 2.5 times diameter of bolt
(b) Minimum edge distance as given in Table 8.6

Plates in compression joint, may separate. To avoid this, it is essential to limit the pitch as below (Clause 10.2.3.2 of IS: 800)

(c) Maximum pitch = $12t$ or 200 mm

where t = thickness of thinner plate

The maximum spacing of the bolts in tension joints shall be as below (Clause 10.2.3.2 of IS: 800)

(d) Maximum pitch = $16t$ or 200 mm

(e) Maximum cross pitch = 32t or 300 mm

Table 8.7 Tensile properties of fasteners and strength per sq mm

Tensile properties of fasteners					*Strength per sq mm*			
Material specification	*Bolt grate*	*Properties*			*Shear strength V_{dsb} in N/ sq mm See note(1)*	*Bearing strength V_{npb} in N/sq mm (see note (2))*	*Tension strength T_{nb} in N/ sq mm(least of 8 & 9) (see note 3)*	
		Yield stress in MPa (Min)	*Ulti-mate ten-sile stress in Mpa (Min)*	*Elon-gation % (Min)*			$0.9\ f_{ub}/1.25$	$f_{yb}\ \gamma_{mb}/\gamma_{mo}$
1	2	3	4	5	6	7	8	9
IS: 1367 (part 3) (ISO 898) specifications of fasteners-threaded steel for technical supply conditions.	3.6	180	330	25	152.42	820	237.6	204.5
	4.6	240	400	22	184.75	820	288	272.7
	4.8	320	420	14	193.99	820	302.4	363.6
	5.6	300	500	20	230.94	820	360	340.9
	5.8	400	520	10	240.18	820	374.4	454.5
	6.8	480	600	8	277.13	820	432	545.4
	8.8 (d<16 mm)	640	800	12	369.5	820	576	727.3
	9.8	720	900	10	415.69	820	648	818.2
	10.9	940	1040	9	480.36	820	748.8	1068.2
	12.9	1100	1220	8	563.49	820	878.4	1250

Note:

(1) This is as per Clause 10.3.3 of IS: 800.

(2) This is as per Clause 10.3.4 of IS: 800. The value of k_b is taken as 1.0, diameter is taken as 1 mm and the thickness of thinner part connected is taken as 1 mm. The value of f_u considered in calculating this is 410 MPa. Multiply the value in column 7 of the above table by ($k_b \times d \times t \times f_u/410$) to get required bearing strength.

(3) This is as per Clause 10.3.5 of IS: 800. To obtain the design strength in tension, multiply the values in column 8 by A_n (net stress area in mm^2) and column 9 by in A_{sb} (shank area of bolt in mm^2). Least out of the two is the design strength in tension.

Table 8.8 Design strength for various property classes of bolts.

Table 8.8(a) Design strength of bolt property class 3.6-As per IS:800:2007

Serial Number	Physical property				Single shear capacity in kN		Tension capacity in kN		Bearing capacity in kN
	Diameter of bolt in mm (Shank)	Gross area in sq mm A_{sb}	Net area in sq mm A_{nb}	A_{nb}/A_{sb}	Thread not in shear plane	Thread in shear plane	$0.9_{fu}\ A_{nb}/\gamma_{mb}$	$f_{yb}\ A_{sb}/\gamma_{mo}$	For t = 10 mm, k_b = 1.0 and f_u = 410 MPa
1	2	3	4	5	6	7	8	9	10
1.	10	78.57	58	0.74	11.98	8.84	13.78	12.86	8.2
2.	12	113.14	84.3	0.75	17.25	12.85	20.03	18.51	9.84
3.	16	201.14	157	0.78	30.66	23.93	37.3	32.91	13.12
6.	20	314.29	245	0.78	47.9	37.34	58.21	51.43	16.4
7.	24	452.57	353	0.78	68.98	53.8	83.87	74.06	19.68
8.	30	707.14	561	0.79	107.78	85.51	133.29	115.71	24.6
9.	33	855.64	694	0.81	130.42	105.78	164.89	140.01	27.06
10.	36	1018.29	817	0.8	155.21	124.53	194.12	166.63	29.52

f_u = 330 N/sq mm
f_{yb} = 180 N/sq mm
$fu/(\sqrt{3}*1.25)$ 152.42 N/sq mm
Tension capacity is the least of value in Column 8 and Column 9
Bearing capacity $2.5 \times k_b \times d \times t \times f_u/1.25$

Note: To get the bearing capacity for any thickness of plate, and for other values of k_b & f_u multiply the bearing capacity in column 10 by $k_b \times$ thickness of plate in mm $\times f_u/410$.

Table 8.8(b) Design strength of bolt property class 4.6–As per IS:800:2007

Serial Number	Physical property				Single shear capacity in kN		Tension capacity in kN		Bearing capacity in kN
	Diameter of bolt in mm (Shank)	Gross area in sq mm A_{sb}	Net area in sq mm A_{nb}	A_{nb}/A_{sb}	Thread not in shear plane	Thread in shear plane	$0.9_{fu}\ A_{nb}/\gamma_{mb}$	$f_{yb}\ A_{sb}/\gamma_{mo}$	For t = 1.0 mm, k_b = 1.0 and f_u = 410 MPa
1	2	3	4	5	6	7	8	9	10
1.	10	78.57	58	0.74	14.52	10.72	16.7	17.14	8.2
2.	12	113.14	84.3	0.75	20.9	15.57	24.28	24.69	9.84
3.	16	201.14	157	0.78	37.16	29.01	45.22	43.89	13.12

Contd...

Contd...

6.	20	314.29	245	0.78	58.06	45.26	70.56	68.57	16.4
7.	24	452.57	353	0.78	83.61	65.22	101.66	98.74	19.68
8.	30	707.14	561	0.79	130.65	103.65	161.57	154.29	24.6
9.	33	855.64	694	0.81	158.08	128.22	199.87	186.69	27.06
10.	36	1018.29	817	0.8	188.13	150.94	235.3	222.17	29.52

f_u = 400 N/sq mm
f_{yb} = 240 N/sq mm
$f_u/(\sqrt{3}*1.25)$ 184.75 N/sq mm
Tension capacity is the least of value in Column 8 and Column 9
Bearing capacity $2.5 \times k_b \times d \times t \times f_u/1.25$

Note: To get the bearing capacity for any thickness of plate, and for other values of k_b & f_u multiply the bearing capacity in column 10 by $k_b \times$ thickness of plate in mmx $f_u/410$.

Table 8.8 (c) Design strength of bolt property class 4.8-As per IS: 800: 2007

Serial Number	*Physical property*				*Single shear capacity in kN*		*Tension capacity in kN*		*Bearing capacity in kN*
	Diameter of bolt in mm (Shank)	*Gross area in sq mm A_{sb}*	*Net area in sq mm A_{nb}*	A_{nb}/A_{sb}	*Thread not in shear plane*	*Thread in shear plane*	$0.9 f_u A_{nb}/\gamma_{mb}$	$f_{yb} A_{sb}/\gamma_{mo}$	*For t = 1.0 mm, k_b = 1.0 and f_u = 410 Mpa*
1	*2*	*3*	*4*	*5*	*6*	*7*	*8*	*9*	*10*
1	10	78.57	58	0.74	15.24	11.25	17.54	22.86	8.2
2	12	113.14	84.3	0.75	21.95	16.35	25.49	32.91	9.84
3	16	201.14	157	0.78	39.02	30.46	47.48	58.51	13.12
6	20	314.29	245	0.78	60.97	47.53	74.09	91.43	16.4
7	24	452.57	353	0.78	87.79	68.48	106.75	131.66	19.68
8	30	707.14	561	0.79	137.18	108.83	169.65	205.71	24.6
9	33	855.64	694	0.81	165.99	134.63	209.87	248.91	27.06
10	36	1018.29	817	0.8	197.54	158.49	247.06	296.23	29.52

f_u = 420 N/sq mm
f_{yb} = 320 N/sq mm
$f_u/(\sqrt{3}*1.25)$ 193.99 N/sq mm
Tension capacity is the least of value in Column 8 and Column 9
Bearing capacity $2.5 \times k_b \times d \times t \times f_u/1.25$

Note: To get the bearing capacity for any thickness of plate, and for other values of k_b & f_u multiply the bearing capacity in column 10 by $k_b \times$ thickness of plate in mmx $f_u/410$.

Table 8.8(d) Design strength of bolt property class 5.6-As per IS: 800: 2007

Serial Number	Physical property				Single shear capacity in kN		Tension capacity in kN		Bearing capacity in kN
	Diameter of bolt in mm (Shank)	Gross area in sqmm A_{sb}	Net area in sqmm A_{nb}	A_{nb}/A_{sb}	Thread not in shear plane	Thread in shear plane	0.9_{fu} A_{nb}/Y_{mb}	f_{yb} A_{sb}/Y_{mo}	For t = 1.0 mm, k_b = 1.0 and f_u = 410 MPa
1	2	3	4	5	6	7	8	9	10
1.	10	78.57	58	0.74	18.15	13.39	20.88	21.43	8.2
2.	12	113.14	84.3	0.75	26.13	19.47	30.35	30.86	9.84
3.	16	201.14	157	0.78	46.45	36.26	56.52	54.86	13.12
6.	20	314.29	245	0.78	72.58	56.58	88.2	85.71	16.4
7.	24	452.57	353	0.78	104.52	81.52	127.08	123.43	19.68
8.	30	707.14	561	0.79	163.31	129.56	201.96	192.86	24.6
9.	33	855.64	694	0.81	197.6	160.27	249.84	233.36	27.06
10.	36	1018.29	817	0.8	235.16	188.68	294.12	277.71	29.52

f_u = 500 N/sqmm
f_{yb} = 300 N/sqmm
$f_u/(\sqrt{3}*1.25)$ 230.94 N/sqmm
Tension capacity is the least of value in Column 8 and Column 9
Bearing capacity $2.5 \times k_b \times d \times t \times f_u/1.25$

Note: To get the bearing capacity for any thickness of plate, and for other values of k_b & f_u multiply the bearing capacity in column 10 by $k_b \times$ thickness of plate in mmx $f_u/410$.

Table 8.8(e) Design strength of bolt property class 5.8-As per IS: 800: 2007

Serial Number	Physical property				Single shear capacity in kN		Tension capacity in kN		Bearing capacity in kN
	Diameter of bolt in mm (Shank)	Gross area in sqmm A_{sb}	Net area in sqmm A_{nb}	A_{nb}/A_{sb}	Thread not in shear plane	Thread in shear plane	0.9_{fu} A_{nb}/Y_{mb}	f_{yb} A_{sb}/Y_{mo}	For t = 1.0 mm, k_b = 1.0 and f_u = 410 MPa
1	2	3	4	5	6	7	8	9	10
1.	10	78.57	58	0.74	18.87	13.93	21.72	28.57	8.2
2.	12	113.14	84.3	0.75	27.17	20.25	31.56	41.14	9.84
3.	16	201.14	157	0.78	48.31	37.71	58.78	73.14	13.12
6.	20	314.29	245	0.78	75.48	58.84	91.73	114.29	16.4

Contd...

Contd...

7.	24	452.57	353	0.78	108.7	84.78	132.16	164.57	19.68
8.	30	707.14	561	0.79	169.84	134.74	210.04	257.14	24.6
9.	33	855.64	694	0.81	205.51	166.68	259.83	311.14	27.06
10.	36	1018.29	817	0.8	244.57	196.23	305.88	370.29	29.52

f_u = 520 N/sq mm
f_{yb} = 400 N/sq mm
$f_u/(\sqrt{3}*1.25)$ 240.18 N/sq mm
Tension capacity is the least of value in Column 8 and Column 9
Bearing capacity $2.5 \times k_b \times d \times t \times f_u/1.25$

Note: To get the bearing capacity for any thickness of plate, and for other values of k_b & f_u multiply the bearing capacity in column 10 by K_b × thickness of plate in mmx $f_u/410$.

Table 8.8 (f) Design strength of bolt property class 6.8-As per IS: 800: 2007

Serial Number	*Physical property*				*Single shear capacity in kN*		*Tension capacity in kN*		*Bearing capacity in kN*
	Diameter of bolt in mm (Shank)	*Gross area in sq mm A_{sb}*	*Net area in sq mm A_{nb}*	*A_{nb}/A_{sb}*	*Thread not in shear plane*	*Thread in shear plane*	*$0.9f_u A_{nb}/\gamma_{mb}$*	*$f_{yb} A_{sb}/\gamma_{mo}$*	*For t = 1.0 mm, k_b = 1.0 and f_u = 410 MPa*
1	*2*	*3*	*4*	*5*	*6*	*7*	*8*	*9*	*10*
1.	10	78.57	58	0.74	21.77	16.07	25.06	34.29	8.2
2.	12	113.14	84.3	0.75	31.36	23.36	36.42	49.37	9.84
3.	16	201.14	157	0.78	55.74	43.51	67.82	87.77	13.12
6.	20	314.29	245	0.78	87.1	67.9	105.84	137.14	16.4
7.	24	452.57	353	0.78	125.42	97.83	152.5	197.49	19.68
8.	30	707.14	561	0.79	195.97	155.47	242.35	308.57	24.6
9.	33	855.64	694	0.81	237.12	192.33	299.81	373.37	27.06
10.	36	1018.29	817	0.8	282.2	226.41	352.94	444.34	29.52

f_u = 600 N/sq mm
f_{yb} = 480 N/sq mm
$f_u/(\sqrt{3}*1.25)$ 277.13 N/sq mm
Tension capacity is the least of value in Column 8 and Column 9
Bearing capacity $2.5 \times k_b \times d \times t \times f_u/1.25$

Note: To get the bearing capacity for any thickness of plate, and for other values of k_b & f_u multiply the bearing capacity in column 10 by k_b × thickness of plate in mmx $f_u/410$.

Table 8.8(g) Design strength of bolt property class 8.8-as per IS: 800: 2007

Serial Number	Physical property				Single shear capacity in kN		Tension capacity in kN		Bearing capacity in kN
	Diameter of bolt in mm (Shank)	Gross area in sq mm A_{sb}	Net area in sq mm A_{nb}	A_{nb}/A_{sb}	Thread not in shear plane	Thread in shear plane	$0.9_{fu} A_{nb}/Y_{mb}$	$f_{yb} A_{sb}/Y_{mo}$	For t = 1.0 mm, k_b = 1.0 and f_u = 410 MPa
1	2	3	4	5	6	7	8	9	10
1.	10	78.57	58	0.74	29.03	21.43	33.41	45.71	8.2
2.	12	113.14	84.3	0.75	41.81	31.15	48.56	65.83	9.84
3.	16	201.14	157	0.78	74.32	58.01	90.43	117.03	13.12
6.	20	314.29	245	0.78	116.13	90.53	141.12	182.86	16.4
7.	24	452.57	353	0.78	167.23	130.43	203.33	263.33	19.68
8.	30	707.14	561	0.79	261.29	207.29	323.14	411.43	24.6
9.	33	855.64	694	0.81	316.16	256.44	399.74	497.83	27.06
10.	36	1018.29	817	0.8	376.26	301.88	470.59	592.46	29.52

f_u = 800 N/sq mm
f_{yb} = 640 N/sq mm
$f_u/(\sqrt{3}*1.25)$ 369.5 N/sq mm
Tension capacity is the least of value in Column 8 and Column 9
Bearing capacity $2.5 \times k_b \times d \times t \times f_u/1.25$

Note: To get the bearing capacity for any thickness of plate, and for other values of k_b & f_u multiply the bearing capacity in column 10 by $k_b \times$ thickness of plate in mmx $f_u/410$.

Table 8.8(h) Design strength of bolt property class 9.8-As per IS: 800: 2007

Serial Number	Physical property				Single shear capacity in kN		Tension capacity in kN		Bearing capacity in kN
	Diameter of bolt in mm (Shank)	Gross area in sq mm A_{sb}	Net area in sq mm A_{nb}	A_{nb}/A_{sb}	Thread not in shear plane	Thread in shear plane	$0.9_{fu} A_{nb}/Y_{mb}$	$f_{yb} A_{sb}/Y_{mo}$	For t = 1.0 mm, k_b = 1.0 and f_u = 410 MPa
1	2	3	4	5	6	7	8	9	10
1.	10	78.57	58	0.74	32.66	24.11	37.58	51.43	8.2
2.	12	113.14	84.3	0.75	47.03	35.04	54.63	74.06	9.84
3.	16	201.14	157	0.78	83.61	65.26	101.74	131.66	13.12

Contd...

Contd...

6.	20	314.29	245	0.78	130.65	101.84	158.76	205.71	16.4
7.	24	452.57	353	0.78	188.13	146.74	228.74	296.23	19.68
8.	30	707.14	561	0.79	293.95	233.2	363.53	462.86	24.6
9.	33	855.64	694	0.81	355.68	288.49	449.71	560.06	27.06
10.	36	1018.29	817	0.8	423.29	339.62	529.42	666.51	29.52

f_u = 900 N/sq mm
f_{yb} = 720 N/sq mm
$f_u/(\sqrt{3}*1.25)$ 415.69 N/sq mm
Tension capacity is the least of value in Column 8 and Column 9
Bearing capacity 2.5 × k_b × d × t × f_u/1.25

Note: To get the bearing capacity for any thickness of plate, and for other values of k_b and f_u multiply the bearing capacity in column 10 by k_b × thickness of plate in mm × f_u/410.

Table 8.8(i) Design strength of bolt property class 10.9-As per IS: 800: 2007

Serial Number	*Physical property*				*Single shear capacity in kN*		*Tension capacity in kN*		*Bearing capacity in kN*
	Diameter of bolt in mm (Shank)	*Gross area in sq mm A_{sb}*	*Net area in sq mm A_{nb}*	*A_{nb}/A_{sb}*	*Thread not in shear plane*	*Thread in shear plane*	*$0.9_{fu}A_{nb}/\gamma_{mb}$*	*$f_{yb}A_{sb}/\gamma_{mo}$*	*For t = 1.0 mm, k_b = 1.0 and f_u = 410 MPa*
1	*2*	*3*	*4*	*5*	*6*	*7*	*8*	*9*	*10*
1.	10	78.57	58	0.74	37.74	27.86	43.43	67.14	8.2
2.	12	113.14	84.3	0.75	54.35	40.49	63.12	96.69	9.84
3.	16	201.14	157	0.78	96.62	75.42	117.56	171.89	13.12
6.	20	314.29	245	0.78	150.97	117.69	183.46	268.57	16.4
7.	24	452.57	353	0.78	217.4	169.57	264.33	386.74	19.68
8.	30	707.14	561	0.79	339.68	269.48	420.08	604.29	24.6
9.	33	855.64	694	0.81	411.01	333.37	519.67	731.19	27.06
10.	36	1018.29	817	0.8	489.14	392.45	611.77	870.17	29.52

f_u = 1040 N/sq mm
f_{yb} = 940 N/sq mm
$f_u/(\sqrt{3}*1.25)$ 480.36 N/sq mm
Tension capacity is the least of value in Column 8 and Column 9
Bearing capacity 2.5 × k_b × d × t × f_u/1.25

Note: To get the bearing capacity for any thickness of plate, and for other values of k_b and f_u multiply the bearing capacity in column 10 by k_b × thickness of plate in mmx f_u/410.

Table 8.8(j) Design strength of bolt property class 12.9-As per IS: 800: 2007

Serial Number	Physical property				Single shear capacity in kN		Tension capacity in kN		Bearing capacity in kN
	Diameter of bolt in mm (Shank)	Gross area in sq mm A_{sb}	Net area in sq mm A_{nb}	A_{nb}/A_{sb}	Thread not in shear plane	Thread in shear plane	$0.9f_u A_{nb}/\gamma_{mb}$	$f_{yb} A_{sb}/\gamma_{mo}$	For t = 1.0 mm, k_b = 1.0 and f_u = 410 MPa
1	2	3	4	5	6	7	8	9	10
1.	10	78.57	58	0.74	44.27	32.68	50.95	78.57	8.2
2.	12	113.14	84.3	0.75	63.76	47.5	74.05	113.14	9.84
3.	16	201.14	157	0.78	113.34	88.47	137.91	201.14	13.12
6.	20	314.29	245	0.78	177.1	138.06	215.21	314.29	16.4
7.	24	452.57	353	0.78	255.02	198.91	310.08	452.57	19.68
8.	30	707.14	561	0.79	398.47	316.12	492.78	707.14	24.6
9.	33	855.64	694	0.81	482.15	391.06	609.61	855.64	27.06
10.	36	1018.29	817	0.8	573.8	460.37	717.65	1018.29	29.52

f_u = 1220 N/sq mm
f_{yb} = 1100 N/sq mm
f_u/
(√3*1.25) 563.49 N/sq mm
Tension capacity is the least of value in Column 8 and Column 9
Bearing capacity 2.5 × k_b × d × t × f_u/1.25

Note: To get the bearing capacity for any thickness of plate, and for other values of k_b and f_u multiply the bearing capacity in column 10 by k_b × thickness of plate in mm× f_u/410.

Bearing Type Bolts

Shear (Clause 10.3.3 of IS: 800)

Factored shear force in bolt (V_{sb}), should be less than or equal to design shear strength V_{dsb} of the bolt as given below:

$$V_{dsb} = \frac{f_u}{\gamma_{mb}\sqrt{3}}(n_n A_{nb} + n_s A_{sb})$$

f_u = ultimate tensile strength of a bolt.

n_n = number of shear planes with threads intercepting the shear plane.

n_s = number of shear planes without threads intercepting the shear plane.

A_{nb} = net shear area of the bolt at threads may be taken as the area corresponding to root diameter at the thread.

A_{sb} = gross cross-sectional area (or) shank area of bolt.

Net area (threaded neck) and shank area (unthreaded portion) for various diameters of bolts are given in Tables 8.4 and 8.7 of this book. The shear strength in N/sq mm for various classes of bolts is given in Table 8.8 of this book. The design shear strength of bolts in kN of different classes and diameters are given in the Table 8.8(a) to (k) of this book.

Bearing (Clause 10.3.4 of IS: 800)

Bearing strength of bolt in bearing on any plate = $V_{dpb} = V_{npb} / \gamma_{mb}$

A_{pb} = nominal bearing area of the bolt on any plate.

V_{npb} = nominal bearing strength of a bolt = $2.5\ k_b.d.tf_u$

k_b is smaller of $\frac{e}{3d_o}; \frac{p}{3d_o} - 0.25; \frac{f_{ub}}{f_u}; 1.0.$

e, p = end and pitch distances of the bolt along the direction of force.

d_o = diameter of the hole

f_{ub}, f_u = ultimate tensile stress of the bolt and the ultimate tensile stress of the plate, respectively.

d = nominal diameter of the bolt

t = summation of the thickness of the connected plates subjected to bearing stress in the same direction.

The bolt materials generally used are of higher strength than the plate material, bearing failure of plate may happen before the bolt fails by bearing.

Bearing stress (strength) of bolt in N/sq mm for various classes of bolts is given in Table 8.7 of this book. Table 8.8 (a) to (k) of this book gives the bearing capacity of different diameters under different product classes of bolts for the value of k_b = 1.0, thickness of plate as 1 mm and f_u = 410 MPa. For all other values of k_b, thickness and f_u the values given in the table shall be multiplied by the value of k_b and thickness of thinner part in mm and (f_u/410) to obtain the bearing strength in kN.

Tension (Clause 10.3.5 of IS: 800)

Factored tensile force of the bolt, should be less than or equal to the tensile strength of the bolt, as given below.

The factored tensile strength of the bolt = T_{db}

T_{db} = design tensile strength of a bolt

$T_{db} = (0.9\ f_{nb}\ A_n / \gamma_{mb}) < f_{yb}\ A_{sb} / \gamma_{mo}$

A_n = net tensile stress area (Table 8.4 of this book).

A_{sb} = shank area of the bolt (Table 8.4 of this book).

Bolts Subjected to Combined Shear and Tension (Clause 10.3.6 of IS: 800)

If the bolt is subjected to shear and tension together, the actual shear and axial tension calculated do not exceed the respective strengths and the expression given below shall be satisfied.

$(V_{sb}/V_{db})^2 + (T_b/T_{db})^2 < 1.0$

8.27 EXAMPLES ON BOLTED JOINTS

Example 8.1 *Determine the bolt value for M20 bolts of grade 4.6 connecting two Fe 410 plates of 10 mm thick each.*

Solution:

Nominal diameter of bolt = 20 mm.

Diameter of hole = 20 + 2 = 22 mm (Table 19 of IS: 800)

Single shear capacity of bolt

$$V_{nsb} = \frac{f_u}{\sqrt{3}}(n_n A_{nb} + n_s A_{sb})/\gamma_{mb} \qquad \text{(Clause 10.3.3 of IS: 800)}$$

n_n = 0 Thread not in shear plane.

A_{nb} = 245 mm^2 (Table 8.4 of this book).

n_s = 1 Only one interface is there since the bolt connects only two plates.

A_{sb} = 314 mm^2 (Table 8.4 of this book).

γ_{mb} = 1.25 (Table 5 of IS: 800)

Bolts-Bearing Type

For property class 4.6 bolts, f_u = 400 MPa.

Capacity in single shear = $V_{dsb} = \left(\frac{400}{\sqrt{3}}\right)\left(\frac{0 + 1 \times 314}{1.25}\right)$ = 58 kN (Table 8.8(b) of this book)

Capacity in bearing = $V_{dpb} = \left(\frac{2.5\, k_b d t f_u}{\gamma_{mb}}\right)$ (Clause 10.3.4 of IS: 800)

k_b is smaller of $\frac{e}{3d_o}, \frac{p}{3d_o} - 0.25, \frac{f_{ub}}{f_u}$, 1.0

e = end or edge distance of fastener along the bearing direction

= 1.5 × 22 = 33 mm or 1.7 × 22 = 37.4 mm. (Clause 10.2.4.2 of IS: 800)

Provide 40 mm

P = pitch distance of fastener along bearing direction (Clause 10.2.2 of IS: 800)

= 2.5 × 20 = 50 mm.

Provide 50 mm pitch.

$$\frac{e}{3d_o} = \frac{40}{3 \times 22} = 0.61$$

$$\frac{p}{3d_o} - 0.25 = \frac{50}{3 \times 22} - 0.25 = 0.51$$

$$\frac{f_{ub}}{f_u} = \frac{400}{410} = 0.97$$

$\therefore \qquad k_b = 0.51$

$$V_{dpb} = \text{Capacity in bearing} = \frac{2.5 \times 0.51 \times 20 \times 10 \times 40}{1000 \times 1.25} = 83.64 \text{ kN}$$

Alternatively, $V_{dpb} = 16.4 \times 0.51 \times 10 = 83.64$ kN (Table 8.8(b) of this book).

Bolt value = Least of V_{dbp} and V_{dsb} = 58 kN.

Example 8.2 *12 mm thick plates are to be joined by double cover butt joint using 10 mm thick cover plates. The steel of main and cover plate conform to f_y = 250 N/mm². Determine the strength and efficiency of the joint per pitch of 60 mm if:*

(i) 20 mm diameter close tolerance and turned bolts of property class 6.8 are used.

(ii) 20 mm diameter close tolerance and turned bolts of property class 8.8 are used.

Solution: Since the total thickness of cover plates is more than the thickness of the main plate, the bolt will be checked for failure by bearing on main plate.

Note:

Thickness of cover plate shall not be less than $(5/8)^{th}$ of main plate.

Diameter of bolt = 20 mm.

Diameter of bolt hole = 20 + 2 = 22 mm (Table 19 of IS: 800).

Strength of joint:

$$\text{Strength of cover plate per pitch} = T_{dn} = \frac{0.9 \times A_n f_u}{\gamma_{m1}} = \frac{0.9 \times (60-22) \times 10 \times 410}{1.25 \times 1000 \text{ NL}}$$

$= 112.18$ kN.

Strength of two cover plates = 224.36 kN (1)

Strength of main plate per pitch

$$T_{dn} = \frac{0.9 \times A_n f_u}{\gamma_{m1}} = \frac{0.9 \times (60 - 22) \times 12 \times 410}{1.25 \times 1000} = 134.61 \text{ kN.} \qquad (2)$$

The least of (1) and (2) is the strength of plate at joint = 134.61 kN

$$\text{Strength of bolt in bearing} = V_{dpb} = \left(\frac{2.5\, k_b\, dtf_u}{\gamma_{m1}}\right) \text{ (Clause 10.3.4 of IS:800)}$$

$$k_b \text{ is smaller of } \frac{e}{3d_o}, \frac{p}{3d_o} - 0.25, \frac{f_{ub}}{f_u}, 1.0$$

e = end or edge distance of fastener along the bearing direction

= 1.5 × 22 = 33 mm or 1.7 × 22 = 37.4 mm. (Clause 10.2.4.2 of IS:800).

Provide 40 mm.

Minimum pitch = P = 2.5 × 20 = 50 mm. (Clause 10.2.2 of IS:800)

Provide 60 mm pitch.

$$\frac{e}{3d_o} = \frac{40}{3 \times 22} = 0.61$$

$$\frac{p}{3d_o} - 0.25 = \frac{60}{3 \times 22} - 0.25 = 0.66$$

$$\frac{f_{ub}}{f_u} = \frac{600}{410} = 1.46$$

$$\therefore \quad k_b = 0.61$$

Strength in bearing = $V_{dph} = \dfrac{2.5 \times 0.61 \times 20 \times 12 \times 410}{1000 \times 1.25} = 120$ kN (3)

Alternatively, $V_{dpb} = 16.4 \times 0.61 \times 12 = 120$ kN (Table 8.8 (b) of this book).

Strength of 20 mm diameter bolt in double shear (shank in shear plane)

$$V_{dsb} = \{(f_u/\sqrt{3})\ [(n_n A_{nb} + n_s A_{sb})/\gamma_{mb}]\} \quad \text{(Clause 10.3.3 of IS: 800)}$$

$n_n = 0$ Thread not in shear plane.

$A_{nb} = 245$ mm^2 (Table 8.4 of this book.)

$n_s = 2$ Two interface for three plates.

$A_{sb} = 314$ mm^2 (Table 8.4 of this book.)

$\gamma_{mb} = 1.25$ (Table 5 of IS:800)

Bolt Property Class 6.8

$$f_u = 600 \text{ (Table 8.7 of this book).}$$

$$V_{dsb} = \left(\frac{600}{\sqrt{3}}\right)\left(\frac{2 \times 314}{1.25 \times 1000}\right) = 174 \text{ kN} \quad (4)$$

Single shear strength = 87.1 kN (Table 8.8(f) of this book).

Double shear strength = 2 × 87.1 = 174.2 kN

Strength of plate in tearing/rupture (Clause 6.3.1 of IS: 800)

$$= \frac{0.9(p - d) \times t \times f_u}{\gamma_{m1}} = \frac{0.9 \times (60 - 22) \times 12 \times 410}{1.25 \times 1000} = 134.61 \text{ kN} \quad (5)$$

The least of 1 to 5 is the strength of the joint = 120 kN.

Strength of solid plate per pitch length = $\dfrac{60 \times 12 \times 410}{1.1 \times 1000}$ = 268.4 kN (Clause 6.2 of IS:800).

Joint efficiency = $\dfrac{120}{268.4} \times 100 = 44.7$

Bolt Property Class 8.8

The strength of joint 120 kN is due to bearing strength which does not depend on the strength of the bolt. Hence, using class 8.8 bolts will not have any impact. This will only increase in strength of the bolt in double shear, which is

$$V_{dsb} = \left(\frac{f_u}{\sqrt{3}}\right)\left(\frac{n_n A_n b + n_s A_s b}{\gamma_{mb}}\right) \quad \text{(Clause 10.3.3 of IS:800)}$$

$$V_{dsb} = \left(\frac{800}{\sqrt{3}}\right)\left(\frac{2 \times 314}{1.25 \times 1000}\right) = 232 \text{ kN}$$

Bolt Property Class 4.6

Let us use class 4.6 bolts and see

$$V_{dsb} = \left(\frac{400}{\sqrt{3}}\right)\left(\frac{2 \times 314}{1.25 \times 1000}\right) = 116 \text{ kN}$$

Since this is less than 120 kN, this governs the design.

Joint efficiency = (116/268.4) × 100 = 43.2%

Bolt Property Class 4.6 Full Threaded Stud Bolts

Let us use Clause 4.6 bolts and the threaded area is in shear plane (full thread stud)

$$V_{dsb} = \left(\frac{400}{\sqrt{3}}\right)\left(\frac{2 \times 245}{1.25 \times 1000}\right) = 90.5 \text{ kN}$$

This will govern the design.

$$\text{Joint efficiency} = \frac{90.5}{268.4} \times 100 = 33.7\%$$

Example 8.3 *Design the bolted joint as shown in Fig. 8.11 to take factored load of 350 kN. Fe 410 plate and class 4.6 bolts are available.*

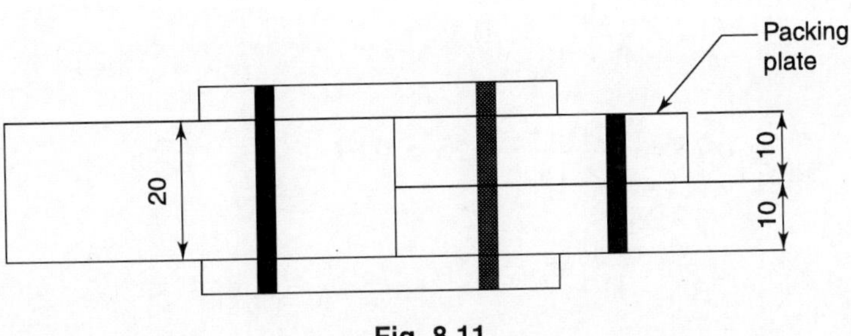

Fig. 8.11

The design shear capacity of bolts carrying shear through a packing plate in excess of 6 mm shall be decreased by a factor β_{pk} given by

$$\beta_{pk} = (1 - 0.0125 t_{pk}) \quad \text{(Clause 10.3.3.3 of IS: 800).}$$

where t_{pk} = thickness of the thicker packing in mm.

In this case t_{pk} = 10 mm

$$\beta_{pk} = (1 - 0.0125 \times 10) = 0.875$$

Providing 16 mm diameter class 4.6 bolts; with threaded portion in shear plane.

Bolt strength in double shear

$$V_{dsb} = \left(\frac{f_u}{\sqrt{3}}\right)\left(\frac{n_n A_n b + n_s A_s b}{\gamma_{mb}}\right) \quad \text{(Clause 10.3.3 of IS:800)}$$

Net area of M16 bolt is 157 mm^2 (Table 8.4 of this book).

$$V_{dsb} = \left(\frac{400}{\sqrt{3}}\right)\left(\frac{2 \times 157}{1.25 \times 1000}\right) = 58 \text{ kN} \quad (1)$$

Single shear strength of bolt = 29 kN (Table 8.8 (b) of this book).

Strength of bolt in bearing = $V_{dpb} = \left(\frac{2.5 k_b d t f_u}{\gamma_{mb}}\right)$ (Clause 10.3.4 of IS: 800)

Diameter of bolt 16 mm.

Diameter of bolt hole = 16 + 2 = 18 mm (Table 19 of IS: 800).

Minimum end/edge distance = 1.5 × 18 = 27 mm or 1.7 × 18 = 30.6 mm (Clause 10.2.4.2 of IS: 800). Provide end/edge distance considering planed edges = e = 30 mm.

Minimum pitch = 2.5 × diameter of bolt = 2.5 × 16 = 40 mm (Clause 10.2.2 of IS: 800).

Provide pitch = p = 50 mm.

k_b is smaller of $\frac{e}{3d_o}, \frac{p}{3d_o} - 0.25, \frac{f_{ub}}{f_u}, 1.0$

$$\frac{e}{3d_o} = \frac{30}{3 \times 18} = 0.55$$

$$\frac{p}{3d_o} - 0.25 = \frac{50}{3 \times 18} - 0.25 = 0.67$$

$$\frac{f_{ub}}{f_u} = \frac{400}{410} = 0.97$$

$$k_b = 0.55$$

$$\text{Strength in bearing} = V_{dph} = \frac{2.5 \times 0.55 \times 16 \times 10 \times 410}{1000 \times 1.25} = 72.16 \text{ kN} \quad (2)$$

= 13.2 × 10 × 0.55 = 72.16 kN (Table 8.8 (b) of this book).

Bolt strength = 58 kN. (Least of 1 and 2)

Reduced strength due to packing = $\beta_{pk} \times 58 = 0.875 \times 58 = 50.75$ kN.

Number of bolts required $= \dfrac{350}{50.75} = 6.9$

Provide 8 Nos. of 16 mm diameter class 4.6 bolts in two rows.

Strength of bolt/pitch = 2 × 50.75 = 101.5 kN since two rows are adopted.

Strength of plate in tearing/rupture

$$= \frac{0.9(p-d)\times t\times f_u}{\gamma_{m1}} = \frac{0.9\times(50-18)\times 10\times 410}{1.25\times 1000} = 94.46 \text{ kN}$$

Strength of solid plate = $T_{dg} = A_g f_y/\gamma_{mo}$ (Clause 6.2 of IS: 800)

$= 2 \times 50 \times 10 \times 250/(1.1 \times 1000) = 227.3$ kN.

Lesser value out of 94.46 kN and 227.3 kN is the strength of plate.

Strength of plate is 94.46 kN and the shear strength of bolts is 101.5 kN.

We can increase the pitch.

Example 8.4 *High tension bolt bearing joint under shear*

Bottom tie of ISA 75 × 75 × 8 of a 16 m span truss is subjected to a tension of 55 kN due to dead load. Design a bearing type connection using class 8.8, high tension bolt between the tie and the gusset plate.

Solution:

Axial tension = P = 55 kN

Factored load = 1.5 × 55 = 82.5 kN

Angle size = ISA 75 × 75 × 8

Edge distance available = 35 mm (SP: 6 (1)). We can use M20 or lesser size bolt.

Since the force and angle size are small, one can select 16 mm diameter bolt.

Minimum edge distance required for rolled edge is 27 mm. (Table 8.6 of this book) and available edge distance is 35 mm.

Provide thickness of the gusset as 8 mm and M16 bolts.

The bolts are under single shear and the shear strength of shank of class 8.8 bolt is V_a = 74.32 kN. (Table 8.8(g) for class 8.8)

Let the end and edge distance be 35 mm.

The Bearing and Tearing Strengths

Bearing strength (clause 10.3.4 of IS: 800)

For M16-class 8.8 bolt = 13.12 × k_b × thickness (Table 8.8(g) of this book).

Diameter of hole = 18 mm (Table 19 of IS: 800) k_b is the least of

(a) $e/3d_o = 35/(3 \times 18) = 0.65$

(b) $(p/3d_o) - 0.25$

Let the pitch be = 3 × diameter of bolt = 3 × 16 = 48 mm. Provide 50 mm

Therefore, $(p/3d_o) - 0.25 = 50/(3 \times 18) - 0.28 = 0.676$

(c) f_{ub}/f_u

f_{ub} = 800 MPa and f_u = 410 MPa

Therefore, $f_{ub}/f_u = 800/410 = 1.95$

(d) 1.0, Hence, $k_b = 0.65$ (the least value is taken)

$$\text{Bearing strength} = V_{dpb} = \left(\frac{2.5k_b dtf_u}{\gamma_{mb}}\right) \text{ (Clause 10.3.4 of IS: 800)}$$

$$= 13.12 \times 0.65 \times 8 = 68.224 \text{ kN}$$

Bolt value is the least of shear strength and bearing strength = 68.224 kN

Number of bolts = 82.5/68.224 = 1.21

Provide 2 numbers M16 bolts at a pitch of 50 mm.

Since the length between centre to centre of extreme bolts are not exceeding 15 times the diameter of bolt, there is no reduction in shear capacity as per Clause 10.3.3.1 of IS: 800.

Block Shear Strength (Clause 6.4.1 of IS: 800)

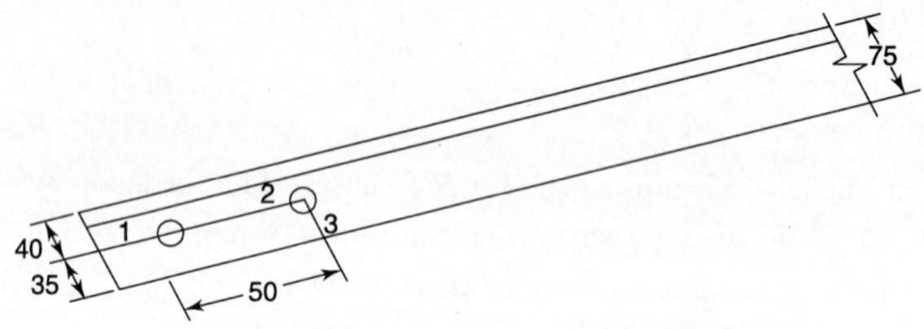

Fig. 8.12

$$T_{db} = A_{vg}\, f_y /(\gamma_{mo} \sqrt{3} + 0.9\, A_{tn}\, f_u/\gamma_{m1}, \text{ (or)}$$

$$= 0.9\, A_{vn}\, f_u/(\gamma_{m1} \sqrt{3} + A_{tg}\, f_y/\gamma_{mo})$$

A_{vg}, A_{vn} = minimum gross and net area in shear along bolt line parallel to external force, (1 – 2 and 3 – 4 as shown in Fig. 3.4 (a), 1 – 2 as shown in Fig. 3.4(b)

A_{tg}, A_{tn} = minimum gross and net area in tension from the bolt line perpendicular to the line of force, respectively (2 – 3 as shown in Fig. 3.4(b)).

$$A_{vn} = (35 + 50 - 18 - (18/2) \times 8 = 464 \text{ mm}^2$$

$$A_{vg} = (35 + 50) \times 8 = 680 \text{ mm}^2$$

$$A_{tg} = 35 \times 8 = 280 \text{ mm}^2$$

$$A_{tn} = (35 - (18/2)) \times 8 = 208 \text{ mm}^2$$

$$T_{db} = 680 \times 250/(1.1 \times \sqrt{3} + 0.9 \times 208 \times 410/1.25) = 150.63 \text{ kN.}$$

(or)

$$= 0.9 \times 464 \times 410/1.25 \times \sqrt{3} + 280 \times 250/1.1 = 142.72 \text{ kN.}$$

Least is 142.72 kN which is greater than 82.5 kN.

Hence, satisfactory.

Example 8.5 *The bracing member of a transmission line tower is to be built using ISA 75 × 75 × 8 and each angle is subjected to a compression of 40 kN. It is to be connected to the tower leg of size ISA 130 × 130 × 12. Design the joint using class 10.9 bolts. The compressive force is inclusive of wind force.*

Solution:

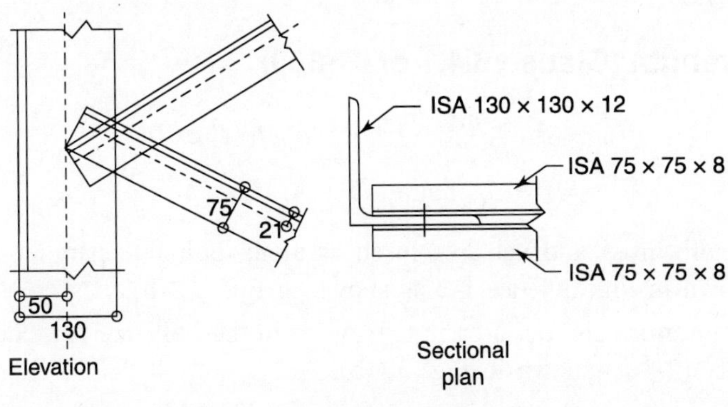

Fig. 8.13

Since the compressive force includes that due to wind, the partial safety factor for load = 1.2

Factored load = 1.2 × 40 = 48 kN (Table 4 of IS: 800).

Thread not in shear plane.

Providing M16 bolt of property class 10.9

Single shear strength = 96.62 kN (Table 8.8(j) of this book)

Double shear strength = 2 × 96.62 = 193.24 kN > 2 × 48 kN

Diameter of bolt hole = 16 + 2 = 18 mm (Table 19 of IS: 800).

Edge distance required = 1.5 × 18 = 27 mm (or) 1.7 × 18 = 30.6 mm (Clause 10.2.4.2 of IS: 800)

Edge distance available = 75 – 40 = 35 mm (SP: 6(1))

Bearing Strength (Clause 10.3.4 of IS:800)

Thickness of ISA75 × 75 × 8 = 8 mm

Thickness of ISA130 × 130 × 12 = 12 mm 8 mm will govern.

$$k_b = \text{least of } e/3d_o, (p/3d_o) - 0.25, f_{ub}/f_u \text{ and } 1$$

$$e/3d_o = 35/(3 \times 18) = 0.648$$

Since only one bolt is used there is no pitch.

$$f_{ub}/f_u = 1040/410 = 2.536$$

Therefore, $k_b = 0.648$

Both the angles are in compression.

Hence, $t = 2 \times 8 = 16$ mm.

Design strength in bearing = 2.5 × 0.648 × 16 × 16 × 410/1.25 × 1000 = 136 kN (or) = 13.12 × 16 × 0.648 = 136 kN. (Table 8.8(j) of this book)

Bolt value = 136 kN.

Factored compressive load = 2 × 48 = 96 kN.

Block Shear Strength (Clause 6.4.1 of IS:800)

$$T_{db} = A_{vg}\, f_y/(\gamma_{mo}\sqrt{3} + 0.9\, A_{tn}\, f_u/\gamma_{m1}) \text{ (or)}$$

$$= 0.9\, A_{vn}\, f_u/(\gamma_{m1}\sqrt{3} + A_{tg}\, f_y/\gamma_{mo})$$

A_{vg}, A_{vn} = minimum gross and net area in shear along bolt line parallel to external force, (1-2 and 3-4 as shown in Fig. 3.4 (a), 1-2 as shown in Fig. 3.4(b).

A_{tg}, A_{tn} = minimum gross and net area in tension from the bolt line perpendicular to the line of force, respectively (2-3 as shown in Fig. 3.4(b).

$$A_{vn} = (35 - (18/2)) \times 8 = 208 \text{ mm}^2$$

$$A_{vg} = 35 \times 8 = 280 \text{ mm}^2$$

$$A_{tg} = 35 \times 8 = 280 \text{ mm}^2$$

$$A_{tn} = (35 - (18/2)) \times 8 = 208 \text{ mm}^2$$

$$T_{db} = 280 \times 250/(1.1 \times \sqrt{3} + 0.9 \times 208 \times 410/1.25) = 98.14 \text{ kN}$$

(or)

$$= 0.9 \times 208 \times 410/1.25 \times \sqrt{3} + 280 \times 250/1.1 = 99.09 \text{ kN.}$$

Least is 98.14 kN which is greater than 40 kN.

One bolt M16 class 10.9 is adequate.

Example 8.6 *Joint between angle tie and gusset plate*

In a 12 m span truss, the bottom chord (ISA 60 × 60 × 6) is subjected to 52.5 kN tension and the rafter consisting of 2 ISA 50 × 50 × 6 is subjected to 60 kN compression. Design a joint between the tie, rafter and gusset plate. Use class 4.6 bolts. The load is due to dead and imposed loads only.

Solution: *Connection between the bottom chord and gusset*

The tension on tie = 1.5 × 52.5 = 78.75 kN (Table 4 of IS: 800).

Size of tie is = ISA 60 × 60 × 6

Since the width of the angle is only 60 mm, only one row of bolts can be provided.

The edge distance available for one row is = 60 – 35 = 25 mm (SP: 6(1))

Using M16 bolts, diameter of the bolt hole = 16 + 2 = 18 mm (Table 19 of IS: 800)

The edge distance for rolled edge = 1.5 × 18 = 27 mm. (Clause 10.2.4.2 of IS: 800).

Available is only 25 mm (SP: 6(1)).

Minimum pitch = 2.5 × 16 = 40 mm.

Provide 50 mm.

The bolt is in single shear.

Single shear strength of M16 bolt of property class 4.6 is 37.16 kN (Table 8.8(b) of this book).

Strength of bolt in bearing against 6 mm thick angle = $\left(\dfrac{2.5k_b dtf_u}{\gamma_{mb}}\right)$ (Clause 10.3.4 of IS: 800).

k_b is the least of

(1) $e/3d_o = 25/3 \times 18 = 0.463$

(2) $(p/3d_o) - 0.25 = (50/3 \times 18) - 0.25 = 0.676$

(3) $f_{ub}/f_u = 400/410 = 0.975$

(4) 1.0

Strength of the bolt in bearing = 2.5 × 0.463 × 16 × 6 × 410/(1.25 × 1000) = 36.451 kN.

Bearing controls the design since it is the lesser value.

Number of bolts required $N = P/F_r = 78.75/36.45 = 2.16$

Provide 3 Numbers M16 Class 4.6 bolts

Block Shear Strength (Clause 6.4.1 of IS: 800)

$$T_{db} = A_{vg}\, f_y/(\gamma_{mo}\sqrt{3} + 0.9\, A_{tn}\, f_u/\gamma_{m1})\ \text{(or)}$$

$$= 0.9\, A_{vn}\, f_u/(\gamma_{m1}\sqrt{3} + A_{tg}\, f_y/\gamma_{mo})$$

A_{vg}, A_{vn} = minimum gross and net area in shear along the bolt line parallel to external force, (1-2 and 3-4 as shown in Fig. 3.4(a), 1-2 as shown in Fig. 3.4(b).

A_{tg}, A_{tn} = minimum gross and net area in tension from the bolt line perpendicular to the line of force, respectively (2-3 as shown in Fig. 3.4(b).

$$A_{vn} = [30 + (2 \times 50) - (2 \times 18) - (18/2)] \times 6 = 510 \text{ mm}^2$$

$$A_{vg} = [30 + (2 \times 50)] \times 6 = 780 \text{ mm}^2$$

$$A_{tg} = 25 \times 6 = 150 \text{ mm}^2$$

$$A_{tn} = (25 - (18/2)) \times 6 = 96 \text{ mm}^2$$

$$T_{db} = 780 \times 250/(1.1 \times \sqrt{3} + 0.9 \times 96 \times 410/1.25) = 130.7 \text{ kN}$$

(or)

$$= 0.9 \times 510 \times 410/1.25 \times \sqrt{3} + 150 \times 250/1.1 = 121 \text{ kN.}$$

Least is 121 kN which is greater than 78.75 kN.

Hence, satisfactory.

Gusset

Thickness of the gusset plate shall not be less than the thickness of the member. Provide 8 mm thick gusset plate.

Edge distance at the edge for flame cut = 1.7 × 18 = 30.6 mm

Minimum length of gusset plate = two edge distances + 2 pitches

$$= 2g + 2p = 2 \times 30.6 + 2 \times 50 = 161.2 \text{ mm}$$

Use end edge distance = g = 30 mm and the pitch = 50 mm

Length required = 2(30 + 50) = 160 mm

The size of the gusset plate has to be arrived at to accommodate the other members.

Rafter and Gusset Joint

Partial safety factor = 1.5 (Table 4 of IS: 800).

Factored axial compression = P_u = 1.5 × 60 = 90 kN.

Edge distance available = 50 – 28 = 22 mm. (SP: 6(1)).

Use M12 property class 4.6 bolts.

Minimum end/edge distance = 1.5 × 13 = 19.5 mm.

Provide end distance of 20 mm.

Diameter of bolt hole = 12 + 1 = 13 mm = d_o (Table 19 of IS: 800)

Minimum pitch = 2.5 × 12 = 30 mm

Provide 30 mm pitch.

The gusset plate is sandwiched between the angles and hence the bolt is in double shear.

Design double shear strength = 2 × 20.9 = 41.8 kN (Table 8.8(b) of this book)

Strength in bearing for 6 mm thick angle

Strength of bolt in bearing against 6 mm thick angle = $\left(\dfrac{2.5k_b dtf_u}{\gamma_{mb}}\right)$ (Clause 10.3.4 of IS: 800)

k_b is the least of

(1) $e/3d_o = 20/(3 \times 13) = 0.513$

(2) $p/3 \times d_o = (30/3 \times 13) - 0.25 = 0.52$

(3) $f_{ub}/f_u = 400/410 = 0.975$

(4) 1.0

Thickness of gusset the plate is 8 mm. Thickness of two angles put together is 12 mm. Hence, thickness of the gusset plate is used in calculating the bearing strength.

Strength of bolt in bearing = 2.5 × 0.513 × 12 × 8 × 410/(1.25 × 1000) = 40.4 kN.

Strength in bearing controls the design since it is the lesser value.

Bolt value = 40.4 kN.

Number of bolts required $n = P/F_r = 90/40.4 = 2.23$

Provide 4 numbers of M12 class 4.6 bolts.

Minimum overlap length of the gusset is = $2g + 3p = 2 \times 20 + 3 \times 30 = 130$ mm.

Provide 135 mm overlap.

The details are shown in Fig. 8.14.

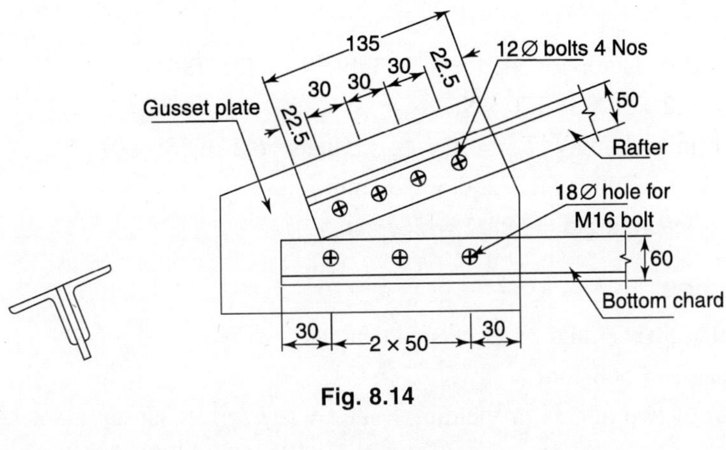

Fig. 8.14

Block Shear Strength (Clause 6.4.1 of IS: 800)

$$T_{db} = A_{vg}\, f_y/(\gamma_{mo}\sqrt{3} + 0.9\, A_{tn}\, f_u/\gamma_{m1}) \text{ (or)}$$

$$= 0.9\, A_{vn}\, f_u/(\gamma_{m1}\sqrt{3} + A_{tg}\, f_y/\gamma_{mo})$$

A_{vg}, A_{vn} = minimum gross and net area in shear along the bolt line parallel to external force, (1-2 and 3-4 as shown in Fig. 3.4(a), 1-2 as shown in Fig. 3.4(b).

A_{tg}, A_{tn} = minimum gross and net area in tension from the bolt line perpendicular to the line of force, respectively (2-3 as shown in Fig. 3.4(b).

$$A_{vn} = [20 + (3 \times 30) - (3 \times 13) - (13/2)] \times 6 = 387 \text{ mm}^2$$

$$A_{vg} = [20 + (3 \times 30)] \times 6 = 660 \text{ mm}^2$$

$$A_{tg} = 20 \times 6 = 120 \text{ mm}^2$$

$$A_{tn} = (20 - (13/2)) \times 6 = 81 \text{ mm}^2$$

$$T_{db} = 660 \times 250/(1.1 \times \sqrt{3} + 0.9 \times 81 \times 410/1.25) = 110.5 \text{ kN}$$

(or)

$$= 0.9 \times 387 \times 410/1.25 \times \sqrt{3} + 120 \times 250/1.1 = 93.23 \text{ kN}.$$

Least is 93.23 kN which is greater than 45 kN which is the force per angle.

Hence, satisfactory.

Example 8.7 *Design of connection between the gusset plate and angle in tension.*

A member is subjected to an axial tension of 100 kN. Design the member using

(a) Two equal angles.

(b) Single unequal angle

Design the bolted connection using M20 bolts of property class 4.6. The load includes load due to wind.

Solution: Patial safety factor for load = 1.2 (Table 4 of IS: 800)

Factored load = 1.2 × 100 = 120 kN.

Tensile strength in yielding = $T_{dg} = A_g f_y/\gamma_{mo}$ (Clause 6.2 of IS:800)

Required A_g = 120 × 1000 × 1.1/250 = 528 mm^2

Required area of each angle = 528/2 = 264 mm^2

Two Equal Angles

Let us try ISA 50 × 50 × 6 and M12 class 4.6 bolt.

Area of each angle = 568 mm^2

Tensile strength of two angles in yielding = $T_{dg} = A_g f_y/\gamma_{mo}$ (Clause 6.2 of IS:800)

$$= 2 \times 568 \times 250/1.1 \times 1000 = 258.2 \text{ kN}.$$

Tensile strength of angle in rupture $T_{dn} = 0.9 \times A_n \times f_u/\gamma_{m1}$ (Clause 6.3.3 of IS:800)

Second term is not used since bolted connection is yet to be designed.

$$A_n = 568 - 1 \times 13 \times 6 = 490 \text{ mm}^2$$

$$T_{dn} = 0.9 \times 490 \times 410/(1.25 \times 1000) = 144.65 \text{ kN per angle}$$

Two angles can take 2 × 144.65 = 289.3 kN

Least of the two is the strength which is = 258.2 kN > Factored load of 120 kN.

Though angle size smaller than this will be sufficient, general engineering practice is to use minimum size of ISA 50 × 50 × 6 for industrial structures.

Hence, satisfactory.

Single Unequal Angle

Area required = 568 mm^2

Using one ISA 75 × 50 × 8

Gross cross-sectional area of angle = 938 mm^2

Longer leg is connected.

Tensile strength of the angle in yielding = $T_{dg} = A_g f_y/\gamma_{mo}$ (Clause 6.2 of IS: 800).

$$= 938 \times 250/1.1 = 213 \text{ kN.}$$

Tensile strength of angle in rupture $T_{dn} = 0.9 \times A_n \times f_u/\gamma_{m1}$ (Clause 6.3.3 of IS: 800).

Second term is not used since bolted connection is yet to be designed.

$$\text{Net area} = A_n = 938 - 1 \times 13 \times 8 = 834 \text{ mm}^2$$

$$T_{dn} = 0.9 \times 834 \times 410/1.25 = 246.2 \text{ kN}$$

Tensile strength = 213 kN > 120 kN

Hence, satisfactory.

Conclusion

Weight/m of 2 ISA 50 × 50 × 6 = 90 N

Weight/m of one ISA 75 × 50 × 8 = 74 N

Using single ISA 75 × 50 × 8 as a tension member which has less weight is advantageous.

Design of Bolted Connection for Single Unequal Angle ISA 75 × 50 × 8 to the Gusset

Minimum pitch = 2.5 × 20 = 50 mm. (Clause 10.2.2 of IS:800).

Provide 50 mm pitch.

Minimum edge distance = 1.5 × 22 = 33 mm. (Clause 10.2.4.2 of IS:800).

Longer leg is connected. Edge distance = 75 – 40 = 35 mm (SP: 6(1)).

Provide 35 mm edge distance.

k_b is the least of

(1) $e/(3 \times d_o) = 35/(3 \times 22) = 0.53$

(2) $p/(3 \times d_o) - 0.25 = (50/(3 \times 22)) - 0.25 = 0.51$

(3) $f_{ub}/f_u = 400/410 = 0.975$

(4) 1.0

Design strength in bearing = 2.5 × 0.51 × 20 × 8 × 410/(1.25 × 1000) = 66.91 kN.

Design strength in single shear = 58 kN (Table 8.8 (b) of this book)

Bolt value = 58 kN

Number of bolts = 120/58 = 2.07.

Provide M20 property class 4.6 bolts 3 numbers.

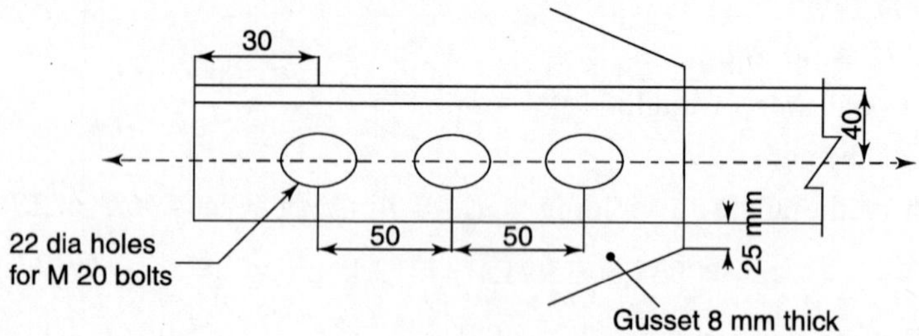

Fig. 8.15

Design strength in rupture (tension) taking into account the second term in the equation

$$T_{dn} = 0.9\, A_{nc}\, f_u/\gamma_{m1} + \beta A_{go}\, f_y/\gamma_{mo} \qquad \text{(Clause 6.3.3 of IS: 800)}$$

$$\beta = 1.4 - 0.076\,(w/t)\,(f_y/f_u)\,(b_s/L_c) \le (f_u \times \gamma_{mo})/(f_y \times \gamma_{m1}) \ge 0.7$$

$$(f_u \times \gamma_{mo})/(f_y \times \gamma_{m1}) = (410 \times 1.1)/(250 \times 1.25) = 1.4432$$

$$(w/t) = 75/6$$

$$L_c = 2 \times 50 = 100 \text{ mm.}$$

$$b_s = w + w_1 - t = 75 + 28 - 8 = 95 \text{ mm.}$$

$$B = 1.4 - 0.076(75/8)\,(250/410)\,(95/100) = 0.98727 \le 1.4432 \text{ and} > 0.7$$

$$T_{dn} = 246.2 - [0.98727 \times 50 \times 8 \times 250/(1.1 \times 1000) = 156.45 \text{ kN}$$

$$> 120 \text{ kN.}$$

Block Shear Strength (Clause 6.4.1 of IS: 800)

$$T_{db} = A_{vg}\, f_y/(\gamma_{mo}\, \sqrt{3} + 0.9\, A_{tn}\, f_u/\gamma_{m1}) \text{ (or)}$$

$$= 0.9\, A_{vn}\, f_u/(\gamma_{m1}\, \sqrt{3} + A_{tg}\, f_y/\gamma_{mo})$$

A_{vg}, A_{vn} = minimum gross and net area in shear along the bolt line parallel to external force, (1-2 and 3-4 as shown in Fig. 3.4(a), 1-2 as shown in Fig. 3.4(b).

A_{tg}, A_{tn} = minimum gross and net area in tension from the bolt line perpendicular to the line of force, respectively (2-3 as shown in Fig. 3.4(b).

$$A_{vn} = [30 + (2 \times 50) - (2 \times 22) - (22/2)] \times 8 = 600 \text{ mm}^2$$

$$A_{vg} = [30 + (2 \times 50)] \times 8 = 1040 \text{ mm}^2$$

$$A_{tg} = 35 \times 8 = 280 \text{ mm}^2$$

$$A_{tn} = (35 - (22/2)) \times 8 = 192 \text{ mm}^2$$

$$T_{db} = 1040 \times 250/(1.1 \times \sqrt{3} + 0.9 \times 192 \times 410/1.25) = 193.14 \text{ kN.}$$

(or)

$$= 0.9 \times 600 \times 410/1.25 \times \sqrt{3} + 280 \times 250/1.1 = 165.9 \text{ kN.}$$

Least is 165.9 kN which is greater than 120 kN.

Hence, satisfactory.

Connecting 2 ISA 50 × 50 × 6

Using M12 class 8.8 bolts.

Minimum pitch = 2.5 × 12 = 30 mm. (Clause 10.2.2 of IS:800).

Provide 40 mm pitch.

Diameter of bolt hole = 12 + 1 = 13 mm (Table 19 of IS:800)

Minimum edge distance = 1.5 × 13 = 19.5 mm. (Clause 10.2.4.2 of IS:800).

Edge distance available = 50 – 28 = 22 mm

Providing the bolts in such a way that the thread is not in shear plane,

Design strength in double shear = 83.62 kN (Table 8.7(g) of this book)

k_b is the least of

(1) $e/(3 \times d_o) = 22/(3 \times 13) = 0.564$

(2) $p/(3 \times d_o) - 0.25 = (40/(3 \times 13)) - 0.25 = 0.776$

(3) $f_{ub}/f_u = 400/410 = 0.975$

(4) 1.0

Therefore, $k_b = 0.564$

Design strength in bearing = 2.5 × 0.564 × 12 × 6 × 410/(1.25 × 1000) = 33.3 kN

Bearing governs.

Therefore, number of bolts = (120/33.3) = 3.6

Provide 4 nos. M12 class 8.8 bolts with an edge distance of 22 mm and pitch of 40 mm.

Design strength in rupture (tension) taking into account the second term in the equation

$$T_{dn} = 0.9\, A_{nc}\, f_u/\gamma_{m1} + \beta A_{go}\, f_y/\gamma_{mo} \text{ (Clause 6.3.3 of IS: 800)}$$

$$\beta = 1.4 - 0.076(w/t)\ (f_y/f_u)\ (b_s/L_c) \leq (f_u \times \gamma_{mo})/(f_y \times \gamma_{m1}) \geq 0.7$$

$$(f_u \times \gamma_{mo})/(f_y \times \gamma_{m1}) = (410 \times 1.1)/(250 \times 1.25) = 1.4432$$

$$(w/t) = 50/6$$

$$L_c = 3 \times 40 = 120 \text{ mm.}$$

$$b_s = w + w_1 - t = 50 + 28 - 6 = 72 \text{ mm.}$$

$$\beta = 1.4 - 0.076\,(50/6)\,(250/410)\,(72/120) = 1.1683 \leq 1.4432 \text{ and} > 0.7$$

$$T_{dn} = 144.65 - [1.1683 \times 50 \times 6 \times 250/(1.1 \times 1000) = 68 \text{ kN.}$$

For two angles 2 × 68 = 136 kN > 120 kN.

Block Shear Strength (Clause 6.4.1 of IS:800)

$$T_{db} = A_{vg}\, f_y/(\gamma_{mo}\sqrt{3} + 0.9\, A_{tn}\, f_u/\gamma_{m1}) \text{ (or)}$$

$$= 0.9\, A_{vn}\, f_u/(\gamma_{m1}\sqrt{3} + A_{tg}\, f_y/\gamma_{mo})$$

A_{vg}, A_{vn} = minimum gross and net area in shear along the bolt line parallel to external force, (1-2 and 3-4 as shown in Fig. 3.4(a), 1-2 as shown in Fig. 3.4(b).

A_{tg}, A_{tn} = minimum gross and net area in tension from the bolt line perpendicular to the line of force, respectively (2-3 as shown in Fig. 3.4(b).

$$A_{vn} = [22 + (3 \times 40) - (3 \times 13) - (13/2)] \times 8 = 772 \text{ mm}^2$$

$$A_{vg} = [22 + (3 \times 40)] \times 8 = 1136 \text{ mm}^2$$

$$A_{tg} = 22 \times 6 = 132 \text{ mm}^2$$

$$A_{tn} = (22 - (13/2)) \times 8 = 124 \text{ mm}^2$$

$$T_{db} = 1136 \times 250/(1.1 \times \sqrt{3} + 0.9 \times 124 \times 410/1.25) = 179.7 \text{ kN.}$$

(or)

$$= 0.9 \times 772 \times 410/1.25 \times \sqrt{3} + 132 \times 250/1.1 = 161.6 \text{ kN.}$$

Least is 161.6 kN which is greater than 120 kN.

Hence, satisfactory.

Note: (1) Note that the use of high tensile bolt of class 8.8 increases the shear strength but not the bearing strength which depends on the thickness and yield strength of the connected parts.

Example 8.8 *Beam to beam connection*

Design the connection between secondary beams MB300 and main beam MB600 using bolts of property class 4.6 and cleat angles on web as shown in Fig. 8.16. Reaction of each secondary beam is = 65 kN. The reaction is due to dead load only.

The webs of the two secondary beams are connected to the web of the main beam by cleat angles on either side of the webs of the secondary beams. The bolts connecting the cleat angles with the web of secondary beam are in double shear since angles are on both sides. The bolts between main beam and cleats also will be in double shear.

Web of Secondary Beam (MB 300)

Thickness of web of secondary beam ISMB 300 = 7.5 mm.

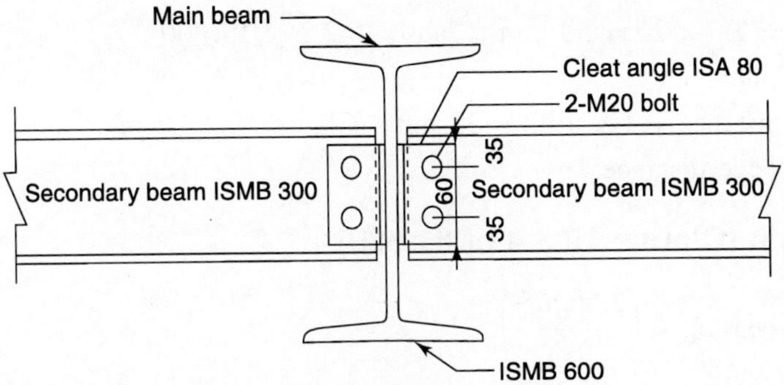

Fig. 8.16

Providing M20 bolts of property class 4.6 with clearance holes.

Diameter of hole = 20 + 2 = 22 mm (Table 19 of IS: 800)

Minimum pitch = 2.5 × 20 = 50 mm (Clause 10.2.2 of IS: 800).

Provide 60 mm pitch.

Minimum edge distance = 1.5 × 22 = 33 mm. (Clause 10.2.4.2 of IS:800).

Provide 35 mm edge distance.

Bearing Strength (Clause 10.3.4 of IS: 800)

Design strength in bearing = $\left(\dfrac{2.5k_b dtf_u}{\gamma_{mb}}\right)$

k_b is the least of

1. $e/(3 \times d_o) = 35/(3 \times 22) = 0.53$
2. $p/(3 \times d_o) - 0.25 = (60/3 \times 22)) - 0.25 = 0.66$
3. $f_{ub}/f_u = 400/410 = 0.975$
4. 1.0

Design strength in bearing = 2.5 × 0.53 × 20 × 7.5 × 410/(1.25 × 1000) = 65.2 kN

Design strength in double shear = 2 × 58.06 = 116.12 kN (Table 8.8(b) of this book)

Bolt value = 65.2 kN

Factored load = 1.5 × 65 = 97.5 kN

Hence, number of bolts = 97.5/65.2 = 1.5

Provide two M20-class 4.6 bolts.

Web of Main Beam (MB 600)

Thickness of web of main beam ISMB 600 = 12 mm.

Providing M20 bolts of property class 4.6 with clearance holes.

Diameter of hole = 20 + 2 = 22 mm (Table 19 of IS: 800)

Minimum pitch = 2.5 × 20 = 50 mm (Clause 10.2.2 of IS: 800).

Provide 60 mm pitch.

Minimum edge distance = 1.5 × 22 = 33 mm. (Clause 10.2.4.2 of IS: 800).

Provide 35 mm edge distance.

Bearing Strength (Clause 10.3.4 of IS: 800)

$$\text{Design strength in bearing} = \left(\frac{2.5k_b dtf_u}{\gamma_{mb}}\right)$$

k_b is the least of

1. $e/(3 \times d_o) = 35/(3 \times 22) = 0.53$
2. $p/(3 \times d_o) - 0.25 = (60/(3 \times 22)) - 0.25 = 0.66$ (Clause 10.2.2 of IS: 800)
3. $f_{ub}/f_u = 400/410 = 0.975$
4. 1.0

Design strength in bearing = 2.5 × 0.53 × 20 × 12 × 410/(1.25 × 1000) = 104.3 kN.

Design strength in double shear = 2 × 58.06 = 116.12 kN (Table 8.8(b) of this book)

Bolt value = 104.3 kN.

Factored load = 2 × 1.5 × 65 = 195 kN.

Number of bolts = 195/104.3 = 1.87

Provide two M20-class 4.6 bolts on each side.

Total will be 4 bolts.

Length of angle that can be accommodated = $D - 2h_2 = 300 - 2 \times 29.25 = 241.5$ mm = h_1

Required length of angle = pitch + 2 × edge distance = 60 + 2 × 35 = 130 mm.

Provided 130 mm < 241.5 mm

To check the adequacy of angle.

Design strength of angles in shear = $A_v \times f_{yw}/(\gamma_{mo} \times \sqrt{3}) = 4 \times 130 \times t \times 250/(1.1 \times \sqrt{3}) = 195 \times 1000$ N.

$$68.232 \times t = 195 \times 1000 \text{ N}$$

$$t = 195 \times 1000/68.232 = 2.86 \text{ mm.}$$

Provide ISA 80 × 80 × 6 for which edge distance = 80 – 45 = 35 mm (SP: 6(1)).

Note:

(1) We can increase the pitch or edge distance or both and increase the length of the angle to reduce the thickness if it is higher. Minimum thickness is 6 mm. However, the length of angle cannot be more than h_1 of the secondary beam.

(2) The reaction of secondary beam will act along bolts connected to its web. The connection to the main beam is at 40 mm from the centre of bolts connected to the web of secondary beam. Hence, there will be moment due to eccentricity. In this case the reaction on both sides of main beam are equal. The moments get balanced.

Example 8.9 *Beam to beam connection*

An ISMB 250 beam is connected to the web of ISMB 500 as shown in Fig. 8.17 and the reaction is 60 kN which is due to dead and imposed loads only. Design the connection.

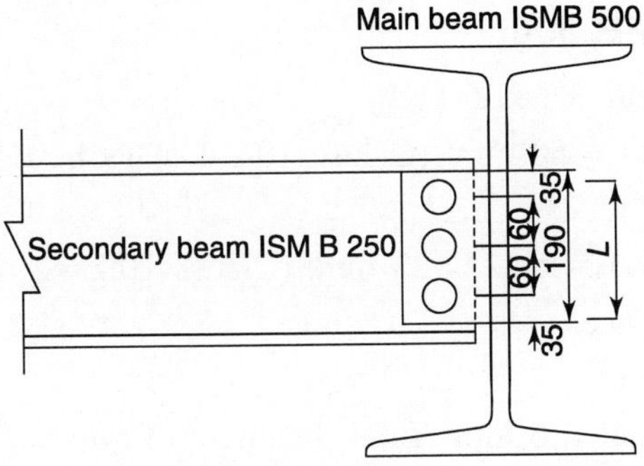

Fig. 8.17

Solution: Reaction from secondary beam = 60 kN.

Partial safety factor of load = 1.5 (Table 4 of IS: 800)

Factored = 1.5 × 60 = 90 kN

Bolts in web of secondary beam are in double shear.

Bolts in web of main beam are in single shear.

Web of Secondary Beam (ISMB 250)

Thickness of web of ISMB 250 = 6.9 mm

Providing M20 bolts, diameter of holes = 22 mm (Table 19 of IS: 800)

Minimum pitch = 2.5 × 20 = 50 mm. (Clause 10.2.2 of IS: 800).

Provide 60 mm pitch.

Minimum edge distance = 1.5 × 22 = 33 mm. (Clause 10.2.4.2 of IS: 800).

Provide 35 mm edge distance.

Bearing strength (Clause 10.3.4 of IS: 800)

$$\text{Design strength in bearing} = \left(\frac{2.5k_b d t f_u}{\gamma_{mb}}\right)$$

k_b is the least of

1. $e/(3 \times d_o) = 35/(3 \times 22) = 0.53$
2. $p/(3 \times d_o) - 0.25 = (60/(3 \times 22)) - 0.25 = 0.66$ (Clause 10.2.2 of IS: 800)
3. $f_{ub}/f_u = 400/410 = 0.975$
4. 1.0

Design strength in bearing = 2.5 × 0.53 × 20 × 6.9 × 410/(1.25 × 1000) = 60 kN.

Design strength in double shear = 2 × 58.06 = 116.12 kN (Table 8.8(b) of this book)

Bolt value = 60 kN.

Number of bolts = 90/60 = 1.5. Provide 3 bolts for similarity.

Web of Main Beam (ISMB 500)

Design strength in single shear = 58.06 kN

Minimum pitch = 2.5 × 20 = 50 mm. (Clause 10.2.2 of IS: 800).

Provide 60 mm pitch.

Minimum edge distance = 1.5 × 22 = 33 mm. (Clause 10.2.4.2 of IS: 800).

Provide 35 mm edge distance.

Thickness of web 10.2 mm.

Thickness of angle proposed = 6 mm. Total thickness of two angles = 2 × 6 = 12 mm.

Bearing strength = 2.5 × 0.53 × 20 × 10.2 × 410/(1.25 × 1000) = 88.66 kN.

Bolt value = 58.06 kN

Number of bolts = 90/58.06 = 1.55. Provide 3 bolts.

Maximum length of angle that can be accommodated = 250 – 2 × h_2

$$= 250 - 2 \times 27.95 = 194.1 \text{ mm}$$

Required length of angle = 2 × 60 + 2 × 35 = 190 mm < 194.1 mm = h_1

Provide 190 mm length of angle.

Design strength in shear = $A_v \times f_{yw}/(\gamma_{mo} \times \sqrt{3}) = 2 \times 190 \times t \times 250/(1.1 \times \sqrt{3}) = 90000$ N

$$60.33t = 90000 \text{ N.}$$

$$t = 90000/60.33 = 1.5 \text{ mm.}$$

The thickness shall not be less than 6 mm.

Provide ISA 80 × 80 × 6 – 190 mm long with 3 bolts of M20.

Note: In practice these connections are made in such a way that the top of secondary beam and main beam are at one level to have floor/platform on top. The arrangement is shown in Fig. 8.18.

L = 250 – (h_2 of MB 500) – (h_2 of ISMB 250) = 250 – 37.95 – 27.95 = 184.1 mm.

Due to notching in ISMB 250; angle length cannot be more than this. Keeping the length of angle as 180 mm, possible pitch = (180 – 2 × 35)/2 = 55 mm > required.

Required length of angle = 2 × 55 + 2 × 35 = 180 mm < 184.1 mm.

k_b is the least of

1. $e/(3 \times d_o) = 35/(3 \times 22) = 0.53$
2. $p/(3 \times d_o) - 0.25 = (55/(3 \times 22)) - 0.25 = 0.583$ (Clause 10.2.2 of IS: 800)
3. $f_{ub}/f_u = 400/410 = 0.975$
4. 1.0

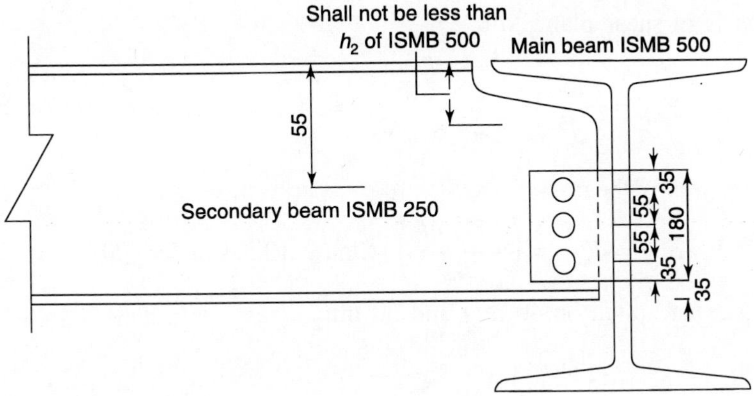

Fig. 8.18

Even now k_b = 0.53.

Hence, there is no change in bolt value.

Example 8.10 *Design the bolted shear connection between the flange of ISMB 500 column and web of ISMB 350 beam. The reaction transferred from the beam is 140 kN. The load is due to dead and imposed loads only. Use class 4.6 bolts.*

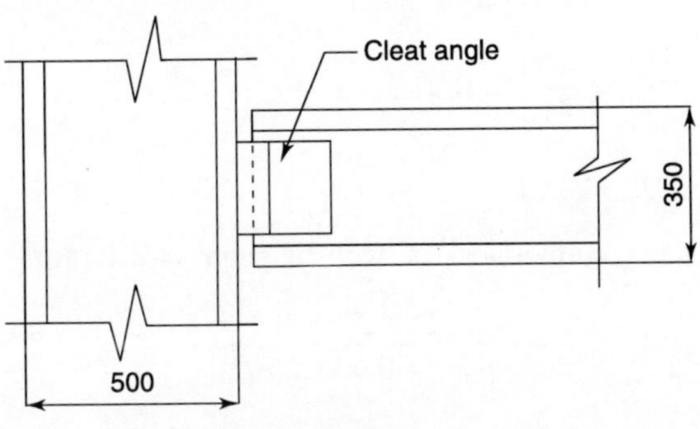

Fig. 8.19

Solution: The width of flange of ISMB 500 is 180 mm.

The cleat angle size can be limited to (180/2) = 90 mm to avoid projection of angle beyond the flange of the column.

Using 90 mm equal angle, the distance of centre of hole will be 40 mm from toe. (SP: 6 (1)).

Minimum edge/end distance for rolled edge = 1.5 × diameter of bolt hole = 40 mm.

Diameter of bolt hole can be = 40/1.5 = 26.7 mm.

Let us use M20 bolts of property class 4.6.

Double shear strength =

$$V_{dsb} = \{(f_u/\sqrt{3})\ [(n_n A_{nb} + n_s A_{sb})/\gamma_{mb}]\} \qquad \text{(Clause 10.3.3 of IS: 800)}$$

Considering thread is in shear plane,

$$V_{dsb} = \left(\frac{400}{\sqrt{3}}\right)\left(\frac{2 \times 245}{1.25 \times 1000}\right) = 90.5 \text{ kN (Table 8.8 (b) of this book).}$$

Bearing capacity on web of beam

Strength of bolt in bearing $= V_{dpb} = \left(\dfrac{2.5k_b dtf_u}{\gamma_{mb}}\right)$ (Clause 10.3.4 of IS: 800)

Let the edge distance and pitch be 40 mm and 50 mm.

k_b is smaller of $\dfrac{e}{3d_o}, \dfrac{p}{3d_o} - 0.25, \dfrac{f_{ub}}{f_u}, 1.0$

e = end distance of fastener along bearing direction = 40 mm (Clause 10.2.4.2 of IS: 800)

P = pitch distance of fastener along bearing direction (Clause 10.2.2 of IS: 800) = 50 mm.

$$\frac{e}{3d_o} = \frac{40}{3 \times 22} = 0.6$$

$$\frac{p}{3d_o} - 0.25 = \frac{50}{3 \times 22} - 0.25 = 0.51$$

$$\frac{f_{ub}}{f_u} = \frac{400}{410} = 0.97$$

$$k_b = 0.51$$

t = thickness of web of beam = 8.1 mm.

Strength in bearing $= V_{dph} = \dfrac{2.5 \times 0.51 \times 20 \times 8.1 \times 410}{1000 \times 1.25} = 67.75$ kN

$$V_{dph} = 16.4 \times 8.1 \times 0.51 = 67.75 \text{ kN (Table 8.8(b) of this book).}$$

Strength of bolt = 67.75 kN

Factored load = 1.5 × 140 = 210 kN (Table 4 of IS: 800).

Number of bolts = n = 210/67.75 = 3.1

Provide four 20 mm diameter bolts.

Let us use web cleats of ISA90 × 90 × 8.

Vertical shear = 210/4 = 52.5 < 67.75 kN

Connection with Column Flange

Bolts are in single shear.

Single shear strength of property class 4.6, M20 bolts (thread in shear plane)

$= 45.26$ kN (Table 8.8(b) of this book).

Bearing strength will be based on the thickness of angle since the thickness of flange of column is more.

Bolt value = 45.26 kN.

$$\text{Number of bolts required} = \frac{210}{45.35} = 4.63$$

Provide similar to the connection with web of beam.

M20 bolts, 4 on each side. Thus, there will be 8 bolts.

For ISMB 500; $g = 100$ mm (as in steel tables). As in SP: 6(1), for ISA 90 angle $c = 50$ mm.

C/C of bolt on the column flange = 2 × 50 + thickness of web of beam = 2 × 50 + 8.1 108.1 mm.

The centre line of the bolt on the web cleat will be at 50 mm away from column face.

Moment due to this = (210 × 50)/1000 = 10.5 kNm

Considering hinge action at bottom most bolt, and r, r_1, r_2, r_3... the distance of bolts from the centre line of bottommost bolt, force in the extreme bolt due to this moment = $\dfrac{M_r}{\Sigma r^2}$

$$\Sigma r^2 = 2(50^2 + 100^2 + 150^2) = 70000 \text{ mm}^2$$

Tension at the topmost bolt = $[(10.5 \times 10^6 \times 150)/(70000 \times 1000)] = 22.5$ kN

Vertical shear = 210/8 = 26.25 kN

Tension capacity:

$$T_{ub} = (0.9 f_{ub} A_n) = \frac{0.9 \times 400 \times 245}{1000} = 882 \text{ kN.}$$

(Clause 10.3.5 of IS:800)

$$T_b = T_{dn} = \frac{T_{ub}}{\gamma_{mo}} = \frac{88.2}{1.25} = 70.56 \text{ kN (Table 8.8(b) of this book).}$$

Shall be less than $f_{yb} \times A_{sb}/\gamma_{mo} = 240 \times 314/1.1 = 68.51$ kN (Tables 8.7 & 8.8(b) of this book).

Hence, tensile capacity of bolt = 68.51 kN.

Combined Shear and Tension (Clause 10.3.6 of IS: 800)

$$\left(\frac{V_{sb}}{V_{db}}\right)^2 + \left(\frac{T_b}{T_{db}}\right)^2 \leq 1.0$$

$$(26.25/45.26)^2 + (22.5/68.51)^2 = 0.44 < 1.0$$

Note:

(1) Since the ratio is only 0.44, we can even reduce the number of bolts (or) diameter of bolts.

(2) But for uniformity with connection to web of beam, the same 4 M20 bolts are used for connecting the cleat angle to flange of column.

(3) In case if such joints are in large numbers, we can optimize the number of bolts to achieve economy.

(4) It is advisable to adopt same class with less number of varieties with respect to diameter and length.

Design of Cleat Angles

Required length of the angle to accomodate bolts = 3 × 50 + 2 × 40 = 230 mm.

This shall be less than (depth of beam – 2 × (thickness of flange + the root radius)).

$$= 350 - 2 \times (14.2 + 14) = 293.6 \text{ mm.}$$

h_1 = 288 mm as in steel table which is greater than the required length of angle 230 mm. Let us provide the length of angle as 230 mm.

Moment capacity = $1.2 \times f_y \times z/\gamma_{mo}$ (Clause 8.2.1.2 of IS:800)

$$= 1.2 \times 250 \times [(8 \times 230^2)/6]/(1.1 \times 10^6) = 19.24 \text{ kNm}$$

Moment per angle = (10.5)/2 = 5.25 < 19.24 kNm

$$V_d = \text{Design shear strength} = (A_v f_{yw})/(\gamma_{mo}\sqrt{3})$$

(Clause 8.4 of IS: 800)

$$= (230 \times 8 \times 250)/(1.1 \times \sqrt{3} \times 1000) = 241.43 \text{ kN} > 210/2 \text{ kN}$$

$$0.6V_d = 0.6 \times 241 = 144.86 \text{ kN} > 210/2 \text{ kN}$$

Applied shear is 210 kN and this is taken by 2 angles. Each angle will take 210/2 = 105 kN. This value is less than $0.6V_d$ and hence the angles are in low shear. The connection with 2 ISA 90 × 90 × 8 with M20 class 4.6 bolts as shown in Fig. 8.20 is satisfactory.

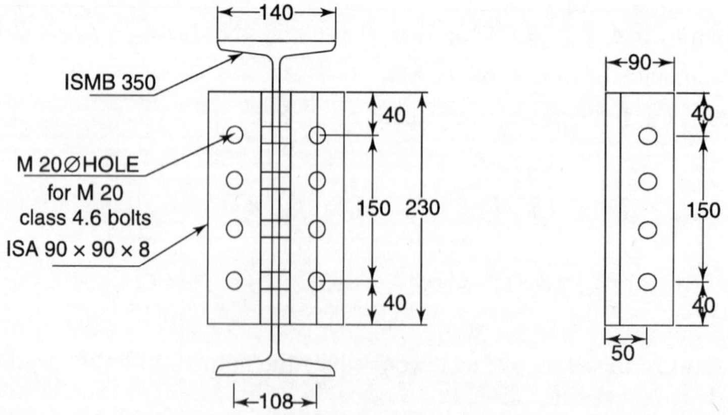

Fig. 8.20

Example 8.11 *For example 8.10; design an end plate connection to transfer only shear.*

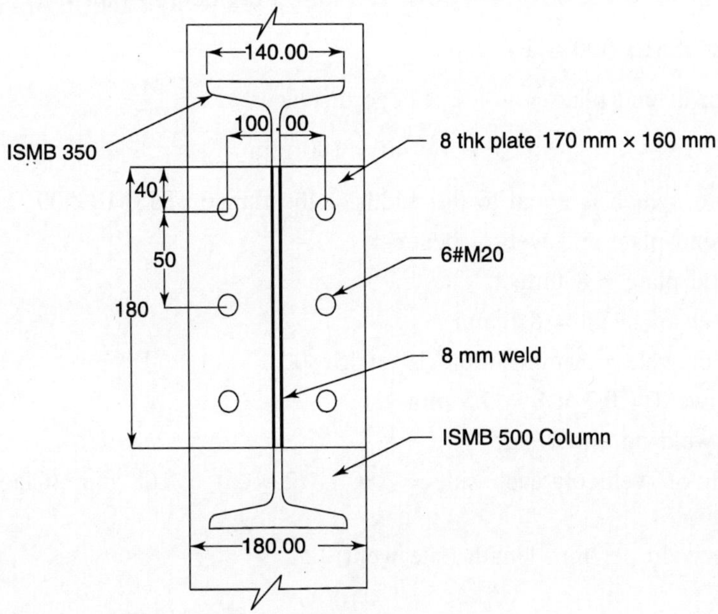

Fig. 8.21

Solution: *Connection to column flange*

Let us adopt the end plate thickness as 8 mm.

Use M20 bolts of property class 4.6.

Let the edge distance and pitch be 40 mm and 50 mm.

$$k_b = 0.51$$

$$\text{Bearing capacity} = \frac{2.5 \times 0.51 \times 20 \times 8 \times 410}{1000 \times 1.25} = 66.91 \text{ kN} \tag{1}$$

Single shear capacity of bolt (shank in shear plane)

$$V_{dsb} = \left(\frac{f_u}{\sqrt{3}}\right)\left(\frac{n_n\, A_n b + n_s\, A_s b}{\gamma_{mb}}\right) \text{ (Clause 10.3.3)}$$

$$V_{dsb} = \left(\frac{400}{\sqrt{3}}\right)\left(\frac{0 + 1 \times 314}{1.25 \times 1000}\right) = 58 \text{ kN} \tag{2}$$

Bolt value = 58 kN (least of 1 and 2) (Table 8.8(b) of this book).

$$\text{Number of bolts required} = \frac{210}{58} = 3.62$$

Provide 6 M20 class 4.6 bolts, three on each side.

Length of end plate = 2 × 40 + 2 × 50 = 180 mm. Provide 180 mm < h_1.

g for ISMB 500 = 100 mm.

Minimum width of end plate = g + 2 × edge distance

= 100 + 2 × 40 = 180 mm

Provide 180 mm, which is equal to the width of the flange of ISMB 500.

Weld between end plate and web of beam

Thickness of end plate = 8 mm.

Thickness of web of beam = 8.1 mm

Minimum size of weld = 3 mm (Table 19 of IS: 800).

Let the size of weld = 0.7 × 8 = 5.6 mm

Provide 6 mm weld on either side.

Effective length of weld on each side = 180 – (2 × 6) = 168 mm (Clause 10.5.4.1 of IS: 800)

Strength of site weld per mm length (site weld)

$$f_{wd} = \frac{f_u}{\sqrt{3} \times \gamma_{mw}} = \frac{410}{\sqrt{3} \times 1.5} = 157.8 \text{ N/mm}^2$$

(Clause 10.5.7.1.1 of IS: 800)

Strength/mm length of 6 mm weld = 157.8 N/mm^2 × 0.707 × 6 = 669 N/mm

Table 7.2 of this book gives the value of design strength of shop weld. To obtain design strength of site weld, multiply the value in the table by (1.25/1.5).

For 6 mm, the strength of site weld = 803.4 × (1.25/1.5) = 669.5 N/mm.

Shear on weld = (210 × 1000)/(2 × 168) = 625 N/mm < 669 N/mm

Hence, satisfactory

Note:

In such joint where no moment is to be transferred, the length of end plate shall be kept as small as possible.

Example 8.12 *Design a fin plate welded to column at shop and bolted to web of beam at site for the Example 8.10.*

Solution: Bolted connection to web of beam

Providing thread only to required length (IS:1367), shank will be in shear plane.

Single shear capacity of bolt M20 class 4.6 = 58 kN (Table 8.8(b) of this book).

Let us provide fin plate of thickness more than the thickness of web of beam. The web will be the thinner part. Thickness of web of MB350 = 8.1 mm. This is considered to calculate the bearing capacity.

$k_b = 0.51$ (Example 8.10 of this book)

Bearing capacity against web of the beam $V_{dpb} = \left(\dfrac{2.5k_b\,dtf_u}{\gamma_{mb}}\right)$

$$= \frac{2.5 \times 0.51 \times 20 \times 8.1 \times 410}{1000 \times 1.25} = 67.75 \text{ kN}$$

Bolt value = 58 kN.

Number of bolts $= \dfrac{210}{58} = 3.62$

Provide 4 bolts M20 class 4.6

Length of plate required = 2 × 40 + 3 × 50 = 230 mm < h_1 of ISMB 350

Let us provide 5 mm gap for rotation and edge distance of 40 mm.

Block Shear Strength (Clause 6.4.1 of IS:800)

Thickness of web of beam = 8.1 mm and the thickness of fin plate = 10 mm. Block shear strength of web will be checked.

$$T_{db} = A_{vg}\, f_y/(\gamma_{mo}\sqrt{3} + 0.9\, A_{tn}\, f_u/\gamma_{m1}) \text{ (or)}$$

$$= 0.9\, A_{vn}\, f_u/(\gamma_{m1}\sqrt{3} + A_{tg}\, f_y/\gamma_{mo})$$

A_{vg}, A_{vn} = minimum gross and net area in shear along bolt line parallel to external force, (1-2 and 3-4 as shown in Fig. 3.4(a), 1-2 as shown in Fig. 3.4(b)).

A_{tg}, A_{tn} = minimum gross and net area in tension from the bolt line perpendicular to the line of force, respectively (2-3 as shown in Fig. 3.4(a) and Fig. 3.4(b)). Also refer Fig. 7 of IS:800

$$A_{vn} = [(3 \times 50) - (3 \times 22)] \times 8.1 = 680.4 \text{ mm}^2$$

$$A_{vg} = [(3 \times 50)] \times 8.1 = 1215 \text{ mm}^2$$

$$A_{tg} = 2 \times 35 \times 8.1 = 567 \text{ mm}^2$$

$$A_{tn} = 2 \times (35 - (22/2)) \times 8.1 = 388.8 \text{ mm}^2$$

$$T_{db} = 1215 \times 250/(1.1 \times \sqrt{3} + 0.9 \times 388.8 \times 410/1.25) = 274.2 \text{ kN}$$

(or)

$$= 0.9 \times 680.4 \times 410/1.25 \times \sqrt{3} + 567 \times 250/1.1 = 244.83 \text{ kN.}$$

Least is 244.83 kN which is greater than 210 kN.

Hence, satisfactory.

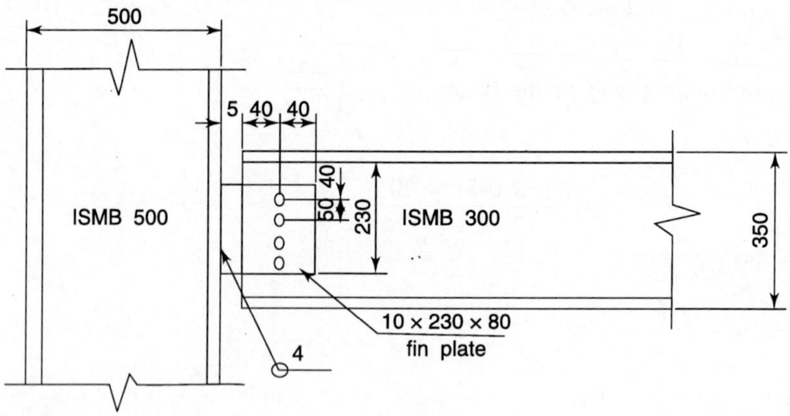

Fig. 8.22

Thickness of Fin Plate

Size of fin plate = 10 × 230 × 80 mm

Moment at the face of column = shear × distance between the face of column and centre of bolt hole = 210 × 40/1000 = 8.4 kNm

Moment capacity of plate = $M_d = 1.2Z_e f_y / \gamma_{mo}$

$$= 1.2(10 \times 230^2/6) \times 250/1.1 \times 10^6 = 24.05 \text{ kNm} > 8.4 \text{ kNm}$$

Design of Fillet Weld

Size of weld = 0.7 × 10 = 7 mm

Since the gap between the column face and the beam is kept as 5 mm, let us provide 4 mm weld.

Shear Capacity of Weld

Effective length of weld = 2(230 – 2 × 4) = 2 × 222 = 444 mm

Shear capacity of site weld = $f_{wd} = f_u/(\sqrt{3} \times \gamma_{mw}) = 410/(\sqrt{3} \times 1.25) = 189.37 \text{ N/mm}^2$

Shear capacity of 4 mm weld of effective length 444 mm

$$= 0.7071 \times 4 \times 444 \times 189.37/1000 = 237.8 \text{ kN} > 210 \text{ kN}$$

Adequate.

If the same weld is done at site, design shear strength of weld

$$= 0.7071 \times 4 \times 444 \times 157.8/1000 = 198.2 \text{ kN} < 210 \text{ kN}.$$

Length of the site weld required to transfer the shear = 210 × 1000/(0.7071 × 4 × 157.8) = 470 mm.

Length of plate = (470/2) + 2 × 4 = 243 mm. Provide 250 mm long plate with same 4 mm size of weld.

Note: Another alternative is to increase the size of the weld and the gap.

Example 8.13 *The column is ISMB 300 and the beam connected to the flange of the column is ISMB 200. The reaction from the beam is 80 kN. Design a seat angle connection using class 4.6 bolts. The load is due to dead and imposed loads only.*

Solution: From steel tables, maximum size of flange bolt is 22 mm for ISMB 300. Let us use class 4.6, M16 bolts.

Flange thickness of ISMB 300 column = 12.4 mm

Strength of Bolt in Single Shear (Thread not in Shear Plane)

$$V_{dsb} = (f_u/\sqrt{3})\ (n_n A_{nb} + n_c A_{sb})/\gamma_{mb} \qquad \text{(Clause 10.3.3 of IS:800)}$$

$$n_n = 0$$

Using class 4.6 bolts, $f_u = 400$ MPa

Using 16 mm bolts $A_{sb} = \dfrac{\pi \times 16^2}{4} = 201\ \text{mm}^2$ (Table 8.4 of this book)

$$V_{dsb} = \frac{400}{\sqrt{3}}\ (1 \times 201)/1.25 = 37.135\ \text{kN} \quad \text{(Table 8.8(b) of this book)}$$

Strength in Bearing (Clause 10.3.4 of IS:800)

M16 bolts of property class 4.6 are proposed.

Diameter of bolt 16 mm.

Diameter of bolt hole = 16 + 2 = 18 mm (Table 19 of IS:800).

Minimum end/edge distance = 1.5 × 18 = 27 mm or 1.7 × 18 = 30.6 mm (Clause 10.2.4.2 of IS:800). Let us keep end/edge distance as 40 mm.

Minimum pitch = 2.5 × diameter of bolt = 2.5 × 16 = 40 mm (Clause 10.2.2 of IS:800).

Let us keep pitch p as = 50 mm.

The value of k_b is the least of

(1) $e/3d_o = 40/(3 \times 18) = 0.74$

(2) $(p/3d_o) - 0.25 = 50/(3 \times 18) - 0.25 = 0.676$

(3) $f_{ub}/f_u = 400/410 = 0.975$

(4) 1.0

The least value is $0.676 = k_b$

Flange thickness of column is 12.4 mm. Flange thickness of beam is 10.8 mm.

Let us provide angle of thickness 10 mm.

$$V_{dsb} = V_{npb}/\gamma_{mb} = 2.5\ k_b\ dt\ f_u/\gamma_{mb} = 2.5 \times 0.676 \times 16 \times 10 \times 410/(1.25 \times 1000) = 88.69\ \text{kN}.$$

Bolt value = 37.135 kN

Factored load = 1.5 × 80 = 120 kN

$$\text{No. of bolts required} = \frac{120}{37.135} = 3.23$$

Four M16 class 4.6 bolts are required for the connecting the angle with the column flange.

Seating Angle Size

Step-by-step design procedure for seating angle connection is given below.

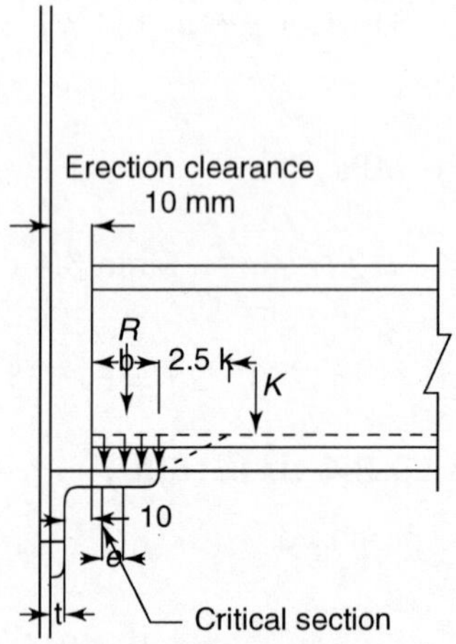

Fig. 8.23

(a) The length of seating angle shall be less than the flange width of column to which the beam is connected and more than the width of the flange of beam. If it is kept less than the width of flange of beam overhead welding will be involved at site.

(b) Leg size of the seating angle is decided based on the web crippling of the beam crippling strength of web $\geq R$

$$(b_1 + n_2) \times t_w \times (f_{yw}/\gamma_{mo}) \geq R$$

$$b_1 \geq (R \times \gamma_{mo}/t_w \times f_{yw}) - n_2$$

R = support reaction of the beam

t_w = web thickness of beam.

(c) Calculate the distance between free end of the angle to root of angle

$$b_2 = b_1 + g - (t_a + r_a)$$

g = erection and rotation gap

t_a = thickness of angle

r_a = root radius of the angle

(d) Assuming that the reaction is uniformly distributed over the bearing area, find the bending moment at the critical section.

(e) By equating this to the bending strength of rectangular section (length and thickness of angle), the thickness of angle can be found out since the length is already decided.

Width of flange of beam ISMB 200 = 100 mm

Width of flange of column ISMB 300 = 140 mm

Length of angle can be 100 mm

Strength at root line of beam = length of bearing × thickness of web × (f_{yw}/γ_{mo}) = Reaction.

Length of bearing × 5.7 × (250/1.1) = Length of bearing × 1295.45 = 120 × 1000 = $(b_1 + n_2)$

$$b_1 = \text{Length of bearing} = \frac{120 \times 1000}{1295.45} - 2.5 \times (10.8 + 11) = 92.63 - 54.5 = 38.13 \text{ mm.}$$

Let the end gap for rotation be 5 mm.

Length of outstanding leg of angle required = 38.13 + 5 = 43.13 mm

Providing ISA 75 × 75 × 8 angle,

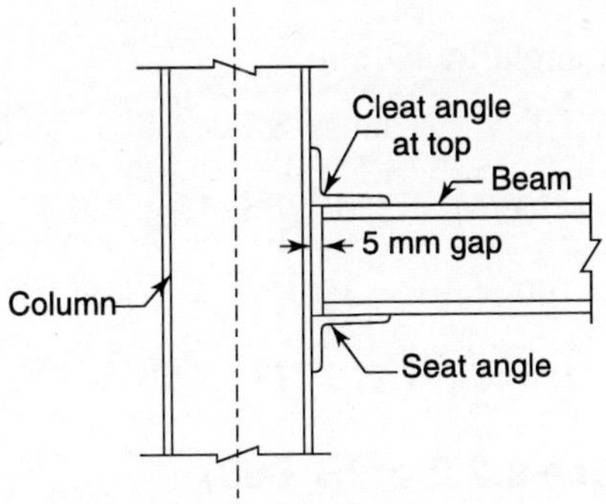

Fig. 8.24

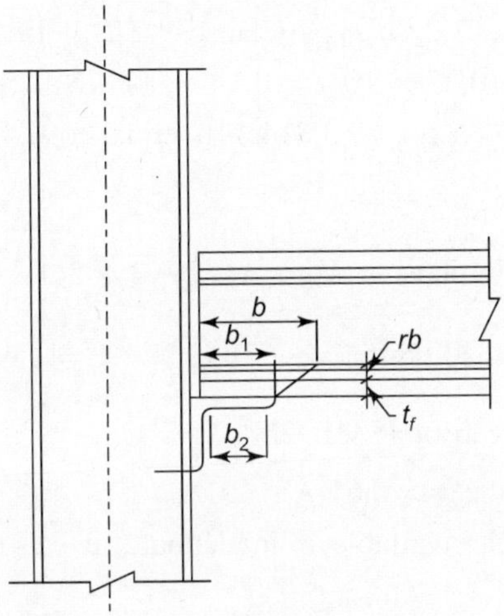

Fig. 8.25

b_2 for ISA 75 × 75 × 8 angle = 75 – (8 + 7) = 60 mm

Uniformly distributed load on overhang length = $\dfrac{120 \times 1000}{100 \times 60}$ = 20 N/mm

Moment at the root of angle = $\dfrac{20 \times 60^2}{2}$ = 0.072 ×10⁶ Nmm

Moment capacity = $M_d = 1.2Z_e f_y/\gamma_{mo}$ (Clause 8.2.1.2 of IS: 800)

$$= 1.2 \times (100 \times t_a^2/6) \times 250/1.1 = 0.072 \times 10^6 \text{ Nmm}$$

t_a = required thickness = 3.98 mm.

Thickness of provided angle ISA 75 × 75 × 8 is adequate.

Shear capacity of the outstanding leg of cleat.

Width × thickness × $f_y/(\sqrt{3} \times 1.1)$ = 100 × 8 × 250/(1.1 × $\sqrt{3}$) = 104.97 kN < 120 kN

Inadequate.

Increase the thickness of the angle to 10 mm.

Moment capacity = $M_d = 1.2Z_e f_y/\gamma_{mo}$ (Clause 8.2.1.2 of IS: 800)

$$M_d = 1.2 \times (100 \times 10^2/6) \times 250/1.1 \times 10^6) = 0.45 \times 10^6 > 0.072 \times 10^6 \text{ Nmm}$$

Design shear strength = V_d = 100 × 10 × 250/($\sqrt{3}$ × 1.1 × 1000) = 131.21 kN > 120 kN

$$0.6 \times V_d = 0.6 \times 131.21 = 78.73 \text{ kN} > 120 \text{ kN}$$

Combined Strength (Clause 9.2.2 of IS: 800)

$$\beta = ((2V/V_d) - 1)^2 = ((2 \times 120/131.21) - 1)^2 = 0.687$$

M_{fd} = 0 since there is no separate shear area.

$M_{dv} = M_d - \beta \times (M_d - M_{fd}) \le 1.2Z_e f_y/\gamma_{mo}$ (Clause 9.2.2 of IS: 800)

$M_{dv} = 0.45 \times 10^6 - 0.687 \times (0.45 \times 10^6) = 0.14 \times 10^6$ kNm > 0.072×10^6 kNm.

This shall be less than $M_d = 1.2Z_e f_y/\gamma_{mo}$ which is equal to 0.45×10^6 kNm.

Strength of Web

Strength of web of ISMB 200 in shear = $V_d = A_v f_{yw}/\sqrt{3} \times 1.1$

$$= 200 \times 5.7 \times 250/\sqrt{3} \times 1.1 = 149.6 \text{ kN} > 120 \text{ kN}$$

No stiffener is required for web of ISMB 200.

The beam is in high shear (shear > $0.6V_d$).

Note: If the thickness of angle available is inadequate, it can be strengthened by providing a plate as shown in Fig. 8.26.

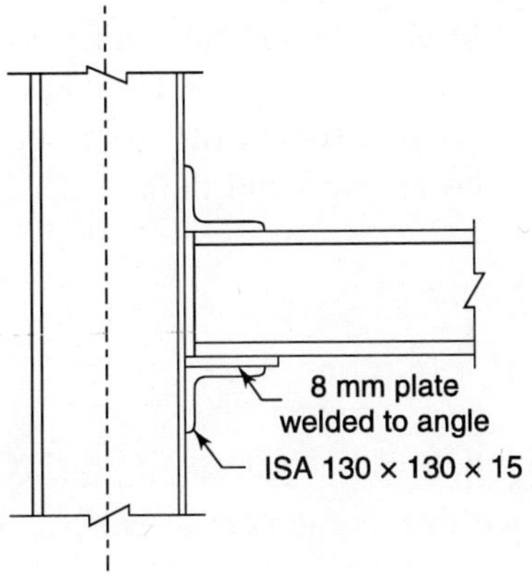

Fig. 8.26

8.28 MOMENT CONNECTIONS

8.28.1 Introduction

When the direct forces do not pass through the centre of the joint, the joints carry moments in addition to direct forces. The fasteners are designed to carry the additional forces (tension or shear) due to bending.

8.28.2 Bolted Joints Subjected to Moment Acting Perpendicular to the Plane of Joints (Bolts in Tension and Shear)

Bolts are subjected to tension and shear both when the applied moment is not in the plane of the bolts as in case of the bolts connecting the bracket to the column flange shown in Fig. 8.27. The external force tends to rotate the connection about a neutral axis. The location of the neutral axis depends upon the initial tension (if any) in the bolts.

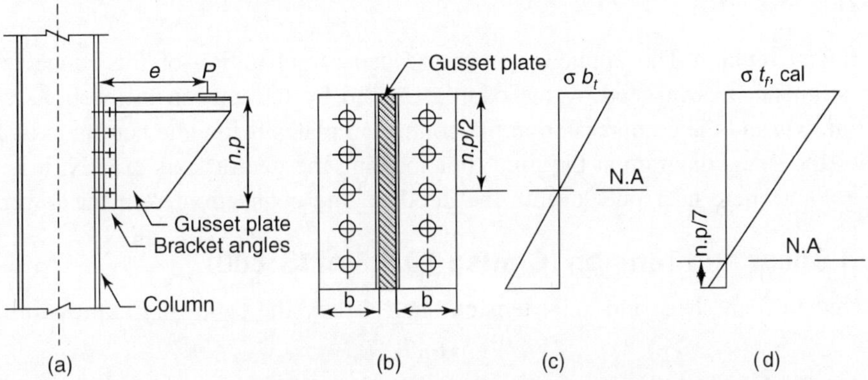

Fig. 8.27 *(a) Bracket, (b) Front view showing bolt location (c) Stress diagram (Initial tension) (d) Stress diagram (No initial tension)*

(i) Initial tension in bolts: Hot-driven bolts have initial tension when they cool and compress the connecting plates together. Figure 8.27 shows the stress variation at the joint due to the applied moment. It is assumed that some of the initial tension in the bolts above the neutral axis will be relieved and the contact of the connected plates will be broken.

If n = number of bolts in each line

m = number of lines of bolts

p = pitch of bolts

Nominal tensile capacity of the bolt = $T_{db} = 0.9\ A_n\ f_{ub}/\gamma_{mb}$ (Clause 10.3.5 of IS: 800)

Actual stress due to tensile force due to moment at extreme fastener

$$= \frac{M \times y}{I} = \frac{M \times \left(\frac{n \times p}{2}\right)}{\frac{1}{12} \times (m \times b) \times (n \times p)^3} = \frac{6M}{m \times b(n \times p)^2}$$

Therefore, the effective tensile load on the extreme fastener,

$$T_b = \text{Actual stress} \times b \times p = \frac{6 \times M \times b \times p}{m \times b(n \times p)^2}$$

or

$$T_b = \frac{6 \times M}{m \times p \times n^2} \tag{1}$$

where A_r = effective cross-sectional area of the fasteners in tension and $A_r = m \times n \times A$ Equation (4) can be used to determine the number of bolts required in eccentric connections. The maximum tension $T_b \not> $ bolt value R. Hence, keeping $R = T_b$ in Equation (1)

Number of bolts
$$n = \sqrt{\frac{6 \times M}{m \times p \times R}} \tag{2}$$

(ii) No initial tension The applied moment M causes separation of the connected plates. The moment M is resisted by the couple formed by the tension in the bolts above the neutral axis and the compression in the connected plates below the neutral axis. Fig. 8.27 shows the stress diagram at the joint. For locating the neutral axis exactly, it is assumed to be somewhere near one-seventh the depth of the connection from the lower end.

Combined Shear and Tension (Clause 10.3.6 of IS: 800)

Bolts subjected to both shear and axial tension shall satisfy the following expression.

$$(V_{sb}/V_{db}) + (T_b/T_{db})$$

where V_{sb} and T_b = the calculated shear and axial tensile forces respectively in the fastener.

V_{db} and T_{db} = design shear and axial tensile strength respectively

8.28.3 Design of Moment Connections

(i) Eccentric load lying in plane of bolts Assume a suitable bolt diameter and find its strength in shear and bearing to determine bolt value R. If the applied moment is small, then determine the number of bolts n required for direct load P by equating n to P/R. Use two to three times the bolts required for direct load, check the individual and combined strength.

If the applied moment is high, then the number of bolts n may be determined by Equation (1) reproduced below.

$$n = \sqrt{\frac{6 \times M}{m \times p \times R}} \tag{1}$$

Arrange the bolts and check the adequacy.

(ii) Eccentric load not lying in the plane of bolts If the fasteners have initial tension, Equation (1) can be used to determine the number of bolts required for the joint. If the fasteners do not have any initial tension, then a slightly less number of bolts is required and Equation (1) is modified to:

$$n = 0.8 \times \sqrt{\frac{6 \times M}{m \times p \times R}} \tag{2}$$

The bolts are then arranged and the force is calculated which should be within the design strength.

Example 8.14 *A concentrated load of 100 kN due to dead load and imposed load is applied on a bracket at a distance of 350 mm from the face of the flange of a column ISMB 400. Design the bracket joint using equal angles and class 4.6 bolts.*

Solution: In this connection, moment is perpendicular to the plane of joint. It is proposed to have cleat angles at top and bottom flanges of beam to transfer moment. This requires design of bracket with cleat angles at top and bottom of beam/bracket and

(a) bolts connecting column to angle cleats; and

(b) bolts connecting cleat to the bracket.

The size of bracket, cleat angle, number and diameter of bolts, edge distances, clearances available, are unknown factors. In such a case, we have to assume sizes by experience and check for the design.

a. Bracket

Factored load = 1.5 × 100 = 150 kN

The load on the bracket introduces shear and bending moment because of eccentricity.

Factored bending moment on the bracket is $M = P_e = 150 \times 0.35 = 52.5$ kNm.

Bending strength = $M_d = 1.2 Z_e f_y / \gamma_{mo}$ (Clause 8.2.1.2 of IS: 800)

Z_e required = $(52.5 \times 10^6 \times 1.1/1.2 \times 250) = 1{,}92{,}500 \text{ mm}^3$

Let us try ISMB 300 for which $Z_e = 573.5 \times 10^3 \text{ mm}^3$

b. Cleat angles

Cleat angles are provided to connect the top and bottom flanges of the bracket with the column as shown in Fig. 8.28.

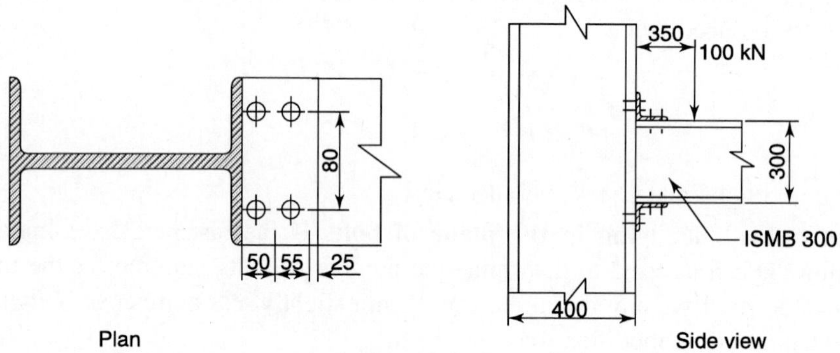

Fig. 8.28

The load on the bracket causes:

- Shear in the bolts connected to column flange,
- Tension at the bolts, and
- Compression on the bottom cleat.

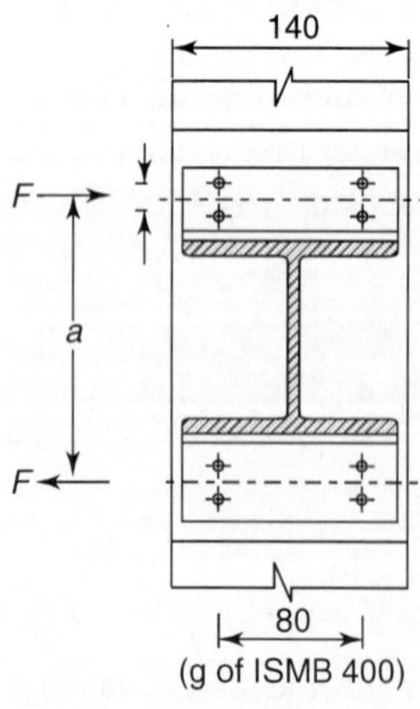

Fig. 8.29

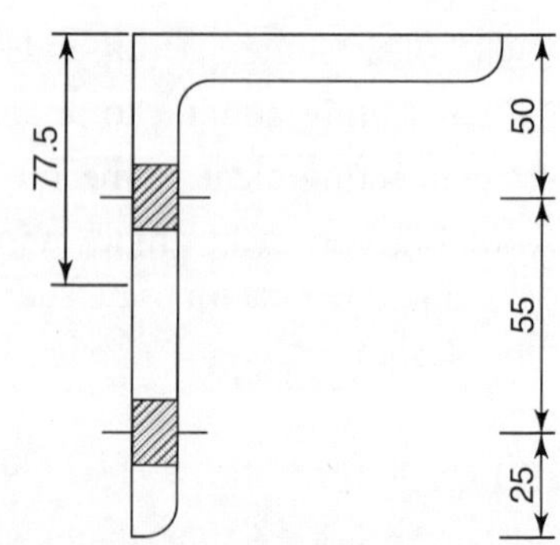

Fig. 8.30

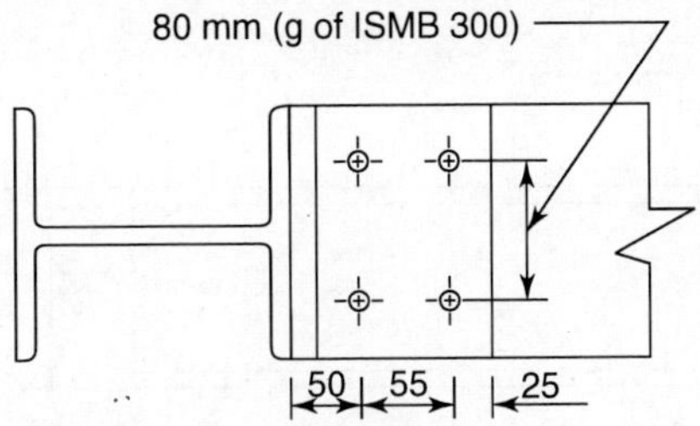

Fig. 8.31

Let us use ISA 130 × 130 × 12 angles as cleats to accommodate two rows of bolts. As per SP: 6(1) the distances from the face of angle to the centre of the hole are 50 mm and 105 mm from the outer face of the leg of angle.

Centre to centre of bolt group = a = 300 + 2 × (50 + (55/2)) = 455 mm

$$M = F \times a;\ F = M/a = 52.5 \times 10^6/455 = 115385 \text{ N}$$

Let us provide four bolts at top.

Tension/bolt = 115385/4 = 28846 N = 28.846 kN.

Net cross-sectional area of each bolt M20 = 245 mm^2 (Table 8.4 of this book).

Tension capacity = $0.9\, f_{ub}\, A_n/\gamma_{mb} < (f_{yb}\, A_{sb}/\gamma_{mo})$ (Clause 10.3.5 of IS: 800)

$$= 0.9 \times 400 \times 245/1.25 = 70.56 \text{ kN}$$

$(f_{yb}\, A_{sb}/\gamma_{mo})$ = 240 × 314/1.1 = 68.51 kN (Table 8.8(b) of this book).

Design strength in tension = 68.51 kN > 28.846 kN.

Thread is not in shear plane.

Design shear strength of M20 bolt of class 4.6

$= \left(\dfrac{f_u}{\sqrt{3}}\right)\left(\dfrac{n_n A_n b + n_s A_s b}{\gamma_{mb}}\right)$ = (400/$\sqrt{3}$) (1 × 314)/(1.25 × 1000) = 58 kN (Table 8.8(b) of this book).

Four bolts connected to column flange at the top and four bolts connected to the column flange at the bottom will take shear.

Hence, shear force per bolt = 150/8 = 18.75 kN < 58 kN.

Combined Strength

$(18.75/58)^2 + (28.846/68.51)^2 = 0.74 < 1.0$ (Clause 10.3.6 of IS: 800)

Hence, the joint between cleat angles and column flange is satisfactory.

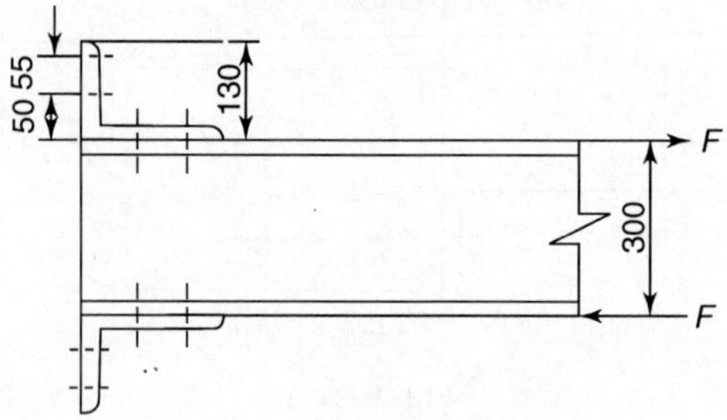

Fig. 8.32

Bolts Connecting the Flange of MB 300

Force = 52.5 × 10^6/300 = 1,75,000 N

Shear force per bolt = 1,75,000/4 = 43,750 N = 43.75 kN < 58 kN.

Hence, provide ISA 130 × 130 × 12 and M20 bolts of property class 4.6.

Note:

(1) It can be noticed that the edge distance is only 25 mm for ISA 130 × 130 × 12 angles with M20 bolts.

(2) Maximum size of bolt M20 (SP: 6(1)).

Example 8.15 *Design the bolted connection for angle bracket as shown in Figs. 8.33 and 8.34 using bolts of property class 4.6.*

Solution: Moment perpendicular to the plane of connections

Moment = 1,50,000 × 200 = 30 × 10^6 Nmm.

The tensile force in the topmost bolt is given by

$$F_1 = \frac{M \times r}{\Sigma r^2}$$

$$\Sigma r^2 = 2 \times (75^2 + 150^2 + 225^2) = 1{,}57{,}500 \text{ mm}^2,$$

$$r = 225 \text{ cm (Distance of extreme bolt)}$$

$$F_1 = (3{,}00{,}00{,}000 \times 225)/(78{,}750 \times 2) = 42{,}857 \text{ N.} = 42.857 \text{ kN.}$$

Providing M24 bolts.

Gross area = 452 mm^2 and net area = 353 mm^2 (Table 8.4 of this book)

Tensile strength = 0.9 f_{ub} $A_n/\gamma_{mb} < f_{yb}$ A_{sb}/γ_{mo} (Clause 10.3.5 of IS: 800).

$$0.9\, f_{ub}\, A_n = 0.9 \times 400 \times 353/(1.25 \times 1000) = 101.7 \text{ kN}$$

$$f_{yb}\, A_{sb}/\gamma_{mo} = 240 \times 452/(1.1 \times 1000) = 98.62 \text{ kN} > 42.857 \text{ kN.}$$

Shear per bolt = 150/8 = 18.75 kN.

Thread is considered in shear plane.

Single shear strength $= V_{nsb} = \dfrac{f_u}{\sqrt{3}} (n_n A_{nb} + n_s A_{sb})/\gamma_{mb}$

$= (400/\sqrt{3})\ (1 \times 353)/(1.25 \times 1000) = 65.2$ kN.

Combined strength $= (18.75/65.2)^2 + (42.857/98.62)^2 = 0.27 < 1.0$

Hence, satisfactory.

Note:

1. Note that the tension is resisted by the neck area. The shear is taken by the shank area if the shear plane is not in threaded area.
2. In this case the neutral axis is considered to be at the centre line of bottommost bolts.

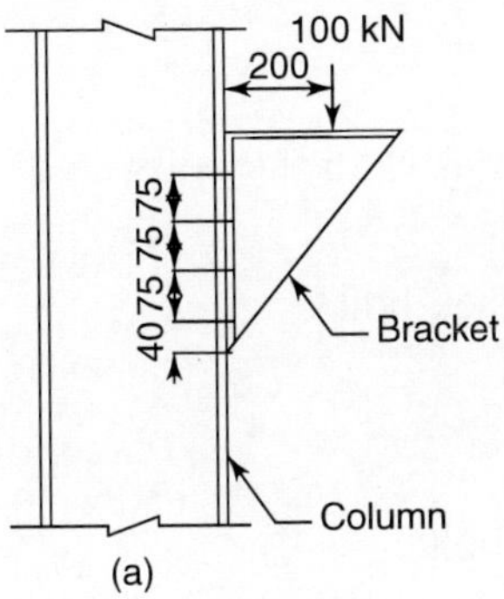

Fig. 8.33

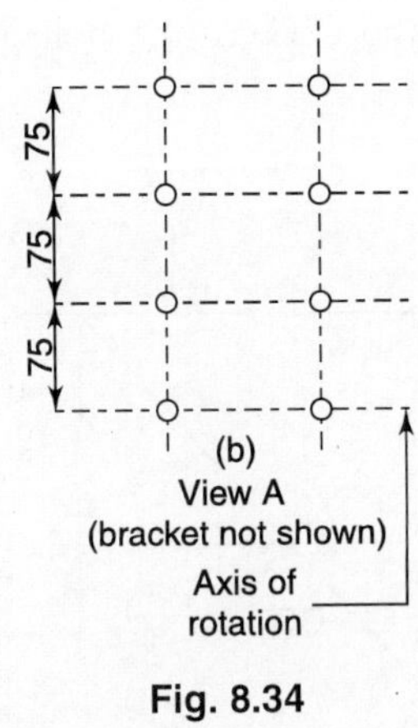

Fig. 8.34

8.28.4 Bolted Joints Subjected to Moment Acting in the Plane of the Joint

Figures 8.35 and 8.36 show the joints subjected to moment in the plane of the joint. In addition to shear stress due to direct load, shear stress due to moment is also developed.

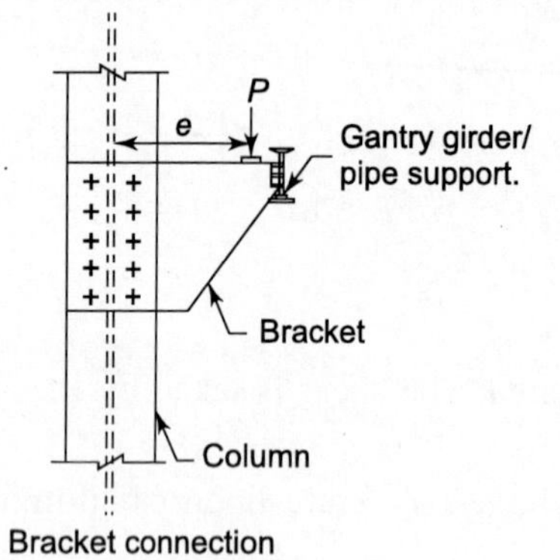

Fig. 8.35

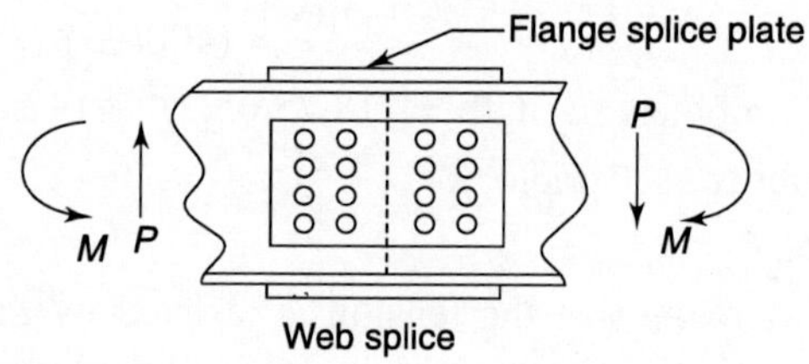

Fig. 8.36

Consider a bolt group shown in Fig. 8.36 the bolts are to resist (i) direct load P, and

(i) moment $M = P \times e$.

(ii) Force due to axial load on each bolt

$$F_a = \frac{P}{n} \tag{4}$$

where, n = total number of bolts.

(iii) Forces due to moment M on any bolt.

F_m is proportional to r if all bolts are of same diameter (since the farthest bolt will get more distortion)

where, r = distance of the bolt from the C.G. of the bolt group

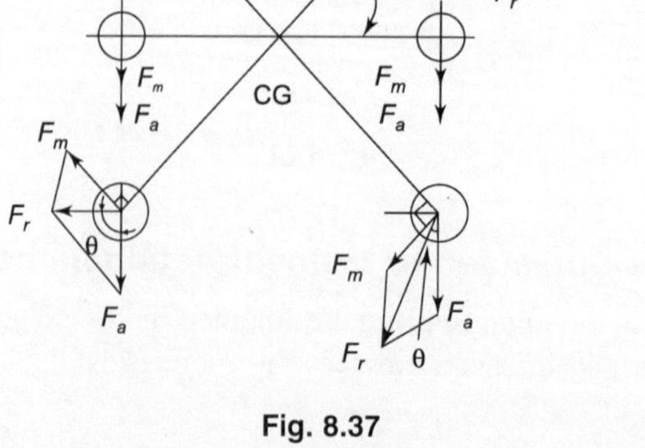

Fig. 8.37

$$\therefore \qquad F_m = k \times r$$

where k = a constant

Moment of force F_m about C.G. of bolt group = $F_m \times r = k \times r^2$

Algebraic sum of the moments = $\Sigma k \times r^2 = k \times \Sigma r^2$

$$= \frac{F_m \times \Sigma r^2}{r}$$

Equating this algebraic sum of moments to the external moment,

$$M = \frac{F_m \times \Sigma r^2}{r}$$

or

$$F_m = \frac{M \times r}{\Sigma r^2} \qquad (5)$$

The direction of F_m is perpendicular to the line joining the bolt under consideration and the C.G. of the bolt group.

The resultant force F_r acting on a bolt is the vectorial sum of the forces F_a due to direct load and F_m due to moment, that is

$$F_r = \sqrt{F_a^2 + F_m^2 + 2 \times F_a \times F_m \times \cos\theta} \qquad (6)$$

where θ = angle between the forces F_a and F_m acting on a bolt.

Example 8.16 *A bracket transmits a factored load of 80 kN at an eccentricity of 300 mm to a column through ten M16 bolts of property class 4.6 in two vertical rows 100 mm apart as shown in Fig. 8.38. The pitch of the bolts is 80 mm. Check the adequacy of the joint.*

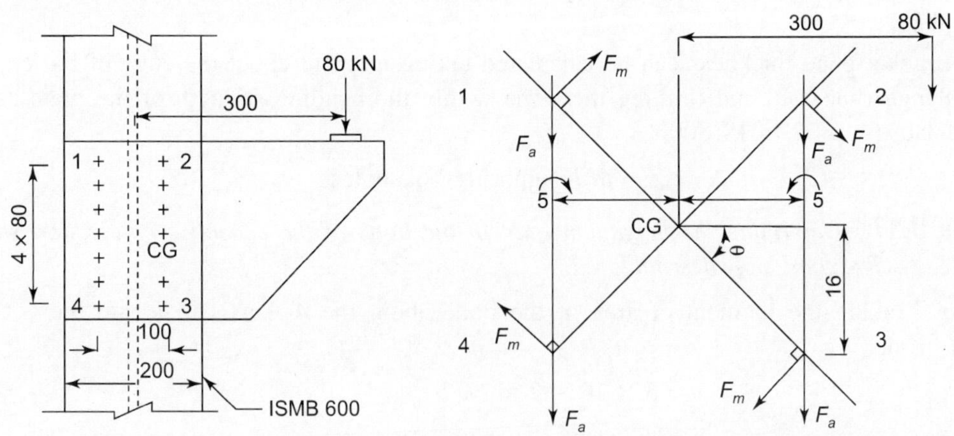

Fig. 8.38

Solution: *Moment lies in the plane of joint.*

The extreme bolts will have maximum shear due to moment. Let F_m is the maximum force. From the force diagram, for extreme bolts, 1, 2, 3 and 4, it is seen that the angle between F_m and F_a is the least for bolts 2 and 3. They will, therefore, have the maximum force.

$$\text{Shear due to direct load, } F_a = \frac{P}{n} = \frac{80}{10} = 8 \text{ kN}$$

$$\text{Shear due to moment, } F_m = \frac{M \times r}{\Sigma r^2} = \frac{(80 \times 30)\sqrt{16^2 + 5^2}}{2 \times 5^2 + 4(5^2 + 8^2) + 4\,(5^2 + 16^2)}$$

$$= 26.29 \text{ kN}$$

$$\text{Angle between } F_m \text{ and } F_a,\ \theta = \tan^{-1}\frac{16}{5} = 72.646°$$

$$\text{Resultant force, } F_r = \sqrt{F_a^{\,2} + F_m^{\,2} + 2 \times F_a \times F_m \times \cos\theta}$$

$$= \sqrt{8^2 + 26.29^2 + 2 \times 8 \times 26.29 \times \cos 72.646°} = 29.675 \text{ kN}$$

The design strength of M16 bolts of property class 4.6 is 37.16 kN (Table 8.8(b) of this book). The resultant force is less than this.

The joint is adequate.

Note:

(1) To decide the diameter of bolts refer steel table which gives the maximum diameter of bolts that can be used for each rolled section.

(2) To find the number of bolts use equation (2 or 3).

(3) Based on the number of bolts, pitch and edge distance decide the depth of bracket plate.

(4) Bracket plate thickness can be calculated taking moment about the edge of the column flange (line 2-3) and limiting the same within the bending strength of the plate as per Clause 8.2.1.2 of IS: 800.

$$Z = td^2/6 \text{ ignoring bolt holes.}$$

Example 8.17 *Determine the maximum load in the bolts of the eccentric connection shown in Fig. 8.39. The load shown is factored.*

Solution: Taking the moment of area of the bolts about the *A-A* axis to locate the C.G. of the bolt group,

$$8 \times \bar{x} = 3 \times 10 + 2 \times 20$$

or

$$\bar{x} = 8.75 \text{ cm}$$

From Fig. 8.39 it is seen that the right-hand top and bottom corner bolts 3 and 8 have the maximum load since their distance from the centre of gravity of the bolt group is maximum.

Thus, for extreme bolts 3 and 8

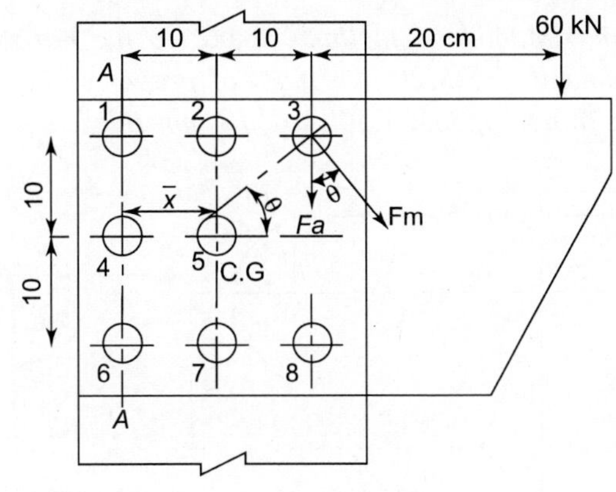

Fig. 8.39

(i) Force due to direct load,

$$F_a = \frac{W}{n} = \frac{60}{8} = 7.5 \text{ kN}$$

(ii) Force due to moment,

$$F_m = \frac{M \times r}{\Sigma r^2}$$

where $M = W \times e = 60 \times (40 - 8.75) = 1875$ kN.cm

$$r = \sqrt{10^2 + 11.25^2} = 15.05 \text{ cm}$$

$$\Sigma r^2 = [2 \times (10^2 + 11.25^2)] + [2 \times (10^2 + 1.25^2)] + (1.25^2) + [2 \times (10^2 + 8.75^2)] + (8.75^2)$$

$$= 1087.5 \text{ cm}^2$$

$$\therefore \quad F_m = \frac{1875 \times 15.05}{1087.5} = 25.95 \text{ kN}$$

The angle between F_m and F_a,

$$\theta = \tan^{-1}\frac{10}{11.25} = 41.63°$$

∴ Resultant force in the right-extreme bolt,

$$F_r = \sqrt{F_a^2 + F_m^2 + 2 \times F_a \times F_m \times \cos\theta}$$

$$= \sqrt{7.5^2 + 25.95^2 + 2 \times 7.5 \times 25.95 \times \cos 41.63°}$$

$$= 31.95 \text{ kN}$$

The design shear strength in single shear for M16 bolts of property class 4.6 is 37.16 kN (Table 8.8(b) of this book). Hence, provide M16 Class 4.6 bolts.

Example 8.18 *Find the maximum load inclined at 60° to the horizontal, which the bracket shown in Fig. 8.40 can transmit if 5 M20 mm bolts of property class 4.6 bolts are used and the plates connected to the flange of ISMB 500 are 10 mm thick.*

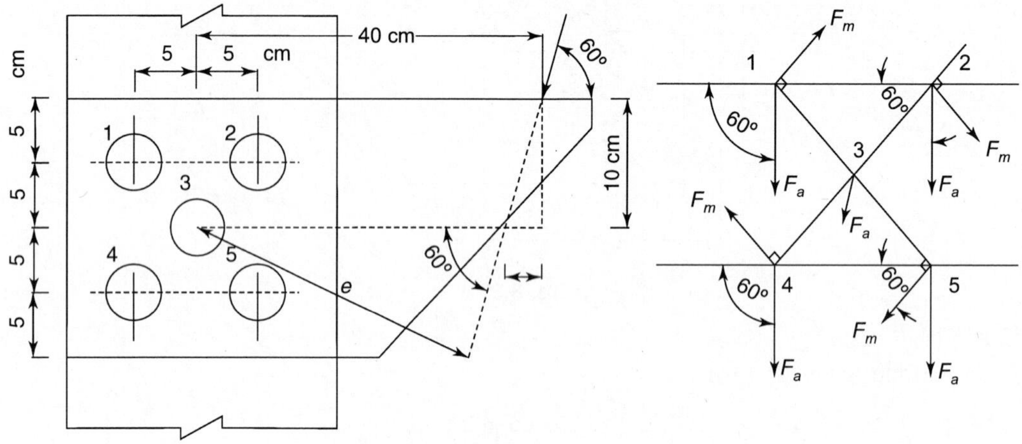

Fig. 8.40

Solution: Eccentricity $e = \left(40 - \dfrac{10}{\sqrt{3}}\right) \sin 60° = 29.64$ cm

From the force diagram it is seen that the angle between F_m and F_a is minimum for the bolt 5. Therefore, bolt number 5 has the maximum force.

Force on bolt 5,

(i) Due t_p direct load, $F_a = \dfrac{W}{5}$

(ii) Due to moment, $F_m = \dfrac{M \times r}{\Sigma r^2} = \dfrac{W \times 29.64 \times \sqrt{5^2 + 5^2}}{4 \times (5^2 + 5^2)} = 1.048$ W

The angle between F_m and F_a, = 15°

Resultant force on the bolt,

$$F_r = \sqrt{F_a^2 + F_m^2 + 2 \times F_a \times F_m \times \cos \theta}$$

$$= \sqrt{\left(\frac{W}{5}\right)^2 + (1.048\ \text{W})^2 + 2 \times \frac{W}{5} \times 1.048\ \text{W} \cos 15°}$$

$$= W \times \sqrt{\frac{1}{25} + 1.048^2 + \frac{2.096}{5} \cos 15°}$$

$$= 1.242\ \text{W}$$

or $$F_r = 1.24\ \text{W}$$

Strength of M20-property class 4.6 bolts,

Strength in Bearing (Clause 10.3.4 of IS: 800)

Diameter of bolt 20 mm.

Diameter of bolt hole = 20 + 2 = 22 mm (Table 19 of IS: 800).

Minimum end/edge distance = 1.5 × 22 = 33 mm or 1.7 × 22 = 37.4 mm (Clause 10.2.4.2 of IS: 800). Provided is 50 mm. Adequate.

Minimum pitch = 2.5 × diameter of bolt = 2.5 × 20 = 50 mm (Clause 10.2.2 of IS: 800).

Provided pitch = 50 mm.

Therefore, k_b = least of

(5) $e/3d_o = 50/(3 \times 22) = 0.76$

(6) $(p/3d_o) \times 0.25 = (50/(3 \times 22)) \times 0.25 = 0.51$

(7) $f_{ub}/f_u = 400/410 = 0.975$

(8) 1.0

Therefore, least is 0.51 = k_b

Plate thickness of column is 10 mm. Flange thickness of beam is 10.8 mm.

Let us provide angle of thickness 10 mm.

$V_{dsb} = V_{npb}/\gamma_{mb} = 2.5\ k_b\ dt\ f_u/\gamma_{mb} = 2.5 \times 0.51 \times 20 \times 10 \times 410/(1.25 \times 1000) = 83.64$ kN.

Design strength in single shear = 58.06 kN.

Bolt value = 58.06 kN

$F_r = 1.24\ W = 58.06$ kN

$W = 58.06/1.24 = 46.82$ kN.

The maximum factored load inclined at 60° to the horizontal shall not be more than 46.82 kN.

Example 8.19 *Design the bolted connection for the bracket as shown.*

(a) With class 4.6 bolts – Bearing type

(b) With class 8.8 bolts – Bearing type

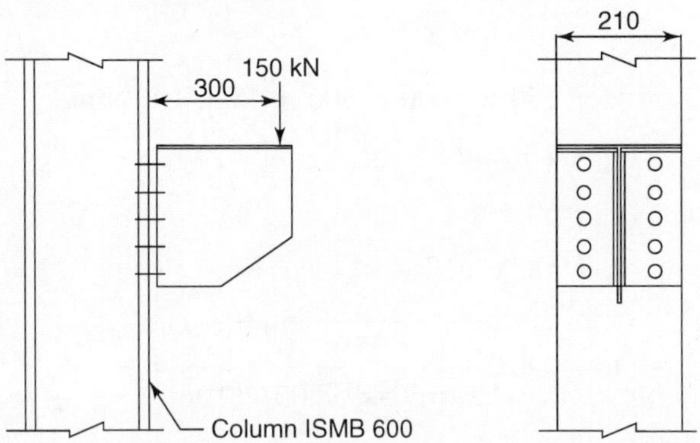

Fig. 8.41

Solution:

(a) Class 4.6 bolts

Using two rows of M20 bolts of property class 4.6 and cleat angles ISA 100 × 100 × 8 Design strength in single shear of M20 class 4.6 bolts = 58.06 kN.

Strength in Bearing (Clause 10.3.4 of IS:800)

Diameter of bolt 20 mm.

Diameter of bolt hole = 20 + 2 = 22 mm (Table 19 of IS:800).

Minimum end/edge distance = 1.5 × 22 = 33 mm (Clause 10.2.4.2 of IS:800).

For ISA 100 × 100 × 8 back mark is 60 mm (SP:6(1))

Hence, edge distance = (100 – 60) = 40 mm. Let us provide end distance also as 40 mm.

Minimum pitch = 2.5 × diameter of bolt = 2.5 × 20 = 50 mm (Clause 10.2.2 of IS:800).

Provide pitch = 60 mm.

Therefore k_b = least of

(9) $e/3d_o = 40/(3 \times 22) = 0.61$

(10) $(p/3d_o) - 0.25 = (60/(3 \times 22)) - 0.25 = 0.66$

(11) $f_{ub}/f_u = 400/410 = 0.975$

(12) 1.0

Therefore, least is 0.61 = k_b

Assuming the bracket plate thickness as 10 mm.

t = least of plate thickness or twice the thickness of angle since angle is there on both sides of the bracket plate.

$V_{dsb} = V_{\eta pb}/\gamma_{mb} = 2.5\ k_b d t f_u/\gamma_{mb} = 2.5 \times 0.61 \times 20 \times 10 \times 410/(1.25 \times 1000) = 100.04$ kN.

Bolts value = R = 58.06 kN.

$$\eta = 0.8\sqrt{\frac{6M}{mPR}}$$

$$M = 150 \times 1000 \times 300 = 45 \times 10^6 \text{ Nmm}$$

$$m = 2 \text{ rows}$$

$$P = 60 \text{ mm}$$

$$R = 58.06 \text{ kN}$$

$$\eta = 0.8\sqrt{\frac{6 \times 45 \times 10^6}{2 \times 60 \times 58.06 \times 10^3}} = 5$$

Provide 7 bolts

Length of cleat angle = 2 × 40 + 6 × 60 = 440 mm.

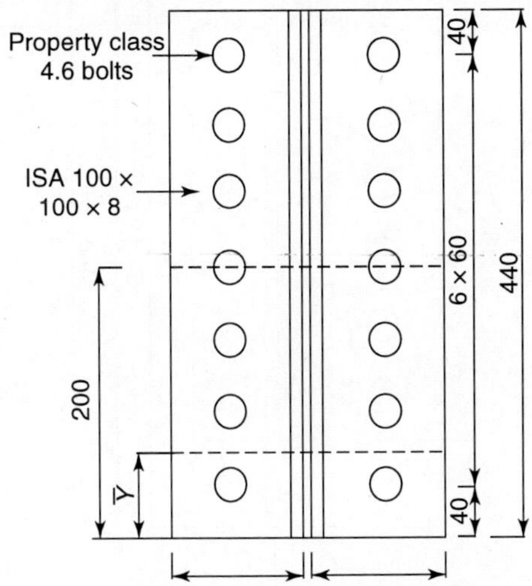

Fig. 8.42

Gross area of cross section of bolt $= \dfrac{\pi d^2}{4} = \dfrac{\pi \times 20^2}{4} = 314 \text{ mm}^2$

Approximate location of *NA* $= \dfrac{440}{7} = 63$ mm

To find exact location of *NA* (considering compression below *NA* and tension above *NA*)

$$2 \times 100 \times \bar{y} \times \frac{\bar{y}}{2} - 2 \times 314 \times (\bar{y} - 40)$$

$$= 2 \times 6 \times 314 \times (250 - \bar{y})$$

$$100 \times \bar{y}^2 - 628\,\bar{y} + 25120 = 942000 - 3768\,\bar{y}$$

$$100\,\bar{y}^2 + 3140\,\bar{y} - 916880 = 0$$

$$\bar{y}^2 + 31.4\,\bar{y} - 9168.8 = 0$$

$$\bar{y} = \frac{-31.4 \pm \sqrt{31.4^2 + 4 \times 9168.82}}{2} = \frac{-31.4 \pm 194.06}{2} = 81.3 \text{ or } -112.73 \text{ mm.}$$

$\bar{y} = 81.3$ mm.

Moment of inertia about *NA*

$$I = \left(\frac{2 \times 100 \times 81.33^3}{3} - 2 \times 314 \times (81.3 - 40)^2\right) + [2 \times 314\,(18.7^2 + 78.7^2 + 138.7^2 + 198.7^2 + 258.7^2 + 318.7^2]$$

$= 3586.4 \times 10^4 - 107.12 \times 10^4 + 21058.59 \times 10^4 = 24537.9 \times 10^4 \text{ mm}^4$

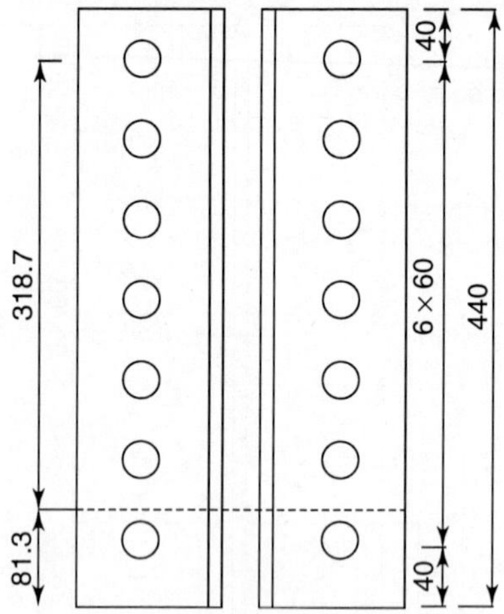

Fig. 8.43

From C.G. to topmost is at = 440 – 81.3 – 40 = 318.7 mm.

Tensile stress in the topmost bolt $\frac{M}{I} \times y$

$$\frac{45 \times 10^6}{24537.9 \times 10^4} \times 318.7 = 58.45 \text{ MPa}$$

Net area of M20 bolt = 245 mm^2

T_b = (58.45 × 245)/1000 = 14.32 kN

Shear per bolt = 150/14 = 10.7 kN.

Design strength in tension = 68.57 kN (Table 8.8(b) of this book).

Design strength in shear = 58.06 kN (Table 8.8(b) of this book)

Combined strength

$$\frac{14.32}{68.57} + \frac{10.7}{58.06} = 0.39 < 1.0$$

Hence, adequate.

(b) Class 8.8 bolts

Design strength in single shear of M20 Class 8.8 bolts = 116.13 kN.

Design strength in bearing = 100.04 kN. (Clause 10.3.4 of IS: 800)

Bolt value = 100.04 kN.

$$\eta = 0.8 \times \sqrt{\frac{6M}{mPR}} = 0.8 \times \sqrt{\frac{6 \times 45 \times 10^6}{2 \times 60 \times 100.04 \times 10^3}} = 3.8$$

Provide 4 bolts. Edge and end distance = 40 mm and pitch = 60 mm.

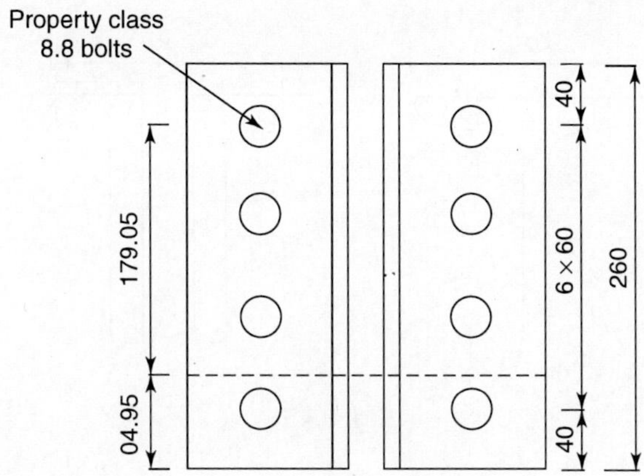

Fig. 8.44

Length of angle = 2 × 40 + 3 × 60 = 260 mm.

$$\bar{y} \text{ (approximate)} = \frac{260}{7} = 37.14 \text{ mm}$$

$$2 \times 100 \times \bar{y} \times \frac{\bar{y}}{2} = 2 \times 3 \times 314\ (130 - \bar{y})$$

$$100\ \bar{y}^2 = 244920 - 1884\ \bar{y}$$

$$100\ \bar{y}^2 + 1884\ \bar{y} - 244920 = 0$$

$$\bar{y}^2 + 18.84\ \bar{y} - 2449.2 = 0$$

$$y = -18.84 \pm \sqrt{\frac{18.84^2 + 4 \times 2449.2}{2}} = \frac{-18.84 \pm 100.76}{2} = 40.95 \text{ mm}$$

$$I = \frac{2 \times 100 \times 40.95^3}{3} - 2 \times 314 \times (40.95 - 40)^2 + 2 \times 314 \times \{(0.95^2 + 60.95^2 + 120.95^2 + 180.95^2)\}$$

$$= 457.8 \times 10^4 - 0.0566677 \times 10^4 + 3208.3 \times 10^4 = 3666 \times 10^4 \text{ mm}^4$$

From C.G. topmost bolt is at 260 – 40.95 – 40 = 179.05 mm.

$$\text{Tensile stress at topmost bolt} = \frac{M}{I} \times y = \frac{45 \times 10^6 \times 179.05}{3666 \times 10^4} = 219.8 \text{ MPa}$$

Actual tension in bolt = T_b = (219.8 × 245)/1000 = 53.85 kN.

$$\text{Shear/bolt} = \frac{150}{2 \times 4} = 18.75 \text{ kN.}$$

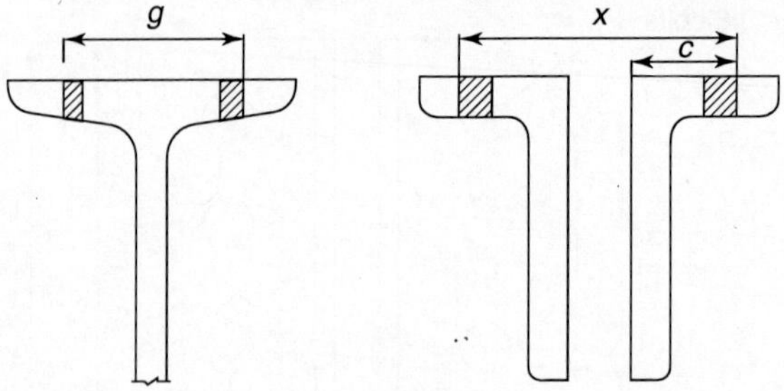

Fig. 8.45

Interaction formula, $\frac{53.85}{141.12} + \frac{18.75}{116.13} = 0.54 < 1.0$

Hence, satisfactory.

The value of g for ISMB 600 is 140 mm for M20. Using ISA 100 × 100 × 8; back mark = 60 mm = C (SP: 6(1))

$$x = 2 \times 60 + 12 = 132 \text{ mm.}$$

Keeping at g; the back mark $C = \frac{100 - 12}{2} = 44$ mm.

Bolts will be at 44 mm instead of 60 mm.

Check for Angle

Design shear strength of angle = $V_d = \frac{A_v f_{yw}}{\gamma_{mo}\sqrt{3}}$ (Clause 8.4.1 of IS: 800)

$$= \frac{260 \times 8 \times 2 \times 250}{1.1 \times \sqrt{3} \times 1000} = 550 \text{ kN.}$$

Shear per angle = 150/2 = 75 kN < 0.6 × 550 kN. Low shear.

Moment capacity per angle = $M_d = 1.2Z_e f_y/\gamma_{mo}$ (Clause 8.2.1.2 of IS: 800)

$$= 1.2 \times (8 \times 260^2/6) \times 250/(1.1 \times 10^6) = 24.58 \text{ kNm.}$$

Moment capacity of two angles = 2 × 24.58 = 49.16 kNm > 45 kNm.

Design of Bracket Plate (Sec AA)

$$e = 240 \text{ mm } M = 150 \times 0.24 = 36 \text{ kNm.}$$

Moment capacity of plate = $M_d = 1.2Z_e f_y/\gamma_{mo}$ (Clause 8.2.1.2 of IS: 800)

$$= 1.2 \times Z_e \times 250/1.1 = 38.4 \times 10^6 \text{ Nmm.}$$

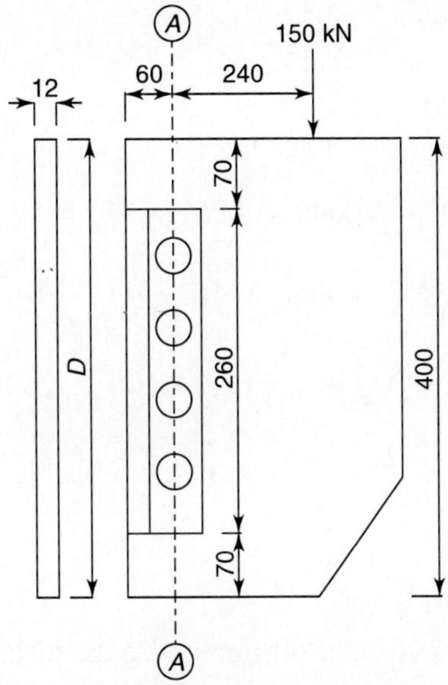

Fig. 8.46

$$Z_e = (36 \times 10^6) \times 1.1/1.2 \times 250 = 132 \times 10^3$$

$$Z_{\text{reqd}} = 140.8 \times 10^3 \text{ mm}^3 = \frac{tD^2}{6}$$

Providing 12 mm thick plate

$$\therefore \qquad D = \sqrt{\frac{6 \times 132 \times 10^3}{12}} = 256.9 \text{ mm}$$

Provide 400 mm.

$$\text{Design shear strength of plate} = V_d = \frac{A_v f_{yw}}{\gamma_{mo}\sqrt{3}} \text{ (Clause 8.4.1 of IS:800)}$$

$$= \frac{400 \times 12 \times 250}{1.1 \times \sqrt{3} \times 1000} = 629.8 \text{ kN.}$$

$$Z \text{ (ignoring holes)} = \frac{12 \times 400^2}{6} = 320000 \text{ mm}^3$$

Moment capacity of plate = $M_d = 1.2 \times 320000 \times 250/1.1 \times 10^6$ (Clause 8.2.1.2 of IS:800)

$$= 87.27 \text{ kNm} > 36 \text{ kNm.}$$

Considering Holes

$$I_{XX} = \frac{12 \times 400^3}{12} \left(\frac{12 \times 22^3}{12} + 12 \times 22\ (2 \times 30^2 + 2 \times 90^2) \right)$$

$$= 64 \times 10^6 - (10648 + 4.752 \times 10^6 = 59.24 \times 10^6 \text{ mm}^4$$

$$Z = \frac{59.24 \times 10^6}{200} = 296.2 \times 10^3 \text{ mm}^3$$

Moment capacity of plate = M_d = 1.2 × 296200 × 250/1.1 × 10^6 (Clause 8.2.1.2 of IS:800)

$$= 80.8 \text{ kNm} > 36 \text{ kNm}.$$

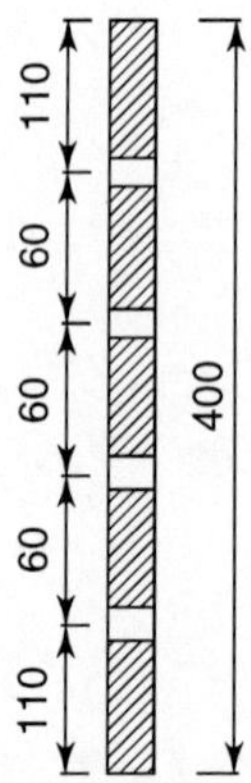

Fig. 8.47

Abstract

(a) Plate: 12 mm × 450 mm

Seven bolts of Class 4.6 at pitch 60 mm and edge distance 40 mm

Angles ISA 100 × 100 × 8 length = 440 mm

(b) Plate 12 mm × 400 mm.

Four bolts of Class 8.8 at P = 60 mm, edge distance = 40 mm

Angles ISA 100 × 100 × 8 length = 260 mm

8.29 BOLTED SPLICES

8.29.1 General Design Principle

(1) At the location of splice, continuity of the member about both axes has to be maintained.

(2) It is considered that the applied moment at the splice is resisted by the flange splice plates.

(3) It is considered that the applied shear force at the splice is resisted by the web splice plates.

(4) In beams, it is considered that the applied axial direct load in compression or tension is resisted equally by the flange splice plates in the case of beams. For columns it is considered that the direct load is shared between the flange and the web splice plates in proportion to their areas.

8.29.2 Flange Splice Plates

The flange forces are given by (Fig. 8.50 of this book)

$$F_f = \pm\ M_z/h + P/2$$

Let us say compressive forces are positive.

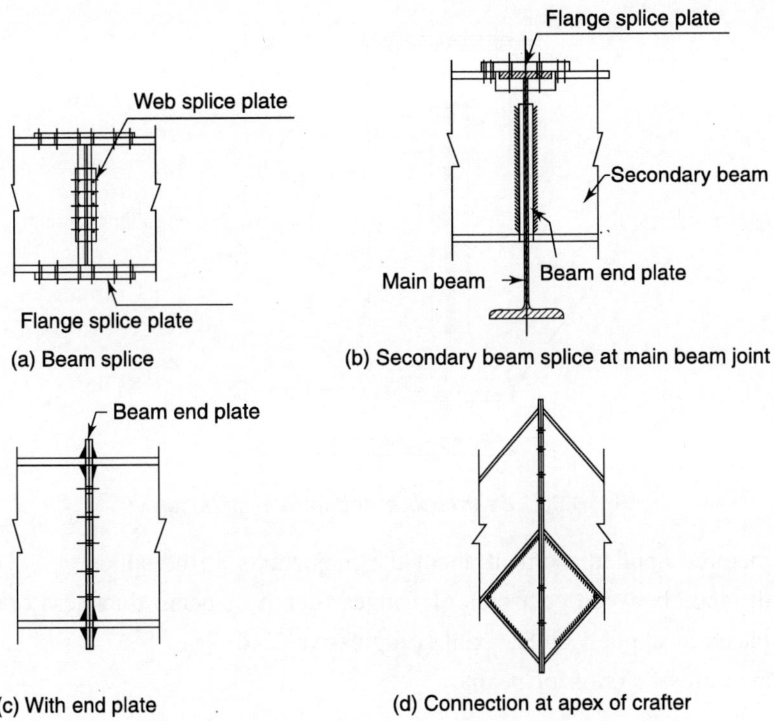

Fig. 8.48 *Details of splice joints for beams*

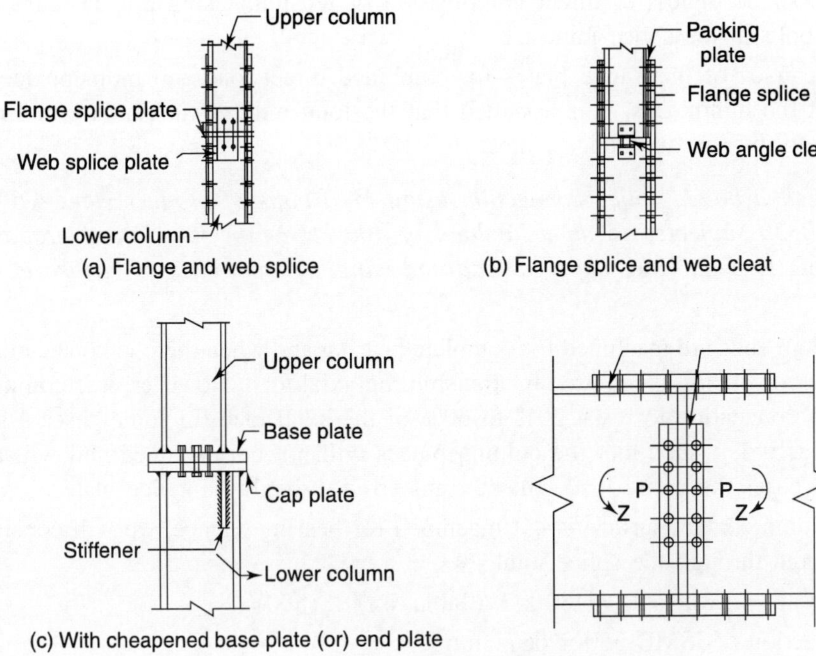

Fig. 8.49 *Splice joints for columns*

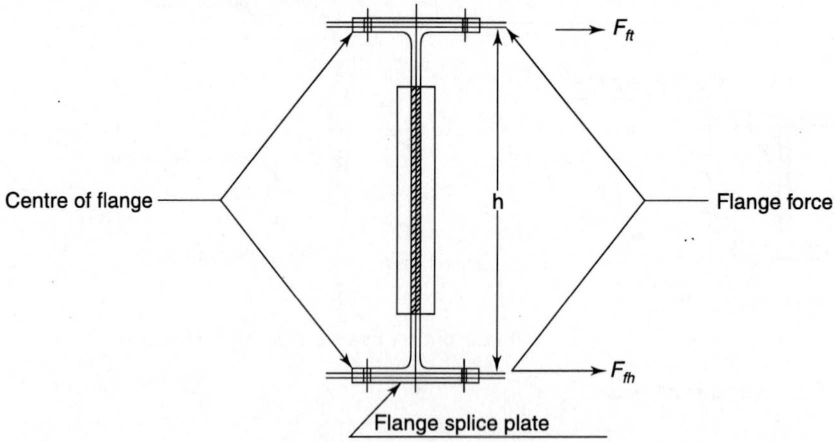

Fig. 8.50 *Beam splice and force distribution*

M_z = factored applied moment about the major axis at the splice

h = distance between centroids of flanges (depth of beam thickness of flange)

P = Factored applied direct axial compressive load

A = area cross section of beam

Design Principle

(1) Design flange splice plates for direct compression or tension, taking into account the holes for the bolts at the splice joint.

(2) Check the adequacy of the flange plates for combined direct load and moment due to bending about the minor axis. It is assumed that the total minor axis bending moment is shared equally by the two flange plates.

Example 8.20 *Design a bolted splice connection using M20 bolts of product class 4.6 for a column section MB350 subjected to an axial load of 400 kN at the location of proposed splice. The load is due to dead load, imposed load and wind. Column contact surface is not machined.*

Solution: If the column ends are machined for complete bearing and when there is no shear and moment, theoretically no splicing is required for transmitting axial load. However, it is common practice to design the connection to resist 20% to 60% of the axial load. In some cases a gap between column pieces is left such that the column pieces will not be machined and will not have contact. In such cases the entire load will be transferred through the splice plates.

In our case the column contact surface is not machined for bearing. Hence, we will consider 100% load is transferred through the splice joint.

Factored axial load = 1.2 × 400 = 480 kN. (Table 4 of IS: 800)

Area of cross section of ISMB 350 = 6671 mm^2

Area of web of ISMB 350 = (350 – 2 × 14.2) × 8.1 = 2605 mm^2

Area of flanges of ISMB 350 = 6671 – 2605 = 4066 mm^2

Load shared by web = (480/6671) × 2605 = 187.44 kN.

Load shared by flanges = 480 – 187.44 = 292.56 kN.

Web Splice

Clear depth of web of ISMB 350 = h_1 = 288 mm.

Providing 220 mm wide splice plate, and equating the area required,

$$2 \times 220 \times t = 2605 \text{ mm}^2$$

$$t = 2605/2 \times 220 = 5.92 \text{ mm.}$$

Providing 6 mm thick and 220 mm wide web splice plates,

Design strength = 2 × 6 × 220 × 250/(1.1 × 1000) = 600 kN > 187.44 kN.

The splice plates are bolted to the web of the column and the bolts will be in double shear.

Design strength of M20 bolts of property class 4.6 in double shear = 2 × 58.06 kN = 116.12 kN (Table 8.8(b) of this book).

Approximate number of bolts = 187.44/116.12 = 1.61

Let us provide two bolts.

Bearing Strength (Clause 10.3.4 of IS: 800)

Design strength in bearing for M20 of property class 4.6 bolt

= 16.4 × k_b × thickness of plate (Table 8.8(g) of this book).

Diameter of hole = 22 mm (Table 19 of IS: 800).

Minimum edge distance = 1.5 × 22 = 33 mm (or) 1.7 × 22 = 37.4 mm.

Providing 60 mm end and edge distance, pitch. 220 – 2 × 60 = 100 mm.

Minimum pitch = 2.5 × 20 = 50 mm.

k_b is the least of

(1) $e/3d_o$ = 60/(3 × 22) = 0.91

(2) $(p/3d_o) - 0.25$ = 100/(3 × 22) – 0.25 = 1.265

(3) f_{ub}/f_u

f_{ub} = 400 MPa and f_u = 410 MPa

Therefore, f_{ub}/f_u = 400/410 = 0.975

(4) 1.0

Hence, k_b = 0.91.

Thickness of two web splice plates put together is greater than that of the web.

Hence, t = thickness of web = 8.1 mm.

Bearing strength = $V_{dpb} = \left(\dfrac{2.5k_b dt f_u}{\gamma_{mb}}\right)$ (Clause 10.3.4 of IS: 800)

$= 16.4 \times 0.91 \times 8.1 = 120.9$ kN

Bolt value is the least of shear strength and bearing strength = 116.12 kN

Number of bolts provided is adequate.

Provide 2 M20 bolts of property class 4.6 as shown in figure.

Since the length between centre to centre of extreme bolts is not exceeding 15 times the diameter of bolt, there is no reduction in shear capacity as per Clause 10.3.3.1 of IS: 800.

Flange splice

Width of flange of ISMB 350 = 140 mm.

Providing 140 mm wide splice plate and equating the area required,

$$2 \times 140 \times t = 4066 \text{ mm}^2$$

$$t = 4066/2 \times 140 = 14.52 \text{ mm.}$$

Let us provide 16 mm thick and 140 mm wide flange splice plates.

Design strength = 2 × 16 × (140 – 2 × 22) × 250/(1.1 × 1000) = 698.2 kN > 292.56 kN.

The splice plates are bolted to the flange of the column and the bolts will be in single shear.

Design strength of M20 bolts of property class 4.6 in single shear = 58.06 kN (Table 8.8(b) of this book).

Approximate number of bolts = 292.56/2 × 58.06 = 2.6

Let us provide four bolts.

Bearing Strength (Clause 10.3.4 of IS: 800)

Design strength in bearing for M20 of property class 4.6 bolt

= 16.4 × k_b × thickness of plate (Table 8.8(g) of this book).

Diameter of hole = 22 mm (Table 19 of IS: 800).

Width of flange of ISMB350 = 140 mm.

Providing *c/c* of bolt hole at 65 mm, edge distance will be (140 – 65)/2 = 37.5 mm.

Minimum edge distance = 1.5 × 22 = 33 mm (or) 1.7 × 22 = 37.4 mm.

Providing 45 mm end distance.

Minimum pitch = 2.5 × 20 = 50 mm.

Provide 55 mm pitch.

k_b is the least of

(1) $e/3d_o = 45/(3 \times 22) = 0.682$

(2) $(p/3d_o) - 0.25 = 55/(3 \times 22) - 0.25 = 0.583$

(3) f_{ub}/f_u

$f_{ub} = 400$ MPa and $f_u = 410$ MPa

Therefore, $f_{ub}/f_u = 400/410 = 0.975$

(4) 1.0

Hence, $k_b = 0.583$.

Thickness of flange is 14.2 mm < 16 mm.

$$\text{Bearing strength} = V_{dpb} = \left(\frac{2.5k_b dtf_u}{\gamma_{mb}}\right) \text{ (Clause 10.3.4 of IS: 800)}$$

$$= 16.4 \times 0.563 \times 14.2 = 131.11 \text{ kN}$$

Bolt value is the least of shear strength and bearing strength = 58.06 kN

Number of bolts calculated is adequate.

Since the length between centre to centre of extreme bolts is not exceeding 15 times the diameter of bolt, there is no reduction in shear capacity as per Clause 10.3.3.1 of IS: 800.

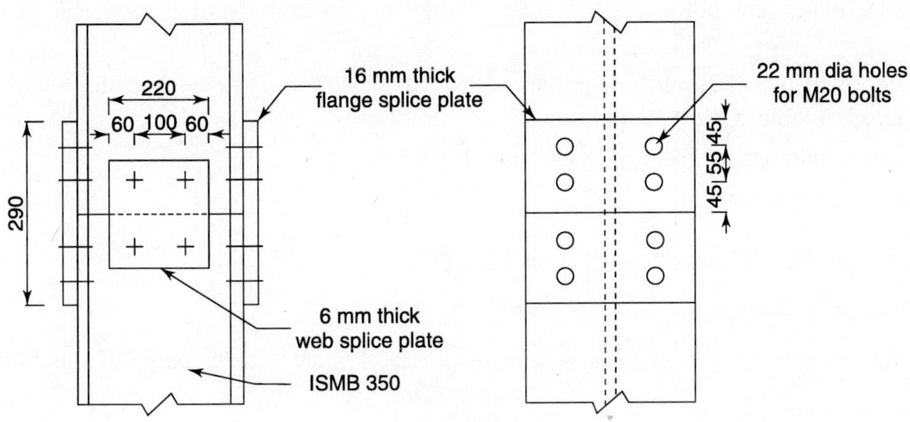

Fig. 8.51

Example 8.21 *Design a bolted splice connection using M20 bolts of product class 4.6 for a column section MB350 subjected to an axial load of 400 kN and a moment 8.20 kNm about its major axis at the location of proposed splice. The load and moment is due to dead load, imposed load and wind. Column contact surface is not machined.*

Solution:

Factored axial load = 1.2 × 400 = 480 kN. (Table 4 of IS: 800)

Factored moment = 1.2 × 8.2 = 9.84 kN.

Let us consider that the axial load is taken by web and flanges and moment is taken by flanges alone.

Axial load to be taken by web = 187.44 kN.

Axial load to be taken by each flange = 292.56/2 = 146.28 kN

Force on each splice due to moment = $\dfrac{\text{Moment}}{\text{Depth of section}} = \dfrac{9.84}{0.35} = 28.11$ kN

Maximum compression on each flange splice = 146.28 + 28.11 = 174.4 kN.

Minimum compression on each flange splice = 146.28 – 28.11 = 118.17 kN. There is no tension.

Web Splice

Clear depth of web of ISMB 350 = h_1 = 288 mm.

Providing 220 mm wide splice plate, and equating the area required,

$2 \times 220 \times t = 2605 \text{ mm}^2$

$t = 2605/2 \times 220 = 5.92$ mm.

Providing 6 mm thick and 220 mm wide web splice plates,

design strength = 2 × 6 × 220 × 250/(1.1 × 1000) = 600 kN > 187.44 kN.

The splice plates are bolted to the web of the column and the bolts will be in double shear.

Design strength of M20 bolts of property class 4.6 in double shear = 2 × 58.06 kN = 116.12 kN (Table 8.8(b) of this book).

Approximate number of bolts = 187.44/116.12 = 1.61

Let us provide two bolts.

Bearing Strength (Clause 10.3.4 of IS:800)

Design strength in bearing for M20 of property class 4.6 bolt

= 16.4 × k_b × thickness of plate (Table 8.8(g) of this book).

Diameter of hole = 22 mm (Table 19 of IS:800).

Minimum edge distance = 1.5 × 22 = 33 mm (or) 1.7 × 22 = 37.4 mm.

Providing 60 mm end and edge distance, pitch 220 – 2 × 60 = 100 mm.

Minimum pitch = 2.5 × 20 = 50 mm.

k_b is the least of

(1) $e/3d_o = 60/(3 \times 22) = 0.91$

(2) $(p/3d_o) - 0.25 = 100/(3 \times 22) - 0.25 = 1.265$

(3) f_{ub}/f_u

$$f_{ub} = 400 \text{ MPa and } f_u = 410 \text{ MPa}$$

Therefore, $f_{ub}/f_u = 400/410 = 0.975$

(4) 1.0

Hence, $k_b = 0.91$.

Thickness of two web splice plates put together is greater than that of the web.

Hence, t = thickness of web = 8.1 mm.

Bearing strength = $V_{dpb} = \left(\frac{2.5k_b dtf_u}{\gamma_{mb}}\right)$ (Clause 10.3.4 of IS:800)

$$= 16.4 \times 0.91 \times 8.1 = 120.9 \text{ kN}$$

Bolt value is the least of shear strength and bearing strength = 116.12 kN

Number of bolts provided is adequate.

Provide 2 M20 bolts of property class 4.6 as shown in Fig. 8.51.

Since the length between centre to centre of extreme bolts is not exceeding 15 times the diameter of bolt, there is no reduction in shear capacity as per Clause 10.3.3.1 of IS: 800.

Flange splice

Width of flange of ISMB 350 = 140 mm.

Providing 140 mm wide splice plate and equating the area required,

$$2 \times 140 \times t = 4066 \text{ mm}^2$$

$$t = 4066/2 \times 140 = 14.52 \text{ mm.}$$

Let us provide 16 mm thick and 140 mm wide flange splice plates.

Design strength of each flange splice plate

$$= 16 \times (140 - 2 \times 22) \times 250/(1.1 \times 1000) = 349.6 \text{ kN} > 174.4 \text{ kN.}$$

The splice plates are bolted to the flange of the column and the bolts will be in single shear.

Design strength of M20 bolts of property class 4.6 in single shear = 58.06 kN (Table 8.8(b) of this book).

Approximate number of bolts = 174.4/58.06 = 3

Let us provide four bolts.

Bearing Strength (Clause 10.3.4 of IS: 800)

Design strength in bearing for M20 of property class 4.6 bolt

$$= 16.4 \times k_b \times \text{thickness of plate (Table 8.8(g) of this book).}$$

Diameter of hole = 22 mm (Table 19 of IS: 800).

Width of flange of ISMB 350 = 140 mm.

Providing *c/c* of bolt hole at 65 mm, edge distance will be (140 – 65)/2 = 37.5 mm.

Minimum edge distance = 1.5 × 22 = 33 mm (or) 1.7 × 22 = 37.4 mm.

Providing 45 mm end distance.

Minimum pitch = 2.5 × 20 = 50 mm.

Provide 55 mm pitch.

k_b is the least of

(1) $e/3d_o = 45/(3 \times 22) = 0.682$

(2) $(p/3d_o) - 0.25 = 55/(3 \times 22) - 0.25 = 0.583$

(3) f_{ub}/f_u

$$f_{ub} = 400 \text{ MPa and } f_u = 410 \text{ MPa}$$

Therefore, $f_{ub}/f_u = 400/410 = 0.975$

(4) 1.0

Hence, $k_b = 0.583$.

Thickness of flange is 14.2 mm < 16 mm.

$$\text{Bearing strength} = \quad V_{dpb} = \left(\frac{2.5k_b\, dtf_u}{\gamma_{mb}}\right) \text{(Clause 10.3.4 of IS: 800)}$$

$$= 16.4 \times 0.563 \times 14.2 = 131.11 \text{ kN}$$

Bolt value is the least of shear strength and bearing strength = 58.06 kN

Number of bolts calculated is adequate.

Since the length between centre to centre of extreme bolts is not exceeding 15 times the diameter of bolt, there is no reduction in shear capacity as per Clause 10.3.3.1 of IS:800.

The splice connection detail is as in Fig. 8.51.

8.29.3 Design of Bolted Splice for Flexural Member Rolled Section

When the span, loading, support conditions and the location of splice are known; then we can calculate the exact bending moment, shear force at the location of splice and design the web and flange splice. In the absence of all the above data, the splice is designed for full capacity. Also refer to Clause 10.7(6) of IS:800 for minimum design action on connections for splicing of flexural members.

The splicing of flexural members can be done by providing flange and web splice plates using bolts. The flange splice is designed to take tensile or compressive force in the flange due to moment. The length of splice plate depends on the number of bolts, pitch and end/edge distances required to take the force in the flange. Splice on web is designed to take up the shear and moment due to eccentricity in web splice connection.

These splice plates when connected using bolts, the holes shall be taken into account while designing the splice connection and the member.

Example 8.22 *Design a splice connection for ISMB 400 beam having f_y = 250 MPa for its full capacity using splice plates and M20 bolts of property class 6.8. Draw a neat sketch showing the connection.*

Solution: Yield stress ratio = $\varepsilon = (250/f_y)^{0.5} = (250/250)^{0.5} = 1$

Shear capacity

$$V_d = \text{Design strength} = (A_v f_{yw})/(\gamma_{mo} \sqrt{3}) \text{ (Clause 8.4 of IS: 800)}$$

$$A_v = \text{Area of web} = 400 \times 8.9 = 3560 \text{ mm}^2$$

$$f_{yw} = 250 \text{ MPa}$$

$$\gamma_{mo} = 1.1 \text{ (Table 5 of IS:800)}$$

$$V_d = 3560 \times 250/(1.1 \times 1000 \times \sqrt{3}) = 467.12 \text{ kN.}$$

When the factored shear force is less than $0.6V_d$, the member will be under low shear. (Clause 8.2.1.2 of IS: 800)

$$0.6V_d = 0.6 \times 467.12 = 280.3 \text{ kN}$$

Moment Capacity

$$M_d = \beta_b Z_p f_y/\gamma_{mo}$$

$$b/t_f = (140/2)/16 = 70/16 = 4.37 < 9.4\varepsilon$$

Clear depth of web = $d = 400 - 2 \times (16 + 14) = 340$ mm.

$$d/t_w = 340/8.9 = 38.2 < 84\varepsilon$$

Section is plastic (Table 2 of IS:800)

$\beta_b = 1.0$ for plastic section. (Clause 8.2.1.2 of IS: 800)

Plastic section modulus = $Z_p = 1176.18 \times 10^3 \text{ mm}^3$ (Table 46 of IS:800)

$$M_d = \beta_b Z_p f_y/\gamma_{mo} = (1.0 \times 1176.18 \times 10^3 \times 250)/(1.1 \times 10^6) = 267.3 \text{ kNm}$$

This shall be less than $1.2\, Z_e f_y/\gamma_{mo}$ for simply supported beam. (Clause 8.2.1.2 of IS:800)

Elastic section modulus = $Z_y = 1020 \times 10^3 \text{ mm}^3$ (Table 46 of IS: 800).

$$1.2\, Z_e f_y/\gamma_{mo} = (1.2 \times 1020 \times 10^3 \times 250)/(1.1 \times 10^6) = 278.2 \text{ kNm}$$

Moment capacity is 267.3 kNm.

Design Data

Moment capacity or design bending strength M_d = 267.3 kNm

Design shear strength = 280.3 kN

Design of Web Splice

Shear area required = A_v = Design shear strength × $(\gamma_{mo} \sqrt{3})/(f_{yw})$

$$= 280.3 \times 10^3 \times 1.1 \times \sqrt{3}/250 = 2136 \text{ mm}^2$$

Clear depth of web = $d = 400 - 2 \times (16 + 14) = 340$ mm.

The depth of web splice plate shall not be more than this.

Depth higher than this will hit against the root of flange.

Let us provide 150 mm depth of web splice.

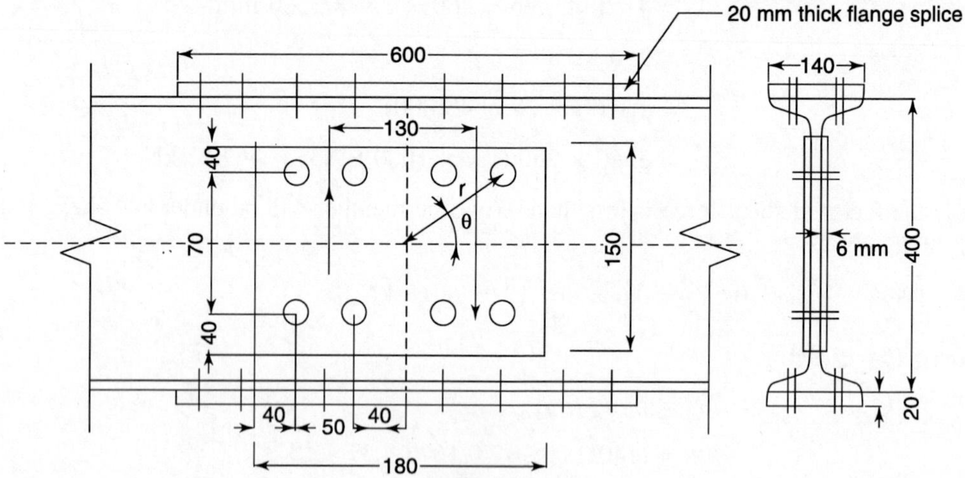

Fig. 8.52

Web splice plate will be provided on both sides of web.

Thickness of each plate required = 2136/(2 × 150) = 7.12 mm

This shall not be less than 50% of thickness of web and 6 mm whichever is more.

Provide 150 × 8 mm splice plates.

The splice plates are bolted to the web of the column and the bolts will be in double shear.

Design strength of M20 bolts of property class 6.8 in double shear

$$= 2 \times 87.1 = 174.2 \text{ kN (Table 8.8(b) of this book).}$$

Approximate number of bolts = 280.3/174.2 = 1.61

Let us provide four bolts on each side.

This itself is adequate as per calculation for taking shear. But there is an eccentricity caused by centre of gravity of bolt group. Due to this there will be moment.

Minimum end/edge distance required = 1.5 × 22 = 33 mm or 1.7 × 22 = 37.4 mm.

Minimum pitch = 2.5 × 20 = 50 mm.

Providing 50 mm pitch and 40 mm end distance, the size of splice plate

$$= 4 \times 40 + 2 \times 50 = 180 \text{ mm.}$$

C.G. of Bolt Group from the Centre of Joint

Since there are two rows and two bolts in each row, eccentricity = 2 × 40 + 2 × (50/2) = 130 mm.

Moment due to this eccentricity = (280.3/2) × (130/1000) = 18.22 kNm

Maximum shear in the extreme bolt due to bending = $M \times r/\Sigma r^2$

r is the distance of extreme bolt from centre of gravity of web splice.

$$r = \sqrt{(90^2 + 35^2} = 96.6 \text{ mm}$$

$$r_1 = \sqrt{(35^2 + 40^2} = 53.15 \text{ mm}$$

$$\Sigma r^2 = 4 \times (53.15^2 + 96.6^2) = 48626 \text{ mm}^2$$

Shear due to moment = $M \times r/\Sigma r^2$ = (18.22 × 10^6 × 96.6)/(48626 × 1000) = 35.2 kN.

Direct shear = 280.3/4 = 70.08 kN.

$$\tan \varphi = 35/90$$

$$\varphi = 21.25$$

Combined shear

$$= \sqrt{(35.2^2 + 70.08^2 + 2 \times 35.2 \times 70.08 \times \cos 21.25°)} = 103.67 \text{ kN} < 174.2 \text{ kN}.$$

Bearing Strength (Clause 10.3.4 of IS: 800)

Design strength in bearing for M20 of property class 6.8 bolt

= 16.4 × k_b × thickness of plate (Table 8.8(f) of this book).

Diameter of hole = 22 mm (Table 19 of IS: 800).

Minimum edge distance = 1.5 × 22 = 33 mm (or) 1.7 × 22 = 37.4 mm.

Provided end and edge distance = 40 mm.

Pitch. 150 – 2 × 40 = 70 mm.

Minimum pitch = 2.5 × 20 = 50 mm.

k_b is the least of

(1) $e/3d_o$ = 40/(3 × 22) = 0.61

(2) $(p/3d_o) - 0.25$ = 70/(3 × 22) – 0.25 = 0.81

(3) f_{ub}/f_u

f_{ub} = 600 MPa and f_u = 410 MPa

Therefore, f_{ub}/f_u = 600/410 = 1.46

(4) 1.0

Hence, k_b = 0.61.

Thickness of two web splice plates put together is greater than that of the web.

Hence, t = thickness of web = 8.9 mm.

$$\text{Bearing strength} = V_{dpb} = \left(\frac{2.5 k_b d t f_u}{\gamma_{mb}}\right) \text{ (Clause 10.3.4 of IS: 800)}$$

$$= 16.4 \times 0.61 \times 8.9 = 89.04 \text{ kN}$$

Bolt value is the least of shear strength and bearing strength = 89.04 kN

Design shear = 280.3 kN

Number of bolts = 280.3/89.04 = 3.14

Provide four M20 bolts of property class 6.8 as shown in Fig. 8.52.

Since the length between centre to centre of extreme bolts is not exceeding 15 times the diameter of bolt, there is no reduction in shear capacity as per Clause 10.3.3.1 of IS:800.

Design of Flange Splice

Moment capacity = 267.3 kNm

Lever arm = centre to centre of flange = 400 – 16 = 384 mm = 0.384 m.

Tensile/compressive force in flange = F = (267.3/0.384) = 696.1 kN

Width of flange = 140 mm

Providing the width of splice plate as 140 mm, thickness required

$$= (1.05 \times 140 \times 16)/140 = 16.8 \text{ mm}$$

Provide 20 mm thick splice plate.

Gross area of each flange splice plate = 140 × 20 = 2800 mm^2

This is greater than the area of flange which is = 140 × 16 = 2240 mm^2.

The tensile force per flange 696.1 kN should be taken by shear in the bolts.

Number of bolts = force/bolt value.

The splice plate is bolted to the flange of the column and the bolts will be in single shear.

Design strength of M20 bolts of property class 6.8 in single shear

= 87.1 kN (Table 8.8(b) of this book).

Approximate number of bolts = 696.1/87.1 = 7.99

Let us provide eight bolts on each side of the joint on each flange.

Let e = 40 mm and p = 50 mm.

Bearing Strength (Clause 10.3.4 of IS:800)

Design strength in bearing for M20 of property class 6.8 bolt

= 16.4 × k_b × thickness of plate (Table 8.8(f) of this book).

Diameter of hole = 22 mm (Table 19 of IS:800).

Minimum edge distance = 1.5 × 22 = 33 mm (or) 1.7 × 22 = 37.4 mm.

Provided end and edge distance = 40 mm.

Minimum pitch = 2.5 × 20 = 50 mm

Let us provide a pitch distance iof 60 mm.

k_b is the least of

(1) $e/3d_o = 40/(3 \times 22) = 0.61$

(2) $(p/3d_o) - 0.25 = 60/(3 \times 22) - 0.25 = 0.66$

(3) f_{ub}/f_u

$$f_{ub} = 600 \text{ MPa and } f_u = 410 \text{ MPa}$$

Therefore, $f_{ub}/f_u = 600/410 = 1.46$

(4) 1.0

Hence, $k_b = 0.61$.

Thickness of flange is the least.

$$\text{Bearing strength} = V_{dpb} = \left(\frac{2.5k_b\, dtf_u}{\gamma_{mb}}\right) \text{ (Clause 10.3.4 of IS:800)}$$

$$= 16.4 \times 0.61 \times 16 = 160.1 \text{ kN}$$

Bolt value is the least of shear strength and bearing strength = 87.1 kN

Number of bolts = 696.1/87.1 = 7.99

Provide eight M20 bolts of property class 6.8 as shown in figure.

Since the length between centre to centre of extreme bolts is not exceeding 15 times the diameter of bolt, there is no reduction in shear capacity as per Clause 10.3.3.1 of IS: 800.

Tensile Strength of Flange Splice Plate

$$T_{dg} = A_g f_y/\gamma_{mo} \quad \text{(Clause 6.2 of IS: 800)}$$

$$= 2800 \times 250/(1.1 \times 1000) = 636.4 \text{ kN.}$$

$$T_{dn} = 0.9A_n\, f_u/\gamma_{m1} \quad \text{(Clause 6.3.1 of IS: 800)}$$

$$= 0.9 \times (2800 - 2 \times 22 \times 20) \times 410/(1.25 \times 1000) = 566.78 \text{ kN.}$$

Number of bolts required = 566.78/87.1 = 6.5.

Provided is eight. Hence, satisfactory.

But this joint can take up only 566.78 kN amounting to 566.78 × 400/1000 = 226.71 kNm.

Example 8.23 *Friction type joint*

Bottom tie having 2 × ISA 60 × 60 × 6 of a truss is subjected to a tension of 50 kN service load due to dead and imposed loads. Design a friction type joint using class 8.8 high tension bolts. Surface is clean mill scale.

Solution:

P_u = Factored axial force on tie = 1.5 × 50 kN = 75 kN (Table 4 of IS: 800).

Member = ISA 60 × 60 × 6

Edge distance available = 60 – 35 = 25 mm (SP: 6(1))

Maximum size bolt admissible based on edge distance = 25/1.5 = 16 mm

Let us use M16 bolts of property class 8.8.

Number of effective friction faces = n_c = 2

Nominal shear capacity = $\mu_f \eta_c K_h F_o$ (Clause 10.4.3 of IS: 800)

μ_f = slip factor = 0.33 = Coefficient of friction (Table 20 of IS: 800).

K_h = 1.0 for clearance holes.

$$F_o = A_{nb} \times f_o = A_{nb} \times 0.7 \times f_{ub} \qquad \text{(Clause 10.4.3 of IS: 800)}$$

$$A_{nb} = 157 \text{ mm}^2 \qquad \text{(Table 8.4 of this book)}$$

$$f_{ub} = 800 \text{ MPa}$$

$$F_o = 157 \times 0.7 \times 800/1000 = 87.92 \text{ kN.}$$

Nominal shear capacity = (0.33 × 2 × 1.0 × 87.92) = 58.03 kN.

Design capacity at service load = 58.03/1.1 = 52.75 kN.

Design capacity at ultimate load = 58.03/1.25 = 46.42 kN.

Number of bolts required = 75/52.75 = 1.42

Use 2 M16 bolts of property class 8.8

Minimum pitch = 2.5 d = 2.5 × 16 = 40 mm

End edge distance = 1.5 × d_o = 1.5 × 18 = 27 mm

Provide 25 mm edge distance (SP: 6(1)) and 50 mm pitch.

EXERCISE

1. Under what circumstances oversize holes are used?
2. Why slotted holes are used?
3. Show by sketch what is end distance and edge distance.
4. Explain (a) bolted simple connection (b) bolted moment connection.
5. How the bolt class is specified?
6. M24 × 14D – 8.8 stands for what?
7. Explain the terminology related to bolt with a sketch of a bolt.
8. What is meant by grip length of a bolt?
9. Explain the failure mechanisms of bearing type bolted joints with sketches.
10. Design a bearing type joint between an angle and gusset plate using class 6.8 bolts to transfer a tensile force of 150 kN. Size of angle is ISA 80 × 80 × 8. The load is due to dead load, imposed load and wind. Also calculate the strength of angle in block shear.
11. A beam is transferring a vertical shear of 180 kN due to dead and imposed loads. The beam is connected by cleats on web to flange of the column ISMB 600 rolled *I* section is used as column and the beam is ISMB 450. Using class 4.6 – M24 bolts, design the shear connection.
12. Using the design data of (8.11), design the connection with end plate using M20-class 6.8 bolts.
13. Write down the procedure for calculating the seating angle size for resting a beam.
14. Design bolted splice joint with plates on flanges and on both sides of web for full strength of a beam ISMB 450.

Chapter 9

Wind Load Calculation (As per IS: 875-Part-3)

9.1 INTRODUCTION

Air in motion is called wind. The area resisting wind is subjected to wind pressure. This induces force on structural members.

IS: 875-Part-3 gives maximum basic wind speed in m/sec based on peak gust velocity averaged over a short time interval of about 3 seconds duration. This is worked out for 50 years return period.

9.2 MULTIPLICATION FACTORS

The following multiplication factors to modify the basic wind speed (velocity) are provided in the code.

Values of k_1 are given in Table 1 of IS: 875-Part-3 based on function of the structure and basic wind speed. This is called probability factor (or) risk coefficient.

Values of k_2 are given in Table 2 of IS: 875-Part-3 which depends on terrain, height of the structure and size of structure. The terrain has four categories which depends on the exposure of the structure. This is to be considered based on site condition. The values of k_2 varies at various heights for the same building/structure. The factor k_2 is called terrain, height and structure size factor.

Values of k_3 is to be worked out as per Clause C_2 of Appendix C. If the slope of the ground is 0° to 3°; $k_3 = 1.0$. for higher values, $k_3 = 1 + C_s$. The value of C is given in Clause C_2 and s is given in Clause $C - 2.1$. The factor k_3 is called topography factor.

9.3 DESIGN WIND SPEED

Considering multification factors k_1, k_2 and k_3 the basic wind speed $V_{b,}$ the design wind speed V_z can be calculated as below.

$$V_z = V_b \times k_1 \times k_2 \times k_3 \text{ (Clause 5.3 of IS: 875-Part-3).}$$

The basic wind speed for any location can be obtained from map or Appendix A of IS: 875-Part-3. If the location of the structure under consideration falls in the boundary; we can take the highest basic wind speed of the adjacent locations of the boundary.

9.4 DESIGN WIND PRESSURE

From the design wind speed (V_z), we can calculate the design wind pressure as below.
$P_d = 0.6V_z^2$ in N/m^2 (Clause 5.4 of IS: 875-Part-3).

9.5 EXTERNAL AND INTERNAL PRESSURE COEFFICIENTS

Wind pressure is always perpendicular to the surface resisting the wind. The force due to wind on the cladding, roof and individual members are calculated by (the area resisting × pressure coefficient × design wind pressure).

Wind load on structural members

$F = (C_{pe} - C_{pi}) A \times P_d$ (Clause 6.2.1 of IS: 875-Part-3).

C_{pe} is the external pressure coefficient.

C_{pi} is the internal pressure coefficient.

The external pressure coefficient for rectangular clad buildings is given in Table 4 of IS: 875-Part-3.

The external pressure coefficients for pitched roof of rectangular clad buildings is given in Table 5 of IS: 875-Part-3.

C_{pi} - The internal pressure coefficient are given in Clause 6.2.3 of IS: 875-Part-3 depending on the percent opening of the wall area.

The + sign denotes the pressure is towards the surface resisting wind.

The – sign indicates that the wind pressure is away from the surface resisting wind. These signs are only to indicate the direction towards/away from the surface.

Number of tables giving pressure coefficients for various types of structures are given in IS: 875-Part-3.

As per Clause 6.3.2.1; Fig. 4 of IS: 875-Part-3 is applicable for flat roofs.

Force coefficients for individual structural members depending on the shape of the member and direction of wind are given in Table 26 of IS: 875-Part-3. If l/h or l/D of the member is equal to or less than 100; reduction factor as in Table 25 of IS: 875-Part-3 shall be applied.

9.6 DYNAMIC EFFECTS

Dynamic effects of closed structures with a height to minimum lateral dimension ratio of more than 5.0 (or) whose natural frequency in the first mode is less than 1.0 Hz shall be considered (Clause 7.1 of IS: 875-Part-3).

Example 9.1 explains the method of calculating wind load on a factory shed.

9.7 WIND LOAD CALCULATION FOR A FACTORY SHED

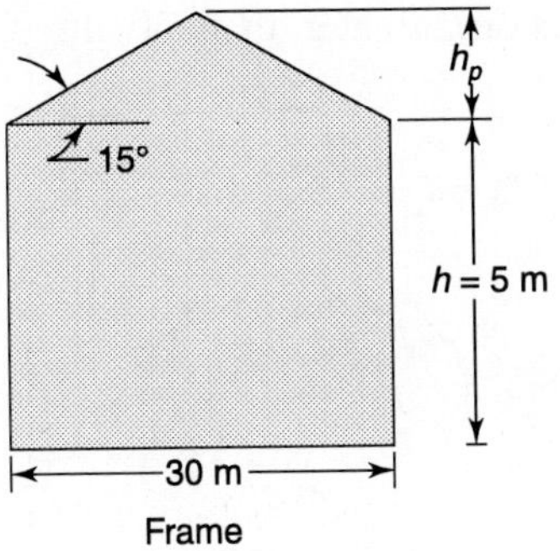

Fig. 9.1

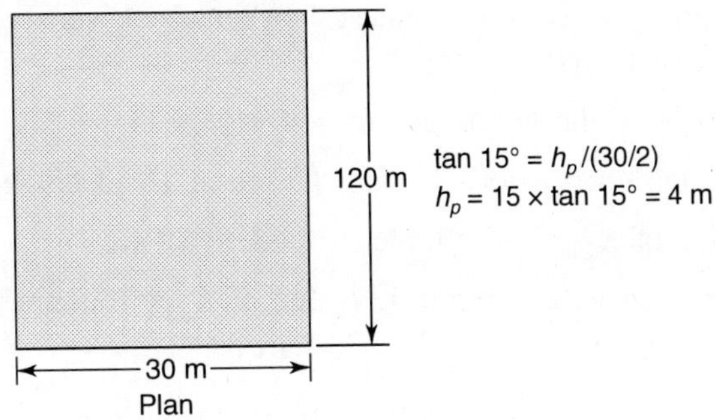

Fig. 9.2

Example 9.1 *Factory shed*

Calculate the wind load for the factory shed of dimension as in Figs. 9.1 and 9.2 based on the following design data.

Location: Trichy.

Slope of roof = 15° to horizontal.

Side covering – Cladding with louver for total height.

Opening: 20% of wall area.

Spacing of frame: 6 m.

Spacing of purlin 1.4 m.

Terrain category = 2.

Length of building = 120 m

Natural ground is almost flat

Solution:

Design pressure

Basic wind speed = 47 m/sec (Appendix-A of IS: 875-Part-3).

V_b = 47 m/sec.

K_1 = probability factor (risk coefficient)

This falls under "all general buildings and structures".

For V_b = 47 m/sec; K_1 = 1.0 (Table 1 of IS: 875-Part-3).

K_2 = terrain, height and structure size factor. (Clause 5.3.2 of IS:875-Part-3).

Category 2 (open terrain with well)

Scattered obstructions having heights generally between 1.5 and 10 m.

Class C (structures and/or their competent such as cladding, glazing. roofing, etc. having maximum dimension (greatest horizontal or vertical dimension) greater than 50 m. (Clause 5.3.2.2 of IS:875-Part-3).

Height of the building 5 m + 4 m = 9 m.

For terrain category 2, class C and at 10 m height,

value of K_2 = 0.93 (Table 2 of IS:875-Part-3).

For almost flat ground the value of K_3 = Topography factor = 1.0 (Clause.5.3.3.1 of IS:875-Part-3).

$$V_z = V_b\, K_1 K_2\, K_3 \text{(Clause 5.3 of IS:875-Part-3).}$$

$$= 47 \times 1 \times 0.93 \times 1 = 43.71 \text{ m/sec}$$

Design wind pressure = $P_z = 0.6\, V_z^2 = 0.6 \times (43.71)^2 = 1146$ N/m^2 (Clause 5.4 of IS:875-Part-3).

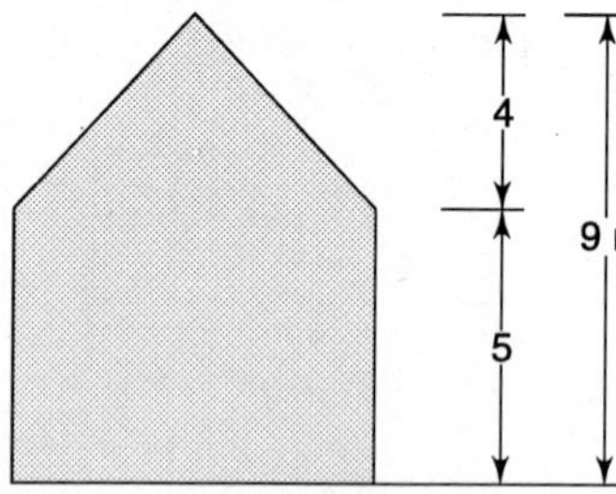

Fig. 9.3

9.7.1 Height to Width & Length to Width Ratio (Table 4 of IS:875-Part-3).

Building height ratio $(h/w) = (5/30) < (1/2)$ $(L/w) = (120/30) = 4$

Note: 'h' is the height between ground and eaves. (Ref. Fig. 9.4)

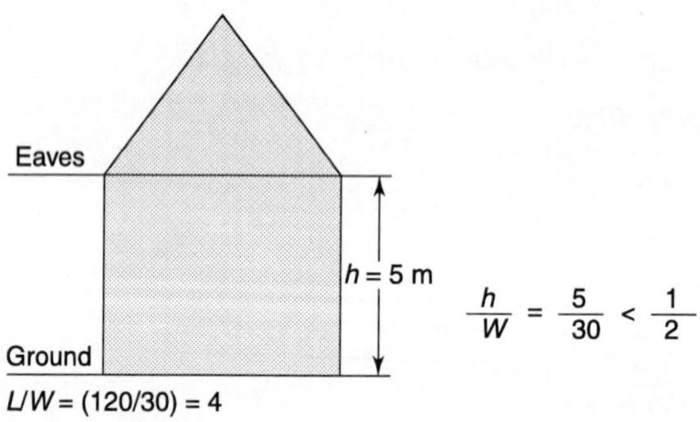

Fig. 9.4

9.7.2 External Pressure Coefficient C_{pe} for Cladding Table 4 of IS:875-Part-3

Since $(L/w) = 4$, we can take coefficient base on $(3/2) < (L/w) < 4$.

Wind angle 0°. Wind is perpendicular to surface A (or) B

Cp_e for surface A = + 0.7

Cp_e for surface B = – 0.25

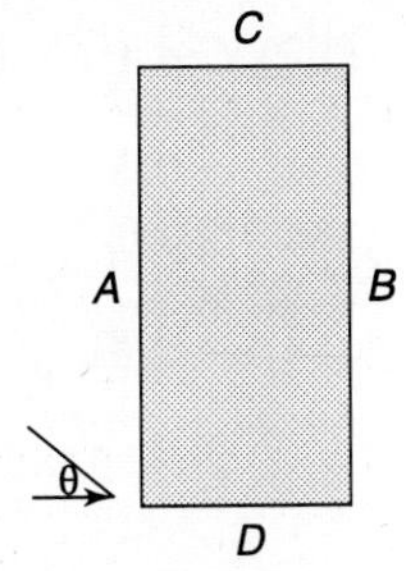

Fig. 9.5

Cp_e for surface C = – 0.6

Cp_e for surface D = – 0.6

Wind angle 90°. Wind is perpendicular to surface C (or) D

Cp_e for surface A = – 0.5

Cp_e for surface B = – 0.5

Cp_e for surface C = + 0.7

Cp_e for surface D = – 0.1

See Note 2 of Clause 6.2.2 of IS: 875-Part-3. Positive wind load indicates the force acting towards the structural elements and negative away from it.

External pressure coefficient (C_{pe})

Case 1 C_{pe} wind direction 0° **Case 2** Cp_e wind direction 90°

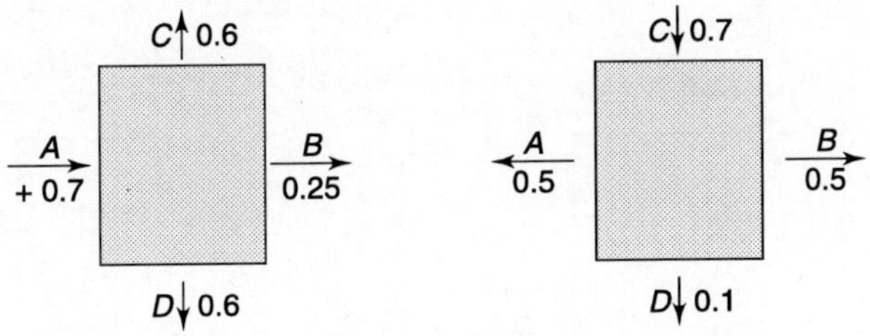

Fig. 9.6

Internal pressure coefficients C_{pi} (Clause 6.2.3 of IS: 875-Part-3).

Since opening is about 20% of wall area, C_{pi} = ± 0.5

Case 1 Wind angle 0°

(1.1) C_{pe} + (–ve C_{pi}) Net coefficient

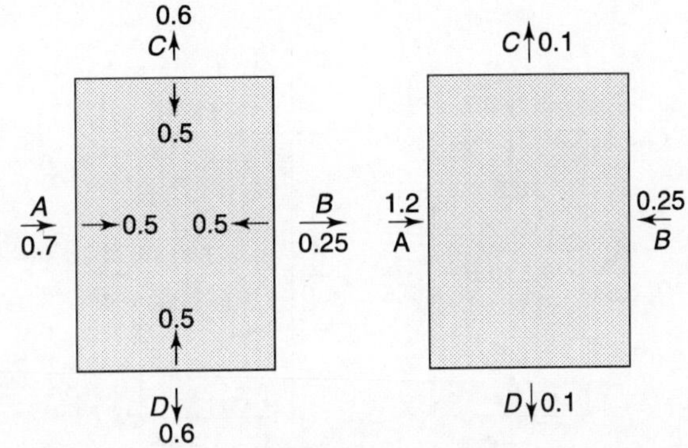

Fig. 9.7

(1.2) C_{pe} + (+ve C_{pi})

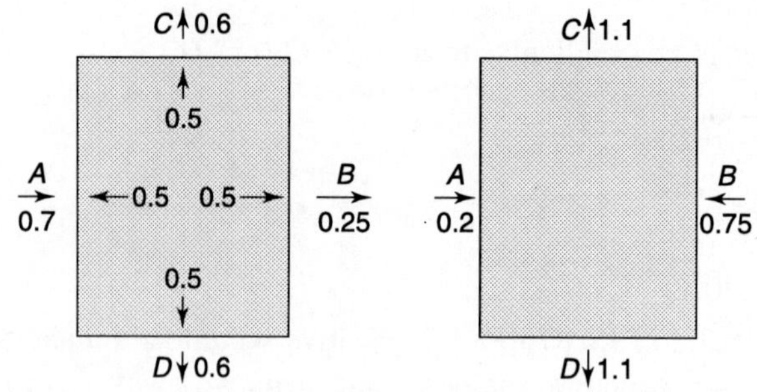

Fig. 9.8

Case 2 (Wind θ = 90°)

(2.1) C_{pe} + (– ve C_{pi})

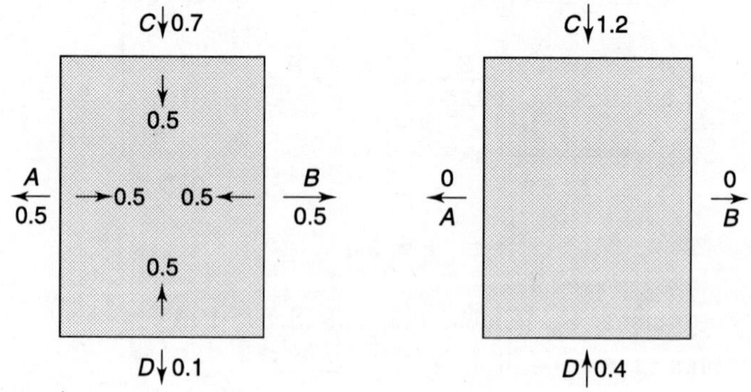

Fig. 9.9

(2.2) C_{pe} + (+ve C_{pi})

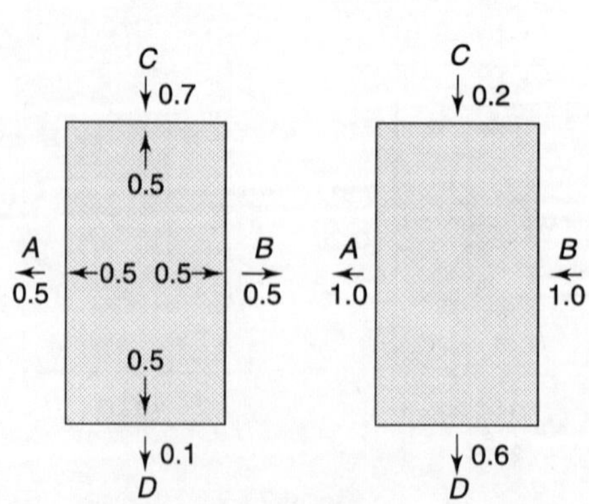

Fig. 9.10

The net coefficient is maximum on side *A* in case 1.1, side *B* is in case 2.2, Side *C* is in case 2.1 and side *D* is in case 1.2. But they are not simultaneously occurring. While analysing through computer package, all the four cases can be considered separately. For manual analysis, it is time consuming to analyse for all the four cases. Hence, one has to select a critical case. For gable columns, the wind force on face *C* or *D* is to be considered. For main columns in the longitudinal direction, wind perpendicular to face *A* or *B* whichever is more can be considered. Let us consider case 1.1 on face *A*. The maximum pressure coefficient for side *B* is 1.0 (case 2.2) which is less than the maximum for side *A*. Columns on side *A* and side *B* are to be of same size because the wind may blow perpendicular to sude *B* also. In such case, the pressure coefficient will reverse.

Wind force on columns from the cladding (other than corner columns) = (Net coefficient × P_d × spacing of column) = (1.2 × 1146 × 6/1000) = 8.25 kN/m.

Note:

1. The corner column will be subjected to forces from two adjacent sides. Consider corner col '*S*' (Fig. 9.11) during wind angle 0° and C_{pi} = +0.5.

   Force from side '*A*' = 6/2 × 1 × 1.2 × 1146 = 4125.6 N/m and from side '*C*' = 6/2 × 1 × 0.1 × 1146 = 343.8 N/m.

   They will act simultaneously which will produce biaxial bending of column.

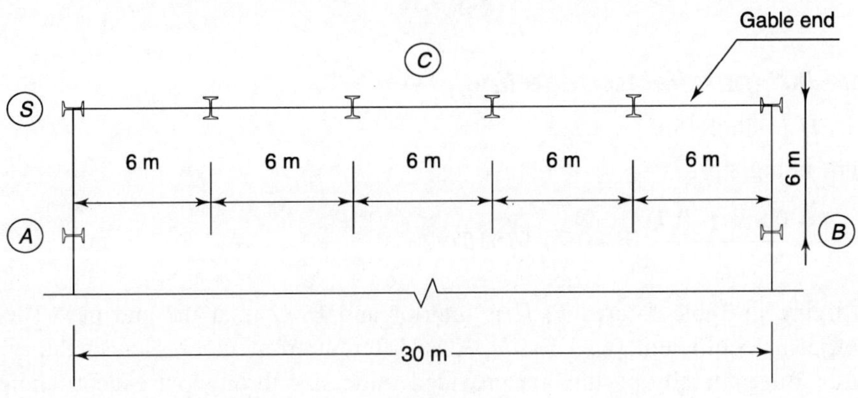

Fig. 9.11

9.7.3 Wind Load on Roof (Table 5 of IS:875-Part-3)

$$h/w = 5/30 < 1/2 \text{ and } \theta = 15°$$

Case 1 Wind angle 0° (perpendicular to ridge line)

Windward side	*Leeward side*
For 10° = –1.2	For 15° = – 0.4
For 20° = –0.4	

By interpolation; for 15° = −0.8

Case 1.1 $C_{pe} + (-C_{pi})$

$C_{pi} = -0.5$ Net coefficient

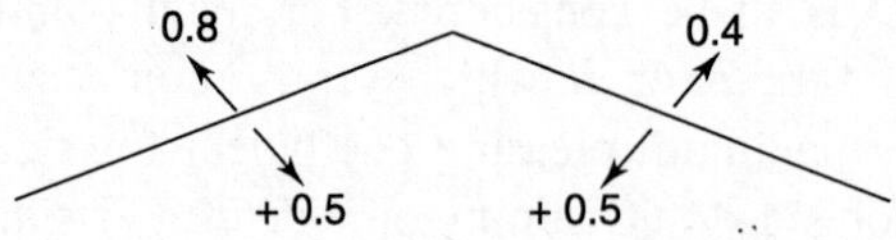

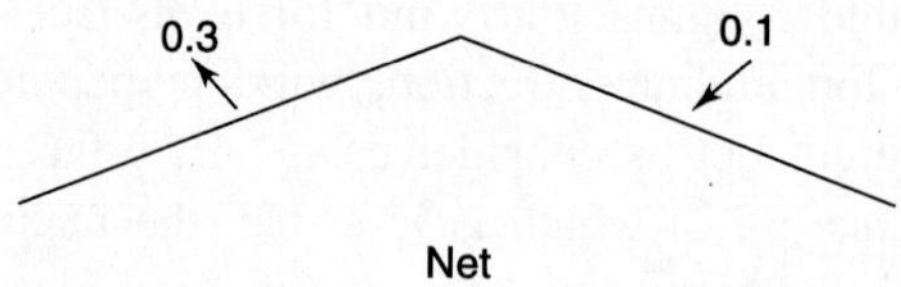

Fig. 9.12

Case 1.2 $C_{pe} + (+C_{pi})$

$C_{pi} = +0.5$

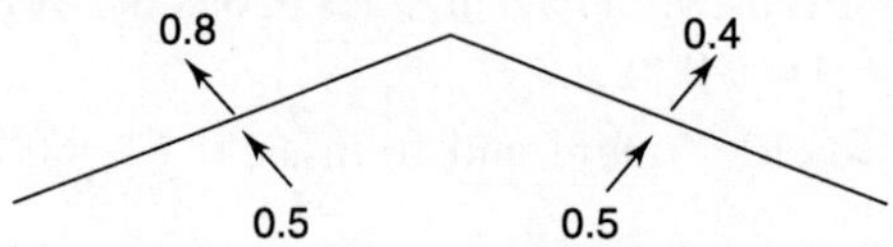

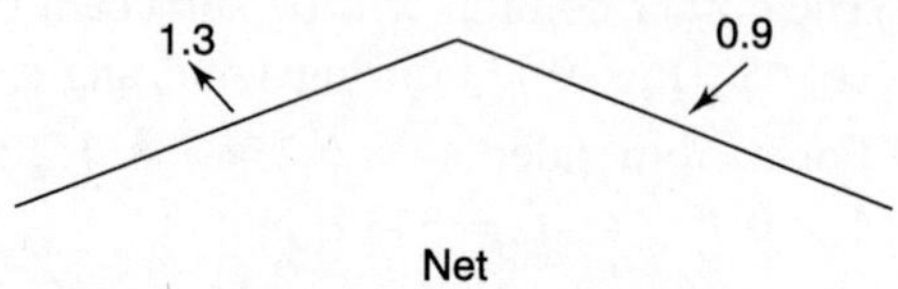

Fig. 9.13

Wind angle 90° (parallel to ridge line)

Refer *EG* & *FH* in Table 5.

EG coefficient is higher.

$$\frac{-0.8 + (-0.7)}{2} = -0.75 \text{ for } 15°$$

Note: From figure in Table 5; area *F*, *H* is interior and *E*, *G* is at the end bay. The external pressure coefficeint coefficient (C_{pe}) for *F*, *H* is different from *E*, *G*. Hence, the wind force will also differ. But generally purlins are provided same size throughout except where span of purlin, i.e., spacing of columns are different. Same way, the rafters are also provided. Hence, we can consider the maximum coefficient, i.e. for area *E*, *G*.

Case 2.1 $C_{pe} + (-C_{pi})$

$C_{pi} = -0.5$

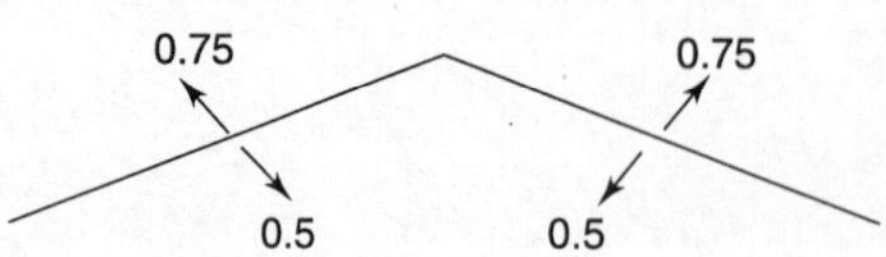

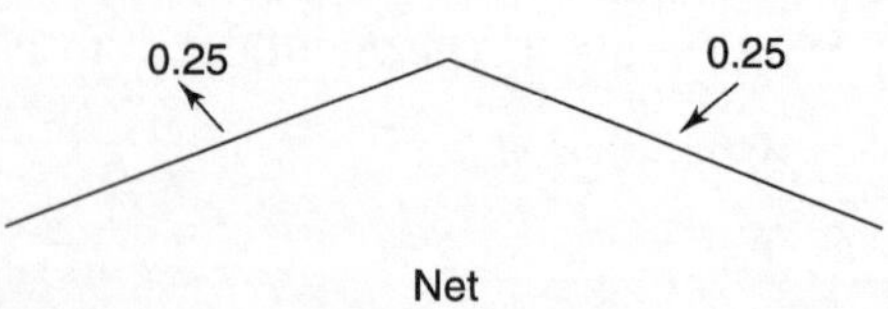

Fig. 9.14

Case 2.2 $C_{pe} + (+C_{pi})$

$C_{pi} = +0.5$

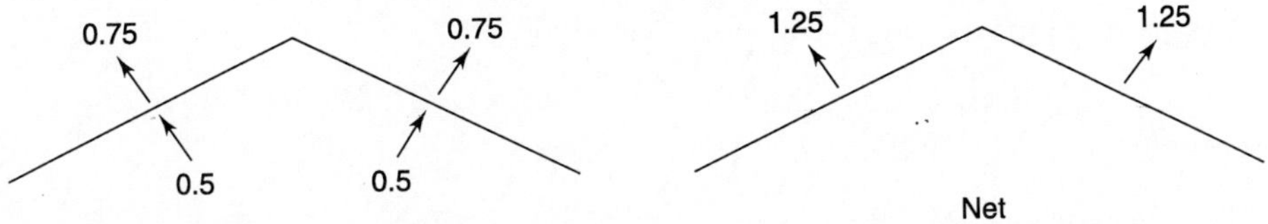

Fig. 9.15

Note the net coefficient diagrams.

Maximum net coefficient for uplift is 1.3 and that for downward is 0.1

The spacing of rafter is 6 m and $P_d = 1146$ N/m^2

(a) Uplift due to wind on rafter = 6 × 1.3 × 1146 = 8939 N/m

(b) Downward force due to wind on rafter = 6 × 0.1 × 1146 = 688 N/m.

(c) Force on purlin/*m* = (Net coefficient × P_d × spacing of purlin) N/m

$$= 1.3 \times 1146 \times 1.4 = 2086 \text{ N/m} = 2.09 \text{ kN/m}$$

Wind force per node is calculated by multiplying the uplift/downward intensity of pressure due to wind in N/m^2 by the spacing of node × spacing of the truss. The nodal force is applied perpendicular to the top rafter.

While considering wind load on cladding (column) and roof, same case has to be taken. This means that if the load on column is considered when $\theta = 0$ and C_{pi} = (–ve); then load on purlin/rafter also to be considered when $\theta = 0$ and C_{pi} = (–ve). There should not be any mix up.

Wind force on purlin due to local coefficient

Local coefficient is applicable for a distance of *y* or 0.15w whichever is less. In this case, 0.15w = 0.15 × 30 = 4.5 < *h*. maximum local coefficient for 15° slope = 1.2 (Table 5 of IS:875-Part-3). Adding C_{pi}, the coefficient = 1.7. Refer Table 5 of IS:875-Part-3.

Force/*m* on Purlin

Spacing of purlin 1.4 m.

Force = 1.4 × 1 × 1.7 × 1146 = 2727.48 N/m say 2.73 kN/m. The local coefficient is applicable at ridge and eaves. Normally, the spacing of purlin at eaves and ridge will be less than those adopted between apex and eaves (1.00 to 1.4 m). Moreover, for uniformity and to maintain the slope, same size of purlins are provided. The purlin at ridge is kept within 150 mm from the ridge point. Hence, it is suggested that local coefficient may be ignored.

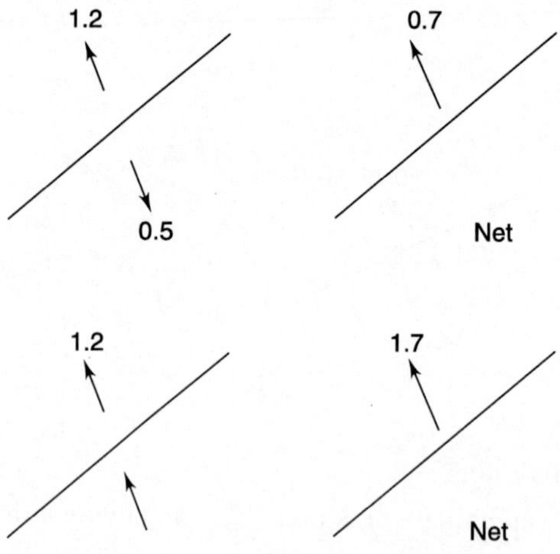

Fig. 9.16

Abstract of forces

1. Design wind force for main column on side '*A*' = 1.2 × 1146 × 6 = 8251 N/m [case 1.1]
2. Design wind force for cable column 1.2 × 1146 × 6 = 8251 N/m [case 2.1] (use appropriate spacing of gable columns)
3. Design wind uplift force for roof rafter 8939 N/m [case 1.2]
4. Design wind downward force for roof rafter 688 N/m [case 1.1]
5. Design wind force for purlin 2086 N/m [case 1.2]
6. Design wind force for side cladding runner on '*A*' [case 1.1] = 1.2 × 1.4 × 1146 = 1925 N/m

For these wind forces with load combinations as in IS: 875-Part-5; the structure shall be designed.

EXERCISE

1. What are the modification factors used to calculate design wind speed?
2. Calculate the wind force on a hoarding board of size 6 m width × 4 m height. The top of the board is at 10 m from ground.
3. Cladding provided for a workshop shed has got opening of 18% of the area. What is the internal pressure coefficient?
4. Explain?

   (a) C_{pe} (b) C_{pi}

   (c) net pressure coefficient

Chapter 10

Design of Roof Trusses

Workshops, storage godowns, school buildings, offices, power plants, process industrial sheds, large gathering areas like cinema theatre, marriage halls require large covered area free of columns. Steel roof trusses are most economical for such covering. Compared to other types of roofing, this is less time consuming for completing the building. Temporary sheds for storing materials, etc., are built with roof trusses. The main advantage is, the same can be removed and reuses elsewhere. Since this is a sloped roof, the rainwater will be drained off quickly. The roof trusses are formed by connecting various members of the truss by using flexible gussets which permit rotation at joints. Hence, it is called a pin jointed frame. The line passing through the centre of gravity of every member at a joint shall meet at on point which is called working point or set off point. Normally, these trusses are loaded at nodes (called panel points) so that all the members are subjected to either axial compression or tension.

In a building the trusses are made to span the shorter span. Ends of these trusses will be supported by R.C.C. columns, steel columns or walls. The trusses are kept at intervals of about 4 to 12 m along the length of the building. Over the top chord of these trusses angles, channels or I sections are kept parallel to the length of the building to support the roof sheeting. These members are called purlins. The roof sheeting is secured to the purlins. The spacing of purlins varies generally between 1 m and 1.4 m depending on the type and strength of the sheet used. Many roof sheeting materials are available in the market. Galvanized iron sheets, asbestoes sheet, resin coated steel sheets and polycarbonate sheets are mostly used. It is advisable to connect the purlins at nodes or panel points of the truss so that all the members of the truss will be subjected to axial force only and there will be no moment. But this may not be possible in all the cases. In many cases purlins are kept away from panel points. Hence, the top chord of the truss will be subjected to bending moment also. This has to be checked as a continuous beam.

Components of roof trusses are rafter or top chord, bottom chord, diagonals, and wind bracing. Refer para 10.03 for further details.

The roof trusses can be pitched roof, pitched roof with monitor, N truss, Pratt truss or lean-to mono slope roof.

Generally, the members of roof trusses are made with single or double angles connected to gusset plates. Tubes and box channels are also used as members of roof truss.

The roof trusses are subjected to the following loads.

(1) **Dead load:** This consists of self weight of the sheet, self weight of the purlin, self weight of the truss itself.

(2) **Imposed load:** The imposed load will be more during erection of sheets only. Since the trusses are sloped, the roof is not easily accesssible. They are very rarely accessed mostly during maintenance. Hence, they are considered inaccessible roof.

(3) **Wind load:** Air in motion is called wind. The wind will impart its force depending on the area resisting. The force due to wind will be perpendicular to the surface. The magnitude depends on the wind speed in the loacality at which the building located, the terrain, shape, size and height of the building. Predominant load on the roof truss is wind load.

10.1 SOME TIPS FOR SPAN DEPTH RATIO AND SPACING

The table below gives type of truss, suggested span and span depth ratio with suggested spacing. This may be useful in selecting the type of truss, its slope and spacing depending on the intended span. However, the slope is decided based on the type of sheet proposed to be used. The resin coated steel sheets and galvalum sheets are used with a slope of 10 degrees to horizontal.

Table 10.1 Types of truss

Type of Truss	*Sketch*	*Span*	*Span-to-depth ratio*	*Spacing*
Pratt	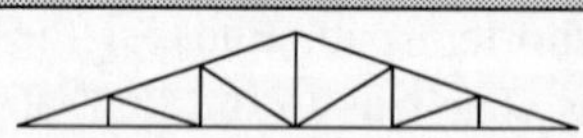	6 m to 12 m	4 to 5	3 m to 5 m
Howe		6 m to 12 m	4 to 5	3 m to 5 m
Fink	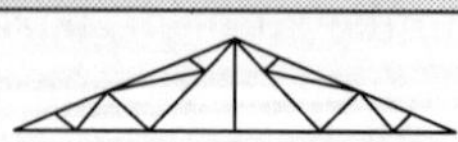	6 m to 15 m	5 to 7	3 m to 5 m
Mansard	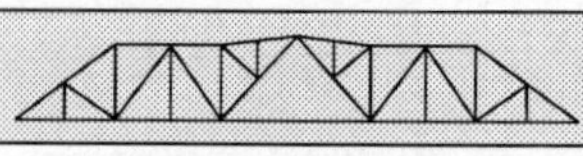	15 m to 30 m	7 to 8	4.5 m to 7.5 m
Pratt (or) *N* Truss		30 m to 50 m	15 to 25	6 m to 12 m
Warren		30 m to 50 m	15 to 25	6 m to 12 m

10.2 SUGGESTED TYPES OF TRUSS ARRANGEMENT FOR VARIOUS SPANS

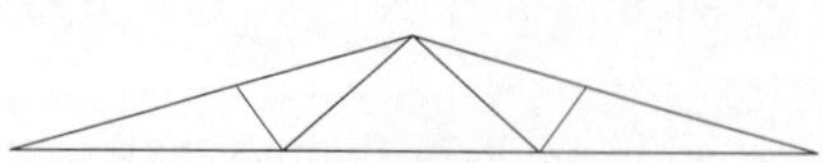

Fink truss (Span < 9 m)
(a)

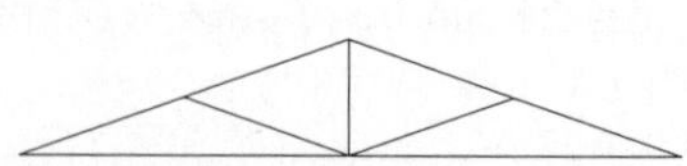

King post truss (Span < 6 m)
(b)

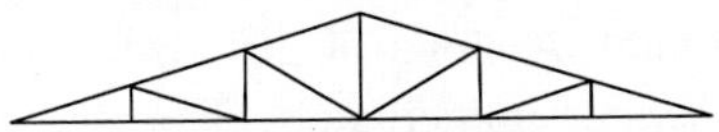

Howe triangular truss
(Span 6 to 24 m)
(c)

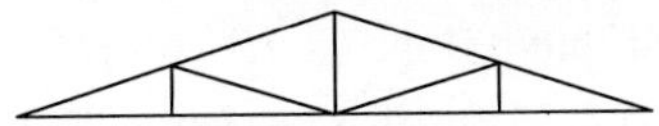

Howe triangular truss
(Span 6 to 24 m)
(d)

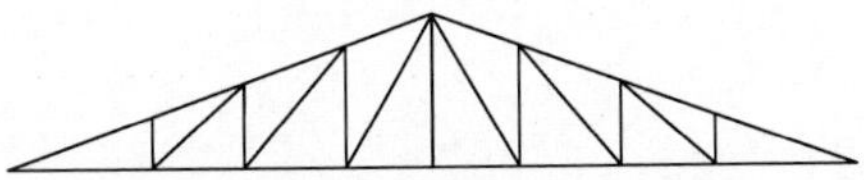

Triangular truss for large span

(e)

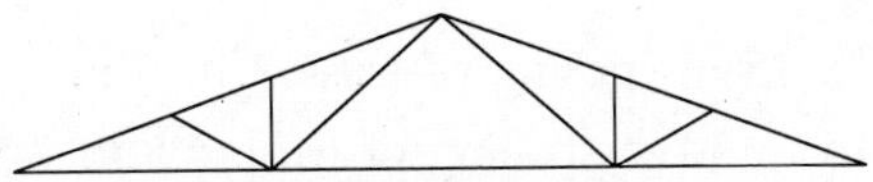

Fan truss (Span < 12 m)

(f)

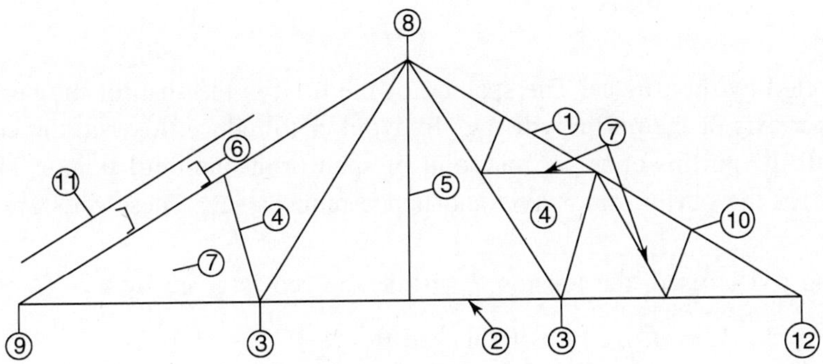

Fink truss (Span 10 to 20 m)

Fig. 10.1

10.3 COMPONENTS OF ROOF TRUSS

1. Principal rafter (or) top chord
2. Bottom chord or main tie.
3. Ties (bottom chord restraints which will be perpendicular to plane of truss)
4. Struts
5. Sag tie
6. Purlins
7. Diagonal member
8. Ridge line.
9. Eaves
10. Panel points
11. Roof covering
12. Base plate anchor

Slope: Generally, adopted slope is 15° to 26° to horizontal. A slope of 26° (1/4 span) is most economical. This slope is most suited to drain the rainwater immediately. This slope is being adoped for tiled or thatched roof. However, for appearance and to drain the rainwater 15° is sufficient. Nowadays even 10° slope is kept for steel roof truss with metal sheeting.

10.4 SPACING OF TRUSS

For spans up to 15 m-spacing of truss 3 to 5 m

For spans 15 to 30 m-spacing of truss 4.5 to 7.5 m

Approximately the spacing can be 1/5 to 1/3 of span

Truss spacing of 4.5 m 5 m and 6 m with a slope 10° to 15° is recommended for asbestos, steel and galvanized sheets.

10.5 SAG RODS

The purlins are supported by the trusses. The spacing of the truss is the span of the purlin. Due to the loads, the weaker axis of the purlin will sag. To avoid or to reduce this sag, threaded rods are used to connect all the purlins either at midpoint of span or at one-third points. These sag rods act as a support for the purlin minor axis and hence reduces sag. These rods are parallel to the truss.

If only one line sag rod is used, the tension T_s in the sag rod is given by

$$T_s = (L_1 L_2 / 4 \cos \theta)\, w_g \sin \theta$$

L_1 = span of the truss

L_2 = spacing of the truss

θ = slope of truss and w_g is the design gravity load.

10.6 WEIGHT OF ROOF COVERING

1. A.C. sheet – 150 N/m² of plan area
2. G.I. sheet – 112.7 N/m² of plan area
3. Metal sheet = (thickness in mm × 78.5) N/mm^2.

10.7 WEIGHT OF PURLIN

For glazed roof – 70 to 120 N/m^2 of plan area

For G.I. sheet – 60 to 90 N/m^2 of plan area

For A.C. sheet – 90 to 150 N/m^2 of plan area

10.8 WEIGHT OF BRACING

12 to 15 N/m^2 of plan area

10.9 WEIGHT OF TRUSS

90 to 150 N/m^2 of plan area (or) the following empirical relations can be used to get the self weight with respect to plan area covered.

$w = 10(L/3 + 5)$ for 4 m spacing.

w = load/m² of plan area due to weight of truss in N/m²

L = span of truss in metres

q = design load in N/m^2

For $L = 18$ m $\quad w = 10[2.2 + q/12.5]$ N/m²

$L = 24$ m $\quad w = 10[2.78 + q/5.42]$ N/m²

$L = 30$ m $\quad w = 10[4.44 + q/3.47]$ N/m²

$L = 36$ m $\quad w = 10[5.27 + q/2.1]$ N/m²

10.10 WELDED ROOF TRUSS

For sheeted roof truss = $w = 10(5.37 + 0.053\text{ A})$ N/m²

For partly glazed truss = $w = 10(5.37 + 0.064\text{ A})$ N/m²

A = plan area in m² covered by each truss.

10.11 DESIGN OF ROOF TRUSS

The dead load, imposed load and wind load per node is to be calculated. Then the forces in the members can be obtained by many methods. Most popular is the method of joints, method of sections and graphical method. Nowadays due to availability of computer software, the analysis is made faster and with various options. It is less time-consuming to optimize the members. An example using graphical method of analysis of truss is worked out below.

Example 10.1 *Design a fink roof truss using angles of material specification IS:2062, f_y = 250 MPa for the following design data*

Span	*20 m*
Slope	*15° to horizontal*
Height of truss (up to eaves from floor)	*10 m*
Spacing	*6 m*
Length of building	*36 m*
Weight of A.C. sheet	*159 N/m^2*
Location	*TRICHY*

Terrain category 3 and the terrain slope is negligible. Opening in walls is arranged 20% of wall area.

Solution: *Purlins and panels.*

$$\sin \theta = 0.2588$$

$$\cos \theta = 0.9659$$

Height of truss = 20/2 × tan 15° = 2.68 m

Length of top chord/rafter = $\sqrt{(10^2 + 2.68^2)}$ = 10.35 m

The top chord is divided into four parts/panels.

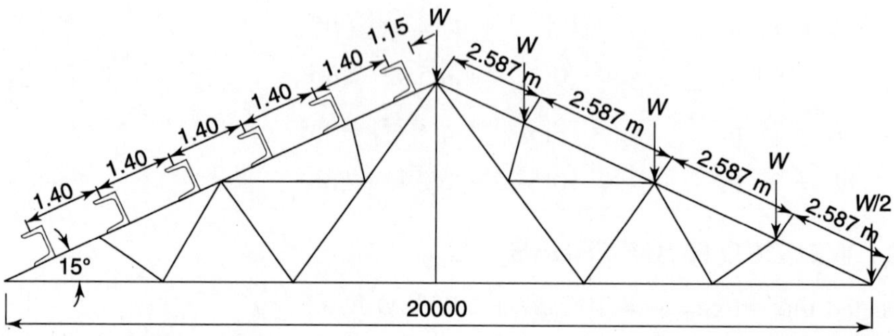

Fig. 10.2

Length of each panel = 10.35/4 = 2.587 m

It is proposed to use 1.5 m long A.C. sheet with 100 mm overlap for roof sheeting.

Spacing of purlin = Length of A.C. sheet-overlap = 1.5 – 0.1 = 1.4 m.

Number of purlins = $\dfrac{10.35}{1.4}$ = 7.39.

Provide 8 purlins on each side.

Number of spacings = 7

Balance = 10.35 – 7 × 1.4 = 0.55 m.

This will be adjusted at ridge and at eaves.

Dead Load

Dead load depends on the density of material as given in IS: 875-Part-1 and the volume of material per unit area/length. For calculating the dead load due to A.C. sheet; weight per sq.m is given in IS: 875-Part-1. If the proposed material is not found in IS: 875-Part-1, we can take the density from supplier data (or) we can find the same. In this problem, the weight is given.

Weight of truss = 150 N/m² (assumed)

A.C. sheet =159 N/m² (given)

A.C. sheet weight/m length of purlin = 159 × 1.4 = 222.6 N/m

Imposed Load on Purlin

The roof is a sloped roof. Slope is 15° to horizontal. Sloped roof can be considered access not provided expect during maintenance. Hence, I.L. = 750 N/m^2 with 20 N/m^2 reduction for every degree of slope more than 10°. (Table 2 serial number (ii) of IS:875-Part-2).

Imposed load on purlin = [750 – (15 – 10) 20] = 650 N/m^2 on plan area.

Imposed load for the design of roof truss

= 2/3 × 650 = 433.33 N/m² (Clause 4.5.1 of IS:875-Part-2).

Wind Load (IS:875-Part-3)

Average height of truss above G.L. = 10 + 2.68/2 = 11.34 m

Basic wind speed = 47 m/sec (Appendix-A of IS:875-Part-3).

K_1 = 1 for general buildings (Table 1 of IS:875-Part-3)

As per Clause 5.3.2.2 of IS:875-Part-3; the structure comes under class B.

Terrain category 3 K_2 = 0.904 at 12 m height. (Clause 5.3.2.1, 5.3.2.2 and Table 2 of IS:875-Part-3)

Since it is plain terrain, K_3 = 1.0

$$Vz = V_b\, K_1\, K_2\, K_3 = 47 \times 1.0 \times 0.904 \times 1.0 = 42.5 \text{ m/sec}$$

(Clause 5.3 of IS:875-Part-3)

$$P_z = 0.6\, V_z^2 = 1083 \text{ N/m}^2$$ (Clause 5.4 of IS:875-Part-3)

$$h = 10 \text{ m}$$

$$w = 20 + 2 \times 0.3 = 20.6$$ (Assuming eaves projection of 0.3 m)

$$h/w = 10/20.6 < 1/2 \text{ and } \theta = 15°$$

C_{pe} as per Table 5 of IS:875-Part-3 is tabulated below.

Table 10.2

	Wind ward	Wind ward	Lee ward	Lee ward
Wind direction	0°	90°	0°	90°
Face	E	E	G	G
C_{pe}	−0.8	−0.75	−0.4	−0.6

Building is with medium permeability C_{pi} = ± 0.5 (Clause 6.2.3.2 of IS:875-Part-3).

$$C_{pe} - C_{pi} = -0.8 - 0.5 = -1.3$$

$$= -0.8 + 0.5 = -0.3$$

Wind intensity = –1.3 × 1083 = –1408 N/m^2

Total wind load on each purlins = 1.4 × 6 × (–1408) = – 11827 N

Wind load per panel = 1408 × 6 × 2.59/1000 = 21.9 kN.

Load on Panel Points of Truss

Dead load

There are 16 purlins on one truss. Consider ISMC 125 purlin for calculating self weight.

Weight of A.C. sheet per panel = 159 × 2.59 × 6 = 2471 N

Self weight of purlin/panel

$$= 1/8(16 \times 6 \times 12.7 \times 9.81) = 1495 \text{ N}$$

Weight of truss = 150 N/m^2

Weight of truss/panel = 1/8(150 × 20 × 6) = 2250 N

Total DL/panel = 2471 + 1495 + 2250 = 6216 N

DL at end panel point = 3108 N

Imposed Load (Live Load)

Imposed load intensity = 433.33 N/m^2 of plan area.

Imposed load per panel = 433.33 × 6 × 2.59 cos 15° = 6504.49 N

Analysis

If the truss is a statically determinate structure; the following expression should be satisfied.

$$m = 2j - r$$

where,

m = number of members

j = number of joints

r = reaction components.

This is applicable for 2*D* truss (planar trusses). Such trusses can be analysed by method of sections (or) joints. Indeterminate trusses are those trusses where number of members will be more than '*m*'. Indeterminate trusses can be analysed by matrix method of analysis for which there are many pieces software available.

These can be solved by graphical method also. We will see how the truss is analysed for gravity load and wind load using graphic static method.

The space between forces are numbered using alphabets and the space between numbers are numbered with numerals to designate the loads and members using Bow's rotation. (Fig.10.3).

Using the graphical method, the forces in members are found and tabulated in Table 10.3. The last two columns of the table gives the design load (+ve for compression and –ve for tension) and the unsupported length of the member. This is the design data for designing the individual member. For designing top chord, the maximum force in one of the four panels is considered since one section profile is used throughout. Same is the case for bottom chord members also.

Forces in Members due to Gravity and Wind Loads

Graphical method-gravity load (Refer Fig. 10.3)

In Bow's notation, to represent spaces between forces, capital letters are used. Figure 10.3 gives the lines of action of the forces which is called the *space diagram* and also the polygon of forces which is called the *force diagram.* To represent a force, usually capital letters (upper case letters) are used on the *space diagram* and the corresponding small letters (lower case letters) are used for the *force diagram.*

It may be noticed that in Fig. 10.3, the space between the loads and reactions are identified with alphabets. The space between the truss membres are given numeric. Around joint *A*, *A-B* denotes load *W*/2 and *B*-1 denotes the part of rafter and 1 – *K* denotes the part of bottom chord. Since loads are due to dead and imposed loads (gravity loads), they are vertically downward. Draw a vertical line. Mark points *A* to *J* to some scale. For example, let us say 1 cm = *W*. Mark *A-B* distance equal to 0.5 cm since load *A-B* = *W*/2. Continue this upto *J*. Note that the

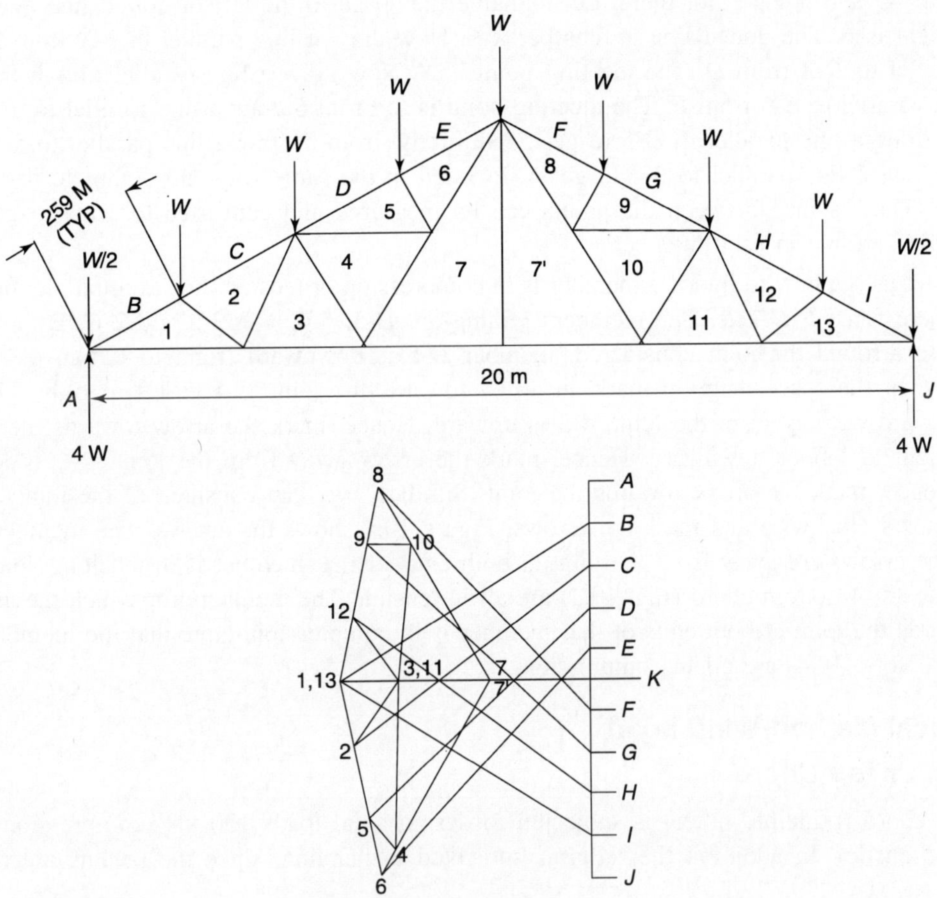

Fig. 10.3

direction is also taken care. Loads *A-B*, *B-C*... are downward and they are marked downard in the vertical line. Now recation *J-K* is vertically upward and mark the same from *J*. Note that the reaction *K-A* is vertically upward. We can measure and check the value.

By locating 1 to 13 in the force diagram, we can measure the force diagram and find the magnitude of force in every member taking into account the scale. Let us consider members *B*-1 and 1-*K* meeting at a joint where force *A-B* and reaction acts. Location *B* is known in the load diagram. From *B* draw a line parallel to *B*-1 and from *K* draw a line parrallel to 1-*K*. Meeting point of these two line is 1. Similarly, draw a line parallel to *C*-2 from *C* and parallel to 1-2 from one. The meeting point of these two lines is 2. Similarly, draw a line parallel to *K*-3 from *K* and parallel to 2-3 from two. The meeting point of these two lines is 3. In the same way we can draw line parallel to memeber 3-4 from 4. But on either side of 4, you may notice that spaces 5 and 7 are there. Since we do not have 5 and 7 on force diagram, we will not be able to draw line parallel to 4-5 or 4-7 to locate 4. Draw a dotted line to join the meeting point of 4-7 on bottom chord and meeting point of 5-6 on top chord. For the present assume that the members 4-5 and 5-6 are not there. Let us name the space to the left of dotted line as x and to the right is 6. The dotted line is member *X*-6. Now draw a line parallel to 3-*X* from 3 and line parallel to *D-X* from *D*. The meeting point is *X*. Now draw a line parallel to *X*-6 from *X* and line parallel to *E*-6 from *E*. The meeting point is 6. From 6 draw a line parallel to 6-5 and from *D* draw a line parallel to *D*-5 to get 5. Similarly, from 5 draw a line parallel to 5-4 and from 3 draw a line parallel to 3-4 to get 4. Proceed in the same way and complete the force diagram. The distance between the points can be measured and converted to scale to get the magnitude of force in each member.

Now let us see how to find if a member is in compression or tension. For example, let us take joint where force *C-D* acts. The members joining are *D*-5, 5-4, 4-3, 3-2 and 2-*C*. By reading clockwise arround the joint considered, member *D*-5 is downward (right to left in the force diagram). In the space diagram mark an arrow towards the joint. 5-4 is left to right. Hence, mark the arrow away from the joint. 4-3 is upward. Hence, mark the arrow towards the joint. 3-2 is right to left or downward. Hence, mark the arrow away from the joint. 2-*C* is left to right. Hence, mark the arrow towards the joint. Similarly, we can consider all the joints, read the members clockwise and mark the arrows. Figure 10.4 shows the arrows. The members for which the arrows are away from the joint at both ends of that member is in tension. Note that the members of bottom chord (Fig. 10.4) are all in tension. The members for which the arrows are towards the joint at both ends of that member is in compression. Note that the members of top chord (Fig. 10.4) are all in compression.

Graphical Method-Wind Load

Design of Top Chord

The top chord (principle rafter) is subjected to design axial loads 130 kN (compression) and 246 kN (tensile). In addition, the rafter is subjected to bending, since the purlins are not at nodes.

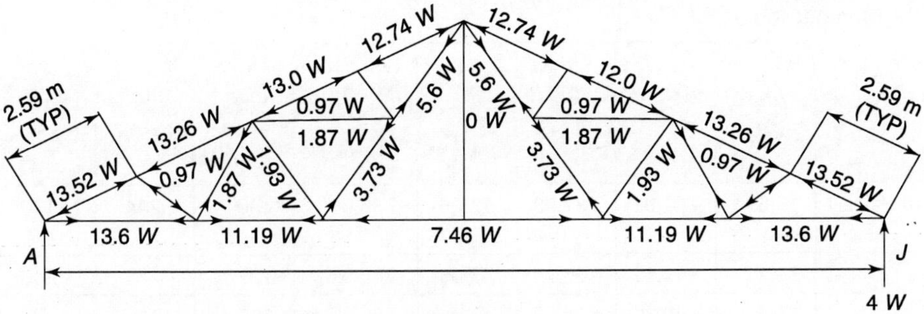

Fig. 10.4

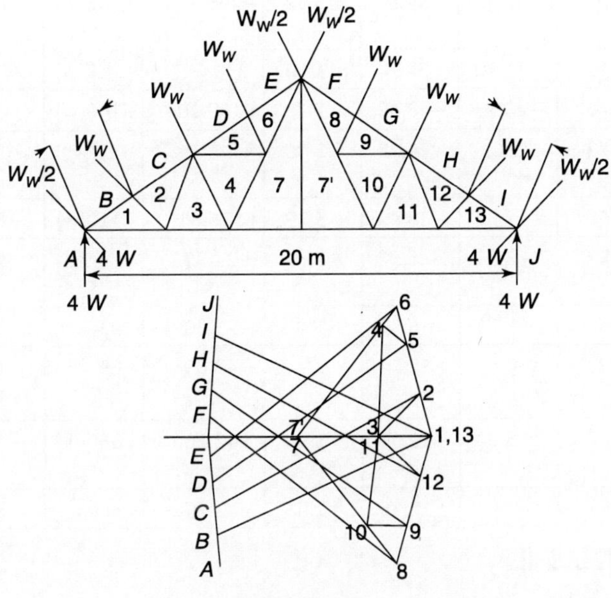

Fig. 10.5

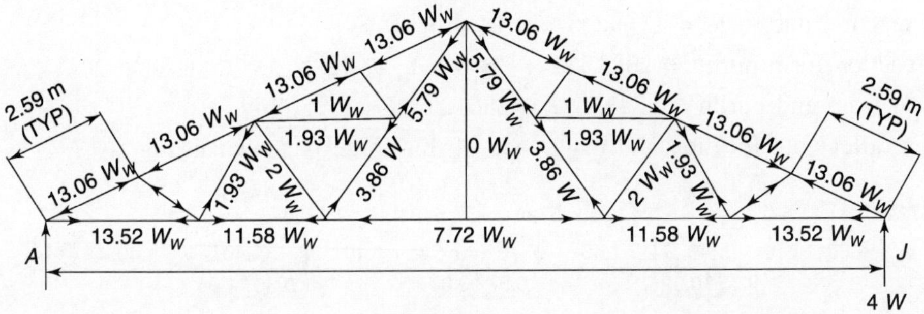

Fig. 10.6

Table 10.3 Member forces

Member Designation (structure)	*Force in terms of W (Fw)*	*DL F_wD.L on node in kN*	*I.L F_wI.L on node in kN*	*Force in terms of W_W (FW_W)*	*Windload (FW_w)W.L on node in kN*	*Maximum compression*	*Maximum tension kN*	*Design load kN*	*Unsupported length in m*
(Top Chord-Rafter) B-1, I-14	13.52	+41.91	+87.88	13.06	−286.01	+130	−244	+130	2.59
C-2, H-13	13.26	+41.11	+86.20	13.06	−286.01	+127	−245	−246	2.59
D-5, G-10	13	+40.30	+84.50	13.06	−286.01	+125	−246	(w)	2.59
E-6, F-9	12.74	39.50	+82.81	13.06	−286.01	122.3	−246		2.59
(Bottom chord) K-1, K-14	13.06	−40.50	−84.89	13.52	+296.1	+256(w)	−125		2.68
K-3, K-12	11.19	−34.70	−72.74	11.58	253.6	+219(w)	−107	+256(w)	2.68
K-7, K-8	7.46	−23.13	−48.49	7.72	+169.10	+146(w)	−72	−125(w)	4.64
Main Slings 4-7,11-8	3.73	−1156	−24.25	3.86	+84.53	+73(w)	−36	+110(w)	2.68
6-7, 9-8	5.60	−17.36	−36.40	5.79	+126.81	+110(w)	−54	−54	2.68
Minor Struts 1-2, 13-14	0.97	+3.01	+ 6.30	1	−21.90	+9.3	−19	+9.3	0.69
5-6, 9-10	0.97	+3.01	+6.30	1	−21.90	+93	−19	−19(w)	0.69
3-4, 11-12	1.93	+6	+12.54	2	− 43.80	+19	− 44	+19 − 44(w)	1.39
Minor Slings 2-3, 13-12	1.87	−5.80	−12.16	1.93	+42.26	+36(w)	−18	+36(w)	2.68
4-5, 10-11	1.87	−580	−12.16	1.93	+42.26	+36(w)	−18	−18	2.68

Moments Due to DL & IL

Purlin reaction due to D.L. = (222.6 + 12.7 × 9.81) × 6 = 2083.122 N

I.L. on purlin = 650 × 1.4 = 910 N/m

Purlin reaction due to IL = 2/3 × (6 × 910) = 3640 N

Total reaction form purlin = 2083.122 + 3640 = 5723.12 N

Reaction from end purlin = 5723.12/2 = 2861.56 N say 2862 N

The top rafter with location and magnitude of this force is shown in Fig. 10.7.

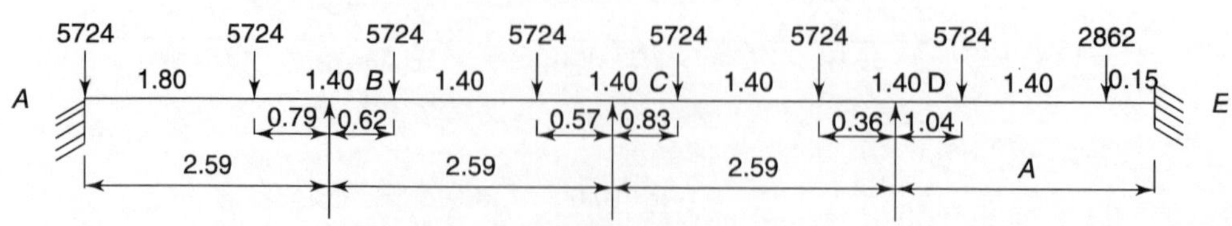

Fig. 10.7

Table 10.4 Moment Distribution (DL + IL)

It is proposed to use only one section throughout the top chord (this is the general engineering practice). Hence, the distribution factor is 0.5. Fixed end moments are calculated and moment distribution is done below.

Joints	*A*	*B*		*C*		*D*		*E*
Members	*AB*	*BA*	*BC*	*CB*	*CD*	*DC*	*DE*	*ED*
Dist Factor	–	0.5	0.5	0.5	0.5	0.5	0.5	-
FEM	−0.96	+2.18	−2.61	+2.63	−2.27	+2.56	−2.16	+1.81
Balance		+0.22	0.22	−0.18	−0.18	−0.2	−0.20	
C/O	+0.11		−0.09	+0.11	−0.1	−0.09		−0.1
Bal		+0.045	+0.045	−0.005	−0.005	+0.045	+0.045	
C/O	+0.0225		−0.003	+0.0225	+0.0225	−0.003		+0.0225
Bal		+0.0015	+0.0015	0	0	+0.0015	+0.0015	
Final	−0.83	+2.45	−2.44	+2.58	−2.53	+2.31	−2.13	+1.73

Moment Due to DL + IL	**Moment Due to DL + WL**
$M_A = 0.83$ kNm	$0.83 \times 1.7 = 1.41$ kNm
$M_B = 2.45$ kNm	$2.45 \times 1.7 = 4.17$ kNm
$M_C = 2.58$ kNm	$2.58 \times 1.7 = 4.39$ kNm
$M_D = 2.31$ kNm	$2.31 \times 1.7 = 3.93$ kNm
$M_E = 1.73$ kNm	$1.73 \times 1.7 = 2.94$ kNm

Moments Due to (DL + WL) Combination

$$W_d = 2083 \text{ N}; \qquad W_w = 11827 \text{ N}$$

Reaction from purlin due to (DL + WL) combination = 11827 – 2083 = 9744 N

Reaction due to end purlins = 1/2 × 9744 = 4872 N

The BM in the rafter due to (DL + WL) combination can be found by multiplying the values in the Table 10.4 by the factor 9744/5724 = 1.70

The moment due to DL + IL and due to DL + WL are tabulated in Table 10.4

Let the stress reduction factor be = 0.4. Therefore, $f_{cd} = 0.4\, f_y = 0.4 \times 250 = 100$ MPa.

Required area = 246 × 1000/100 = 2460 mm^2

Try section ISA 90 × 90 × 8 having an area of 2 × 1379 = 2758 mm^2

$$A = 2 \times 13.79 \text{ cm}^2 = 27.58 \text{ cm}^2$$

$$C_{yy} = 2.51 \text{ cm}$$

$$I_{zz} = 2 \times 104.2 = 208.4 \text{ cm}^4$$

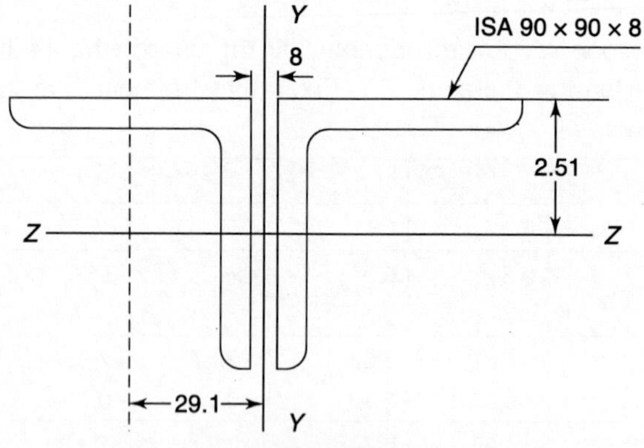

Fig. 10.8

$$C_{yy} + 1/2 \text{ gap} = 25.1 + (8/2) = 29.1 \text{ mm}$$

$$I_{yy} = 2\{104.2 + 13.79 \times 2.91^2) = 442 \text{ cm}^4$$

$$r_{zz} = \sqrt{(208.4/27.58)} = 2.75 \text{ cm} = 27.5 \text{ mm.}$$

$$r_{yy} = \sqrt{(442/27.58)} = 4 \text{ cm} = 40 \text{ mm.}$$

ZZ is the axis of bending.

$$Z_{zt} = 208.4/(9.00 - 2.51) = 32.11 \text{ cm}^3;\ Z_{zc} = 208.4/2.51 = 83 \text{ cm}^3$$

Effective length for *ZZ* axis = 0.7 × 2.59 × 1000 = 1813 mm (Clause 7.5.3 and 7.2.4 of IS: 800).

Effective length for *yy* axis = 1.0 × 1400 = 1400 mm.

Purlins will act as restraints on *YY* axis of truss.

Purlin spacing = 1400 mm

$$\lambda_z = 1813/27.5 = 66$$

$$\lambda_y = 1400/40 = 35$$

Since this is a built up member, this comes under buckling class c (Table 10 of IS: 800).

For KL/r = 66 and f_y = 250 MPa; f_{cd} = 158 MPa (Table 9c of IS: 800).

Check for Section at B

Partial safety factor for load = 1.5 (Table 4 of IS: 800)

M = 2450 Nm. Factored moment = 1.5 × 2450 = 3675 Nm.

P = 130 kN (comp). Factored load = 1.5 × 130 = 195 kN.

Design compressive strength = $P_d = A_e \times f_{cd}$ (Clause 7.1.2 of IS: 800).

$$P_d = 2758 \times 158/1000 = 436 \text{ kN.} > 195 \text{ kN.}$$

$$b/t = 90/8 = 11.25 < 15.7\ \varepsilon$$

$$d/t = 90/8\ 11.25 < 15.7\ \varepsilon$$

$$(b + d)/t = 22.5 < 25\ \varepsilon$$

Semi-compact. (Table 2 of IS:800).

$$Z_{pz} = 2 \times (90 \times 8 \times 21.1 + 17.1 \times 8 \times (17.1/2)) + 2 \times (64.9 \times 8 \times (64.9/2)) = 66419.36 = 66.42 \times 10^3\ \text{mm}^3$$

$$\beta_p = Z_{ez}/Z_{pz} = 83/66.42 = 1.25$$

$$M_d = \beta_p\ Z_p f_y/\gamma_{mo} \quad \text{(Clause 8.2.1.2 of IS:800).}$$

$$= 1.25 \times 66.42 \times 10^3 \times 250/(1.1 \times 10^6) = 18.87\ \text{kNm.}$$

This should be less than $M_d = 1.2Z_e f_y/\gamma_{mo}$ (Clause 8.2.1.2 of IS:800)

$$= 1.0 \times 83 \times 10^3 \times 250/(1.1 \times 10^6) = 22.64\ \text{kNm.}$$

$$M_{dz} = 18.87\ \text{kNm} > 3.675\ \text{kNm}$$

Combined strength = (195/436) + (3.675/18.87) = 0.65 < 1.0 (Clause 9.3.1.1 of IS:800).

If wind is considered

Partial safety factor for load = 1.2 (Table 4 of IS:800)

M = 4170 Nm. Factored moment = 1.2 × 4170 = 5 kNm.

P = 246 kN (Tension). Factored load = 1.2 × 246 = 295.2 kN.

Using M16 mm bolt, the diameter of hole will be 18 mm. (Table 19 of IS:800)

Effective area of angle = 2758 – 2 × 18 × 8 = 2470 mm^2

$$T_{dg} = A_g f_y/\gamma_{mo} = (2758 \times 250/1.1 \times 1000) = 626.8\ \text{kN}$$

(Clause 6.2 of IS:800)

$$T_{dn} = 0.9 \times A_n \times f_u/\gamma_{m1} = 0.9 \times 2470 \times 410/1.25 \times 1000 = 729.1\ \text{kN.}$$

(Clause 6.3 of IS:800)

$$M_{dz} = 18.87\ \text{kNm}$$

Combined strength = (295.2/626.8) + (5/18.87) = 0.73 < 1.0

Hence, 2 ISA 90 × 90 × 8 as top chord is satisfactory.

Design of Bottom Chord (Main Tie)

Partial safety factor for load = 1.5 (Table 4 of IS:800)

Maximum tension = 1,25,000 N. Factored load = 1.5 × 125 = 187.5 kN.

Partial safety factor for load when wind is considered = 1.2 (Table 4 of IS:800)

Maximum compression due to wind = 256 kN.

Factored load = 1.2 × 256 = 307.2 kN.

Let f_{cd} = 100 MPa.

Area required = 307.2 × 1000/100 = 3072 mm^2

Try 2 – 90 × 90 × 10

$$A = 3406 \text{ mm}^2;\ I_{zz} = 2 \times 126.7 = 253.4 \times 10^4 \text{ mm}^4$$

$$C_{yy} = 2.59 \text{ cm}$$

$$I_{yy} = 2[126.7 + 17.03 \times 2.99^2] = 558 \text{ cm}^4,\ r_{zz} = 2.73 \text{ cm: } r_{yy} = 4.05 \text{ cm}$$

Check for Maximum Tension

Providing M16 bolts; Diameter of holes = 18 mm (Table 19 of IS: 800).

$$A_n = 3406 - 2 \times 18 \times 10 = 3046 \text{ mm}^2$$

$$T_{dg} = A_g f_y/\gamma_{mo} = 3406 \times 250/(1.1 \times 1000) = 774 \text{ kN.}$$

(Clause 6.2 of IS: 800)

$$T_{dn} = 0.9 \times 3046 \times 410/(1.25 \times 1000) = 899 \text{ kN.}$$

Design strength in tension = 774 kN > 187.5 kN.

Hence, satisfactory in tension.

Check for Maximum Compression

Let us provide bottom chord tie (restraint for *YY* axis of bottom chord) at mid span and at 4.64 m on either side of mid span as shown in Fig. 10.9.

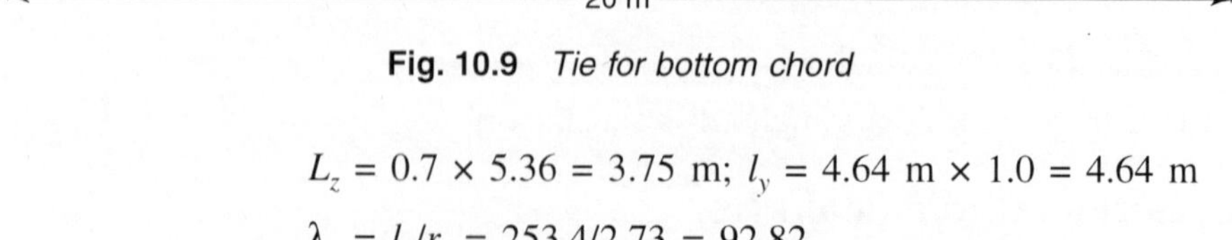

Fig. 10.9 *Tie for bottom chord*

$$L_z = 0.7 \times 5.36 = 3.75 \text{ m};\ l_y = 4.64 \text{ m} \times 1.0 = 4.64 \text{ m}$$

$$\lambda_x = l_z/r_z = 253.4/2.73 = 92.82$$

$$\lambda_y = l_y/r_y = 558/4.05 = 137.8$$

For f_y = 250 MPa; and λ = 137.8; f_{cd} = 68 MPa (Table 9c of IS: 800).

Compressive strength = 68 × 3406/1000 = 231.6 kN < 307.2 kN.

Two angles ISA 90 × 90 × 10 as bottom chord is not adequate.

We can try with higher size angles (or) box made out of channels (or) by providing bottom chord ties at the closer interval than 5.36 m to restrain *YY* axis.

Design of Main Slings (4-7, 6-7)

Maximum compressive force (due to wind) = 110 kN.

Factored = 1.2 × 110 = 132 kN.

Let f_{cd} = 100 MPa

Area of cross section required = 132*x* 1000/100 = 1320 mm^2

Maximum tensile force = 54 kN = 54 kN.

Factored = 1.5 × 54 = 81 kN.

Length of member on *Z* axis is 2.68 m and on *Y* axis is 5.36 m

Try 2 numbers 75 × 75 × 10 placed back to back with 8 mm gap to accommodate gusset plate of 8 mm thickness.

$$A = 2 \times 14.02 = 28.04 \text{ cm}^2 = 2804 \text{ mm}^2$$

$$C_{zz} = C_{yy} = 22.2 \text{ mm}$$

$$I_{zz} = 2 \times 71.4 \text{ mm}^4 = 142.8 \text{ cm}^4$$

$$I_{yy} = 2(71.4 + 14.2 \times 2.62^2) = 335.28 \text{ cm}^4$$

$$r_{zz} = 22.6 \text{ mm}; \; r_{yy} = \sqrt{(335.28/23.04} = 3.46 \text{ cm} = 34.6 \text{ mm}.$$

Check for Maximum Tension

Let us use M16 bolts of property class 4.6.

$$A_n = 2804 - 2 \times 18 \times 10 = 2444 \text{ mm}^2$$

$$T_{dg} = A_g f_y/\gamma_{mo} \text{ (Clause 6.2 of IS:800)}$$

$$T_{dg} = 2804 \times 250/(1.1 \times 1000) = 637.3 \text{ kN}.$$

$$T_{dn} = 0.9\, A_{nc}\, f_u/\gamma_{m1} = 0.9 \times 2444 \times 410/(1.25 \times 1000) = 721.5 \text{ kN}.$$

Design strength in tension = 637.3 kN > 81 kN.

Hence, satisfactory in tension.

Check for Maximum Compression

$$L_z = 0.7 \times 2.68 = 1.876 \times 10^3 \text{ mm}$$

$$L_y = 1 \times 5.36 = 5.36 \times 10^3 \text{ mm}$$

$$\lambda_Z = 1876/22.6 = 83; \; \lambda_y = 5360/34.6 = 154.9$$

For f_y = 250 MPa; and λ = 154.9; f_{cd} = 56.25 MPa (Table 9c of IS: 800).

Design compressive strength = 56.25 × 2804/1000 = 157.725 kN > 132 kN.

Hence, 2 ISA 75 × 75 × 10 is satisfactory.

Design of Minor Sling (2-3, 4-5)

Check for tension

Maximum tension = 18 kN.

This is due to DL + IL

Factored load = 1.5 × 18 = 27 kN.

Maximum compression = 36000 N.

This is inclusive of wind.

Factored load = 1.2 × 36 = 43.2 kN.

Try 2 ISA 50 × 50 × 6

$$A = 2 \times 568 = 1136 \text{ mm}^2$$

$$I_{zz} = 2 \times 12.9 = 25.8 \text{ cm}^4$$

Thickness of gusset = 8 mm

$$C_{yy} = 14.5 \text{ mm}; x = C_{yy} + (8/2) = 18.5 \text{ mm}$$

$$I_{yy} = 2[12.9 + 5.68 \times 1.85^2] = 64.7 \text{ cm}^4$$

$$r_{zz} = 15.1 \text{ cm}$$

$$r_{yy} = \sqrt{(64.7/11.36)} = 2.4 \text{ cm}$$

Providing M16 bolt; the diameter of hole = 18 mm (Table 19 of IS: 800)

Therefore, A_n = 1136 – 2 × 18 × 6 = 920 mm^2

$$T_{dg} = A_g f_y/\gamma_{mo} = (1136 \times 250)/(1.1 \times 1000) = 258.2 \text{ kN}$$

$$T_{dn} = 0.9 \times A_n \times f_u/\gamma_{m1} = 0.9 \times 920 \times 410/(1.25 \times 1000) = 271.6 \text{ kN.}$$

Tensile strength = 258.2 kN > 27 kN

Check for Compression Stress

$$L_z = 0.7 \times 2.68 = 1.876 \text{ m}$$

$$L_y = 1.0 \times 2.68 = 2.68 \text{ m}$$

$$\lambda_z = 1876/15.1 = 124$$

$$\lambda_y = 2680/24 = 112$$

For f_y = 250 MPa and λ = 124; f_{cd} = 80 MPa (Table 9c of IS: 800).

Design strength in compression = 80 × 1136/1000 = 90.9 kN > 43.2 kN

Hence, 2 ISA50 × 50 × 6 is satisfactory.

Design of Main Strut (3-4)

Maximum compression = 19 kN.

This is due to DL + IL

Factored load = 1.5 × 19 = 28.5 kN.

Maximum tension = 44 kN

This is inclusive of wind.

Factored load = 1.2 × 44 = 52.8 kN.

Length of member= 1390 mm

Try single angle ISA 50 × 50 × 6

$$r_{\min} = 9.6 \text{ mm} \quad A = 568 \text{ mm}^2$$

Providing M16 bolts, A_n = net effective area = 568 – 18 × 8 = 460 mm^2

$$L/r = 1390/9.6 = 144.8 \text{ Say } 145$$

$$T_{dg} = A_g f_y/\gamma_{mo} = 568 \times 250/(1.1 \times 1000) = 129 \text{ kN.}$$

$$T_{dn} = 0.9 \times 460 \times 410/(1.25 \times 1000) = 135.8 \text{ kN.}$$

Design strength in tension = 135.8 kN > 52.8 kN.

Single angle. Buckling class c. (Table 10 of IS: 800).

For f_y = 250 MPa and λ = 145; f_{cd} = 62.7 MPa (Table 9c of IS: 800).

Design strength in compression = (62.7 × 568/1000) = 35.6 kN > 28.5 kN

Hence, single angle 50 × 50 × 6 is adequate.

Design of Minor Struts (Mem 1-2, 5-6)

Maximum compression 0.930 kN.

This is due to DL + IL

Factored load = 1.5 × 0.93 = 1.4 kN.

Maximum tension 19 kN.

This includes wind.

Factored load = 1.2 × 19 = 22.8 kN.

Length = 690 mm

Try single angle ISA 50 × 50 × 6

$$L = 690 = 690 \text{ mm}$$

$$\lambda = L/r_{\min} = 690/9.6 = 72$$

$$A_n = \text{net area} = 568 - 18 \times 6 = 460 \text{ mm}^2.$$

$$T_{dg} = A_g f_y / \gamma_{mo} = 568 \times 250/(1.1 \times 1000) = 129 \text{ kN}.$$

$$T_{dn} = 0.9 \times 460 \times 410/(1.25 \times 1000) = 135.8 \text{ kN}.$$

Design strength in tension = 129 kN > 22.8 kN

For f_y = 250 MPa; and λ = 72; f_{cd} = 151 MPa (Table 9c of IS: 800).

Design strength in compression = 151 × 568/1000 = 85.8 kN > 1.4 kN.

Hence, single angle 50 × 50 × 6 is adequate for members 1-2 and 5-6

Note: As general engineering practice angle of size less than 50 × 50 × 6 is not used as structural members in industrial structures.

Design of Central Hanger

This takes no force. It is only a restraint for bottom chord on *ZZ* axis.

Maximum compression 256 kN in bottom chord.

Force in the restraint = 2.5% of maximum compression in bottom chord

= (2.5/100) × 256 = 6.4 kN (Clause 8.2.1 of IS:800).

Factored load = 1.2 × 6.4 = 7.68 kN.

Unsupported length of central hanger

Tan 15° = h/10.

$$h = 10 \times \tan 15° = 2.7 \text{ m}; \ L = 1.0 \times 2.7 = 2.7 \text{ m}$$

Try single angle ISA 60 × 60 × 6 single

$$L/r_{\min} = 2700/11.5 = 235$$

For f_y = 250 MPa; and λ = 235; f_{cd} = 27.25 MPa (Table 9c of IS: 800).

Design strength in compression = 27.25 × 568/1000 = 15.5 kN > 7.68 kN.

Hence, single angle 60 × 60 × 6 is satisfactory.

Lateral Ties between Trusses

Truss spacing = 6 m = Length of ties

There are three ties (Refer Fig. 10.9)

Maximum compression in bottom chord = 256 kN.

Factored load = 1.2 × 256 = 307.2 kN.

2.5% of this = 7.68 kN

Force in each tie = 7.68/3 = 2.56 kN

$$L = 1.0 \times 6.0 = 6 \text{ m}$$

Provide 2 ISA 50 × 50 × 6 star angle as shown in Fig. 10.11.

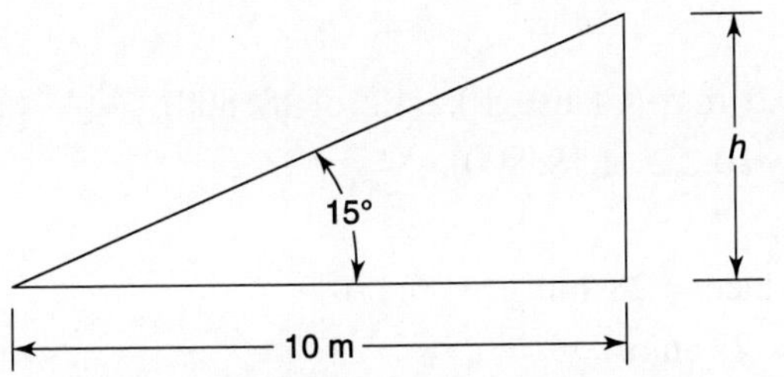

Fig. 10.10

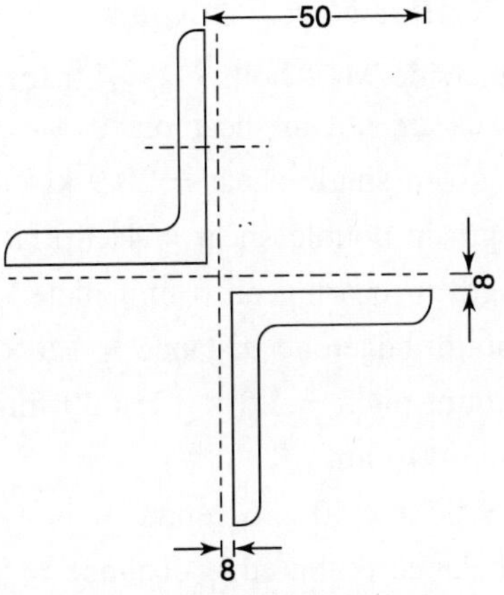

Fig. 10.11

$$I = 2[12.9 + 5.68 \times (1.45 + 0.4)^2] = 64.7 \text{ cm}^4$$

$$A = 2 \times 5.68 = 11.36 \text{ cm}^2$$

$$r = \sqrt{\frac{64.7}{11.36}} = 2.38 \text{ cm} = 23.8 \text{ mm}$$

$$Z = \frac{I}{Y} = \frac{64.7}{5.4} = 11.98 \text{ cm}^3$$

$$r_{\text{min}} = 23.8 \text{ mm}; \ L/r_{\text{min}} = 6000/23.8 = 252.$$

For $\lambda = 250$ and $f_y = 250$ N/mm^2 $f_{cd} = 24.3$ MPa (Table 9c of IS: 800)

Design compressive strength = 24.3 × 1136/1000 = 27.6 kN. > 2.56 kN.

Self weight of the tie = 4.5 kg/m = 44 N/m. Factored self weight = 66 N/m.

Bending due to self weight = $w1^{2/8}$ = (66 × 2 × 6^2)/8 = 594 Nm = 0.6 kNm

Moment capacity = $M_d = 1.2Z_e f_y/\gamma_{mo}$ (Clause 8.2.1.2 of IS: 800)

= 1.2 × 11.98 × 10^3 × 250/(1.1 × 10^6) = 3.267 kNm > 0.6 kNm

Combined Strength

$$1 = \frac{2.56}{27.6} + \frac{0.6}{3.267} = 0.276 < 1$$

Hence, 2 × ISA 50 × 50 × 6 (star) is satisfactory.

Design of Connection

For ISA 50 × 50 × 6 angles

Let us provide M12 bolts class 4.6 for connecting ISA 50 × 50 × 6 angles in such away that the threads are not in shear plane.

Strength in single shear = 20.9 kN (Table 8.8(b) of this book).

Strength in double shear = 41.8 kN

Strength in bearing on 6 mm plate

Minimum edge/end distance = 1.5 × 13 = 19.5 mm (Clause 10.2.4.2 of IS: 800).

Minimum pitch = 2.5 × 12 = 30 mm (Clause 10.2.2 of IS: 800).

Provide 40 mm.

For ISA 50 × 50 × 6 centre of bolt to angle face = 28 mm (SP: 6(1)).

The balance is the edge distance = 50 – 28 = 22 mm.

k_b is the least of

$$(e/3do) = (22/3 \times 13) = 0.564$$

$$(p/3do) - 0.25 = 40/(3 \times 13) - 0.25 = 0.775$$

$$f_{ub}/f_u = 400/410 = 0.975$$

For single ISA50 × 50 × 6

Design strength in bearing strength 2.5 × 0.564 × 12 × 6 × 410/1000 = 41.62 kN

Bolt value = 20.9 kN.

For 2 ISA 50 × 50 × 6

There are two angles and the thickness of both put together is more than the thickness of gusset plate.

Design strength in bearing based on thickness of gusset 2.5 × 0.564 × 12 × 8 × 410/1000 = 55.5 kN

Bolt value = 41.8 kN.

For angles other than ISA 50 × 50 × 6

Let us provide M16 bolts class 4.6 for all other angles.

Threads are not in shear plane.

Strength in single shear = 37.16 kN (Table 8.8(b) of this book).

Strength in double shear = 74.32 kN

Strength in bearing on 6 mm plate

Minimum edge/end distance = 1.5 × 18 = 27 mm (Clause 10.2.4.2 of IS: 800).

Minimum pitch = 2.5 × 16 = 40 mm (Clause 10.2.2 of IS: 800).

Provide 50 mm.

In this truss, we are using the following sizes of angles for which the bolt location as per SP: 6(1) are written against each.

Size of angle	Back mark as in SP: 6(1) in mm	Edge distance available in mm (e)	Diameter of hole for M16 bolt in mm (d_o)	$e/3d_o$
2 ISA 90 × 90 × 8	50	40	18	0.74
2 ISA 90 × 90 × 10	50	40	18	0.74
2 ISA 75 × 75 × 10	40	35	18	0.65
2 ISA 60 × 60 × 6	35	25	18	0.46

We will keep end distance equal to or greater than edge distance.

$$(p/3d_o) - 0.25 = 50/(3 \times 18) - 0.25 = 0.676$$

$$f_{ub}/f_u = 400/410 = 0.975$$

For 2 ISA 90 × 90 × 8 and for 2 ISA 90 × 90 × 10;

$$k_b = 0.676$$

Thickness of gusset plate = 8 mm.

Therefore, bearing strength = 2.5 × 0.676 × 16 × 8 × 410/1000 = 88.7 kN

Bolt value = 74.32 kN

For single ISA 50 × 50 × 6, $k_b = 0.41$

Therefore, bearing strength = 2.5 × 0.41 × 16 × 6 × 410/1000 = 40.34 kN

Bolt value = single shear strength = 37.16 kN

For 2 ISA 75 × 75 × 10, $k_b = 0.65$

Bearing strength = 2.5 × 0.65 × 16 × 8 × 410/1000 = 85.3 kN

Bolt value = 74.32 kN

Table 10.5 Abstract and number of bolts class 4.6-16 mm dia

Sl No.	*Member & size*	*Force kN*	*Bolt value kN*	*No. of bolts required*	*Number and diameter of bolts to be provided*
1.	Rafters 2 ISA 90 × 90 × 8	+195 −295(w)	74.32	3.97	4- M16 bolts
2.	Bottom chord 2 ISA 90 × 90 × 10	+307.2(w) −187.5 (w)	74.32	4.13	5-M16 bolts
3.	Main struts (3-4) ISA 50 × 50 × 6	+28.5 −52.8(w)	20.9	2.53	3 -M12 bolts
4.	Minor struts (1-2, 5-6) ISA 50 × 50 × 6	+1.4 −22.8(w)	20.9	1.09	2- M12 bolts
5.	Main slings (4-7, 6-7) 2 ISA 75 × 75 × 10	+132(w) − 81	74.32	1.78	2- M16 bolts
6.	Minor slings(2-3, 4-5) 2 ISA 50 × 50 × 6	+43.2(w) −27	41.8	1.03	2- M12 bolts

Note: (1) The connections are designed. The number and diameter of bolts are known. The adequacy of the angle with net area is to be checked. (2) Block shear strength is to be checked.

Design of End Support

The end supports should satisfy the following:

1. The size of base plate should be sufficient to accommodate the connection between the top chord and bottom chord.
2. Anchor bolts should be capable of taking the uplift due to wind and shear.
3. Sliding joint should be provided at one end to enable thermal expansion of the truss.

Design of Bearing Plate

$$W_d = 8 \times 6.216 = 49.73 \text{ kN.}$$

$$W_L = 8 \times 6.504 = 52.03 \text{ kN.}$$

$$W_d + W_L = 49.73 + 52.03 = 101.762 \text{ kN.}$$

End reaction due to DL and IL = 101.762/2 = 50.88 kN.

Factored load = 1.5 × 50.88 = 76.32 kN.

Vertical component of wind load = 21.9 × sin (90° – 15°) = 21.15 kN.

Bearing stress in M20 concrete = 12 MPa. (Clause 7.4.1 of IS: 800)

Provide plate of 250 × 400 mm

$$\text{Bearing stress} = \frac{76.32 \times 1000}{250 \times 400} = 0.76 \text{ MPa} < 12 \text{ MPa.}$$

Tension is Taken by Anchor Bolts

Net reaction due to wind + DL = 4(6.216 – 21.15) = (–) 59.74 kN.

Factored load = 1.2 × 59.74 = 71.7 kN.

Providing M16 thread on 20 mm diameter M.S. rods.

Net cross section area = 157 mm^2 (Table 8.3 of this book).

Design rupture strength = $T_{dn} = 0.9 \times A_n \times f_{ub} \times /\gamma_{m1}$ (Clause 6.3.2 of IS: 800)
= 0.9 × 157 × 410/(1.25 × 1000) = 46.35 kN.

$$\text{Number of bolts} = \frac{71.7}{46.35} = 1.55.$$ Provide 4 bolts.

Length of Anchorage

The bond strength for M.S. rod and M20 concrete is 1.2 MPa (Clause 26.2.1 – IS: 456)

$$L = \frac{71.7 \times 1000}{\pi \times 20 \times 1.2 \times 4} = 238 \text{ mm.}$$

Embed 250 mm inside concrete.

10.12 ANALYSIS OF TRUSS BY THE METHOD OF JOINTS

If the relation $m = 2j - 3$ is satisfied, the truss is statically determinate.

m = number of members

j = number of joints

The member forces of a determinate truss can be obtained by using the laws of statics/equilibrium condition. By using free body diagrams, the forces in each member can be obtained.

The following methods are adopted for analysing trusses.

(a) Method of joints

(b) Method of sections

(c) Graphical method

(d) Matrix method

In the method of joints, each and every joint is considered and by using equilibrium conditions, the forces are obtained in the selected sequence.

In the method of sections, equilibrium conditions are applied by considering the truss part by part to obtain member forces.

In the graphical method force diagrams are drawn to scale and forces in members are obtained by measuring. An example using the analysis of a plane truss by graphical method is presented in the Example 10.1. Though the principles of statics is valid for indeterminate trusses (where $m \neq 2j - 3$), they cannot be analyzed by method of joints/sections. These can be analyzed by matrix method. Using matrix method determinate and indeterminate trusses are analyzed through computer software pieces available in the market. These analysis and design software pieces reduce the time taken by a design engineer to a greater extent.

An example using the analysis of a plane truss by the method of joints is presented in Example 10.2.

Example 10.2 *Find the forces in members of truss for the loads shown in Fig. 10.12 by the method of joints.*

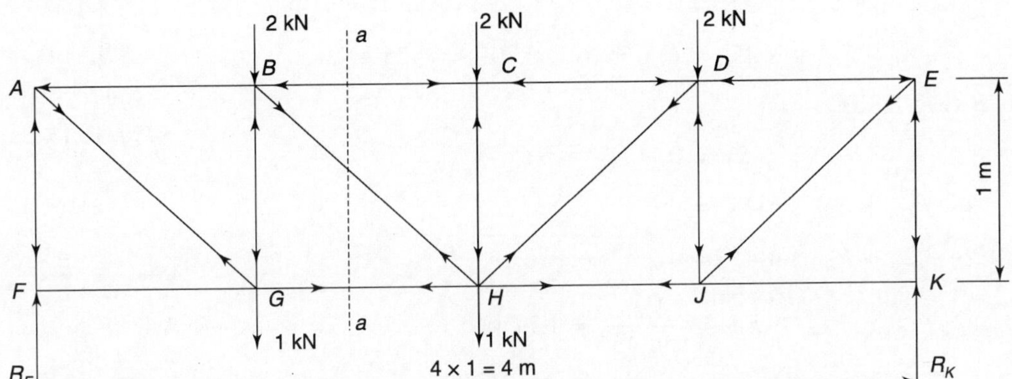

Fig. 10.12

To find reaction at $F(R_F)$, taking moment about K,

$R_F \times 4 = 2 \times 3 + 2 \times 2 + 2 \times 1 + 1 \times 3 + 1 \times 2 = 6 + 4 + 2 + 3 + 2 = 17$ kNm.

$$R_F = 17/4 = 4.25 \text{ kN.}$$

Total load = $3 \times 2 + 2 \times 1 = 8$ kN.

$$R_K = 8 - 4.25 = 3.75 \text{ kN}$$

Let us take joint by joint to obtain forces in members.

Joint *F* (Fig. 10.13)

$$\Sigma H = 0; F_{FG} = 0; \Sigma V = 0; F_{FA} = R_F = 4.25 \text{ kN}$$

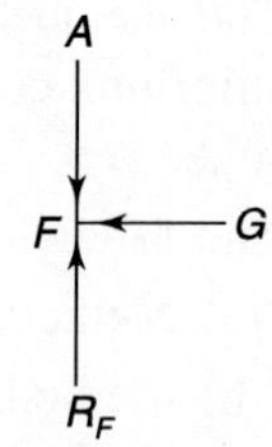

Fig. 10.13

Joint *A* (Fig. 10.14)

θ is the inclination of the member to the horizontal plane.

$$\Sigma V = 0$$

$$F_{FA} - F_{AG} \sin 45° = 0$$

$$F_{AG} = \frac{F_{FA}}{\sin 45°} = \frac{4.25}{0.707} = 6 \text{ kN}$$

$$\Sigma H = 0$$

$$F_{AB} = F_{AG} \cos 45° = 6 \times 0.707 = 4.25 \text{ kN}$$

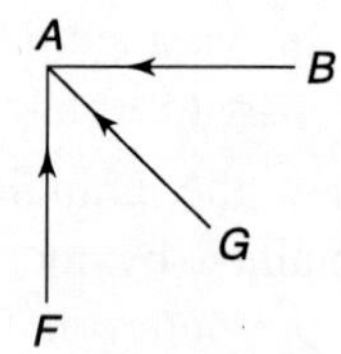

Fig. 10.14

Joint *G* (Fig. 10.15)

$$\Sigma V = 0$$

$$F_{GB} = F_{GA} \sin 45° + 1 = 0$$

$$F_{GB} - (6 \times 0.707) + 1 = F_{GB} - 4.25 + 1 = 0$$

$$F_{GB} = 3.25 \text{ kN}$$

$$\Sigma H = 0$$

$$F_{GA} \cos 45° = F_{GH} = (6 \times 0.707) = 4.25 \text{ kN.}$$

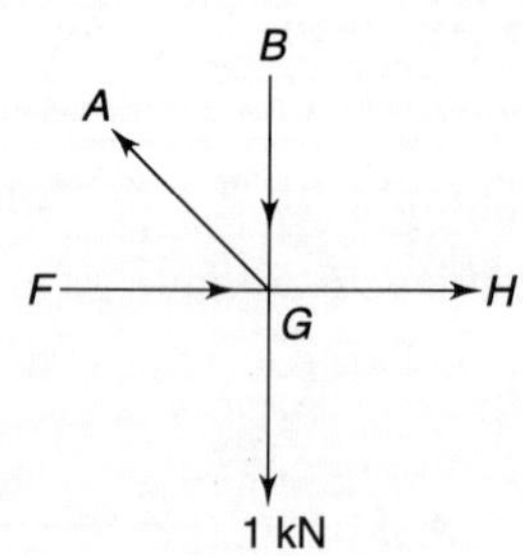

Fig. 10.15

Joint *B* (Fig. 10.16)

$$\Sigma V = 0$$

$$2 - F_{GB} + F_{BH} \sin 45° = 0$$

$$2 - 3.25 + F_{BH} \sin 45° = 0$$

$$F_{BH} = \frac{1.25}{\sin 45°} = 1.77 \text{ kN}$$

$$\Sigma V = 0$$

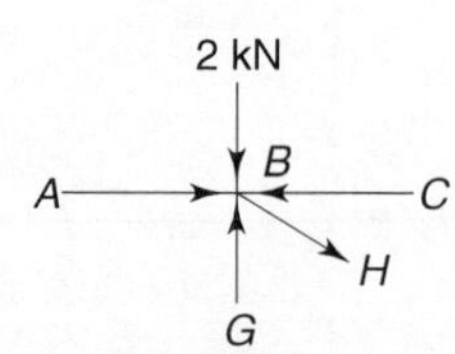

Fig. 10.16

$$F_{AB} + F_{BH} \cos 45° = F_{BC}$$

$$4.25 + 1.77 \times \cos 45° = F_{BC} = 4.25 + 1.25 = 5.5 \text{ kN}$$

Joint *C* (Fig. 10.17)

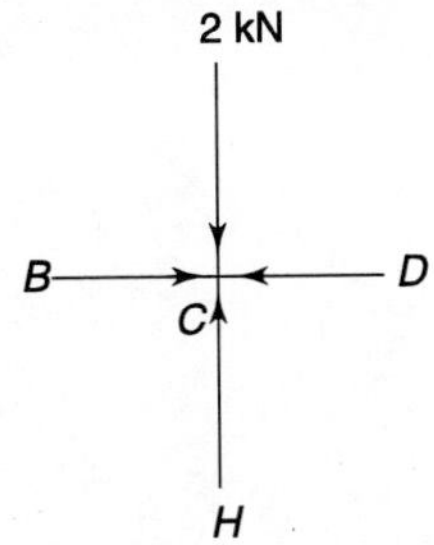

Fig. 10.17

$$\Sigma V = 0 \quad F_{HC} = 2 \text{ kN}$$

$$\Sigma H = 0 \quad F_{CB} = F_{CD} = 5.5 \text{ kN}$$

Joint *K* (Fig. 10.18)

$$\Sigma H = 0 \quad F_{JK} = 0$$

$$F_{KE} = 3.75 \text{ kN}$$

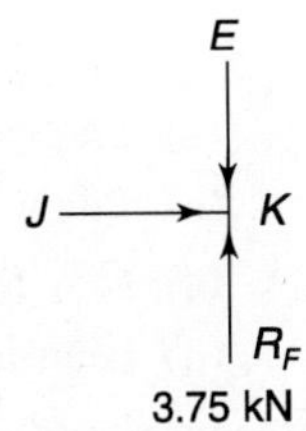

Fig. 10.18

Joint *E* (Fig. 10.19)

$$\Sigma V = 0$$

$$F_{KE} = F_{EJ} \sin 45°$$

$$F_{EJ} = \frac{3.75}{0.707} = 5.3 \text{ kN}$$

$$\Sigma H = 0$$

$$F_{ED} = F_{EJ} \cos 45° = 3.75 \text{ kN}$$

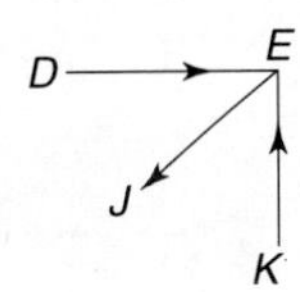

Fig. 10.19

Joint *J* (Fig. 10.20)

$$\Sigma V = 0$$

$$F_{JD} = F_{JE} \sin 45° = 5.3 \times 0.707 = 3.75 \text{ kN}$$

$$\Sigma H = 0 \quad F_{JH} = F_{JE} \cos 45° = 3.75 \text{ kN}$$

Joint *D* (Fig. 10.21)

$$F_{DH} \sin 45° + 2 - F_{DJ} = 0$$

$$F_{DH} = \frac{3.75 - 2}{0.707} = \frac{1.75}{0.707} = 3.75 \text{ kN}$$

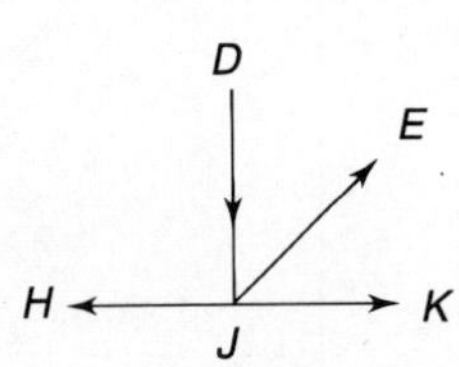

Fig. 10.20

Check with Joint *H* (Fig. 10.22)

$$\Sigma V = 0$$

$$F_{HC} - F_{HD} \sin 45° - F_{HB} \sin 45° + 1 = 0$$

$$2 - 1.75 - 1.25 + 1 = 0$$

The directions are marked in the figure. Arrow towards the joints indicates that the member is in compression.

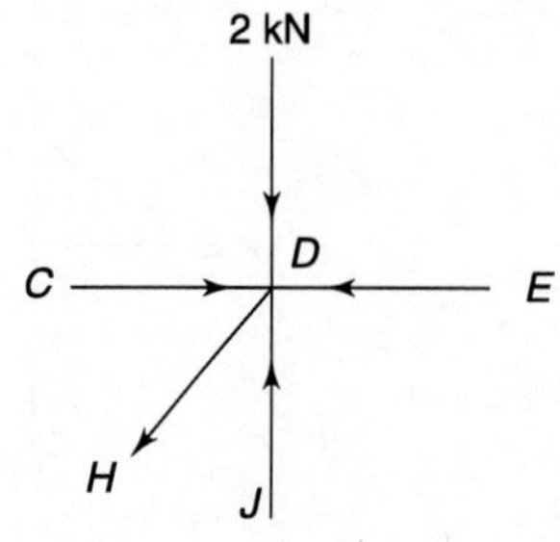

Fig. 10.21

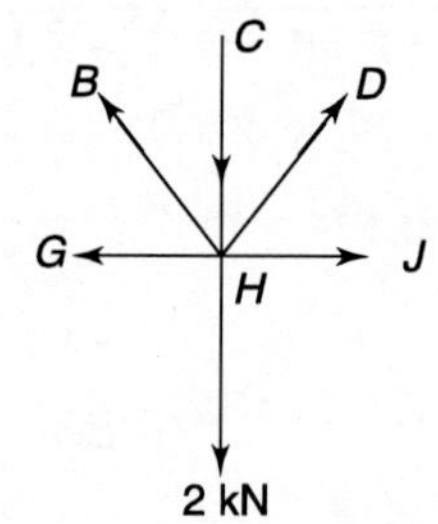

Fig. 10.22

EXERCISE

1. Sketch the following:
   (a) Pitched roof
   (b) Pitched roof with monitor
   (c) *N* – Truss
   (d) bean-to roof
   (e) North light roof
2. Write down the components of a pitched roof truss.
3. What is meant by sag rod and explain its use?
4. By using method of joints find the forces in members *JA* and *JB*.

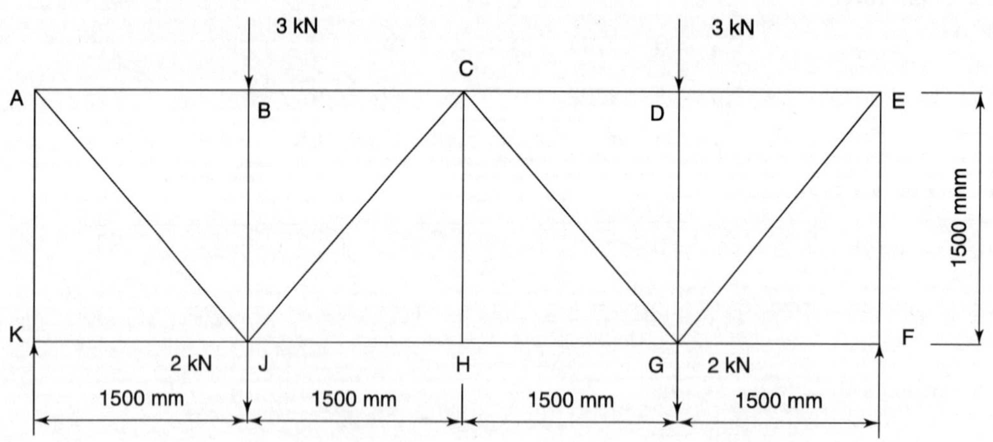

Chapter 11

Calculation of Earthquake Load on Structures (As per IS:1893-Part-1)

Ground in motion is called earthquake. The magnitude and intensity of earthquake is important to the structural design engineer. The effect of earthquake on structure is influenced by the soil layers below the structure, height of structure and the weight it carries at various heights.

The force due to the effects of earthquake on the structure is to be calculated based on the seismic coefficients given in IS:1893-Part-1. The seismic coefficient used in the design of any structure depends on many factors. It is difficult to determine the exact seismic coefficient.

IS:1893-Part-1 deals with a assessment of seismic loads on structures and earthquake resistant design of buildings.

11.1 DESIGN SPECTRUM

India is classified into four seismic zones for the purpose of determining seismic forces. The design horizontal seismic coefficient A_h for a structure shall be determined by the following expression

$$A_h = ZIS_a/(2.\ R.g) \text{ (Clause 6.4.2 of IS:1893-Part-1)}$$

Structure with $T = 0.1$ second, the value of A_h shall not be taken less than $Z/2$ whatever be the value of I/R.

where Z = zone factor given in Table 2 of IS:1893-Part-1 is for the maximum considered earthquake (MCE) and service life of structure in a zone. The number 2 in the denominator of Z is used so as to reduce the MCE zone factor to the factor for design basis earthquake (DBE).

I = Importance factor (Table 6 of IS:1893-Part-1)

R = Response reduction factor (Table 7 of IS:1893-Part-1),

Value of I/R should not exceed 1.0

S_a/g = Average response acceleration coefficient as in Fig. 2 of IS:1893-Part-1.

The random earthquake, which causes the structure to vibrate, can be resolved in any three mutually perpendicular directions. The predominant direction of ground vibration is horizontal.

Vertical acceleration should be considered in structures with large spans, those in which stability is a criteria for design, or for overall stability analysis of structures and for cantilevers.

11.2 DESIGN HORIZONTAL EARTHQUAKE LOAD

The structure shall be designed for the effects due to design earthquake load in one horizontal direction at a time, when the lateral load resisting members are oriented along orthogonal horizontal directions.

11.3 DESIGN VERTICAL EARTHQUAKE LOAD

The design acceleration spectra for vertical motions when required, may be taken as 2/3 of the design horizontal acceleration spectrum.

$$(2/3) \times A_h = (2/3) \times \frac{Z\,IS_a}{2 \times R \times g} \quad \text{(Clause 6.4.5 of IS:1893-Part-1).}$$

11.4 LOAD FACTOR AND REDUCTION IN IMPOSED LOAD

When earthquake forces are considered along with other normal design forces, the permissible stresses in material, in the elastic method of design may be increased by 1/3 in working stress method. The load factors takes care of this in limit state design (Table 4 of IS:800).

Imposed load for various loading classes as specified in IS:875-Part-2 may be used. The earthquake force shall be calculated for the full dead load plus the % of imposed load as in Table 8 of IS:1893-Part-1.

For calculating the design seismic forces of the structure, the imposed load on roof need not be considered (Clause 7.3.2 of IS:1893-Part-1).

No further reduction of the imposed load will be used as envisaged in IS:875-Part-2 for number of story's above the one under consideration or for large spans of beams or floors.

11.5 SEISMIC WEIGHT OF FLOORS

The seismic weight of each floor is its full dead load plus appropriate imposed load as in Table 8 of IS:1893-Part-1).

11.6 SEISMIC WEIGHT OF BUILDING

The seismic weight of the whole building is the sum of the seismic weights of all the floors.

11.7 DESIGN LATERAL FORCE

Design Seismic Base Shear (V_b)

$$V_b = A_h \times W$$

W = Seismic weight of the building.

A_h = Horizontal seismic coefficient.

11.8 FUNDAMENTAL NATURAL PERIOD

The fundamental natural period of vibration T_a in seconds, of a moment resisting frame building without brick infill panels may be calculated using the following expression.

$T_a = 0.085\ h^{0.85}$ for steel fame building. This is applicable for rigid frame without wall. The connection shall be rigid to transfer moment. This is not applicable for structure with bracing.

h = height of the building in metre aboveground. This excludes the basement storey, where basement walls are connected with the ground floor deck or fitted between the building columns. But, it includes the basement storey, when they are not connected.

The fundamental natural period of vibration T_a in seconds, of all other buildings, including moment-resisting fame buildings with brick infill panels, many be estimated by the empirical expression

$T_a = 0.09h/\sqrt{d}$ (Clause 7.6.2 of IS:1893-Part-1)

d = base dimension of the building at the plinth level in metre along the considered direction of the lateral force and h is as above.

This is applicable for rigid frames with brick infill and building structures with vertical bracing members. For a structure having different plan dimension in 'z' and 'x' directions (mutually perpendicular directions) the value of T_a will vary. Due to this, S_a/g will vary and thereby we will get two different values of forces depending on the direction considered.

11.9 DISTRIBUTION OF DESIGN FORCE

Vertical Distribution of Base Shears to Different Floor Levels (Clause 7.7.1 of IS:1893-Part-1)

The design base shear V_b shall be distributed along the height of the building as per the following expression

$$Q_i = V_b \frac{W_i h_i^2}{\sum_{i=1}^{n} W_i h_i^2}$$

Q_i = Design lateral force at floor i

W_i = Seismic weight of floor i

h_i = Height of floor i measured from base

n = Number of storeys in the building is the number of levels at which the masses are located.

Q_i is the design lateral force at floor i. This is the total lateral seismic force at that floor. This has to be distributed at various nodes judiciously according to the distribution of DL and IL on that floor. In a vertically braced structure; these lateral forces are taken by the vertical bracing system to the foundation. Heavily loaded beams will be subjected to lateral loads and thereby lateral bending will be there.

The magnitude of Q_i depends on seismic weight and the height at which it acts. For example, if the seismic weight at a lower floor is equal to the seismic weight at a higher floor, the lateral force at higher floor will be many times more than that at the lower floor. The example worked out in this chapter will explain this.

11.10 VALUE OF DAMPING

The value of damping for steel buildings can be taken as 2% of critical. (Clause 7.8.2.1 of IS:1893-Part-1).

The values of S_a/g obtained from Fig. 2 of IS:1893-Part-1 is for 5% spectra. The value of S_a/g obtained from Fig. 2 of IS:1893 Part-1 is to be multiplied by multiplying factors as given in Table 3 of IS:1893-Part-1 for obtaining values of S_a/g for other damping.

For steel structure, the multiplying factor is 1.4.

11.11 CANTILEVER PROJECTION

Vertical Projections

Tower, tanks, parapets, chimneys and other vertical projections attached to buildings and projection above the roof, shall be designed and checked for stability for five times the design horizontal seismic coefficient A_h. The weight of these projecting elements above the roof will be lumped with the roof weights.

Horizontal Projection

All horizontal projections shall be designed and checked for stability for five times the design vertical coefficient specified in Clause 6.4.5 of IS:1893-Part-1.

This means $5 \times (2/3)\ A_h = 10/3\ A_h$.

The increased design forces specified above are only for designing the projected parts and their connections with the main structures. For the design of main structure, such increase need not be considered.

11.12 CONNECTION BETWEEN PARTS

All connections between beams to columns and columns to their footings, should be made capable of transmitting a force, in all possible directions, of magnitude (Q_i/W_i) times but not less than 0.05 times the weight of the smaller part (or) the total of dead and imposed load reaction. Frictional resistance between such connected parts shall not be relied upon for fulfilling these requirements.

Example *Calculate the design lateral force due to earthquake at every floor for the structure with the following design data.*

Location:	*Chennai.*
Soil type:	*soft soil (from soil report)*
Height:	*36 metre*
Damping:	*2% critical*
Steel Building Braced.	
Number of floors:	*Ten*
Structure:	*Power station building*

Dead load including self weight is 1 kN/sq·m on all floors.

Imposed load 5 kN/m^2 in all floors

The floor area and height of each floor from ground is given below.

Area in square metres	*Height of each floor from ground in metres*
10	3
20	6
25	9
30	13
20	17
10	20
40	24
20	28
30	32
10	36

$$A_h = ZIS_a/(2 \times R \times g) \text{ (Clause 6.4.2 of IS:1893-Part-1)}$$

Zone 3 from Fig. 1 and Annexe *E* of IS:1893-Part-1.

Chennai lies in seismic zone III.

Z = 0.16 (Table 2 of IS:1893-Part-1)

I = 1.5 (Table 6 of IS:1893-Part-1)

R = 4 (Table 7 item iii of IS:1893-Part-1 for steel frame with concentric braces)

To find S_a/g

$T_a = 0.085\ h^{0.75}$ (Clause 7.6 of IS:1893-Part-1)

$= 0.085 \times 36^{0.75} = 1.25;\ 0.67 < T < 4.0$

For soft soil sites $S_a/g = 1.67/T = 1.67/1.25 = 1.34$ (Clause 6.4.5 of IS:1893-Part-1).

This is for 5% damping.

For 2% damping, multiplying factor as per Table 3 of IS:1893-Part-1 = 1.4

$S_a/g = 1.34 \times 1.4 = 1.876.$

$A_h = 0.16 \times 1.5 \times 1.876/2 \times 4 = 0.05626.$

Imposed load in all floors 5 kN/m^2 > 3 kN/m^2. Hence, 50% of imposed load to be considered in seismic weight calculation (Table 8 of IS:1893-Part-1).

Distribution of Design Base Shear to Different Floor Levels (Clause 7.7 of IS:1893-Part-1)

The design base shear $V_b = A_h\, w$ shall be distributed along the height of the building as per the following expression

$$Q_I = \frac{V_b W_i H_i^2}{\sum_{n=1}^{i} W_i H_i^2} \quad \text{(Clause 7.7.1 of IS:1893-Part-1)}$$

Dead load is 1 kN/m^2 on all floors.

The table below gives the floor area in each floor in column 2. DL and LL are given in columns 3 and 4 separately. DL + 50% of IL is given in column 5 which is the seismic weight.

Height of the respective floor from ground is (h_i) given in column 6. Column 7 gives Wh_i^2, the total of which is ΣWh_i^2 used in column 8 of the table. The values in column 9 are Q_i which is the horizontal seismic load at level 'i' obtained by multiplying the content in column 8 and V_b. To obtain V_b, the total seismic weight in column 5 is multiplied by horizontal seismic coefficient A_h.

Floor	Area in M^2	DL kN	IL kN	DL + 50% IL kN	h_i	$W_i\, h_i^2$	$(W_i h_i^2)/(\Sigma W_i h_i^2)$	Q_i
1	2	3	4	5	6	7	8	9
1	10	10	50	35	3	315	8.99×10^{-4}	0.04
2	20	20	100	70	6	2520	7.19×10^{-3}	0.30
3	25	25	125	87.5	9	7087.5	0.02	0.85
4	30	30	150	105	13	17745	0.051	2.159
5	20	20	100	70	17	20230	0.058	2.45
6	10	10	50	35	20	14000	0.04	1.69
7	40	40	200	140	24	80640	0.23	9.74
8	20	20	100	70	28	54880	0.16	6.77
9	30	30	150	105	32	107520	0.31	13.12
10	10	10	50	35	36	45360	0.13	5.5
		215		752.5		350297.5		42.62 kN

$$V_b = A_h \times W = 0.05626 \times 752.5 = 42.33 \text{ kN}.$$

In the above example, the seismic weight for floors 4 and 9 are 105 kN.

Note that the distribution Q_i at the 4th floor is 2.159 kN, whereas at 9th floor it is 13.12 kN. Though the seismic weight is same. As the height increases, the same seismic weight induces higher lateral force. The lateral force Q_i may be applied at bracing (or) column nodes at the appropriate elevations as an approximation. This distribution may be judiciously done depending on the load pattern on the platform at that elevation.

11.13 LIMITATION OF SWAY

Though there are no generally accepted limits for permissible lateral deflection (sway) of buildings, some building codes limit sway of any storey of a building to a maximum of 0.25% of the storey height for wind and 0.50% of the storey height for earthquake loads. Buildings less than 14 storeys height and with height width ratio of 2.5 or less are exempted. Three methods are used to control the sway and transmit the loads to the foundations. These can be used in combination also.

Method 1

Introduce shear walls in a building. Example core wall structure.

Method 2

Providing resistance to lateral loads by introducing diagonal members called bracing. These are provided in vertical planes to take care of lateral (horizontal) forces due to wind and seismic. Horizontal bracings are also provided at platform elevations to transfer the the lateral loads to nodes of vertical bracings. Such bracing (diagonal members in vertical and horizontal planes) reduces the sway and increases the stability of structures particularly when the structure is tall and slender.

Method 3

Rigid frames are structurally efficient. But for tall steel structures, vertically braced structures will be economical.

Designers should provide vertical bracing system for tall buildings. It is better to provide these bracings to form vertical trusses. In many industrial buildings, providing bracings in a particular pattern may not be possible due to interference.

EXERCISE

1. What is meant by seismic weight of (a) floors (b) building?
2. When multiplying factor is used to find S_a/g?
3. What are the factors influencing the seismic force on a structure?

Chapter 12

Design Basis for Analysis and Design of Boiler Supporting Steel Structure

12.1 DESIGN BASIS

In this chapter a model of design basis is presented. The user may modify as per the requirement. This template can be used for any type of steel structure. Preparation of such design basis at the initial stage of the design will help to understand the scope, requirement and documents available. Using this avoid problems croping up at a latter stage.

ANALYSIS AND DESIGN OF BOILER SUPPORTING STEEL STRUCTURE AND CHIMNEY

CLIENT: ______________________________

OWNER: ______________________________

LOCATION: ______________________________

____

CONSULTANT: ______________________________

DOCUMENT NUMBER: ____________________ **DATE:** __________

PREPARED BY: __________ **CHECKED BY:** __________ **APPROVED BY:** __________

REVISION STATUS: ______________________________

12.2 SCOPE

This design criteria covers the design of structural steel supporting boiler.

12.3 CODES AND STANDARDS

Design of steel structure is in accordance with Indian Standards.

The following Indian Standards are followed:

IS:800–2007	General construction in steel—Code of practice.
IS:875–1987	Code of practice for design loads (other than earthquake) for buildings and structures—Part 1-Dead Loads
IS:875–1987	Code of practice for design loads (other than earthquake) for buildings and structures–Part 2-Imposed Loads
IS:875–1987	Code of practice for design loads (other than earthquake) for buildings and structures–Part 3-Wind Load
IS:1893–2002	Criteria for earthquake resistant design of structures—Part 1-General provisions and buildings.
IS:1893–2006	Criteria for earthquake resistant design of structures—Part 4-Industrial structures including stack like structures.
IS:456–2000	Plain and Reinforced Concrete-Code of practice
IS:2062–2006	Hot rolled low, medium and high tensile structural steel.
IS:6533–1989	Code of practice for design and construction of steel chimneys-Part 2–Structural aspects.

12.4 DESIGN PRINCIPLE

Structural System

Boiler steel structure is a framed unbraced construction.

The framed structure is to transmit vertical load (dead load, equipment load, imposed load) and lateral load (wind load, earthquake load) to the foundation.

Analysis Modeling

Beam-to-beam and beam-to-column connection are considered fixed since the ends of beams are connected to take shear, axial force, and moment.

The support of the steel boiler columns are considered pinned on the R.C.C. columns supporting the boiler columns which acts as foundation for the boiler columns.

12.5 DEFLECTION LIMITS

The deflection slimits of beams due to the dead and imposed loads is as per Table 6 of IS:800/2007

12.6 DESIGN LOADS

Dead Load

Self weight of boiler pressure parts, drum, piping, dust blowers, wall blowers, air pre-heater and valves fall in this category.

Dead load includes the self weight of all structural components such as columns, beams, chimney, burners, floor grills, platforms, handrails, stairs, step tread, ladders, casing and refractory/insulation. R.C.C. floor and its finish if any.

Equipment Load

Equipment load including boiler pressure parts, ducts, integral pipes, main pipes, valve and auxiliary equipment.

Hydraulic testing loads for all pressure parts considered.

Imposed Loads

Imposed load is shown in table.

Load category	*Load value (kN/m²)*
	Imposed load
Floor Load	2.5
Platform Load (operating floor)	5 to 10

Imposed load is considered only where floor grating/chequred plate/R.C.C. slab is proposed. Apart from the above, weight of water and weight of ash is also considered.

12.6.1 Wind Load

Wind loading is in accordance with IS:875-Part-3.

The following is the basis for calculating wind load.

Basic wind speed: m/sec.

Exposure category: Category-, Class-

$$\text{Design wind speed} = V_z = V_b k_1 k_2 k_3$$

V_b = Basic wind speed in m/s

k_1 = Probability factor (risk coefficient)

k_2 = Terrain, height and structure size factor.

k_3 = Topography factor.

$$\text{Design wind pressure} = P_z = 0.6\ V_z^2$$

P_z = Design wind pressure in N/m² at height z

V_z = Design wind velocity in m/s at height z.

The height to minimum width ratio of the boiler structure is greater than 5 and the natural frequency is greater than 1. Hence, the dynamic effect due to wind is considered.

12.6.2 Seismic Load

Seismic loading is in accordance with IS:1893-Part-1. The total design lateral seismic force is determined from the following formula:

$A_h = ZIS_a/2Rg$

A_h = Design horizontal seismic coefficient.

Z = Zone factor as in Table 2 of IS:1893-Part-1

I = Importance factor as in Table 6 of IS:1893-Part-1.

S_a/g = Average response acceleration coefficient

R = Response reduction factor as in Table 7 of IS:1893-Part-1.

Soft soil is assumed for design calculation since the type of soil is not known and no soil test report is available. Since the site is seismic prone, earthquake of worst magnitude is considered. Zone V for which $Z = 0.36$. Importance factor of 2 is considered. Chimney is supported on top of boiler structure and the structure itself acts as a smoke stack (chimney) right from base. Hence, the structure will be designed for five times the horizontal seismic coefficient as per Clause 7.12.2.1 of IS:1893-Part-1.

12.6.3 Temperature Load

Since the structure is not subjected to equipment heat load, temperature load is not taken into account.

12.6.4 Snow Load

This is not applicable since there is no snowfall in the region.

12.6.5 Settlement Load

Settlement loads are those loads associated with differential settlement of the foundation. For steel design, a relative displacement equal to 3 mm between two adjacent columns is used to account for settlement. Settlement of this magnitude has no impact on design of the structural steel. Moreover, this structure is supported on R.C.C. columns.

12.6.6 Combination Loads: (100) SW + DL + LL

(1) SW + DL + LL

(2) SW + DL + WL (+ x)

(3) SW + DL + WL (– x)

(4) SW + DL + WL (+ z)

(5) SW + DL + WL (– z)

(6) SW + DL + EQL (+ x)

(7) SW + DL + EQL (– x)

(8) SW + DL + EQL (+ *z*)

(9) SW + DL + EQL (– *z*)

Legend:

SW - Self weight
DL - Dead load
LL - Imposed load
WL - Wind load
EQL - Earthquake load
X - GLOBAL *X*-AXIS
Y - GLOBAL *Y*-AXIS

Note:

(1) Wind and earthquake are not to occur simultaneously.

(2) The combined load cases 2 to 9 will be repeated adding partial imposed load as per Clause 3.2.1, Appendix-A of IS:875-Part-2 and Table 8 of IS:1893-Part-1. But all reductions will not be considered simultaniously.

12.7 MATERIAL

12.7.1 Structure Steel

Description	*Material*
I Shape rolled steel	IS – 2062 f_y = 250 MPa
L Shaped rolled steel	IS – 2062 f_y = 250 MPa
[Shaped rolled steel	IS – 2062 f_y = 250 MPa
Built up-*I* shape steel and steel plate	IS – 2062 f_y = 250 MPa

Other than the above, high strength steel having f_y = 275 MPa, 335 MPa are also used. It is preferable to use high strength steel for plate girders.

12.8 SECTION CONFIGURATION

(a) Columns will be fabricated using rolled sections or plate formed built-up sections.

(b) Column connecting beams, platform beams, cladding runners, etc., will be by using *I* shaped rolled steel or rolled channel sections or channel box sections or built-up sections using plates.

(c) Chimney will be rolled using plates and stiffened with angle or flats.

12.9 FOUNDATION ANCHOR BOLTS

Foundation anchor bolt material proposed is IS:2062/IS:432-Part-1.

All welded connections are proposed to be as per IS:816.

The size of fillet weld will be 0.8 times the thickness of thinner of the connecting plates/parts and the size of the weld shall not be less than 4 mm. However, the size given in the drawing will govern.

12.10 DESIGN METHOD

The columns, column connecting beams, secondary beams are incorporated into a three-dimensional space frame model and analyzed using computer software. Material of construction of columns, beam and tie beams are as in 12.07.1 Design will be by the limit state method is as per IS:800/2007. Chimney will be designed as per IS:6533. Loads considered are as per IS:875 and IS:1893.

12.11 UNITS

Design calculations are carried out in the following measuring system.

Quantity	*Unit*	*Abbreviation*
Length	Metre	m
Area	Square metre	Sq. m
Volume	Cubic metre	cum
Force	Kilo Newton	kN
Stress & intensity of load	Kilo Newton/metre square Newton/square metre	kN/m^2 N/mm^2
Moment	Kilo Newton metre	kNm

12.12 MECHANICAL PROPERTY OF MATERIAL (CLAUSE 2.2.4.1 OF IS:800)

Rolled section (*I*, *L*,) – IS:2062

Yield stress 250 N/mm^2

Plates: IS:2062

Young's modulus $E = 2 \times 10^5$ N/mm^2

Poisson's ratio = 0.3

Density = 78.5 kN/m^3

Coefficient of thermal expansion = $\alpha_t = 12 \times 10^{-6}$/degree centigrade.

12.13 DIMENSION OF ROLLED SECTION INCLUDING PLATES

The dimension as in IS:808, SP:6(1) will be used or as per the manufacturers property table.

12.14 SECTION STRENGTH AND OVERALL STRENGTH

This will be as per IS:800/2007–Limit State method.

12.15 ELECTRODES

Electrodes to be used are *E* 6013 & *E* 7018

12.16 BOLT FOR SIDE CLADDING RUNNERS

It is proposed to use property class 4.6 bolts as per IS:1367 and size not less than M12.

12.17 THICKNESS OF STIFFENER PLATE

The thickness of stiffener plate for columns base shall not be less than 12 mm. The web stiffener (vertical) for beams shall not be less than 8 mm and to be provided wherever concentrated load acts on the beam in addition to required web stiffeners. The diaphragm plates inside the built-up columns shall be provided wherever beams are connected to column flange/web. Thickness of such diaphragm plates shall not be less than the thickness of the flange of the connecting beam.

12.18 DESIGN OF BASE PLATE

Base plate thickness will be arrived at using "Timoshenko" coefficient for various end conditions. No welded joints will be permitted in the base plate. The base plate shall be a single plate. The top of base plate and base of column shall be machined to achieve a minimum of 80% contact. The bottom of base plate need not be machined. The thickness of base plate given in the drawing is after machining.

12.19 FOUNDATION CONCRETE

Grade of concrete for foundation/columns supporting the boiler structure shall not be less than M20.

12.20 STEEL CHIMNEY DESIGN

The design is based on IS:6533.

Project

12.21 LIST OF REFERENCE DRAWINGS

Drawing number	*Title of the drawing*	*Project*

12.22 LIST OF DOCUMENTS/DELIVERABLES

12.22.1 Load Calculation

Dead load and imposed load

Wind load

Earthquake load

Wind load calculation

Equipment load

12.22.2 Foundation Load Table

12.22.3 Arrangement Drawings with Member Sizes Including Size & Thickness of Base Plate and Size & Length of Anchor Bolt

Chapter 13

Codes and Templates

13.1 LIST OF IS CODES

1. IS:456–2000 Plain and reinforced concrete–Code of practice (third revision)
2. IS:800–2007 General Construction in steel–Code of practice (third revision).
3. IS:806-1968 Code of practice for use of steel tubes in general building construction
4. IS:808–1989 Dimensions for hot rolled steel beams, channels and angle.
5. IS:812–1957 Glossary of terms relating to welding and cutting of metals.
6. IS:813–1986 Scheme of symbols for welding (Revised).
7. IS:816–1969 Code of practice for use of metal arc welding for general construction in mild steel) (first revision)
8. IS:814-2004 Covered electrodes for manual metal arc welding of carbon and carbon manganese steel–specification
9. IS:819–1957 Code of practice for resistance spot welding for light assemblies in Mild steel.
10. IS:875–1987 Code of practice for Design loads (other than earthquake) for building and structures.
11. IS:1024–1999 Code of practice for use of welding in bridges and structures subject to dynamic loading
12. IS:1161–1998 Steel tubes structural purposes—specification (fourth revision)
13. IS:1261–1959 Code of practice for seam welding in mild steel.
14. IS:1278–1972 Filler rods and wires for gas welding (second revision)
15. IS:1323–1982 Code of practice for oxy-acetylene welding for structural work in mild steels (second revision)
16. IS:1363–2002 Hexagon head bolts, screws and nuts of product grade-*C*-Part-1-Hexagon head bolts.
17. IS:1364–2002 Hexagon head bolts, screws and nuts of product grade-*A*&*B*-Part-1-Hexagon head bolts.
18. IS:1730-1989 Dimensions for steel plates, sheets and strips for structural and general engineering purposes. (second revision)
19. IS:1732-1989 Dimensions for round and square steel bars for structural and general engineering purposes. (second revision)

20. IS:1911-1967 schedule of unit weight of materials (first revision)
21. IS:2062–2006 hot rolled low, medium and high tensile structural steel (sixth revision)
22. IS:3640–1982 specification for hexagon fit bolts.
23. IS:3757–1985 Specification for high strength structural bolts (second revision)
24. IS:4000–1992 Code of practice for high strength bolts in steel structures (first revision).
25. IS:6623–2004 High strength structural nuts—specification (first revision)
26. IS:6639–1972 Specification for hexagon bolts for steel structures.
27. IS:6649–1985 Specification for hardened and tempered washers for high strength structural bolts and nuts (first revision)
28. IS:7205–1974 safety code for erection of structural steel work
29. IS:9595–1996 metal arc welding of carbon and carbon manganese steels—Recommendations.

    SP:6 Handbook for structural engineers

13.2 TEMPLATES FOR COLUMN DETAILS

The following templates are given for the designer to write the values in the table. These templates will help the designer to ensure that the sizing of all the parts of column is completed and to give complete detail for the preparation of fabrication drawings.

Column Section

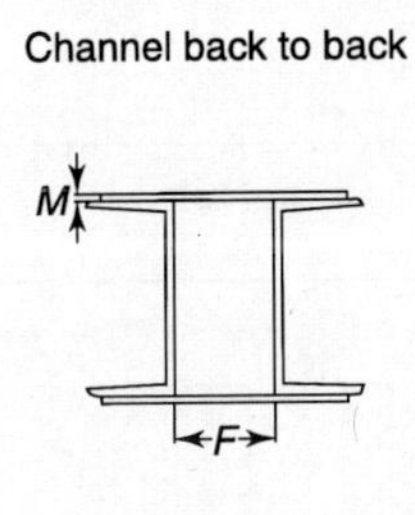

Column Base Plate

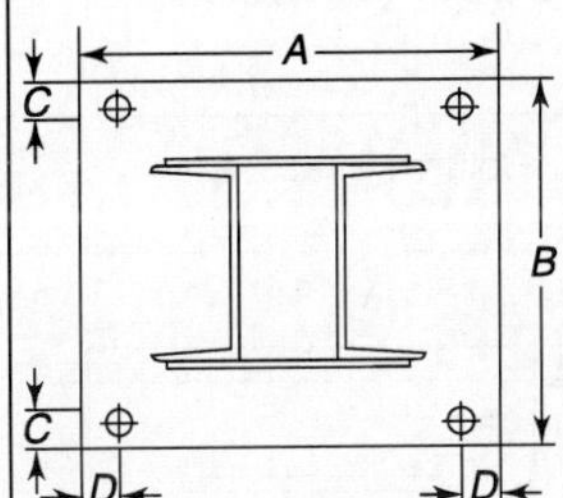

Batten Plate

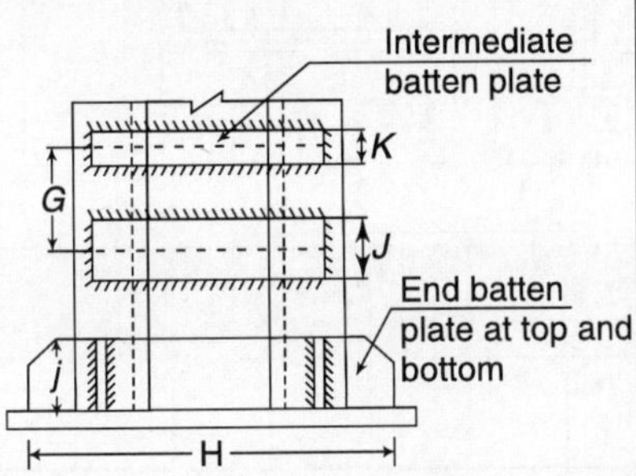

Anchor Bolt Details

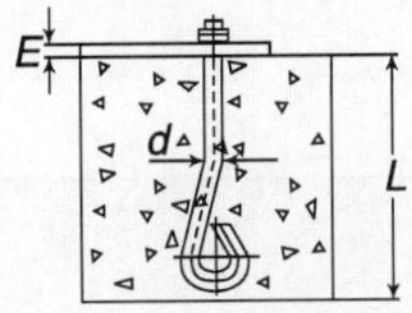

13.2.1 ISMC Back to Back

Col Mark	*Column Size*	*A*	*B*	*C*	*D*	*E*	*F*	*G*	*H*	*J*	*K*	*L*	*M*	Φ_d
	ISMC ____ with batten plate													
	ISMC ____ with batten plate													
	ISMC ____ with batten plate													
	ISMC ____ with batten plate													
	ISMC ____ with batten plate													
	ISMC ____ with batten plate													
	ISMC ____ with batten plate													
	ISMC ____ with batten plate													

Note: *N* = Number of bolts. All dimensions are in mm.

13.2.2 Double *I* Section (ISMB)

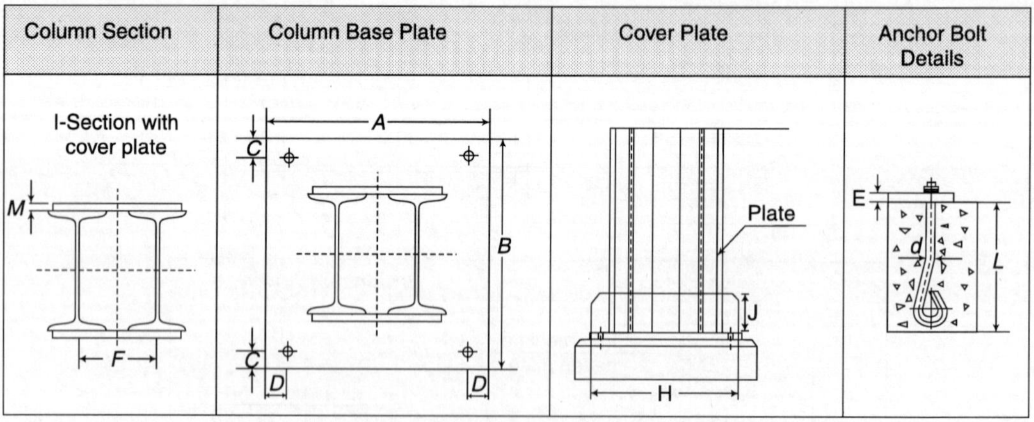

Col Mark	*Column Size*	*A*	*B*	*C*	*D*	*E*	*F*	*H*	*J*	*K*	*L*	*M*	Φ_d
	ISMB with cover plate												

	ISMB with cover plate												
	ISMB with cover plate												
	ISMB with cover plate												
	ISMB with cover plate												
	ISMB with cover plate												

Note: *N* = Number of bolts. All dimensions are in mm.

13.2.3 Single Channel (ISMC)

Column Section	Column Base Plate	Anchor Bolt Details
Single Channel Section 	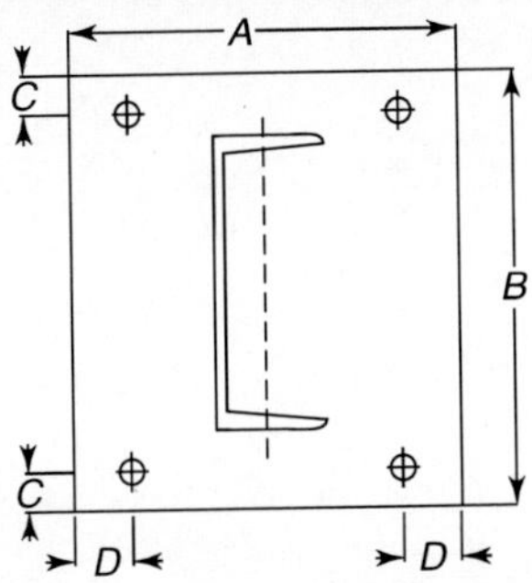	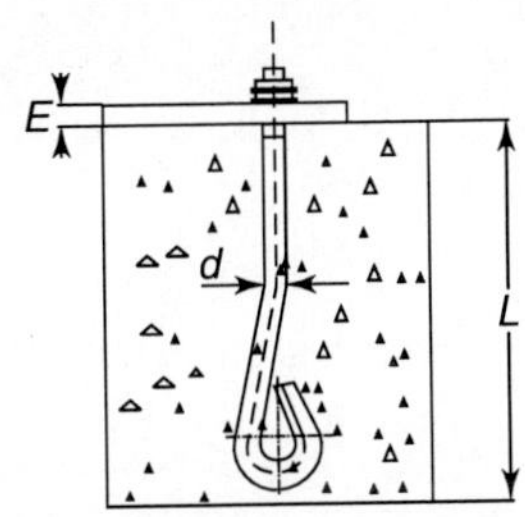

Col Mark	*Column Size*	*A*	*B*	*C*	*D*	*E*	Φ_d	*L*
	ISMC							
	ISMC							
	ISMC							
	ISMC							
	ISMC							
	ISMC							
	ISMC							

Note: *N* = Number of bolts. All dimensions are in mm.

References

1. S. Kanthimathinathan, *Manual for Detailing of Steel Structures,* I. K. International Publishing House Private Ltd., New Delhi.
2. S. S. Bhavikatti, *Design of Steel Structures by Limit State Method as per IS : 800 : 2007,* I. K. International Publishing House Private Ltd., New Delhi.
3. S. Kanthimathinathan, *Design of Connections for Steel Structures*, J. M. Jaina & Brothers, New Delhi.
4. S. K. Duggal, *Limit State Design of Steel Structures,* Tata McGraw-Hill Education Private Limited, New Delhi.
5. N. S. Trahair M. A. Brad Ford, D. A. Nethercot and L. Gardner, *The Behavior and Design of Steel Structures to EC3*, Taylor Francis, London & New York.
6. N. Subramanian, *Design of Steel Structures*, Oxford University Press, New Delhi.
7. Stephen P. Timoshenko, S. Woinowsky-Krieger, *Theory of Plates and Shells*, McGraw Hill International Editions, New York.

Reference Codes

IS : 800/2007

IS : 875/1987

IS : 1893/2002

IS : 1367

IS : 816

IS : 808

SP : 6 (1)

Index